Advanced Biology A2

Advanced Biology A2 2005

Student Resource and Activity Manual

Previous annual editions 1999-2004
This seventh edition 2005

ISBN 1-877329-22-3

Published by **BIOZONE International Ltd**
Printed in New Zealand

About the Writing Team

Tracey Greenwood joined the staff of BIOZONE at the beginning of 1993. She has a Ph.D in biology, specialising in lake ecology, and taught undergraduate and graduate biology at the University of Waikato for four years.

Richard Allan has had 11 years experience teaching senior biology at Hillcrest High School in Hamilton, New Zealand. He attained a Masters degree in biology at Waikato University, New Zealand.

Lyn Shepherd joined the writing team this year, bringing her 20 years experience as a secondary school biology teacher to Biozone.

The authors acknowledge and thank our graphic artist **Dan Butler** for general assistance with production and his tireless pursuit of the perfect 3-D landscape.

Purchases of this manual may be made direct from the publisher:

www.biozone.co.uk

UNITED KINGDOM:
BIOZONE Learning Media (UK) Ltd
P.O. Box 16710, Glasgow G12 9WS
Telephone: (0141) 337 3355
FAX: (0141) 337 2266
E-mail: info@biozone.co.uk

AUSTRALIA:
BIOZONE Learning Media Australia
P.O. Box 7523, GCMC 4217 QLD, Australia
Telephone: +61 (7) 5575 4615
FAX: +61 (7) 5572 0161
E-mail: sales@biozone.com.au

NEW ZEALAND:
BIOZONE International Ltd
P.O. Box 13-034, Hamilton, New Zealand
Telephone: +64 (7) 856 8104
FAX: +64 (7) 856 9243
E-mail: sales@biozone.co.nz

Preface to the 2005 Edition

This seventh edition of the manual is designed to meet the needs of students enrolled in the following biology courses: **AQA specifications A** and **B**, **EDEXCEL**, **OCR**, and Cambridge International Examinations (CIE), as well as senior biology courses for **Wales**, **Northern Ireland**, and **Scotland**. Previous editions of this manual have received very favourable reviews (see our web site: **www.biozone.co.uk** for details). This year's manual has been revised to meet two main objectives: to expand areas of core material to offer greater depth and content coverage, and to remove material duplicated between this manual and its companion volume, *Advanced Biology AS*. We have continued to refine the stimulus material in the manual to improve its accessibility and interest level. Biozone's supplementary product, the **Teacher Resource Handbook** (on CD-ROM), which was released in 2003, has been further developed as a supplementary resource and now provides the topic material that is specific to particular exam boards, but not common to all. In previous editions, this material has had to be included in the manual at the expense of developing core material. A guide to using the Teacher Resource Handbook to best effect has been included in the introductory section of the manual, and can be used in conjunction with the revised course guides. As in previous years, we have updated all of the listed resources: comprehensive and supplementary textbooks, video documentaries, computer software, periodicals, and internet sites. As in previous editions, changed and new activities are indicated in the contents pages. These annual upgrades are in keeping with our ongoing commitment to providing up-to-date, relevant, interesting, and accurate information to students and teachers.

A Note to the Teacher

This manual has been produced as a student-centred resource, enabling students to become more independent in seeking out information. By keeping the amount of reading required to a minimum and providing a highly visual presentation, it is hoped that students will be both motivated and challenged. They are provided with a clear map of where they are going through the course and become aware of the resources available to them. Today, many teachers are finding that a single textbook does not provide all of the information they need. This manual is **not a textbook**. It is a generic resource and, to this end, we have made a point of referencing texts from other publishers. Above all, we are committed to continually revising and improving this resource **each and every year**. We have managed to keep the price below £11 for student purchase. This is a reflection of our commitment to provide high-quality, cost effective resources for biology. Please do not treat this valuable resource as a *photocopy master* for your own handouts. We simply cannot afford to supply single copies of manuals to schools and continue to provide annual updates as we intend. Please **do not photocopy** from this manual. If you think it is worth using, then get your students to purchase what is a very good value-for-money resource. Because it is upgraded annually, it is encouraged that the students themselves purchase this manual. A free model answer book is supplied with your **first order** of 5 or more manuals.

How Teachers May Use this Manual

This manual may be used in the classroom to guide the student through each topic. Some activities may be used to introduce topics while others may be used to consolidate and test concepts already covered by other means. The manual may be used as the primary tool in teaching some topics, but it should not be at the expense of good, 'hands-on' biology. Students may attempt the activities on their own or in groups. The latter provides opportunities for healthy discussion and peer-to-peer learning. Many of the activities may be set as homework exercises. Each page is perforated, allowing for easy removal of pages that must be submitted for formal marking. Teachers may prescribe the specific activities to be attempted by the students (using the check boxes next to the objectives for each topic), or they may allow students a degree of freedom with respect to the activities they attempt. The objectives for each topic will allow students to keep up to date even if they miss lessons. Teachers who are away from class may set work easily in their absence.

I thank you for your support.
Richard Allan

Acknowledgements

We would like to thank those who have contributed towards this edition: • Campus Photography at the University of Waikato for photographs of monitoring instruments • Corel Corporation, for use of their eps clipart of plants and animals from the Corel MEGAGALLERY collection • Genesis Research and Development Corp. Auckland, for the photo used on the HGP activity • Human Genome Sciences, Inc. for photos of large scale DNA sequencing • Kurchatov Inst., for the photographs of Chornobyl • Mary McDougall for proofreading • Myles McInnes for his drawings • Jan Morrison for her line diagrams • Sirtrack NZ Ltd, for the photo of penguin radiotracking • TechPool Studios, for their clipart collection of human anatomy: Copyright ©1994, TechPool Studios Corp. USA (some of these images were modified by Richard Allan and Tracey Greenwood) • Totem Graphics, for their clipart collection of plants and animals • Janice Windsor, for her photograph of a baboon in East Africa • The 3D modelling by Dan Butler using Poser IV, Curious Labs and Bryce.

Photo Credits

Royalty free images, purchased by Biozone International Ltd, are used throughout this manual and have been obtained from the following sources: Corel Corporation from various titles in their Professional Photos CD-ROM collection; IMSI (International Microcomputer Software Inc.) images from IMSI's MasterClips® and MasterPhotosTM Collection, 1895 Francisco Blvd. East, San Rafael, CA 94901-5506, USA; ©1996 Digital Stock, Medicine and Health Care collection; ©Hemera Technologies Inc, 1997-2001; ArtToday ©1999-2001 www.arttoday.com; ©Click Art, ©T/Maker Company; ©1994., ©Digital Vision; Gazelle Technologies Inc.; PhotoDisc®, Inc. USA, www.photodisc.com • Human 3D models were created using Poser IV, Curious Labs. Photos kindly provided by individuals or corporations have been indentified by way of coded credits as follows: **BH**: Brendan Hicks (University of Waikato), **BOB**: Barry O'Brien (Uni. of Waikato), **CDC**: Centers for Disease Control and prevention, Atlanta, USA, **COD**: Colin O'Donnell, **EII**: Education Interactive Imaging, **EW**: Environment Waikato, **HGSI**: Dena Borchardt, at Human Genome Sciences Inc., **JDG**: John Green (University of Waikato), **JR-PE**: Jane Roskruge, **NASA**: National Aeronautics and Space Administration, **RA**: Richard Allan, **RCN**: Ralph Cocklin, **TG**: Tracey Greenwood, **WBS**: Warwick Silvester (Uni. of Waikato), **UCSF**: University of California San Francisco, **WMU**: Waikato Microscope Unit.

Special thanks to all the partners of the Biozone team for their support.

Cover Photographs

Main photograph: The jaguar (*Panthera onca*) is the only member of this family found in the Americas and it is the biggest cat on the continent. Its range is much more restricted now than at the beginning of last century, when it spanned from the southern states of the USA to the tip of South America. It is now found in the northern and central parts of South America and is primarily a forest dweller, with the highest population densities occurring in the lowland forests of the Amazon basin. Melanistic forms are common. PHOTO: Natural Selection Inc. ©Bruce Coleman Collection.

Background photograph: SEM of the villi lining the ileum. PHOTO: ©1996 Digital Stock Corporation, Medicine and Healthcare collection.

Contents

CODES: Δ **Upgraded** this edition ☆ **New** this edition **Activity** is marked: ☐ to be done; ☑ when completed

CONTENTS *(continued)*

CODES: Δ **Upgraded** this edition ☆ **New** this edition **Activity** is marked: ☐ to be done; ☑ when completed

Supplementary Material on the **Teacher Resource Handbook** *on CD-ROM*

AQA-A: Biology and Human Biology (HB)

- ☆ Ecosystems and Energy
  Biology: Module 5: 14.5, 14.7, 14.9
- ☆ Physiology and the Environment
  Biology: Module 6: 15.1-15.2, 15.4-15.7
- ☆ The Human Life Span
  HB: Module 7: 16.1-16.4, 16.6, 16.9, 16.11

AQA-B: Biology

- ☆ Thermoregulation
  Module 4: 13.5
- ☆ Muscles and Movement
  Module 4: 13.8
- ☆ Energy Flow and Nutrient Cycles
  Module 5a: 14.1-14.2
- Δ Applied Ecology ....Option Module 6
- Δ Microbes and Disease ....Option Module 7
- Δ Behaviour and Populations ....Option Module 8

OCR

- Δ Growth, Development & Reproduction .....*Option 01*
- Δ Applications of Genetics ....*Option 02*
- Δ Environmental Biology ....*Option 03*
- Δ Microbiology and Biotechnology ....*Option 04*
- Δ Mammalian Physiology & Behaviour ....*Option 05*

EDEXCEL: Biology and Human Biology

- ☆ Farming and Conservation
  Biology: *Unit 5B.3*
  Human Biology: *Unit 5H.4*
- ☆ Gene Technology
  Biology: *Unit 5B.4*
  Human Biology: *Unit 5H.1*
- ☆ Human Evolution
  Human Biology: *Unit 5H.2*
- Δ Microbiology and Biotechnology ....*Option A*
- Δ Food Science ....*Option B*
- Δ Human Health and Fitness ....*Option C*

Cambridge International Examinations

- ☆ Gene Technology
  Core: section O
- ☆ Mammalian Physiology ....*Option 1*
- ☆ Microbiology and Biotechnology ....*Option 2*
- ☆ Growth, Development & Reproduction .....*Option 3*
- ☆ Applications of Genetics ....*Option 4*

CODES: Δ **Upgraded** this edition ☆ **New** this edition **Activity** is marked: • to be done; ✓ when completed

How to Use this Manual

This manual is designed to provide you with a resource that will make the subject of biology more enjoyable and fun to study. The manual addresses the second year requirements (**A2 schemes**) for the following biology courses: **AQA specifications A** and **B**, **Edexcel**, **OCR**, and **CIE** (Cambridge International Examinations) This manual is also highly suitable for senior biology courses in **Northern Ireland**, **Wales**, and **Scotland (Higher Still)**. Consult the Syllabus Guides on pages 14-19 of this manual (or on the Teacher Resource Handbook) to find out where material for your syllabus is covered. It is hoped that this manual will reinforce and extend the ideas developed by your teacher. It must be emphasised that this manual is **not a textbook**. It is designed to complement the biology textbooks written for your course. The manual provides the following useful resources for each topic:

Guidance Provided for Each Topic

Learning objectives:
These provide you with a map of the topic content. Completing the learning objectives relevant to your course will help you to satisfy the knowledge requirements of your syllabus. Your teacher may decide to leave out points or add to this list.

Topic outcomes:
This panel provides details of the learning objectives that need to be completed to satisfy the requirements (relevant to that topic) for each exam board. Attempt to meet the objectives that relate your exam board only. See pages 14-19 for a synopsis of the syllabus requirements for your course.

Key words:
Key words are displayed in **bold** type in the learning objectives and should be used to create a glossary as you study each topic. From your teacher's descriptions and your own reading, write your own definition for each word.
Note: Only the terms relevant to your learning objectives should be used to create your glossary. Free glossary worksheets are also available from our web site.

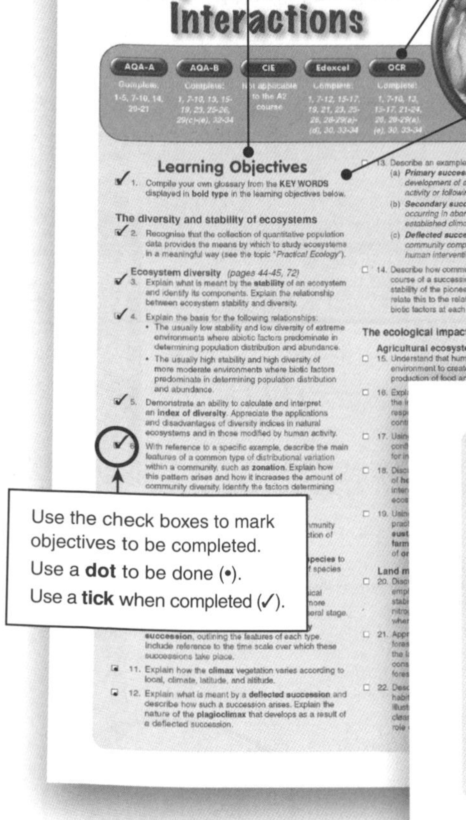

Use the check boxes to mark objectives to be completed.
Use a **dot** to be done (•).
Use a **tick** when completed (✓).

Textbook references:
Provides a list of current texts appropriate for this course. Go to the *Textbook Reference Grid* on page 8 to see the comprehensive textbooks listed (these are texts covering the whole of the course). The grid provides the page numbers from each text relevant to each topic in the manual. Page numbers for supplementary texts, which have only a restricted topic coverage, are provided as appropriate in each topic.

Chapter title tabs:
These are arranged to allow easy identification of different chapters in the manual.

Periodical articles:
Ideal for those seeking more depth or the latest research on a specific topic. Articles are sorted according to their suitability for student or teacher reference. Visit your school, public, or university library for these articles.

Internet addresses:
Access our database of links to more than **800** web sites (updated regularly) relevant to topics covered in the manual. Go to Biozone's own web site: **www.biozone.co.uk** and link directly to listed sites using the *'BioLinks'* button.

Video documentaries and computer software:
Listings of computer software and videos relevant to each topic in the manual are provided in the **"Teacher Resource Handbook"** on CD-ROM (which may be purchased separately). See pages 12-13 for details.

Some of the titles listed may already be available in your school. If not, and you are interested in purchasing them, full supplier details are provided via the resource hub at Biozone's website (see pages 10-11 for details).

Activity Pages

The activities and exercises make up most of the content of this book. They are designed to reinforce the concepts you have learned about in the topic. Your teacher may use the activity pages to introduce a topic for the first time, or you may use them to revise ideas already covered. They are excellent for use in the classroom, and as homework exercises and revision. In most cases, the activities should not be attempted until you have carried out the necessary background reading from your textbook. Your teacher should have a model answers book with the answers to each activity. Because this manual caters for more than one exam board, you will find some activities that may not be relevant to your course. Although you will miss out these pages, our manuals still represent exceptional value.

Introductory paragraph:
The introductory paragraph sets the 'scene' for the focus of the page. Note any words that appear in **bold**, as they are 'key words' worthy of including in a glossary of biological terms for the topic.

Tear-out pages:
Each page of the book has a perforation that allows easy removal. Your teacher may ask you to remove activity pages from this manual for marking. Your teacher may also request that you to tear out the pages and place them in a ring-binder folder with other work on the topic.

Easy to understand diagrams:
The main ideas of the topic are represented and explained by clear, informative diagrams.

Write-on format:
Your understanding of the main ideas of the topic is tested by asking questions and providing spaces for your answers. Where indicated, your answers should be concise. Questions requiring explanation or discussion are spaced accordingly. Answer the questions adequately according to the questioning term used (see the facing page).

175

The Founder Effect

Occasionally, a small number of individuals from a large population may migrate away, or become isolated from, their original population. If this colonising or 'founder' population is made up of only a few individuals, it will probably have a *non-representative sample* of alleles from the parent population's gene pool. As a consequence of this **founder effect**, the colonising population may evolve differently from that of the parent population, particularly since the environmental conditions for the isolated population may be different. In some cases, it may be possible for certain alleles to be missing altogether from the individuals in the isolated population. Future generations of this population will not have this allele.

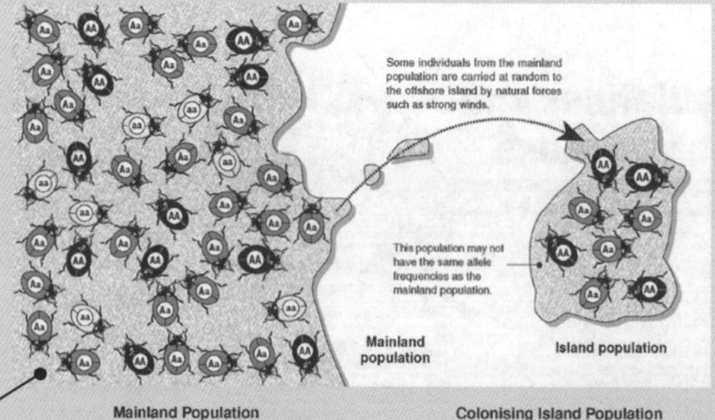

Mainland Population

	Allele frequencies	
	Actual numbers	Calculate %
Allele A		
Allele a		
Total		

Phenotype frequencies		
Black	Dark	Pale

AA Aa aa

Colonising Island Population

	Allele frequencies	
	Actual numbers	Calculate %
Allele A		
Allele a		
Total		

Phenotype frequencies		
Black	Dark	Pale

AA Aa aa

1. Compare the mainland population to the population which ended up on the island (use the spaces in the tables above):
   (a) Count the **phenotype** numbers for the two populations (ie. the number of black, dark and pale beetles).
   (b) Count the **allele** numbers for the two populations: the number of dominant alleles (A) and recessive alleles (a). Calculate these as a percentage of the total number of alleles for each population.
2. Describe how the allele frequencies of the two populations are different: ____________
3. Describe some possible ways in which various types of organism can be **carried** to an offshore island:
   (a) Plants: ____________
   (b) Land animals: ____________
   (c) Non-marine birds: ____________
4. Since founder populations are often very small, **describe** another process that may further alter the allele frequencies:

Code: DA 3

Population Genetics & Evolution

Activity code:
To assist you in identifying the type of activities in this manual and the skills they require, activities are coded. Note that most activities will require some knowledge recall. Unless this is all that is required, this code is excluded from the coding list.

Code: DA 3

Activity Level

1 = Simple questions not requiring complex reasoning
2 = Some complex reasoning may be required
3 = More challenging, requiring integration of concepts

Type of Activity

D = Data handling and/or interpretation
P = Paper practical
R = Research outside the information on the page*
A = Application of knowledge to solve a problem
K = Knowledge recall from information on the page
E = Extension material

* Material to assist with the activity may be found on other pages of the manual or in textbooks.

Explanation of Terms

Questions come in a variety of forms. Whether you are studying for an exam or writing an essay, it is important to understand exactly what the question is asking. A question has two parts to it: one part of the question will provide you with information, the second part of the question will provide you with instructions as to how to answer the question. Following these instructions is most important. Often students in examinations know the material but fail to follow instructions and do not answer the question appropriately. Examiners often use certain key words to introduce questions. Look out for them and be clear as to what they mean. Below is a description of terms commonly used when asking questions in biology.

Commonly used Terms in Biology

The following terms are frequently used when asking questions in examinations and assessments. Students should have a clear understanding of each of the following terms and use this understanding to answer questions appropriately.

Analyse: Interpret data to reach stated conclusions.

Annotate: Add **brief** notes to a diagram, drawing or graph.

Apply: Use an idea, equation, principle, theory, or law in a new situation.

Appreciate: To understand the meaning or relevance of a particular situation.

Calculate: Find an answer using mathematical methods. Show the working unless instructed not to.

Compare: Give an account of similarities and differences between two or more items, referring to both (or all) of them throughout. Comparisons can be given using a table. Comparisons generally ask for similarities more than differences (see contrast).

Construct: Represent or develop in graphical form.

Contrast: Show differences. Set in opposition.

Deduce: Reach a conclusion from information given.

Define: Give the precise meaning of a word or phrase as concisely as possible.

Derive: Manipulate a mathematical equation to give a new equation or result.

Describe: Give a detailed account, including all the relevant information.

Design: Produce a plan, object, simulation or model.

Determine: Find the only possible answer.

Discuss: Give an account including, where possible, a range of arguments, assessments of the relative importance of various factors, or comparison of alternative hypotheses.

Distinguish: Give the difference(s) between two or more different items.

Draw: Represent by means of pencil lines. Add labels unless told not to do so.

Estimate: Find an approximate value for an unknown quantity, based on the information provided and application of scientific knowledge.

Evaluate: Assess the implications and limitations.

Explain: Give a clear account including causes, reasons, or mechanisms.

Identify: Find an answer from a number of possibilities.

Illustrate: Give concrete examples. Explain clearly by using comparisons or examples.

Interpret: Comment upon, give examples, describe relationships. Describe, then evaluate.

List: Give a sequence of names or other brief answers with no elaboration. Each one should be clearly distinguishable from the others.

Measure: Find a value for a quantity.

Outline: Give a brief account or summary. Include essential information only.

Predict: Give an expected result.

Solve: Obtain an answer using algebraic and/or numerical methods.

State: Give a specific name, value, or other answer. No supporting argument or calculation is necessary.

Suggest: Propose a hypothesis or other possible explanation.

Summarise: Give a brief, condensed account. Include conclusions and avoid unnecessary details.

In Conclusion

Students should familiarise themselves with this list of terms and, where necessary throughout the course, they should refer back to them when answering questions. The list of terms mentioned above is not exhaustive and students should compare this list with past examination papers / essays etc. and add any new terms (and their meaning) to the list above. The aim is to become familiar with interpreting the question and answering it appropriately.

Resources Information

Your set textbook should always be a starting point for information. There are also many other resources available, including scientific journals, magazine and newspaper articles, supplementary texts covering restricted topic areas, dictionaries, computer software and videos, and the internet.

A synopsis of currently available resources is provided below. Access to the publishers of these resources can be made directly from Biozone's web site through our resources hub: **www.biozone.co.uk/resource-hub.html**, or by typing in the relevant addresses provided below. Most titles are also available through amazon.co.uk. Please note that our listing any product in this manual does not, in any way, denote Biozone's endorsement of that product.

Comprehensive Biology Texts Referenced

Appropriate texts for this course are referenced in this manual. Page or chapter references for each text are provided in the text reference grid on page 8. These will enable you to identify the relevant reading as you progress through the activities in this manual. Publication details of texts referenced in the grid are provided below and opposite. For further details of text content, or to make purchases, link to the relevant publisher via Biozone's resources hub or by typing: **www.biozone.co.uk/resources/uk-comprehensive-pg1.html**

Bailey, M. and K. Hirst, 2001
Collins Advanced Modular - Biology A2, 2/edn
Publisher: HarperCollins Publishers Ltd.
Pages: 250
ISBN: 0-00-327752-6
Comments: *Appropriate to AQA A2-level Biology (B) specification. Student support packs for AQA B and Edexcel specifications are available at extra cost.*

Baker, M., B. Indge, and M. Rowland, 2001
AQA Biology Specification A: Further Studies in Biology (A2)
Publisher: Hodder & Stoughton
Pages: 276
ISBN: 0-340-80244-8
Comments: *Written specifically for the AQA biology specification A. The text covers, in full, the first two modules of the A level course. Extension material is provided.*

Boyle, M. and K. Senior, 2002
Collins Advanced Science: Biology, 2/e
Publisher: HarperCollins Publishers Ltd.
Pages: 675
ISBN: 0-00-713600-5
Comments: *Fully revised to reflect changes in the specifications. Includes a new section on plant biology and a new chapter on biotechnology.*

Bradfield, P., J. Dodds, J. Dodds, and N. Taylor, 2002
A Level Biology
Publisher: Longman
Pages: 464
ISBN: 0-582-42945-5
Comments: *Written to match A2 level specifications, providing synoptic questions and core content.*

Clegg, C.J., 2000
Introduction to Advanced Biology
Publisher: John Murray
Pages: 528
ISBN: 0-7195-7671-7
Comments: *Matches new AS/A2 specifications, but aims to provide a bridge from GCSE (grade C pass) to A level. Detailed specification matches are available on CD-ROM, FREE on demand.*

Clegg, C.J. and D.G. McKean, 2000
Advanced Biology: Principles and Applications 2/e
Publisher: John Murray
Pages: 712
ISBN: 0-7195-7670-9
Comments: *Student study guide also available. A general text with specific references for use by biology students of the AS and A2 curricula in the UK.*

Fullick, A., 2000
Heinemann Advanced Science: Biology, 2 edn
Publisher: Heinemann Educational Publishers
Pages: 760
ISBN: 0-435-57095-1
Comments: *Suitable as a general text for use with all specifications. The earlier edition (1994) is still available.*

Hanson, M.,1999
New Perspectives in Advanced Biology
Publisher: Hodder & Stoughton
Pages: 880
ISBN: 0-340-66443-6
Comments: *Black and white format. A general text suitable for students of biology in the UK and elsewhere.*

Jones, M., R. Fosbery, D. Taylor, and J. Gregory, 2003
CIE Biology AS and A Level
Publisher: Cambridge University Press
Pages: 352
ISBN: 0-521-53674-X
Comments: *This compact text caters for the Cambridge International Examinations core syllabus for Biology AS and A level. It includes questions and learning objectives.*

Jones, M., and J. Gregory, 2001
Biology 2
Publisher: Cambridge University Press
Pages: 152
ISBN: 0-521-79714-4
Comments: *Provides coverage of the core material for the second year of Advanced Level Biology (OCR), and covers the Advanced Level module, 'Central Concepts'.*

Jones, M. and G. Jones, 1997
Advanced Biology
Publisher: Cambridge University Press
Pages: 560
ISBN: 0-521-48473-1
Comments: *Provides full coverage of the core material included in all advanced biology syllabuses. Suitable for students studying for a range of qualifications.*

Kent, N. A. 2000
Advanced Biology
Publisher: Oxford University Press
Pages: 624
ISBN: 0-19-914195-9
Comments: *Each book comes with a free CD-ROM to help with specification planning. Book is formatted as a series of two page concept spreads.*

Lea, C., P. Lowrie, and S. McGuigan, (S. McGuigan, ed), 2001
Advanced Level Biology for AQA: A2
Publisher: Heinemann Educational Publishers
Pages: 315
ISBN: 0-435-58081-7
Comments: *Appropriate for AQA specification B. Accompanying Resource Pack, providing support material, and extra questions and tasks, is also available.*

Roberts, M., G. Monger, M. Reiss, 2000
Advanced Biology
Publisher: NelsonThornes
Pages: 780
ISBN: 0-17-438732-6
Comments: *Provides complete coverage of the new Advanced Level specifications. Additional support provided from a complementary website.*

Taylor D.J., N.P.O. Green, & G.W. Stout, 1997
Biological Science 1 & 2, 3 ed. (hardback)
Publisher: Cambridge University Press
Pages: 984
ISBN: 0-521-56178-7
Comments: *Revised and updated to cover the latest AS/A2 syllabuses. Also available as a softback two-volume set: Biological Science 1: 0-521-56721-1 and Biological Science 2: 0-521-56720-3.*

Toole G. and S. Toole, 2000
New Understanding Biology for Advanced Level
Publisher: NelsonThornes
Pages: 672
ISBN: 0-7487-3964-5
Comments: *Fully revised to match AS/A2 specifications. Also available with study guide, which provides support material.*

Toole G. and S. Toole, 2004
Essential A2 Biology for OCR
Publisher: NelsonThornes
Pages: approx 170
ISBN: 0-7487-8518-3
Comments: *An appealing and student-friendly text written to meet the core A2 specifications for OCR. A comprehensive glossary is included.*

Williams, G., 2000
Advanced Biology for You
Publisher: NelsonThornes
Pages: 464
ISBN: 0-7487-5298-6
Comments: *Covers the current AS level specifications and the core topics of all A2 specifications.*

Supplementary Texts

For further details of text content, or to make purchases, link to the relevant publisher via Biozone's resources hub or by typing: **www.biozone.co.uk/resources/uk-supplementary-pg1.html**

Barnard, C., F. Gilbert, F., and P. McGregor, 2001
Asking Questions in Biology: Key Skills for Practical Assessments & Project Work, 208 pp.
Publisher: Prentice Hall
ISBN: 0130-90370-1
Comments: *Coverage of design, analysis, and presentation of practical work in senior biology.*

Cadogan, A. and Ingram, M., 2002
Maths for Advanced Biology
Publisher: NelsonThornes
ISBN: 0-7487-6506-9
Comments: *Fully covers the new maths requirements of the AS/A2 biology specifications. Includes worked examples.*

Freeland, P., 1999
Hodder Advanced Science: Microbes, Medicine, and Commerce, 160 pp.
Publisher: Hodder and Stoughton
ISBN: 0340731036
Comments: *Combines biotechnology, microbiology, pathology, and immunity in a comprehensive text.*

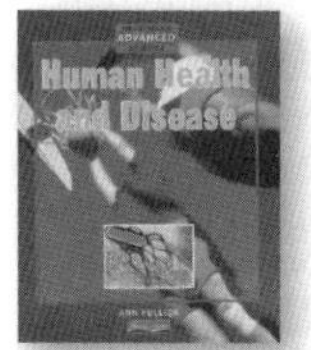

Fullick, A., 1998
Human Health and Disease, 162 pp.
Publisher: Heinemann Educational Publishers
ISBN: 0435570919
Comments: *An accompanying text for courses with modules in human health and disease. Covers both infectious and non-infectious disease.*

Jones, A., R. Reed, and J. Weyers, 2nd edn, 1998
Practical Skills in Biology, 292 pp.
Publisher: Longman
ISBN: 0-582-29885-7
Comments: *Good, accurate guidance on study design, implementation, and data analysis. The third edition in hardback is also available (2002, ISBN: 0-130-45141-X). Available through Amazon.*

Morgan, S., 2002
Advanced Level Practical Work for Biology 128 pp.
Publisher: Hodder and Stoughton
ISBN: 0-340-84712-3
Comments: *Caters for the practical and investigative requirement of A level studies. Covers experimental planning, technique, observations and measurement, and interpretation and analysis.*

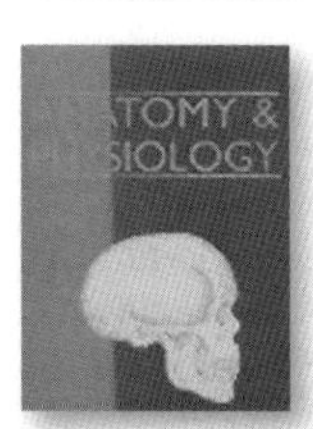

Rowett, H.G.Q, 1999
Basic Anatomy & Physiology, 4 ed. 132 pp.
Publisher: John Murray
ISBN: 0-7195-8592-9
Comments: *A revision of a well established reference book for the basics of human anatomy and physiology. Accurate coverage of required AS/A2 content with clear, informative diagrams.*

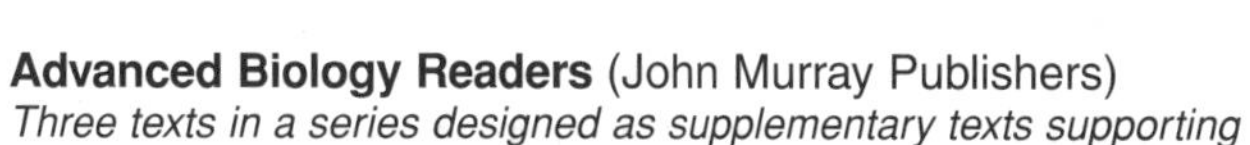

Advanced Biology Readers (John Murray Publishers)
Three texts in a series designed as supplementary texts supporting a range of Advanced Biology options. They are also useful as teacher reference and student extension reading for core topics.

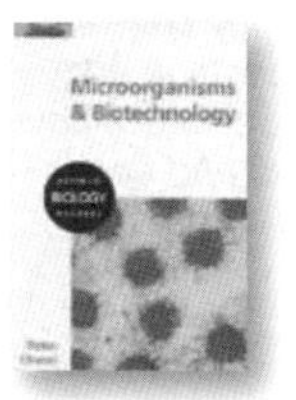

Chenn, P., 1997.
Microorganisms and Biotechnology, 176 pp.
ISBN: 0-71957-509-5
Thorough coverage of the nature of microorganisms, their culture and growth, and their various roles in biotechnology. It includes chapters on the genetic engineering of microbes and enzyme technology.

Chenn, P., 1999.
Ecology, 224 pp.
ISBN: 0-7195-7510-9
A useful, well organised supplementary resource covering all basic areas of ecology (including practical work). Chapters offer considerable depth and many illustrative examples.

Jones, N., A. Karp., & G. Giddings, 2001.
Essentials of Genetics, 224 pp.
ISBN: 0-7195-8611-9
Thorough supplemental for genetics and evolution. Comprehensive coverage of cell division, molecular genetics, and genetic engineering is provided, and the application of new gene technologies to humans is discussed in a concluding chapter.

Bath Advanced Sciences Series (NelsonThornes)
The Bath Advanced Sciences series provides a number of titles (including a full text) to cover Advanced Level material for biology.

Taylor, J., 2001.
Microorganisms and Biotechnology, 192 pp.
ISBN: 0-17-448255-8
Comments: *Provides information for microbiology and biotechnology topics at A2 level. Other older titles in this series include Applied Genetics, 1990 (ISBN: 0174385110) and Biochemistry and Molecular Biology, 1994 (ISBN: 0174482078).*

Cambridge Advanced Sciences / Advanced Biology Topics (Cambridge University Press)
Modular titles. Those in the Advanced Sciences Series cover the A2 options for OCR, but are suitable as teacher reference and student extension reading for core topics in other AS/A2 biology courses.

Dockery, M. and M. Reiss, 1999. **Advanced Biology Topics: Behaviour**, 120 pp.
ISBN: 0521597544
An introductory text covering the basics of human and animal behaviour including environmental influence, behavioural development, courtship, and social interaction.

Gregory, J. 2000.
Applications of Genetics, 82 pp.
ISBN: 0521787254
Topics covered include the basis of variation, selective breeding and heritability (including AI in humans and livestock), genetic diversity and cloning, genetic engineering (techniques and ethics), and a chapter on human genetics.

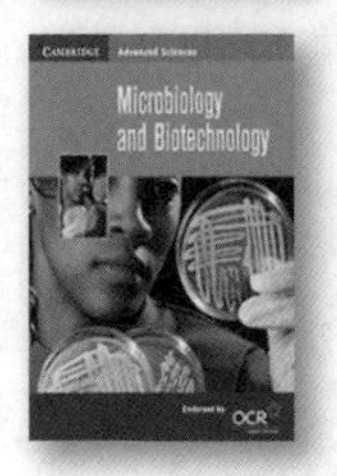

Jones, M. and G. Jones, 2002.
Mammalian Physiology and Behaviour, 104 pp.
ISBN: 0521797497
Covers mammalian nutrition, the structure and function of the liver, support and locomotion, the nervous system, and senses and behaviour. Questions and exercises are provided and each chapter includes an introduction and summary.

Lowrie, P. & S. Wells, 2000.
Microbiology and Biotechnology, 112 pp.
ISBN: 0521787238
This text covers the microbial groups important in biotechnology, basic microbiological techniques, and the various applications of microbes. The chapter "Biotechnology in industry and public health" covers the industrial-scale production and use of microbial enzymes.

Reiss, M. & J. Chapman, 2000.
Environmental Biology, 104 pp.
ISBN: 0521787270
An introduction to environmental biology covering agriculture, pollution, resource conservation and conservation issues, and practical work in ecology. Questions and exercises are provided and each chapter includes an introduction and summary.

Taylor, D. 2001.
Growth, Development and Reproduction, 120 pp.
ISBN: 0521787211
Includes coverage of asexual reproduction, sexual reproduction in flowering plants and mammals, and regulation of growth and development. Questions and exercises are provided and each chapter includes an introduction and summary.

Collins Advanced Modular Sciences (HarperCollins)
Modular-style texts covering material for the A2 options for AQA specification B, but suitable as teacher reference and student extension reading for core topics in other A2 biology courses.

Allen, D, M. Jones, and G. Williams, 2001.
Applied Ecology, 104 pp.
ISBN: 0-00-327741-0
Includes coverage of methods in practical ecology, the effects of pollution on diversity, adaptations, agricultural ecosystems and harvesting (including fisheries), and conservation issues. Local examples are emphasised throughout.

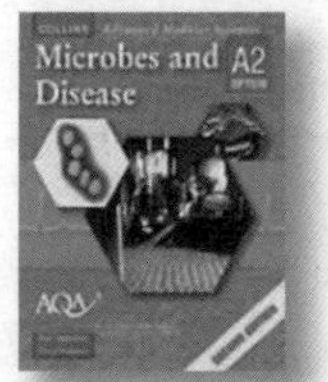

Hudson, T. and K. Mannion, 2001.
Microbes and Disease, 104 pp.
ISBN: 0-00-327742-9
Coverage of selected aspects of microbiology including the culture and applications of bacteria, and the role of bacteria and viruses in disease. Immunity, vaccination, and antimicrobial drug use are covered in the concluding chapter.

Murray, P. & N. Owens, 2001.
Behaviour and Populations, 82 pp.
ISBN: 0-00-327743-7
This text covers an eclectic range of topics including patterns of behaviour, reproduction and its control, human growth and development, human populations, aspects of infectious disease, and issues related to health and lifestyle.

Nelson Advanced Sciences (NelsonThornes)
Modular-style texts covering material for the A2 specifications for Edexcel, but suitable as teacher reference and student extension reading for core topics in other AS/A2 biology courses.

Adds, J., E. Larkcom & R. Miller, 2000.
Exchange and Transport, Energy and Ecosystems, 216 pp.
ISBN: 0-17-448294-9
Includes exchange processes (gas exchanges, digestion, absorption), transport systems, adaptation, sexual reproduction, energy and the environment, and human impact. Practical activities are included in several of the chapters.

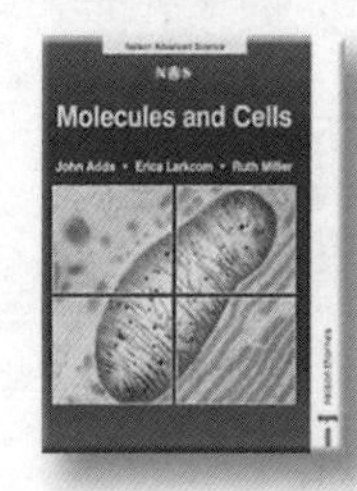

Adds, J., E. Larkcom & R. Miller, 2000.
Molecules and Cells, 112 pp.
ISBN: 0-17-448293-0
Includes coverage of the basic types of biological molecules, with extra detail on the structure and function of nucleic acids, enzymes, cellular organisation, and cell division. Practical activities are provided for most chapters.

Adds, J., E. Larkcom & R. Miller, 2001.
Genetics, Evolution, and Biodiversity, 200 pp.
ISBN: 0-17-448296-5
A range of topics including photosynthesis and the control of growth in plants, classification and quantitative field ecology, populations and pest control, conservation, Mendelian genetics and evolution, gene technology and human evolution. Practical activities are included in many chapters.

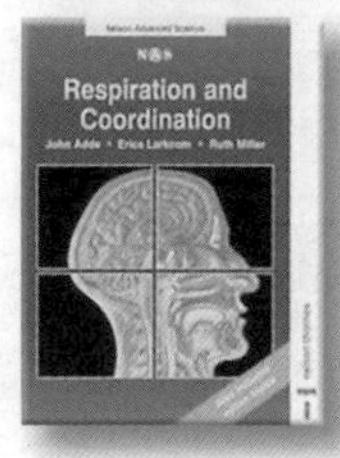

Adds, J., E. Larkcom & R. Miller, 2001.
Respiration and Coordination, 200 pp.
ISBN: 0-17-448295-7
This text covers Unit 4 and all option modules for Edexcel, providing material on metabolic pathways, internal regulation and nervous coordination, microbiology and biotechnology, food science, and health and exercise physiology.

Adds, J., E. Larkcom, R. Miller, & R. Sutton, 1999. **Tools, Techniques and Assessment in Biology**, 160 pp.
ISBN: 0-17-448273-6
A course guide covering basic lab protocols, microscopy, quantitative techniques in the lab and field, advanced DNA techniques and tissue culture, data handling and statistical tests, and exam preparation. Includes useful appendices.

Illustrated Advanced Biology (John Murray Publishers)
Modular-style texts aimed as supplements to students of AS and A2 level biology courses.

Clegg, C.J., 2002.
Microbes in Action, 92 pp.
ISBN: 0-71957-554-0
Microbes and their roles in disease and biotechnology. It includes material on the diversity of the microbial world, microbiological techniques, and a short, but useful, account of enzyme technology.

Clegg, C.J., 1999.
Genetics and Evolution, 96 pp.
ISBN: 0-7195-7552-4
Concise but thorough coverage of molecular genetics, genetic engineering, inheritance, and evolution. An historical perspective is included by way of introduction, and a glossary and a list of abbreviations used are included.

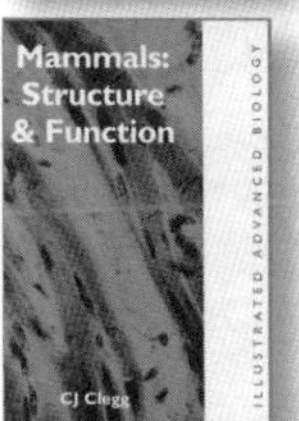

Clegg, C.J., 1998.
Mammals: Structure & Function, 96 pp.
ISBN: 0-7195-7551-6
Includes an introduction to classification and diversity followed by a section on basic mammalian histology. Aspects of mammalian physiology are presented, including nutrition, exchange and transport, and reproduction.

Clegg, C.J., 2003
Green Plants: The Inside Story, approx. 96 pp.
ISBN: 0-7195-7553-2
The emphasis in this text is on flowering plants. Topics include leaf, stem, and root structure in relation to function, reproduction, economic botany, and sensitivity and adaptation.

Biology Dictionaries

A good biology dictionary is of great value when studying biology. Some that you may wish to locate or purchase are listed below. They can usually be sourced directly from the publisher or they are all available (at the time of printing) from www.amazon.co.uk. For further details of text content, or to make purchases, link to the relevant publisher via Biozone's resources hub or by typing: **www.biozone.co.uk/resources/dictionaries-pg1.html**

Clamp, A.
AS/A-Level Biology. Essential Word Dictionary, 2000, 161 pp. Philip Allan Updates.
ISBN: 0-86003-372-4.
Carefully selected essential words for AS and A2. Concise definitions are supported by further explanation and illustrations where required.

Hale, W.G., J.P. Margham, & V.A. Saunders
Collins: Dictionary of Biology 3 ed. 2003, 672 pp. HarperCollins.
ISBN: 0-00-714709-0.
Completely revised and updated to take in the latest developments in biology from the Human Genome Project to advancements in cloning.

McGraw-Hill (ed). **McGraw-Hill Dictionary of Bioscience**, 2 ed., 2002, 662 pp. McGraw-Hill.
ISBN: 0-07-141043-0
22 000 entries encompassing more than 20 areas of the life sciences. It includes synonyms, acronyms, abbreviations, and pronunciations for all terms. Accessible, yet comprehensive.

Thain, M. and M. Hickman. **Penguin Dictionary of Biology** 10 ed., 2000, 704 pp. Penguin.
ISBN: 0-14-051359-0
Concise reference with definitions to more than 7500 terms, including more than 400 new entries. It covers fundamental concepts and recent advances.

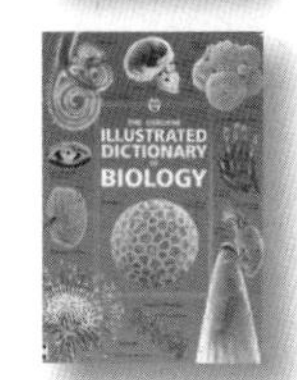

Stockley, C., 2000.
The Usborne Illustrated Dictionary of Biology 2nd ed., 128 pp. Usborne Publishing Ltd.
ISBN: 0-7460-3792-9
User-friendly and thematically organised, with terms explained in context and definitions well supported by illustrations. Available through selected retailers.

Periodicals, Magazines and Journals

Articles in *Biological Sciences Review (Biol. Sci. Rev.)*, *New Scientist*, and *Scientific American* can be of great value in providing current information on specific topics. Periodicals may be accessed in your school, local, public, and university libraries. Listed below are the periodicals referenced in this manual. For general enquiries and further details regarding subscriptions, link to the relevant publisher via Biozone's resources hub or by typing: **www.biozone.co.uk/resources/resource-journal.html**

Biological Sciences Review: *An informative and very readable quarterly publication for teachers and students of biology.* Enquiries: Philip Allan Publishers, Market Place, Deddington, Oxfordshire OX 15 OSE.
Tel: 01869 338652
Fax: 01869 338803
E-mail: sales@philipallan.co.uk
or subscribe from their web site.

New Scientist: *Widely available weekly magazine. Provides summaries of research in articles ranging from news releases to 3-5 page features on recent research.*
Subscription enquiries:
Reed Business Information Ltd
151 Wardour St. London WIV 4BN
Tel: (UK and intl):+44 (0) 1444 475636
E-mail: ns.subs@qss-uk.com
or subscribe from their web site.

Scientific American: *A monthly magazine containing mostly specialist feature articles. Articles range in level of reading difficulty and assumed knowledge.*
Subscription enquiries:
415 Madison Ave. New York. NY10017-1111
Tel: (outside North America): 515-247-7631
or subscribe from their web site.

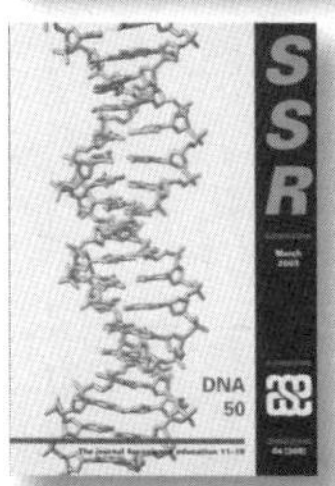

School Science Review: *A quarterly journal published by the ASE for science teachers in 11-19 education. SSR includes articles, reviews, and news on current research and curriculum development. Free to Ordinary Members of the ASE or available on subscription.* Subscription enquiries:
Tel: 01707 28300
Email: info@ase.org.uk *or visit their web site.*

Biologist: *Published five times a year, this journal from the IOB includes articles relevant to teachers of biology in the UK. Articles referenced in the manual can be identified by title and volume number. The IOB also publish the Journal of Biological Education, which provides articles and reviews relevant to those in the teaching profession. Archived articles from both journals are available online at no cost to IOB members and subscribers. Visit their web site for more information.*

Textbook Reference Grid

Guide to use: Page numbers given in the grid refer to the material provided in each text relevant to the stated topic in the manual.

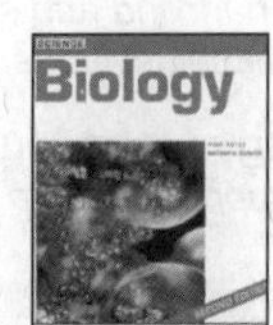

TOPIC IN MANUAL	Bailey & Hirst 2001	Baker *et al.* 2001	Boyle & Senior 2002	Bradfield *et al.* 2002
Cellular Metabolism	2-17	93-103	133-186	Chpt 1-2
Populations and Interactions	168-188	79-87, 108 (symbiosis)	663-689, 695-696	379-402, 409-416
Practical Ecology	156-167	75-79	635-645 (in part)	388, 402-408
Sources of Variation	118-124	7-13, 31-40	Chpt. 24, 27 (in part)	105-109
Inheritance	96-117	13-27	435-444, 503-520	57-99
Population Genetics and Evolution	124-134	41-51, 54-59, 68	521-546	110-150
Biodiversity and Classification	134-137	59-69	N/A	155-182
Homeostasis	18-33	174-192	269-312	293-341
Responses and Coordination	48-85	249-270	345-382	Chpt. 6-7 & 11

Figures refer to page numbers unless indicated otherwise

TOPIC IN MANUAL	Jones & Jones 1997	Kent 2000	Lea *et al.* 2001	Roberts *et al.* 2000
Cellular Metabolism	169-204	86-110	10-13, 156-183	138-155, 197-213
Populations and Interactions	433-443, 463-468	512-525, 546-547	216-231	663-667, 690-717
Practical Ecology	428-432	570-572	232-242	Parts of chpt 37-39 as required
Sources of Variation	88-94, 107-12, 147-50	78-79, 414-415	14-19, 38-45	461-465, 585-586, 588-594
Inheritance	127-131, 133-143	416-431	6-9, 14-35	571-583, 587-590
Population Genetics and Evolution	132-133, 147-168	436-437, 442-454	46-71	Chpt. 36, 42-44 as required
Biodiversity and Classification	497-524	464-489, 494-501	See companion AS text	79-102
Homeostasis	323-348	136-158, 184-187	79-100	267-315 as required
Responses and Coordination	297-321, 349-356	188-208, 226-33, 308-13	74-78, 101-143	345-399

Figures refer to page numbers unless indicated otherwise

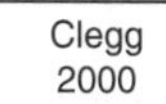

Clegg 2000	Clegg & McKean 2000	Fullick 2000	Hanson 1999	Jones & Gregory 2001	Jones *et al.* 2003
169-204	211, 246-269, 314-319	177-196, 224-235	76-93, 186-87, 606-638	1-27, 63-64	196-220
374-391	48-55, 64-71, 80-83	510-513, 524-557, 570-572	687-724, 727, 742-745	28-36, 38-44	N/A
376-377	68-69	NA	711-712	36-38	N/A
103-106, 115-16, 436-38	196-99, 616-17, 624-27	368-371, 440-42, 452-455, 461-464	116-22, 133-39, 158-59	47-52, 61-64	221-226, 235-237, 244-245
439-456	604-625	429-449	122-152	51-60	226-234
458-471	656-667	495-505	222-254	67-79	245-256
344-368	18-43	479-492	158-159, 257-294	80-85	N/A
252-272	372-391 (in part)	287-294, 314-323	431-460	87-105	257-276
274-303	418-467, 516-522	241-285, 298-312	303-347, 586-605	106-123	276-294

Taylor *et al.* 1997	Toole & Toole 2000	Toole & Toole 2004	Williams 2000
197-226, 264-277, 806	43, 269-284, 316-329	6-31	284-309
312-326, 332-333	347-348, 354-361, 366-368, 378-383	32-49	406-415, 424, 431
349-374	348-354	N/A	416-422
783-90, 814-17, 828-33	139-145, 161-162, 179-181, 186-195	50-53, 68-71, 82-83	119-122, 361, 370-373
808-828	151-172	54-67	354-369
884-927	199-210	84-97	377-384, 388
3-76	74-97	76-81	385-387
598-608, 665-698	476-86, 497-511, 514-16	98-109, 126-131	310-325
532-597, 609-622	526-561, 577-589	110-125, 132-135	326-348

Using the Internet

The internet is a vast global network of computers connected by a system that allows information to be passed through telephone connections. When people talk about the internet they usually mean the **World Wide Web** (WWW). The WWW is a service that has made the internet so simple to use that virtually anyone can find their way around, exchange messages, search libraries and perform all manner of tasks. The internet is a powerful resource for locating information. Listed below are two journal articles worth reading. They contain useful information on what the internet is, how to get started, examples of useful web sites, and how to search the internet.

- **Click Here: Biology on the Internet** Biol. Sci. Rev., 10(2) November 1997, pp. 26-29.
- **An A-level biologists guide to The World Wide Web** Biol. Sci. Rev., 10(4) March 1998, pp. 26-29.

Using the Biozone Website: www.biozone.co.uk

The **Back** and **Forward** buttons allow you to navigate between pages displayed on a WWW site

The current **internet address (URL)** for the web site is displayed here. You can type in a new address directly into this space.

Tool bar provides a row of buttons with shortcuts for some commonly performed tasks, such as printing a page or 'refreshing' the page (i.e. making the page load again).

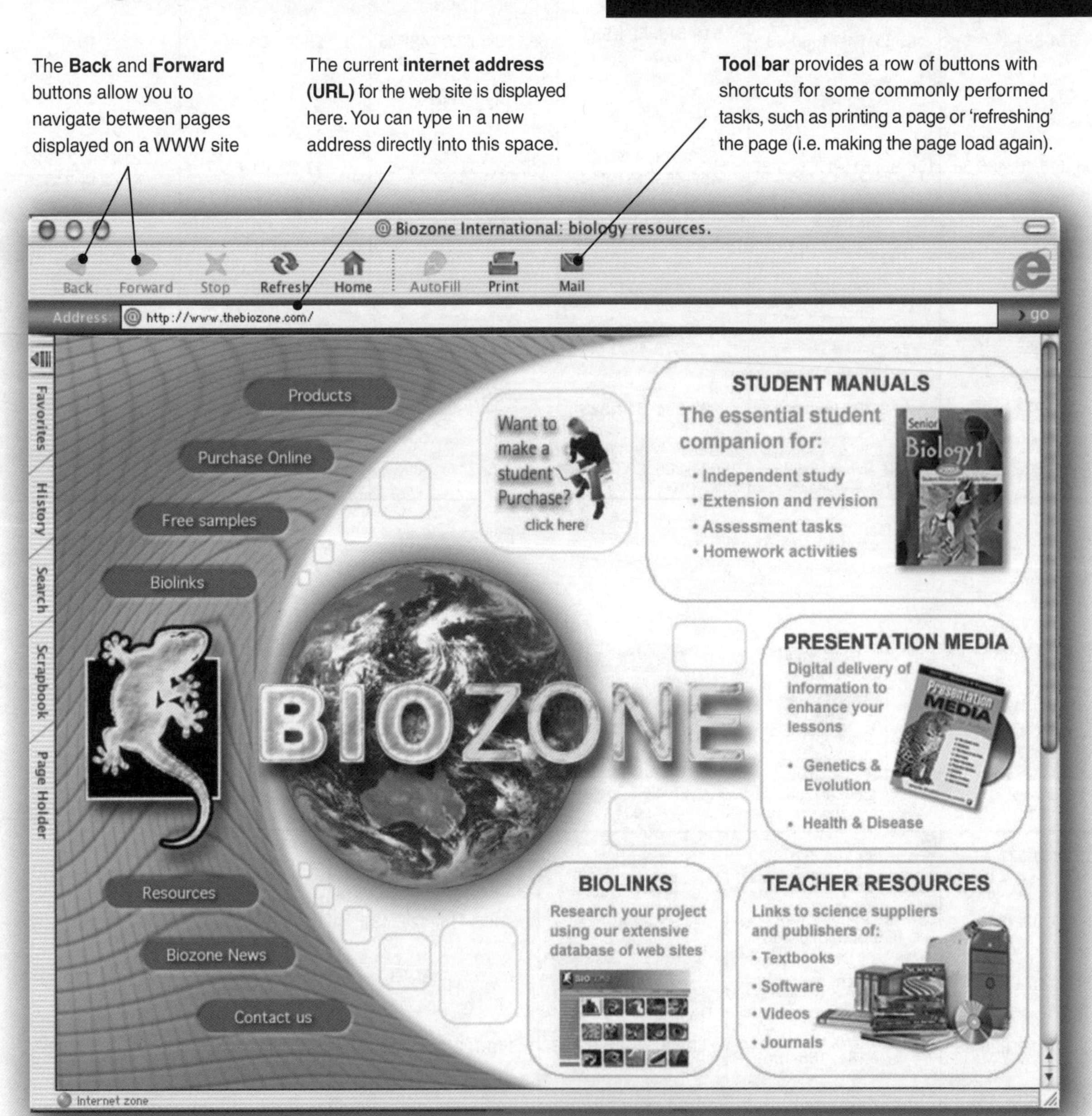

Searching the Net

The WWW addresses listed throughout the manual have been selected for their relevance to the topic in which they are listed. We believe they are good sites. Don't just rely on the sites that we have listed. Use the powerful 'search engines', which can scan the millions of sites for useful information. Here are some good ones to try:

Alta Vista:	**www.altavista.com**
Ask Jeeves:	**www.ask.com**
Excite:	**www.excite.com/search**
Google:	**www.google.com**
Go.com:	**www.go.com**
Lycos:	**www.lycos.com**
Metacrawler:	**www.metacrawler.com**
Yahoo:	**www.yahoo.com**

Biozone International provides a service on its web site that links to all internet sites listed in this manual. Our web site also provides regular updates with new sites listed as they come to our notice and defunct sites deleted. Our **BIO LINKS** page, shown below, will take you to a database of regularly updated links to more than 800 other quality biology web sites.

The **Resource Hub**, accessed via the homepage or resources, provides links to the supporting resources referenced in the manual and the ***Teacher Resource Handbook***. These resources include *comprehensive* and *supplementary texts, biology dictionaries, computer software, videos*, and *science supplies*. These can be used to supplement and enhance your learning experience.

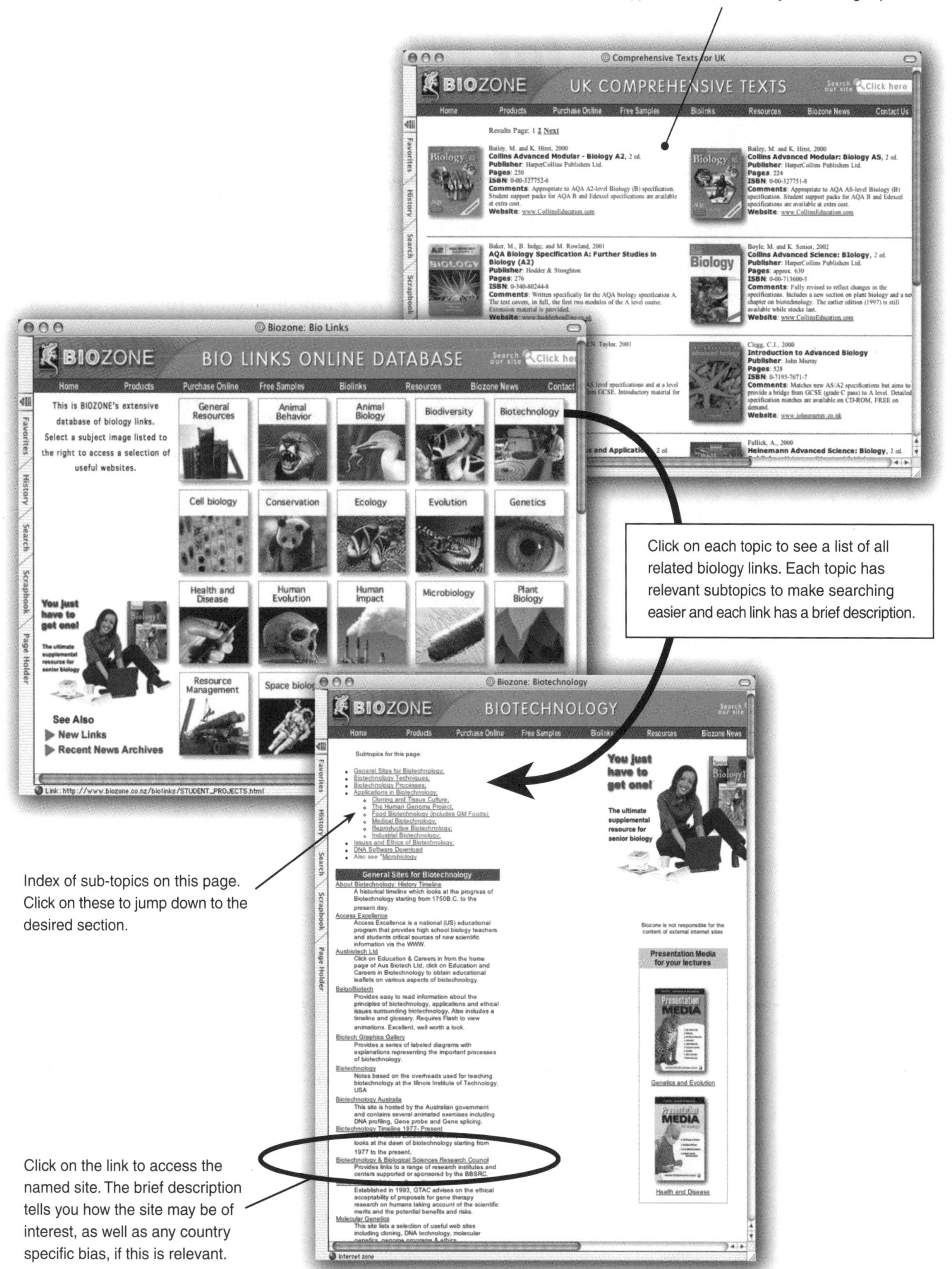

Using the Teacher Resource Handbook

A great collection of supporting resources on CD-ROM for this manual

Acrobat PDF files supplied on CD-ROM provide hyperlinks to an extensive collection of resources.

Details of this product, which can be purchased for just £30, can be viewed at: **www.biozone.co.uk/Products_UK.html**

NOTE: Photocopy licence EXPIRES on **30 June 2005**

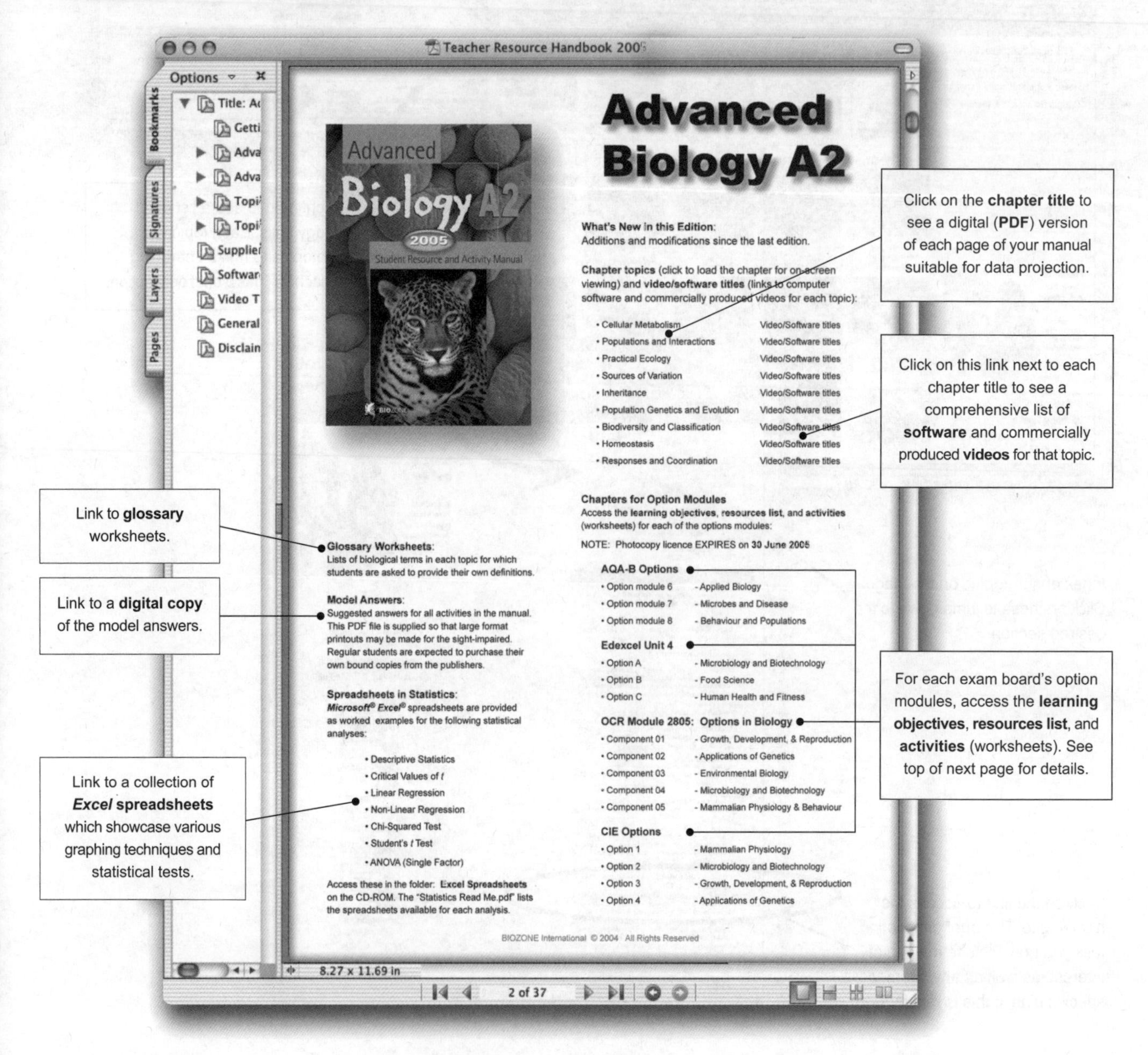

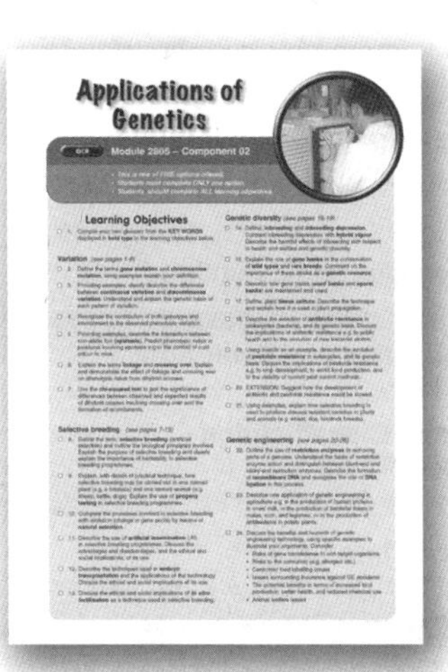

Chapters for Option Modules

Access the **learning objectives**, **resources list**, and **activities** (worksheets) for each of the options modules. The purchase of this package entitles the school to print and photocopy these pages under a limited licencing agreement. NOTE: Photocopy licence EXPIRES on **30 June 2005**

AQA-B Options
- Option module 6: Applied Biology
- Option module 7: Microbes and Disease
- Option module 8: Behaviour and Populations

OCR Module 2805: Options in Biology
- Component 01: Growth, Development, & Reproduction
- Component 02: Applications of Genetics
- Component 03: Environmental Biology
- Component 04: Microbiology and Biotechnology
- Component 05: Mammalian Physiology & Behaviour

Edexcel Unit 4
- Option A: Microbiology and Biotechnology
- Option B: Food Science
- Option C: Human Health and Fitness

CIE Options
- Option 1: Mammalian Physiology
- Option 2: Microbiology and Biotechnology
- Option 3: Growth, Development, & Reproduction
- Option 4: Applications of Genetics

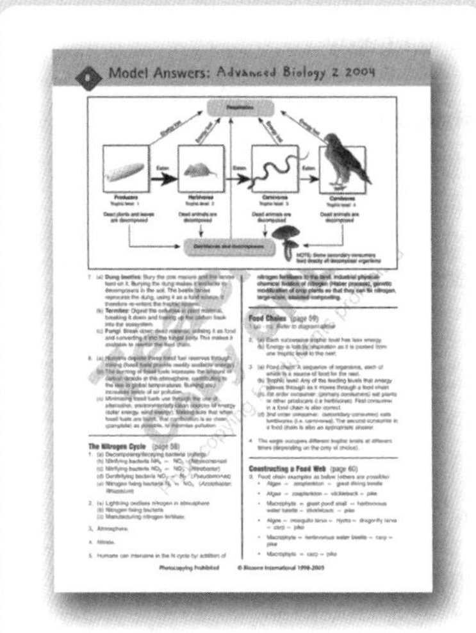

Model Answers

Provide suggested answers for all the questions in each topic.

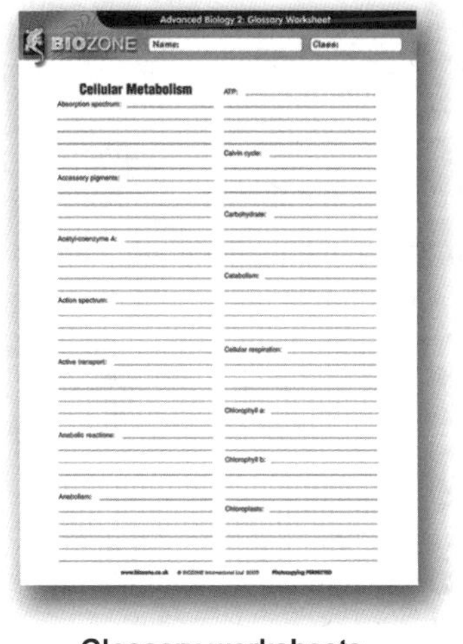

Glossary worksheets

Provide a means to test your understanding of the scientific terms used in each topic.

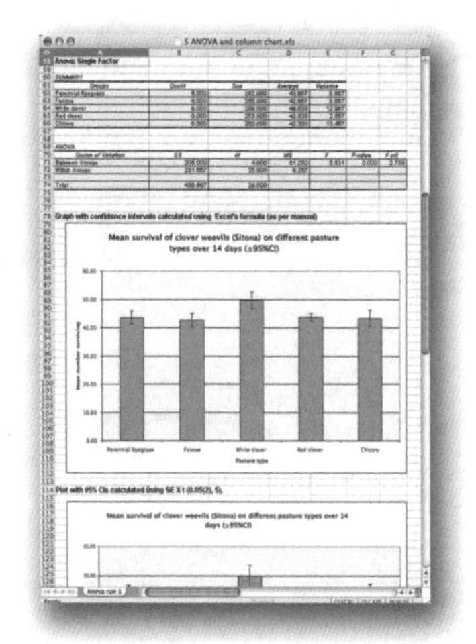

***Microsoft® Excel®* spreadsheets** with worked examples for various kinds of statistical analysis.

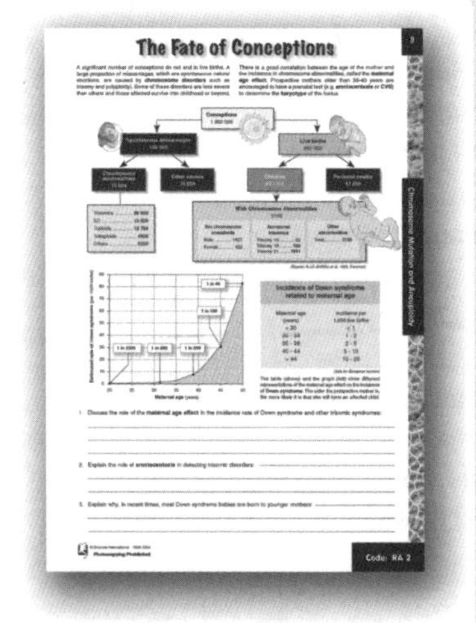

Supplementary Material

Additional material specific to particular exam boards, but not common to all.

Biology Video Titles

Access a database of more than 300 **biology video** titles

Biology Software Titles

Access a database of more than 200 **biology software** titles.

AQA Specification A: A2

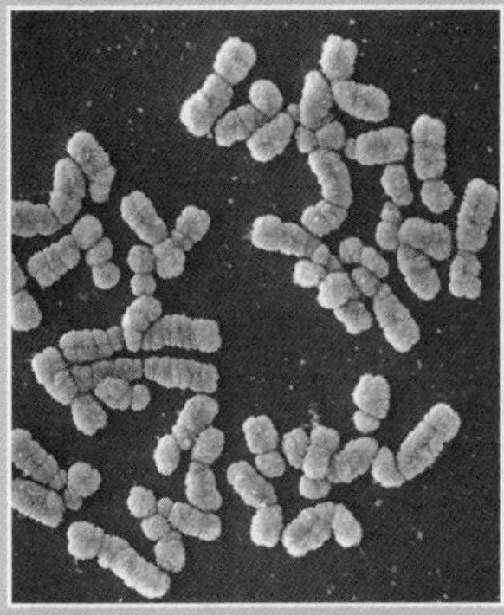

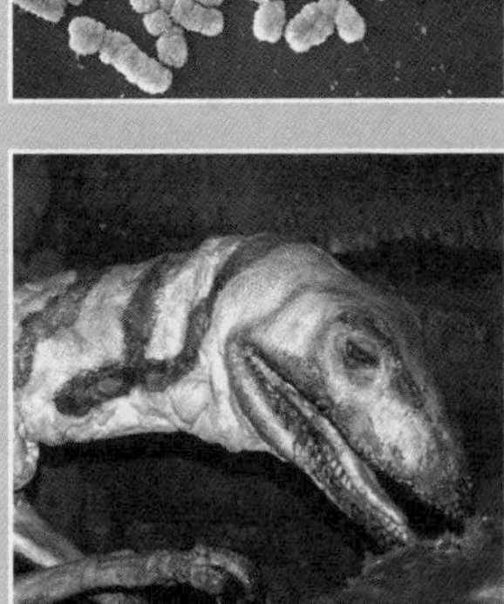

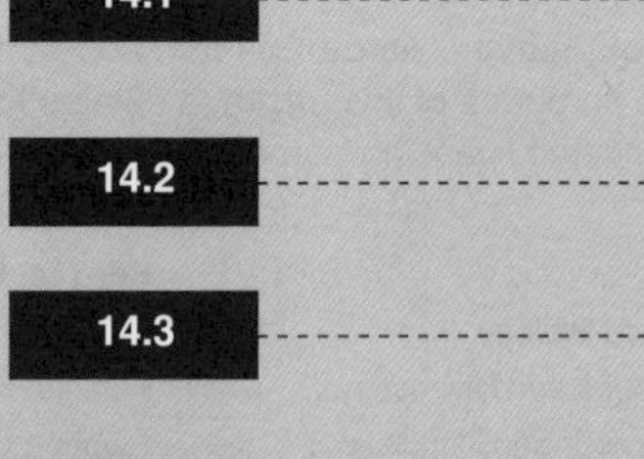

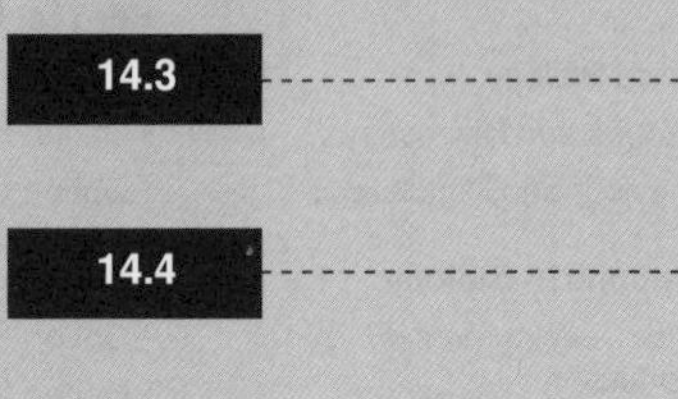

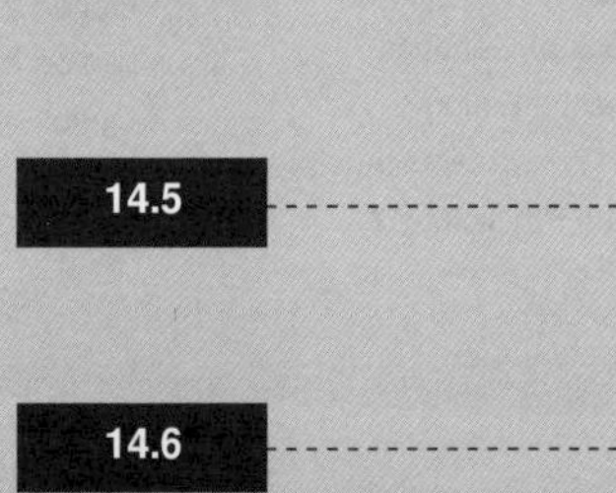

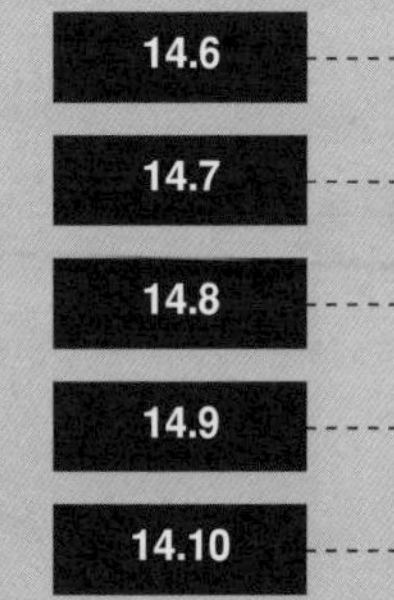

A2 Module	Module section	Manual and topic where material is covered
A2 Module 5: Inheritance, Evolution and Ecosystems	14.1	A2: **Sources of Variation** *(meiosis)* A2: **Inheritance** *(Mendelian inheritance, use of χ^2)*
	14.2	A2: **Sources of Variation** *(causes of variation)* A2: **Inheritance** *(polygeny, variation displayed graphically)*
	14.3	A2: **Population Genetics and Evolution**
	14.4	A2: **Population Genetics and Evolution** *(the species concept)* A2: **Biodiversity and Classification** *(five kingdom classification)*
	14.5	A2: **Populations and Interactions** *(density and distribution, diversity indices, succession)* A2: **Practical Ecology** *(investigating population numbers and distribution)* TRH: **Ecosystems and Energy**
	14.6	A2: **Cellular Metabolism**
	14.7	TRH: **Ecosystems and Energy**
	14.8	A2: **Cellular Metabolism**
	14.9	TRH: **Ecosystems and Energy**
	14.10	A2: **Populations and Interactions**
A2 Module 6: Physiology and the Environment (Biology only)	15.1 - 15.2	TRH: **Physiology and the Environment**
	15.2 - 15.3	A2: **Homeostasis** *(except thermoregulation)*
	15.4 - 15.7	TRH: **Physiology and the Environment**
	15.8 - 15.10	A2: **Responses and Coordination**
A2 Module 7: The Human Life Span (Human Biology only)	16.1 - 16.3	TRH: **The Human Life Span**
	16.4	TRH: **The Human Life Span**
	16.5	A1: **Aspects of Human Health**
	16.6	TRH: **The Human Life Span**
	16.7 - 16.8	A2: **Responses and Coordination**
	16.9	TRH: **The Human Life Span**
	16.10	A2: **Responses and Coordination**
	16.11	A2: **Homeostasis** *(control of blood glucose)* TRH: **The Human Life Span**
A2 Modules 8a/9a, 8b/9b	Candidates are assessed on practical work, statistics, interpretation and evaluation of results, and communication.	A1: **Skills in Biology** A2: **Practical Ecology**

Note: There is no material to cover 16.12 of Module 7

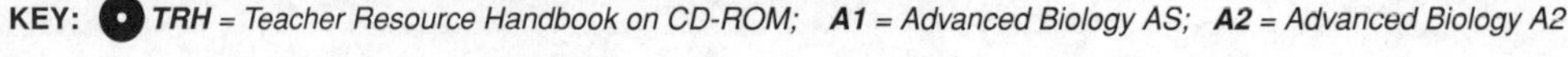

KEY: *TRH = Teacher Resource Handbook on CD-ROM;* ***A1*** *= Advanced Biology AS;* ***A2*** *= Advanced Biology A2*

AQA Specification B: A2

A2 Module	Module section	Manual and topic where material is covered
A2 Module 4: Energy, Control and Continuity	13.1	A2: **Cellular Metabolism** *(role of ATP)*
	13.2	A2: **Cellular Metabolism** *(photosynthesis)*
	13.3	A2: **Cellular Metabolism** *(cellular respiration)*
	13.4	A2: **Homeostasis** *(role of hormones)* A2: **Responses and Coordination** *(stimuli and receptors, simple reflex arcs)*
	13.5	A2: **Homeostasis** *(control of blood glucose, osmoregulation & excretion)* TRH: **Thermoregulation**
	13.6-13.7	A2: **Responses and Coordination**
	13.8	TRH: **Muscles and Movement**
	13.9	A2: **Sources of Variation** *(meiosis)* A2: **Inheritance** *(sex determination, genetic crosses)*
	13.10	A2: **Sources of Variation** *(causes of variation, discontinuous variation)* A2: **Inheritance** *(continuous variation)*
	13.11	A2: **Population Genetics and Evolution**
	13.12	A2: **Population Genetics and Evolution**
A2 Module 5(a): Environment	14.1-14.2	TRH: **Energy Flow & Nutrient cycles**
	14.3	A2: **Practical Ecology**
	14.4	A2: **Populations and Interactions**
A2 Module 5(b): Coursework	Candidates will be assessed on planning, implementing, analysing evidence, evaluation, and synthesis of principles and concepts.	A1: **Skills in Biology** A2: **Practical Ecology**
A2 Option Modules	Option Module 6	TRH: **Option Module 6** *(applied biology)*
	Option Module 7	TRH: **Option Module 7** *(microbes and disease)*
	Option Module 8	TRH: **Option Module 8** *(behaviour and populations)*

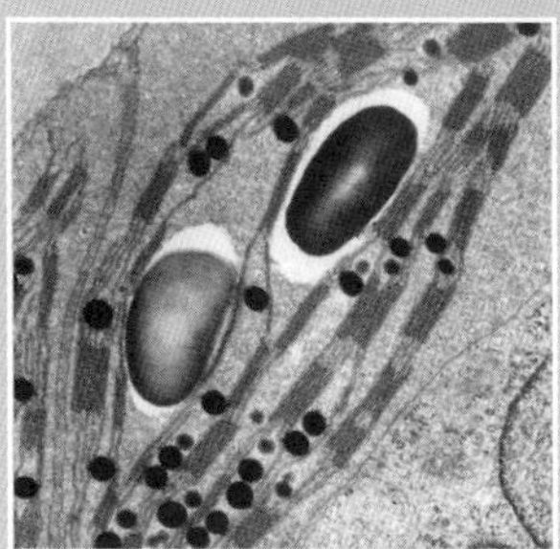

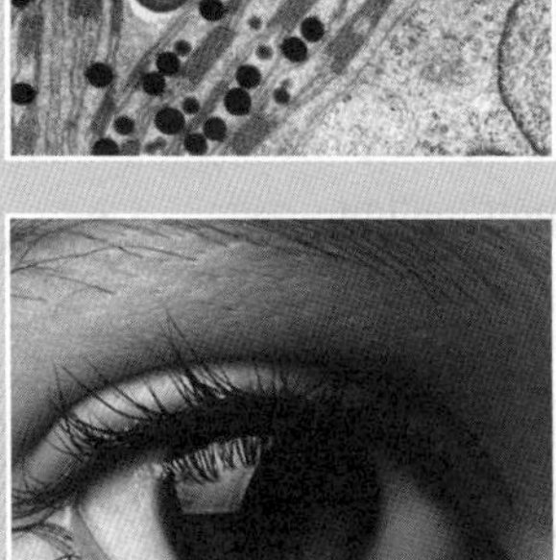

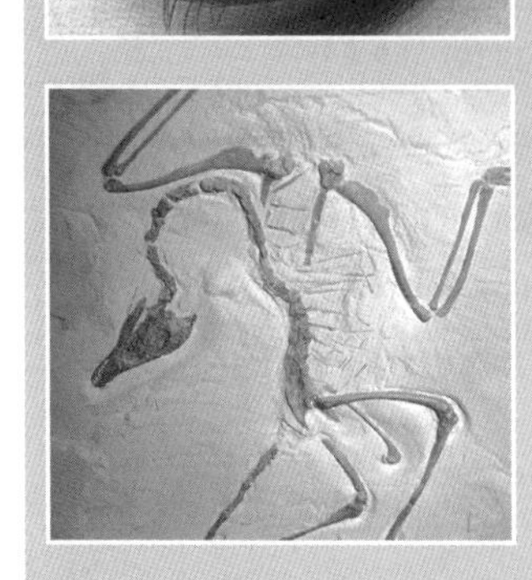

KEY: ● ***TRH*** *= Teacher Resource Handbook on CD-ROM;* ***A1*** *= Advanced Biology AS;* ***A2*** *= Advanced Biology A2*

Cambridge International Exam A2

Advanced Level Syllabus	Section	Manual and topic where material is covered

CORE SYLLABUS

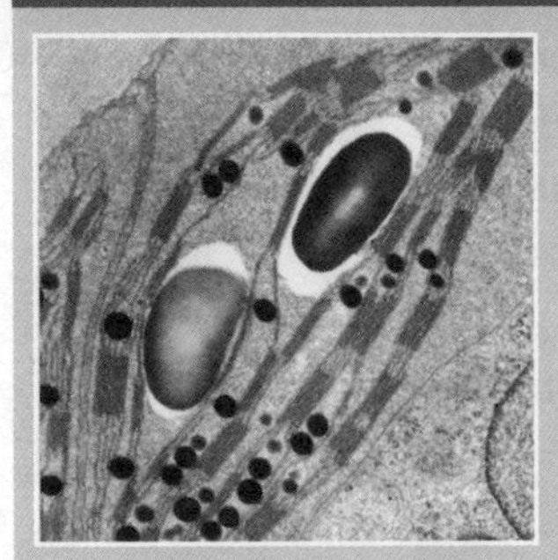

(sections A-K are in Advanced Biology AS)

L: Energy and Respiration

A2: Cellular Metabolism
- *energy in living organisms*
- *aerobic respiration and fermentation*
- *RQ and energy values of substrates*

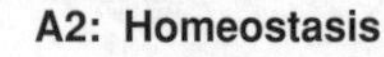

M: Photosynthesis

A2: Cellular Metabolism
- *the biochemistry of photosynthesis*
- *limiting factors in photosynthesis*

N: Regulation and Control

A2: Homeostasis
- *importance of homeostasis*
- *excretion and osmoregulation*
- *nervous and hormonal regulatory mechanisms*

A2: Responses and Coordination
- *nervous system and responses*
- *communication and control in flowering plants*

O: Inherited Change and Gene Technology

A2: Sources of Variation
- *meiosis, mutation, environment and phenotype*

A2: Inheritance
- *monohybrid and dihybrid crosses, test crosses*
- *sex linkage, codominance, multiple alles, use of c2*

TRH: Gene Technology (or see A1)
- *techniques and applications of genetic engineering*
- *ethical issues associated with genetic engineering*

P: Selection and Evolution

A2: Population Genetics and Evolution
- *variation, natural selection, and evolution*
- *allele frequencies in populations*
- *artificial selection*

Options Syllabus

candidates study one option

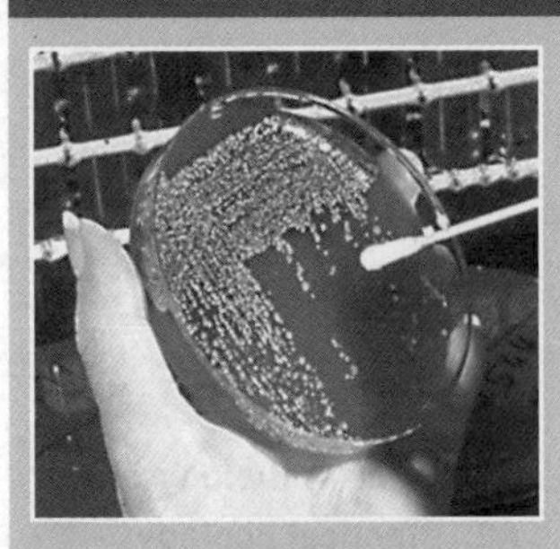

Option 1 — **TRH: Option 1** *(mammalian physiology)*

Option 2 — **TRH: Option 2** *(microbiology and biotechnology)*

Option 3 — **TRH: Option 3** *(growth, development, and reproduction)*

Option 4 — **TRH: Option 4** *(applications of genetics)*

Meeting assessment objectives

A: Knowledge with understanding

B: Handling information and solving problems

C: Experimental skills and investigations

A1: Skills in Biology *(data handling and analysis)*

A2: Practical Ecology *(ecological methodology)*

KEY: *TRH* = *Teacher Resource Handbook on CD-ROM;* ***A1*** = *Advanced Biology AS;* ***A2*** = *Advanced Biology A2*

Edexcel A2

A2 Module	Module section	Manual and topic where material is covered
6104 Unit 4: Respiration and Coordination, and Options	4.1	A2: Cellular Metabolism
	4.2	A2: Homeostasis *(homeostasis, excretion, hormonal coordination)* A2: Responses and Coordination *(nervous control, the mammalian eye, phytochromes)*
	Option A	TRH: Option Module A *(microbiology and biotechnology)*
	Option B	TRH: Option Module B *(food science)*
	Option C	TRH: Option Module C *(human health and fitness)*
6105 Unit 5B: Genetics, Evolution, and Biodiversity (Biology only)	5B.1	A2: Cellular Metabolism
	5B.2	A2: Responses and Coordination
	5B.3	A2: Biodiversity and Classification *(taxonomy and the five kingdom classification system)* A2: Populations and Interactions *(population dynamics, predator-prey, succession, farming & woodlands)* A2: Practical Ecology *(investigating distribution)* TRH: Farming and Conservation *(grasslands, national conservation issues)*
	5B.4	A2: Sources of Variation *(sources of variation including point mutations)* A2: Inheritance *(sex determination and inheritance)* TRH: Chromosome Mutations & Aneuploidy A2: Population Genetics and Evolution *(species, gene pools, natural selection, and speciation)* TRH: Gene Technology
6115 Unit 5H: Genetics, Human Evolution, and Biodiversity (Human Biology only)	5H.1	A2: Sources of Variation *(sources of variation including point mutations)* A2: Inheritance *(sex determination and inheritance)* TRH: Chromosome Mutations & Aneuploidy A2: Population Genetics and Evolution *(species, gene pools, natural selection, and speciation)* TRH: Gene Technology
	5H.2	TRH: Human Evolution
	5H.3	A2: Populations and Interactions
	5H.4	A2: Populations and Interactions *(density and distribution, succession)* A2: Practical Ecology *(investigating distribution)* TRH: Farming and Conservation
Paper 01: T1 or Paper 02: W2. Paper 03 (Synoptic paper)	Two components to assess skills in investigation, evaluation, and communication.	A1: Skills in Biology

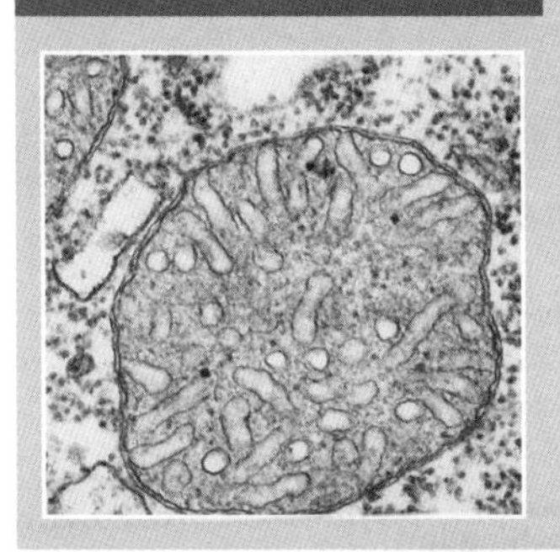

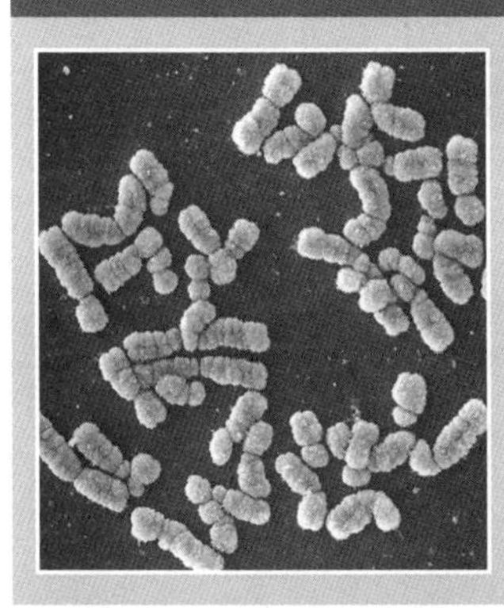

KEY: *TRH = Teacher Resource Handbook on CD-ROM;* ***A1*** *= Advanced Biology AS;* ***A2*** *= Advanced Biology A2*

OCR A2

A2 Module	Module section	Manual and topic where material is covered
5.4 Module 2804: Central Concepts	5.4.1 - 5.4.2	A2: **Cellular Metabolism**
	5.4.3	A2: **Populations and Interactions** *(population dynamics, succession, ecosystem management)* A2: **Practical Ecology** *(techniques for measuring distribution & abundance)*
	5.4.4	A2: **Cellular Metabolism** *(regulation of protein synthesis in bacteria: Lac operon)* A2: **Sources of Variation** *(meiosis, mutation, environment and phenotype, HGP)* A2: **Inheritance** *(inheritance, including sex linkage, use of χ^2)*
	5.4.5	A2: **Biodiversity and Classification** *(five kingdom classification, classification & phylogeny)* A2: **Population Genetics and Evolution** *(gene pools, natural selection, speciation, adaptation)*
	5.4.6	A2: **Homeostasis** *(endocrine control, blood glucose, kidney function)* A2: **Responses and Coordination** *(nervous control & sensory reception, plant hormones)*

A2 Module	Module section	Manual and topic where material is covered
5.5 Module 2805: Options in Biology	Component 01	TRH: **Option Component 01** *(growth, development, and reproduction)*
	Component 02	TRH: **Option Component 02** *(applications of genetics)*
	Component 03	TRH: **Option Component 03** *(environmental biology)*
	Component 04	TRH: **Option Component 04** *(microbiology and biotechnology)*
	Component 05	TRH: **Option Component 05** *(mammalian physiology and behaviour)*

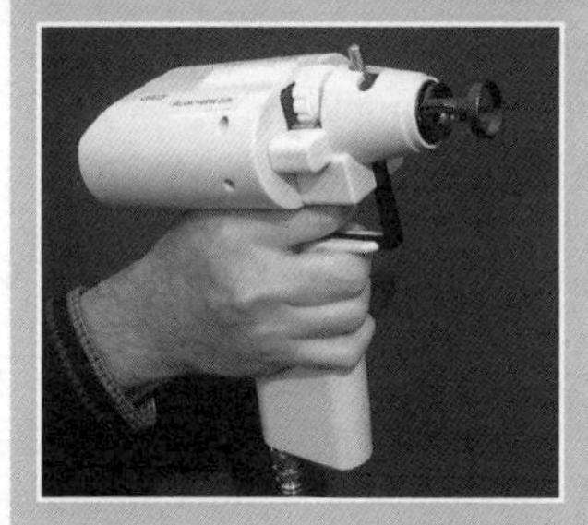

5.10 Module 2806

Unifying Concepts in Biology (compulsory) Component 01

Component 01: Unifying concepts in Biology

Examination questions based on the concepts covered in modules 2801, 2802, 2803 component 01 of the AS scheme, and module 2804 of the A2 scheme.

Experimental Skills 2
Candidates take either Component 02 OR component 03

Component 02: Coursework 2

Internal assessment of experimental and investigative work. Skills include planning an implementing a study, recording, analysing, and evaluating data, and drawing conclusions.

- A1: **Skills in Biology**
- A2: **Practical Ecology**

Component 03: Practical examination 2

External assessment of experimental and investigative skills, including planning an implementing a study, recording, analysing, and evaluating data, and drawing conclusions. Components consist of planning a task, experimentation, and microscopy.

- A1: **Cell Structure** *(microscopy)*
- A1: **Skills in Biology** *(experimental design & data analysis)*
- A2: **Practical Ecology**

KEY: ***TRH*** *= Teacher Resource Handbook on CD-ROM;* ***A1*** *= Advanced Biology AS;* ***A2*** *= Advanced Biology A2*

WJEC A2

A2 Module	Module section	Manual and topic where material is covered
A2 Module BI4: Biochemistry and Health	4.1-4.3	A2: Cellular Metabolism
	4.4	TRH: Digestion and Absorption
	4.5	TRH: Microbiology
	4.6	A1: Infectious Disease
	4.7	A1: Defence and the Immune System
	4.8	A1: Defence and the Immune System *(monoclonal antibodies, vaccination)* A2: Sources of Variation *(antibiotic resistance)* A2: Population Genetics and Evolution *(antibiotic resistance)* TRH: Microbiology *(industrial fermentation)*
A2 Module BI5: Variety and Control	5.1	A2: Sources of Variation *(mutagens, gene mutations, random assortment)* A2: Inheritance *(alleles and Mendelian inheritance, sex linkage)* TRH: Chromosome Mutations & Aneuploidy *(chromosome mutations)*
	5.2	A2: Inheritance *(continuous variation)* A2: Popuation Genetics and Evolution *(gene pools, natural and artificial selection, speciation)*
	5.3-5.4	TRH: Sexual Reproduction *(sexual reproduction in angiosperms and humans)*
	5.5	A1: Cell Division *(cloning in plants and animals)*
	5.6	A2: Biodiversity and Classification
	5.7	A2: Homeostasis *(homeostasis and excretion)*
	5.8	A2: Responses and Coordination *(the ear)*
	5.9	A2: Responses and Coordination *(nervous system)* TRH: Muscles and Movement
A2S Module BI6 Practical Work	Planning and implementing practicals: observations and recording, data interpretation and analysis, evaluation and communicaton of results.	A2: Practical Ecology *(ecological methodology)* A1: Skills in Biology *(experimental design, data analysis, statistical tests)*

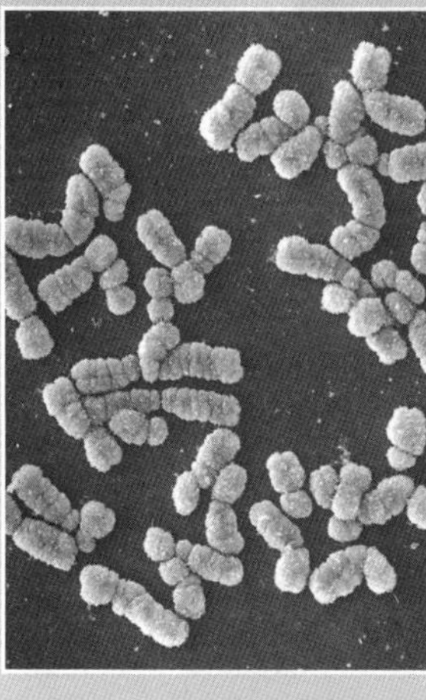

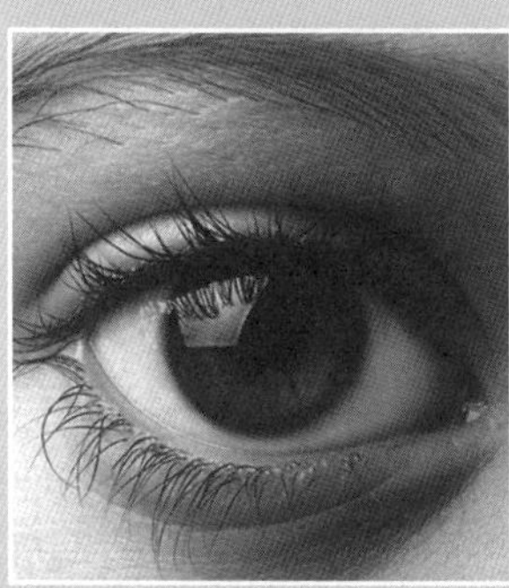

KEY: ● *TRH = Teacher Resource Handbook on CD-ROM;* ***A1*** *= Advanced Biology AS;* ***A2*** *= Advanced Biology A2*

Cellular Metabolism

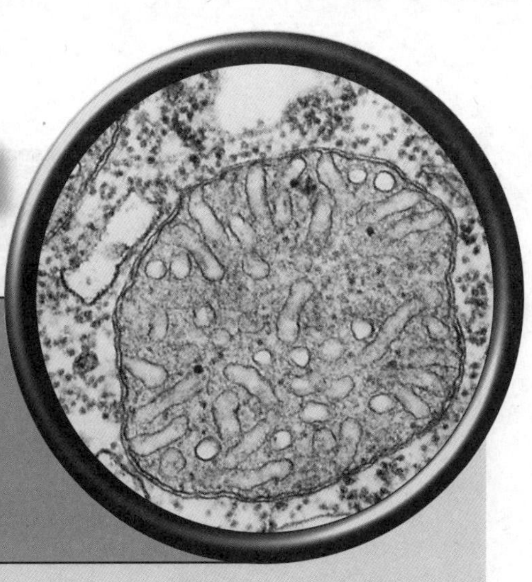

AQA-A	AQA-B	CIE	Edexcel	OCR
Complete:	Complete:	Complete:	Complete:	Complete:
1-3, 5-6, 10-18, 21, 24	*1-3, 5-6, 10-17, 21, 24*	*1-3, 5-21, 24-25*	*1-6, 10-17, 20-27*	*1-3, 5-21, 24-25*

Learning Objectives

☐ 1. Compile your own glossary from the **KEY WORDS** displayed in **bold type** in the learning objectives below.

ATP and metabolic pathways *(pages 22-23, 26-27)*

☐ 2. Explain the need for energy in living things and the universal role of ATP in metabolism, as illustrated by specific examples e.g. **glycolysis**, **active transport**, **anabolic reactions**, movement, and **thermoregulation**.

☐ 3. Describe the structure of **ATP** as a phosphorylated nucleotide. Describe its synthesis from ADP and inorganic phosphate and explain how it stores and releases its energy.

☐ 4. Explain what is meant by a **metabolic pathway**. Describe the role of enzymes in the control of metabolic pathways as illustrated by specific examples, e.g. **oxidoreductases** and **hydrolases**, **anabolism** and **catabolism**, control of phenylalanine metabolism.

☐ 5. Outline the principles involved in **photosynthesis** and **cellular respiration**, explaining in which way the two processes can be considered opposites.

☐ 6. Appreciate that both photosynthesis and cellular respiration involve the molecule **ATP** and a hydrogen carrier molecule (**$NADPH_2$** in photosynthesis, and **$NADH_2$** in respiration).

The operon hypothesis *(pages 24-25)*

☐ 7. Understand the term **operon** as being a unit of genes that function in a coordinated way under the control of an operator gene.

☐ 8. Explain how simple metabolic pathways are regulated in bacteria, as illustrated by the ***lac* operon** in *E. coli*. Outline the principles involved in **gene induction** in the *lac* operon, identifying how lactose activates transcription and how metabolism of the substrate is achieved. Explain the adaptive value of gene induction.

☐ 9. Appreciate that the end-product of a metabolic pathway can activate a repressor and switch genes off (**gene repression**). Appreciate the adaptive value of gene repression for the control of a metabolic end-product.

Cellular respiration

Introduction to respiration *(pages 26-27, 29)*

☐ 10. Describe the role of mitochondria in cellular respiration. Draw and label the structure of a **mitochondrion**, including the **matrix**, the outer and inner membrane and the **cristae**. Briefly state the function of each.

☐ 11. Identify the main steps in cellular respiration: **glycolysis**, **Krebs cycle** (*tricarboxylic acid cycle*) **electron transport system** (*electron transport chain, ETS, or respiratory chain*). On a diagram of a mitochondrion, indicate where each stage occurs. Recognise glycolysis as first stage in cellular respiration and the major anaerobic pathway in cells.

☐ 12. Identify **glucose** as the main respiratory substrate. Appreciate that other substrates can, through conversion, act as substrates for cellular respiration.

☐ 13. Outline **glycolysis** as the phosphorylation of glucose and the subsequent splitting of a 6C sugar into two triose phosphate molecules (2 X **pyruvate**). State the net yield of ATP and $NADH_2$ from glycolysis and appreciate that the subsequent metabolism of pyruvate depends on the availability of oxygen.

Aerobic respiration *(pages 30-31)*

☐ 14. Describe the complete oxidation of glucose to CO_2, with reference to:
- The conversion of pyruvate to **acetyl-coenzyme A**.
- The entry of acetyl CoA into the Krebs cycle by combination with **oxaloacetate**.
- The **Krebs cycle** (as a series of oxidation reactions involving release of CO_2, the production of $NAD.H_2$ or $FAD.H_2$, and the regeneration of oxaloacetate).
- The *role* of the coenzymes NAD and FAD.
- Synthesis of **ATP** by **oxidative phosphorylation** in the electron transport chain.
- The role of oxygen as the terminal electron acceptor and the formation of water.
- The net yield of ATP from aerobic respiration compared to the yield from glycolysis.

☐ 15. Understand the terms **decarboxylation** and **dehydrogenation** as they relate to the Krebs cycle.

Fermentation *(page 32)*

☐ 16. Understand the situations in which the pyruvate formed in glycolysis may not undergo complete oxidation. Describe the following examples of **fermentation**, identifying the H^+ acceptor to each case:

(a) Formation of **lactic acid** in muscle.
(b) Formation of **ethanol** in yeast.

NOTE: Appreciate that, although fermentation is often used synonymously with anaerobic respiration, they are not the same. Respiration always involves hydrogen ions passing down a chain of carriers to a terminal acceptor, and this does not occur in fermentation. In anaerobic respiration, the terminal H^+ acceptor is a molecule other than oxygen, e.g. Fe^{2+} or nitrate.

☐ 17. Compare and explain the differences in the yields of ATP from aerobic respiration and from fermentation.

Respiratory quotients *(page 28)*

☐ 18. Explain the relative energy values of carbohydrate, lipid, and protein as respiratory substrates. Explain the term **respiratory quotient** (RQ). Explain what RQ reveals about the substrate being respired.

☐ 19. Use a simple respirometer to measure RQ and the effect of temperature on respiration rate.

Photosynthesis

The structure of the dicot leaf *(pages 33-35)*

☐ 20. Recognise the leaf as the main photosynthetic organ in plants. Describe the internal and external structure of a dicotyledonous leaf with reference to: the location of the palisade tissue and the distribution of chloroplasts. If required, compare dicot leaf structure with monocot leaf structure and describe adaptations of the leaf to maximising photosynthesis in different environments.

Chloroplasts *(pages 36-37)*

☐ 21. Describe the structure and role of **chloroplasts**, identifying the **stroma**, **grana**, lamellae (**thylakoids**), and location of the chlorophylls and other pigments.

☐ 22. Describe the role of **chlorophyll a** and **b**, and **accessory pigments** (e.g. carotenoids) in light capture. Explain what is meant by the terms **absorption spectrum** and **action spectrum** with respect to the light absorbing pigments.

☐ 23. Investigate chloroplast pigments using chromatography.

Photosynthesis in C3 plants *(pages 38-39)*

☐ 24. Describe, using diagrams, the reactions of photosynthesis in a C3 plant with reference to:

The light dependent phase (LDP) with reference to:

- Where in the chloroplast the LDP occurs and the location and role of the photosystems.
- The **photoactivation** of chlorophyll.
- The splitting of water (**photolysis**) to produce protons and electrons.
- The production of O_2 as a result of photolysis.
- The transfer of energy to ATP (photophosphorylation) and the formation of $NADPH_2$ (reduced NADP).

The light independent phase (LIP) with reference to:

- Where in the chloroplast the LIP occurs.

The **Calvin cycle** including:

- The fixation of carbon dioxide into a 5C compound, **ribulose bisphosphate** (RuBP).
- The reduction of **glycerate-3-phosphate** (PGA) to **carbohydrate** and the role of **ATP** and **$NADPH_2$** (formed in the light dependent phase) in this.
- The regeneration of the ribulose bisphosphate.

Photosynthetic rate *(pages 34, 40)*

☐ 25. Understand the term **photosynthetic rate**. With reference to **limiting factors**, describe the effect of each of the following on photosynthetic rate:

- Light intensity and wavelength.
- Temperature.
- Carbon dioxide concentration.

☐ 26. Explain what is meant by the (light) **compensation point**. Interpret the compensation point for plants in different environments.

Plant mineral nutrition *(page 41)*

☐ 27. Describe the uptake of mineral ions in plant roots. Describe the function of some important plant mineral ions, including: nitrate (NO_3^-), phosphate (PO_4^{3-}), and magnesium (Mg^{2+}) ions.

See the 'Textbook Reference Grid' on pages 8-9 for textbook page references relating to material in this topic.

Supplementary Texts

See pages 4-6 for additional details of this text:

■ Adds, J. *et al.*, 2001. **Respiration and Coordination** (NelsonThornes), pp. 1-8 (respiration).

See pages 10-11 for details of how to access **Bio Links** from our web site: **www.biozone.co.uk**. From Bio Links, access sites under the topics:

GENERAL BIOLOGY ONLINE RESOURCES > **Online Textbooks and Lecture Notes**: • S-Cool! A level biology revision guide • Learn.co.uk • Mark Rothery's biology web site *... and others*

CELL BIOLOGY AND BIOCHEMISTRY: • Cell and molecular biology online • MIT biology hypertextbook *... and others* > **Biochemistry and Metabolic Pathways**: • Calvin cycle (C3 cycle) • Cellular energy references • Cellular respiration • Cycle (Krebs cycle, Citric Acid Cycle) • Electron transport chain • Energy, enzymes, and catalysis problem set • Glycolysis • Learning about photosynthesis • Chapter 7: Metabolism and biochemistry *... and others*

GENETICS > **Molecular Genetics (DNA)**: • Prokaryotic genetics and gene expression chapter • Model of Lac operon (animation) • Induction of the Lac operon • Molecular genetics of prokaryotes

See page 6 for details of publishers of periodicals:

STUDENT'S REFERENCE

Metabolic pathways and their control

■ **Tyrosine** Biol. Sci. Rev., 12 (4) March 2000, pp. 29-30. *The central metabolic role of the amino acid tyrosine (includes errors in tyrosine metabolism).*

■ **Leptin** Biol. Sci. Rev., 15(3) February 2003, pp. 30-32. *The role of the hormone leptin in the regulation of body mass and the control of obesity. The negative feedback mechanisms regulating gain and loss in mass are illustrated.*

■ **Gene Structure and Expression** Biol. Sci. Rev., 12 (5) May 2000, pp. 22-25. *The nature of genes and the control of gene expression. Includes gene induction in the Lac operon in E. coli.*

■ **Genes that Control Genes** New Scientist, 3 November 1990 (Inside Science). *The control of gene expression in prokaryotes by gene induction and repression. The operon model is explained.*

Chlorophyll and photosynthesis

■ **Growing Plants in the Perfect Environment** Biol. Sci. Rev., 15(2) November 2002, pp. 12-16. *To manipulate the growth of plants in controlled environments, one must understand how plants grow and what influences photosynthetic rate.*

■ **Chemistry that Comes Naturally** New Scientist, 31 July 1993, pp. 24-28. *The processes of photosynthesis and respiration.*

■ **Green Miracle** New Scientist, 14 August 1999, pp. 26-30. *The mechanism by which plants split water to make oxygen remains a mystery.*

■ **Chlorophyll** Biol. Sci. Rev., 8(3) January 1996, pp. 28-30. *The chlorophyll molecule: how it absorbs light and its role in photosynthesis.*

■ **Why Don't Plants Wear Sunhats?** Biol. Sci. Rev., 9(3) Jan. 1997, pp. 32-35. *Plants need light, but too much is damaging - how do they cope?*

ATP and cellular respiration

■ **Fat Burns in the Flame of Carbohydrate** Biol. Sci. Rev., 15(3) February 2003, pp. 37-41. *A thorough account of both carbohydrate metabolism and how fatty acid oxidation feeds into the Krebs cycle. Starvation and ketosis are described and several points for discussion are included.*

■ **Fuelled for Life** New Scientist, 13 January 1996 (Inside Science). *Energy and metabolism: ATP, glycolysis, electron transport system, Krebs cycle, and enzymes and cofactors.*

■ **Glucose Catabolism** Biol. Sci. Rev., 10(3) Jan. 1998, pp. 22-24. *Glucose in cells: oxidative phosphorylation and the role of mitochondria.*

TEACHER'S REFERENCE

■ **Molecular Machines that Control Genes** Scientific American, February 1995, pp. 38-45. *Regulation of gene action by protein complexes that assemble on DNA.*

■ **How Mitochondria Respond to Hormones** Biol. Sci. Rev., 8(3) January 1996, pp. 21-24. *The effects of hormones on mitochondrial activity (includes the role of cAMP in the response).*

■ **The Bigger Picture** New Scientist, June 2000, pp. 54-61. *How to understand the complexity of biochemical pathways and the effects of altering genetic constitution.*

■ **Measuring the Metabolism of Small Organisms** Scientific American, December 1995, pp. 84-85. *Methods of measuring and monitoring respiration and metabolic rate in small organisms.*

Software and video resources are provided on the Teacher Resource Handbook on CD-ROM

Metabolic Pathways

Metabolism is all the chemical activities of life. The myriad enzyme-controlled **metabolic pathways** that are described as metabolism form a tremendously complex network that is necessary in order to 'maintain' the organism. Errors in the step-wise regulation of enzyme-controlled pathways can result in metabolic disorders that in some cases can be easily identified. An example of a well studied metabolic pathway, the metabolism of phenylalanine, is described below.

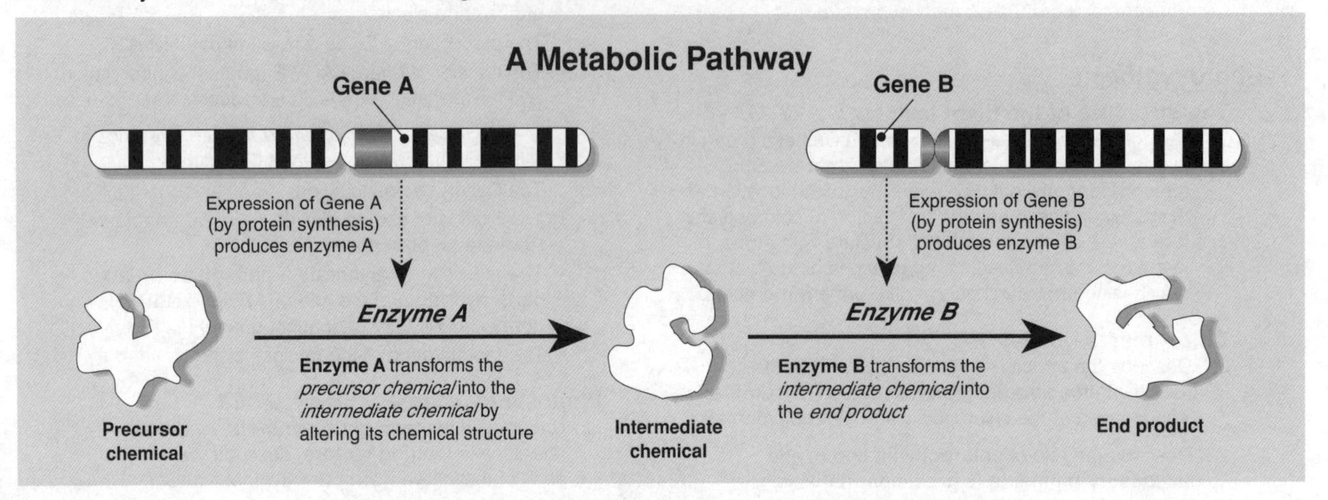

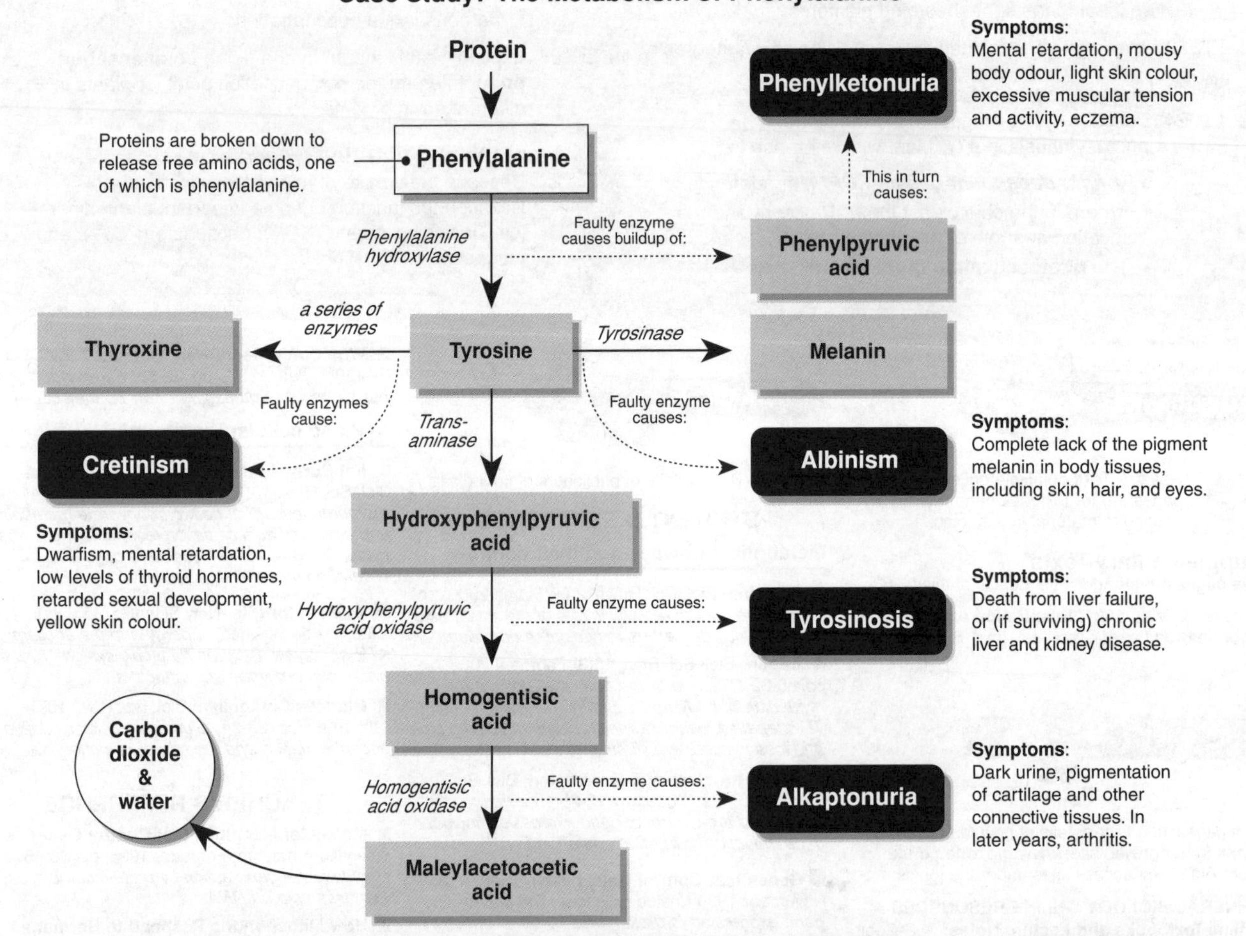

A well-studied metabolic pathway is the metabolic breakdown of the essential amino acid **phenylalanine**. The first step is carried out by an enzyme produced in the liver, called phenylalanine hydroxylase. This enzyme converts phenylalanine to the amino acid **tyrosine**. Tyrosine, in turn, through a series of intermediate steps, is converted into **melanin**, the skin pigment, and other substances. If phenylalanine hydroxylase is absent, phenylalanine is in part converted into phenylpyruvic acid, which accumulates, together with phenylalanine, in the blood stream. Phenylpyruvic acid and phenylalanine are toxic to the central nervous system and produce some of the symptoms of the genetic disease **phenylketonuria**. Other genetic metabolic defects in the tyrosine pathway are also known. As indicated above, absence of enzymes operating between tyrosine and melanin, is a cause of **albinism**. **Tyrosinosis** is a rare defect that causes hydroxyphenylpyruvic acid to accumulate in the urine. **Alkaptonuria** makes urine turn black on exposure to air, causes pigmentation to appear in the cartilage, and produces symptoms of arthritis. A different block in another pathway from tyrosine produces thyroid deficiency leading to goiterous **cretinism** (due to lack of thyroxine).

1. Using the metabolism of phenylalanine as an example, discuss the role of enzymes in **metabolic pathways**:

2. Identify three **products** of the metabolism of phenylalanine:

3. Identify the enzyme failure (faulty enzyme) responsible for each of the following conditions:

   (a) Albinism:

   (b) Phenylketonuria:

   (c) Tyrosinosis:

   (d) Alkaptonuria:

4. Explain why people with **phenylketonuria** have light skin colouring:

5. Discuss the consequences of disorders in the metabolism of **tyrosine**:

6. The five conditions illustrated in the diagram are due to too much or too little of a chemical in the body. For each condition listed below, state which chemical causes the problem and whether it is absent or present in excess:

   (a) Albinism:

   (b) Phenylketonuria:

   (c) Cretinism:

   (d) Tyrosinosis:

   (e) Alkaptonuria:

7. If you suspected that a person suffered from phenylketonuria, how would you test for the condition if you were a doctor:

8. The diagram at the top of the previous page represents the normal condition for a simple metabolic pathway. A starting chemical, called the **precursor**, is progressively changed into a final chemical called the **end product**.

   Consider the effect on this pathway if **gene A** underwent a mutation and the resulting **enzyme A** did not function:

   (a) Name the chemicals that would be present in **excess**:

   (b) Name the chemicals that would be **absent**:

Controlling Metabolic Pathways

The **operon** mechanism was proposed by **Jacob and Monod** to account for the regulation of gene activity in response to the needs of the cell. Their work was carried out with the bacterium *Escherichia coli* and the model is not applicable to eukaryotic cells where the genes are not found as operons. An operon consists of a group of closely linked genes that act together and code for the enzymes that control a particular **metabolic pathway**. These may be for the metabolism of an energy source (e.g. lactose) or the synthesis of a molecule such as an amino acid. The structural genes contain the information for the production of the enzymes themselves and they are transcribed as a single **transcription unit**. These structural genes are controlled by a **promoter**, which initiates the formation of the mRNA, and a region of the DNA in front of the structural genes called the **operator**. A gene outside the operon, called the **regulator gene**, produces a **repressor** molecule that can bind to the operator, and block the transcription of the structural genes. It is the repressor that switches the structural genes on or off and controls the metabolic pathway. Two mechanisms operate in the operon model: gene induction and gene repression. **Gene induction** occurs when genes are switched on by an inducer binding to the repressor molecule and deactivating it. In the ***Lac* operon model** based on *E.coli*, lactose acts as the **inducer**, binding to the repressor and permitting transcription of the structural genes for the utilisation of lactose (an infrequently encountered substrate). **Gene repression** occurs when genes that are normally switched on (e.g. genes for synthesis of an amino acid) are switched off by activation of the repressor.

Control of Gene Expression Through Induction: the *Lac* Operon

Structure of the operon

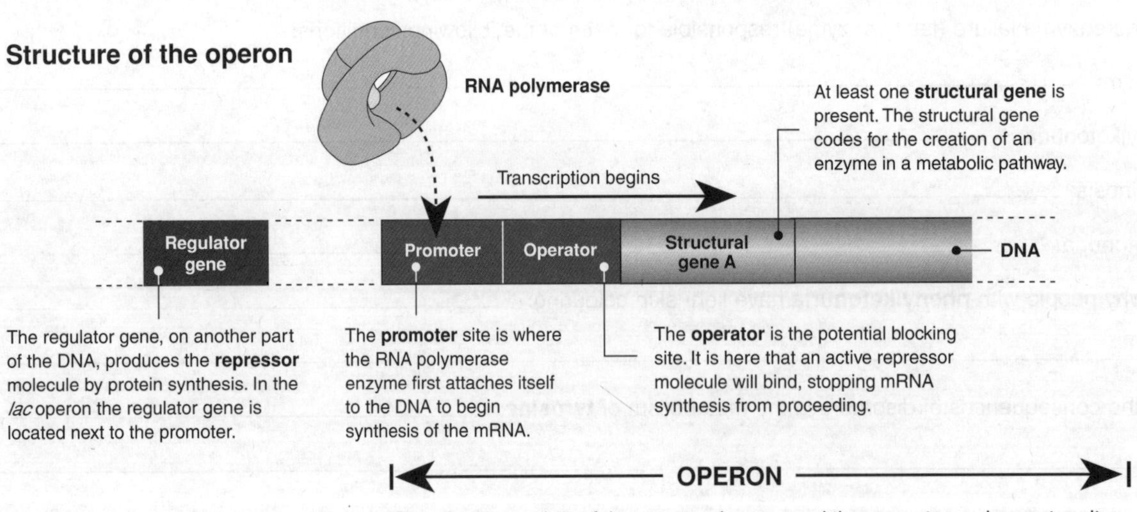

The operon consists of the structural genes and the promoter and operator sites

Structural genes switched off

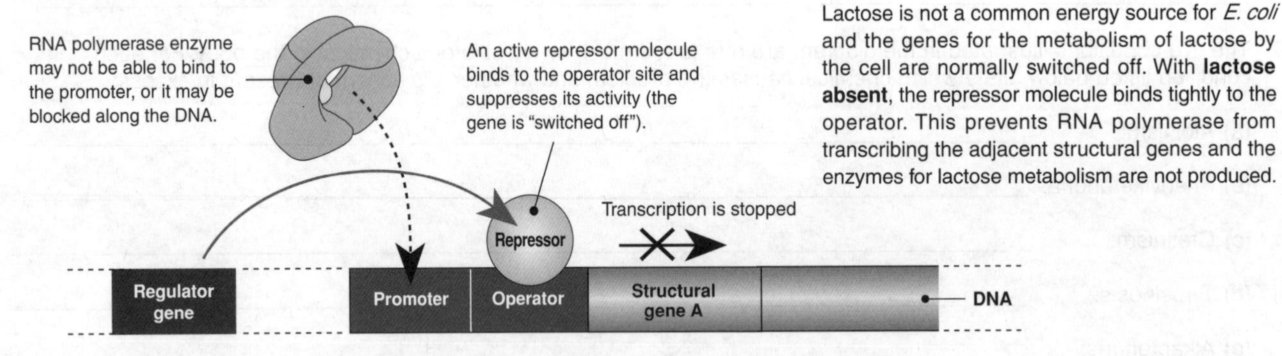

Lactose is not a common energy source for *E. coli* and the genes for the metabolism of lactose by the cell are normally switched off. With **lactose absent**, the repressor molecule binds tightly to the operator. This prevents RNA polymerase from transcribing the adjacent structural genes and the enzymes for lactose metabolism are not produced.

Gene induction

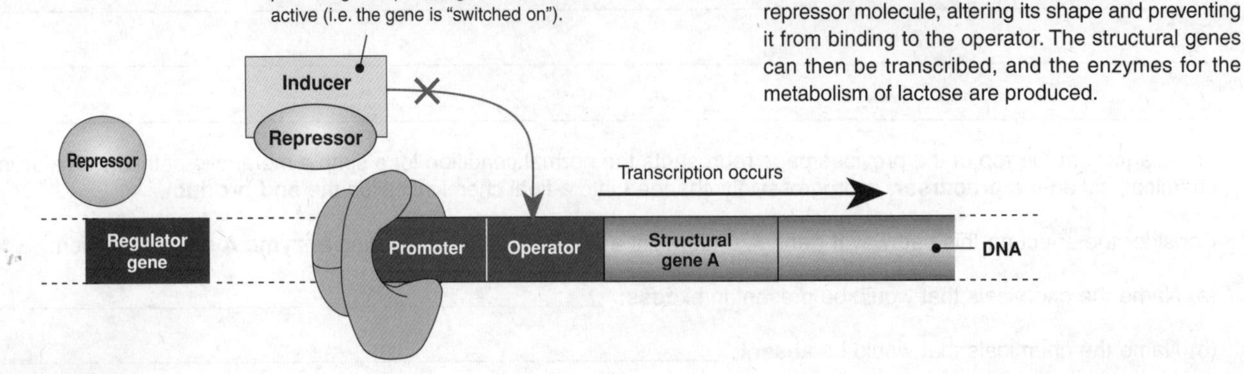

When lactose is available, some of it is converted into the **inducer** allolactose. Allolactose binds to the repressor molecule, altering its shape and preventing it from binding to the operator. The structural genes can then be transcribed, and the enzymes for the metabolism of lactose are produced.

Control of Gene Expression Through Repression

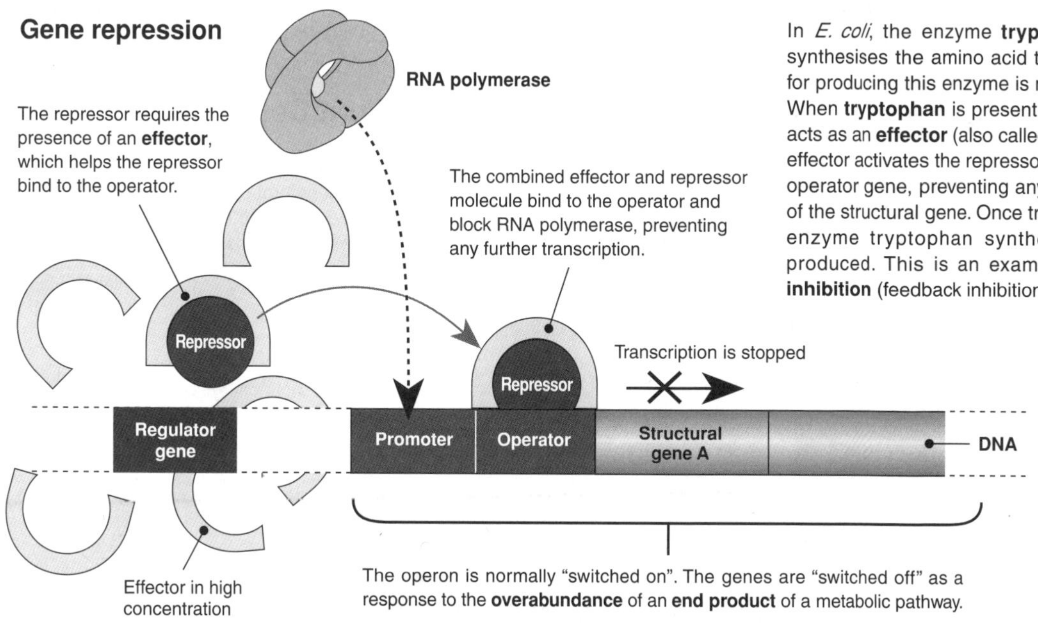

Gene repression

The repressor requires the presence of an **effector**, which helps the repressor bind to the operator.

RNA polymerase

The combined effector and repressor molecule bind to the operator and block RNA polymerase, preventing any further transcription.

Repressor

Repressor

Transcription is stopped

Regulator gene

Promoter

Operator

Structural gene A

DNA

Effector in high concentration

The operon is normally "switched on". The genes are "switched off" as a response to the **overabundance** of an **end product** of a metabolic pathway.

In *E. coli*, the enzyme **tryptophan synthetase** synthesises the amino acid tryptophan. The gene for producing this enzyme is normally switched on. When **tryptophan** is present in excess, some of it acts as an **effector** (also called a co-repressor). The effector activates the repressor, and they bind to the operator gene, preventing any further transcription of the structural gene. Once transcription stops, the enzyme tryptophan synthetase is no longer produced. This is an example of **end-product inhibition** (feedback inhibition).

1. Explain the functional role of each of the following in relation to gene regulation in a prokaryote, e.g. *E. coli*:

   (a) Operon:

   (b) Regulator gene:

   (c) Operator:

   (d) Promoter:

   (e) Structural genes:

2. (a) Explain the advantage in having an inducible enzyme system that is regulated by the presence of a substrate:

   (b) Suggest when it would **not** be adaptive to have an inducible system for metabolism of a substrate:

   (c) Giving an example, outline how gene control in a non-inducible system is achieved through **gene repression**:

3. Describe how the two mechanisms of gene control described here are fundamentally different:

Energy in Cells

A summary of the flow of energy within a plant cell is illustrated below. Animal cells have a similar flow except the glucose is supplied by eating instead of photosynthesis. Note that energy is ultimately used to make energy rich molecules or is lost as heat energy. The role that ATP molecules play in the energy conversion processes is illustrated on the next page.

Energy Transformations in a Photosynthetic Plant Cell

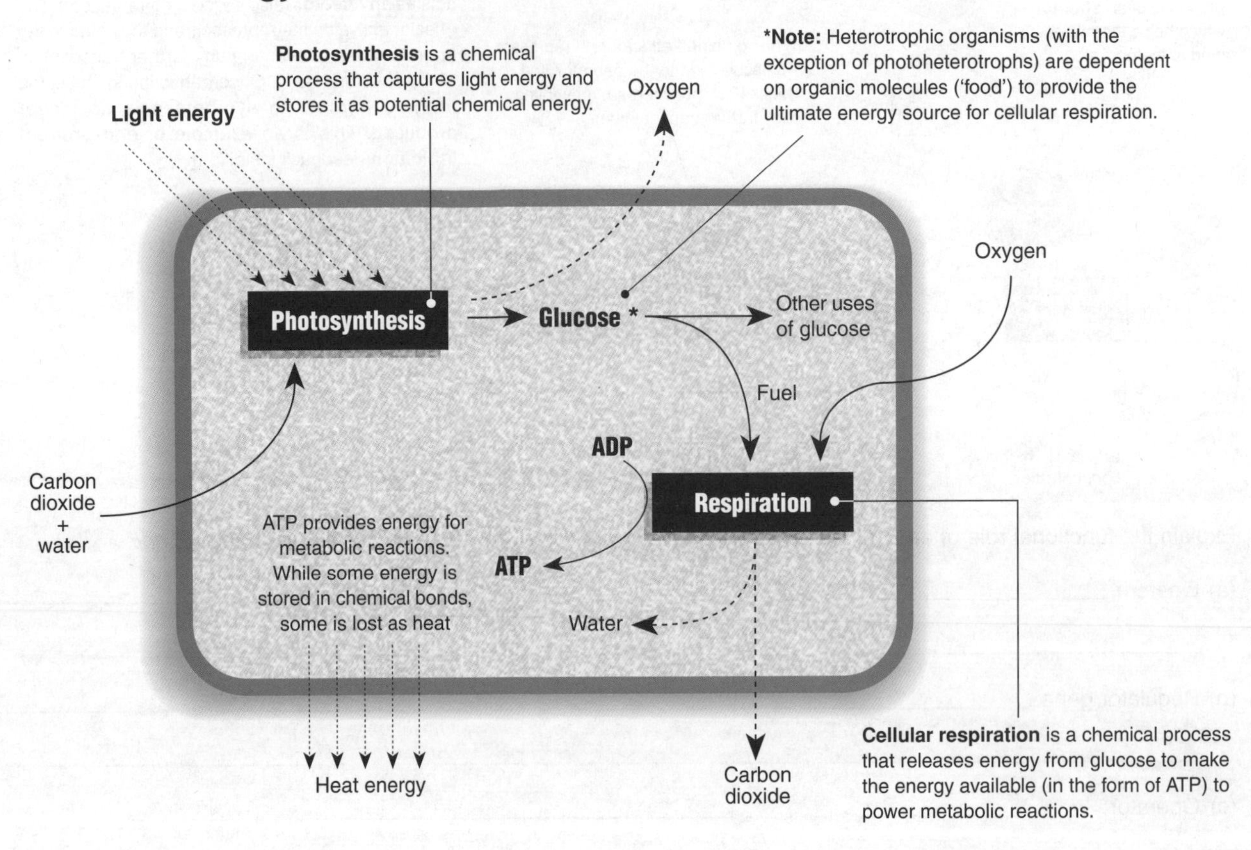

1. Define the following terms that classify how organisms derive their source of energy for metabolism:

   (a) Heterotrophs: ______________________________

   (b) Photosynthetic autotrophs: ______________________________

   (c) Chemosynthetic autotrophs: ______________________________

2. In 1977, scientists working near the Galapagos Islands in the equatorial eastern Pacific found warm water spewing from cracks in the mid-oceanic ridges 2600 metres below the surface. Clustered around these hydrothermal vents were strange and beautiful creatures new to science. The entire community depends on sulphur-oxidising bacteria that use hydrogen sulphide dissolved in the venting water as an energy source to manufacture carbohydrates. This process is similar to photosynthesis, but does not rely on sunlight to provide the energy for generating ATP and fixing carbon:

   (a) Explain why a community based on photosynthetic organisms is not found at this site: ______________________________

   (b) Name the ultimate energy source for the bacteria: ______________________________

   (c) This same chemical that provides the bacteria with energy is also toxic to the process of cellular respiration; a problem that the animals living in the habitat have resolved by evolving various adaptations. Explain what would happen if these animals did not possess adaptations to reduce the toxic effect on cellular respiration:

   ____

   (d) Name the energy source classification for these sulphur-oxidising bacteria: ______________________________

The Role of ATP in Cells

The molecule ATP (adenosine triphosphate) is the universal energy carrier for the cell. ATP can release its energy quickly; only one chemical reaction (hydrolysis of the terminal phosphate) is required. This reaction is catalysed by the enzyme ATPase. Once ATP has released its energy, it becomes ADP (adenosine diphosphate), a low energy molecule that can be recharged by adding a phosphate. This requires energy, which is supplied by the controlled breakdown of respiratory substrates in cellular respiration. The most common respiratory substrate is glucose, but other molecules (e.g. fats or proteins) may also be used.

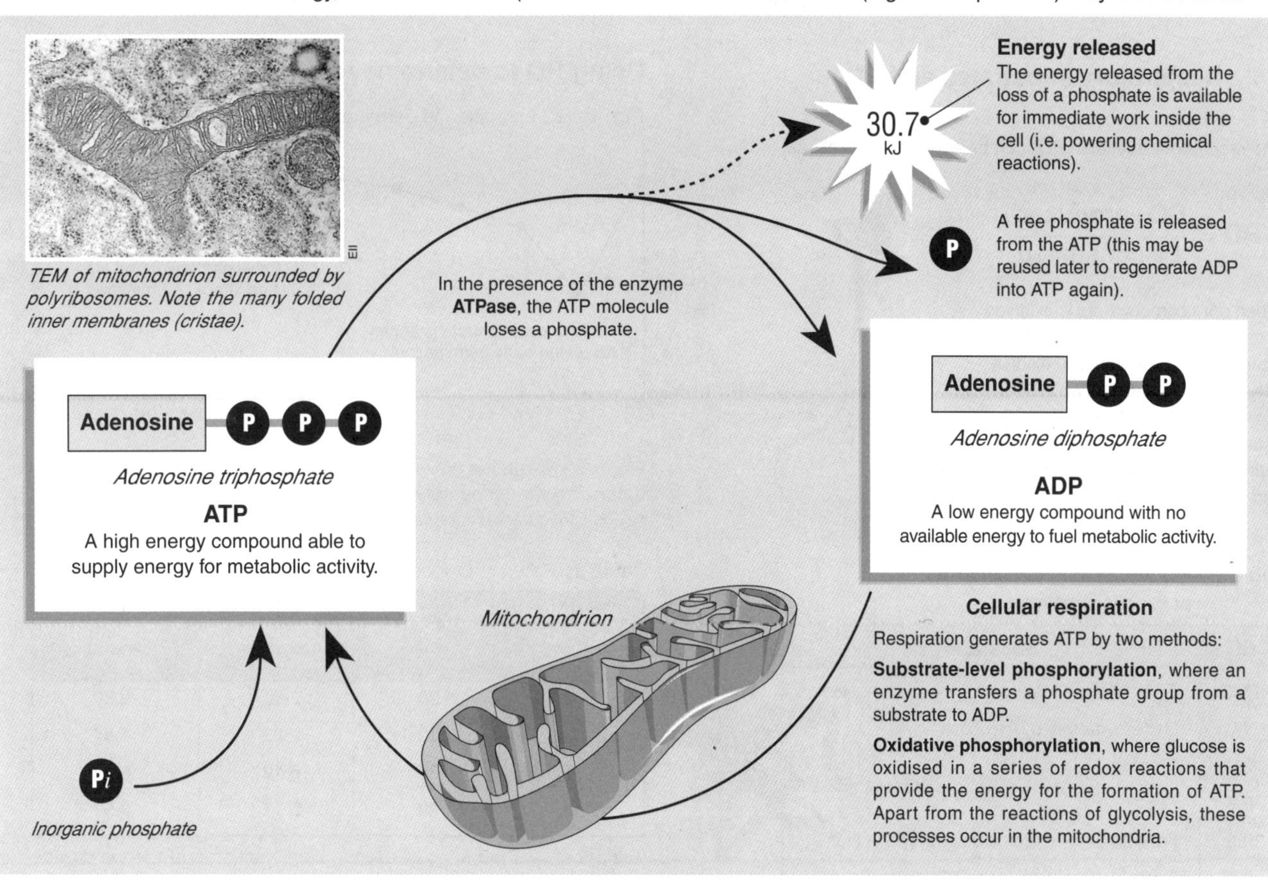

1. Describe how ATP acts as a supplier of energy to power metabolic reactions: ____________________

____

2. Name the immediate source of energy used to reform ATP from ADP molecules: ____________________

3. Name the process of re-energizing ADP into ATP molecules: ____________________

4. Name the ultimate source of energy for plants: ____________________

5. Name the ultimate source of energy for animals: ____________________

6. Explain in what way the ADP/ATP system can be likened to a rechargeable battery: ____________________

____

7. In the following table, use brief statements to contrast photosynthesis and respiration in terms of the following:

Feature	Photosynthesis	Cellular respiration
Starting materials		
Waste products		
Role of hydrogen carriers: NAD, NADP		
Role of ATP		
Overall biological role		

Measuring Respiration

In small animals or germinating seeds, the rate of cellular respiration can be measured using a simple respirometer: a sealed unit where the carbon dioxide produced by the respiring tissues is absorbed by soda lime and the volume of oxygen consumed is detected by fluid displacement in a manometer. Germinating seeds are also often used to calculate the **respiratory quotient** (RQ): the ratio of the amount of carbon dioxide produced during cellular respiration to the amount of oxygen consumed. RQ provides a useful indication of the respiratory substrate being used.

Respiratory substrates and RQ

The respiratory quotient (RQ) can be expressed simply as:

$$RQ = \frac{CO_2 \text{ produced}}{O_2 \text{ consumed}}$$

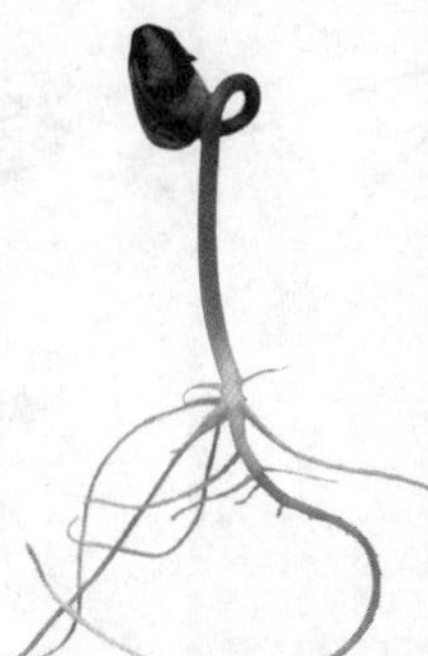

When pure carbohydrate is oxidised in cellular respiration, the RQ is 1.0; more oxygen is required to oxidise fatty acids (RQ = 0.7). The RQ for protein is about 0.9. Organisms usually respire a mix of substrates, giving RQ values of between 0.8 and 0.9 (see table 1, below).

Table 1: RQ values for the respiration of various substrates

RQ	Substrate
> 1.0	Carbohydrate with some anaerobic respiration
1.0	Carbohydrates, e.g. glucose
0.9	Protein
0.7	Fat
0.5	Fat with associated carbohydrate synthesis
0.3	Carbohydrate with associated organic acid synthesis

Using RQ to determine respiratory substrate

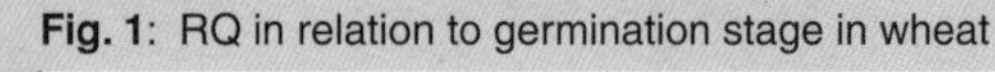

Fig. 1: RQ in relation to germination stage in wheat

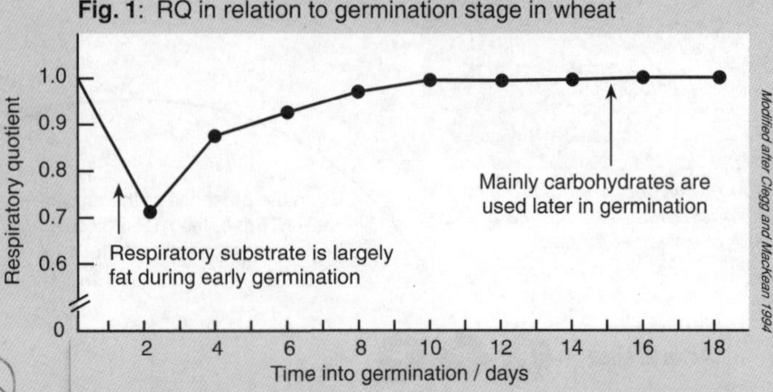

Fig. 1, above, shows how experimental RQ values have been used to determine the respiratory substrate utilised by germinating wheat seeds (*Triticum sativum*) over the period of their germination.

Table 2: Rates of O_2 consumption and CO_2 production in crickets

Time after last fed/ h	Temperature/ C	Rate of O_2 consumption/ $mlg^{-1}h^{-1}$	Rate of CO_2 production/ $mlg^{-1}h^{-1}$
1	20	2.82	2.82
48	20	2.82	1.97
1	30	5.12	5.12
48	30	5.12	3.57

Table 2 shows the rates of oxygen consumption and carbon dioxide production of crickets kept under different experimental conditions.

1. Table 2 above shows the results of an experiment to measure the rates of oxygen consumption and carbon dioxide production of crickets 1 hour and 48 hours after feeding at different temperatures:

   (a) Calculate the RQ of a cricket kept at 20°C, 48 hours after feeding (show working): ____________________

   (b) Compare this RQ to the RQ value obtained for the cricket 1 hour after being fed (20°C). Explain the difference:

2. The RQs of two species of seeds were calculated at two day intervals after germination. Results are tabulated to the right:

   (a) Plot the change in RQ of the two species during early germination:

   (b) Explain the values in terms of the possible substrates being respired:

Days after germination	RQ Seedling A	RQ Seedling B
2	0.65	0.70
4	0.35	0.91
6	0.48	0.98
8	0.68	1.00
10	0.70	1.00

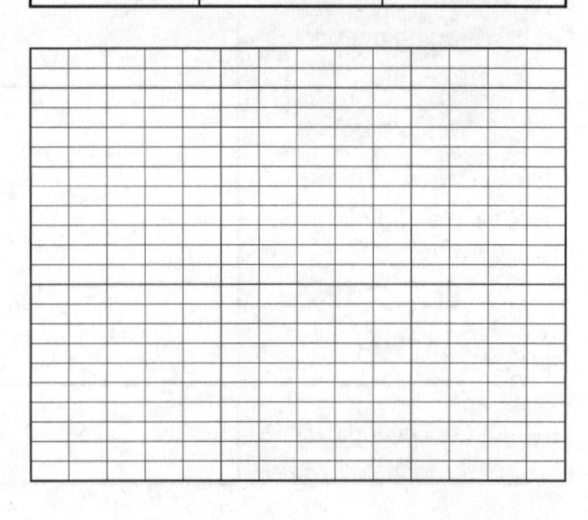

Cellular Respiration

Cellular respiration is the process by which organisms break down energy rich molecules (e.g. glucose) to release the energy in a useable form (ATP). All living cells respire in order to exist, although the substrates they use may vary. **Aerobic respiration** requires oxygen. Forms of cellular respiration that do not require oxygen are said to be **anaerobic**. Some plants and animals can generate ATP anaerobically for short periods of time. Other organisms use only anaerobic respiration and live in oxygen-free environments. For these organisms, there is some other final electron acceptor other than oxygen (e.g. nitrate or Fe^{2+}).

An Overview of Cellular Respiration

Respiration involves three metabolic stages, summarised below. The first two stages are the catabolic pathways that decompose glucose and other organic fuels. In the third stage, the electron transport chain accepts electrons from the first two stages and passes these from one electron acceptor to another. The energy released at each stepwise transfer is used to make ATP. The final electron acceptor in this process is molecular oxygen.

1. **Glycolysis**. This occurs in the cytoplasm and involves the breakdown of glucose into two molecules of pyruvate.
2. **The Krebs cycle**. This occurs in the mitochondrial matrix, and decomposes a derivative of pyruvate to carbon dioxide.
3. **Electron transport and oxidative phosphorylation**. This occurs in the inner membranes of the mitochondrion and accounts for almost 90% of the ATP generated by respiration.

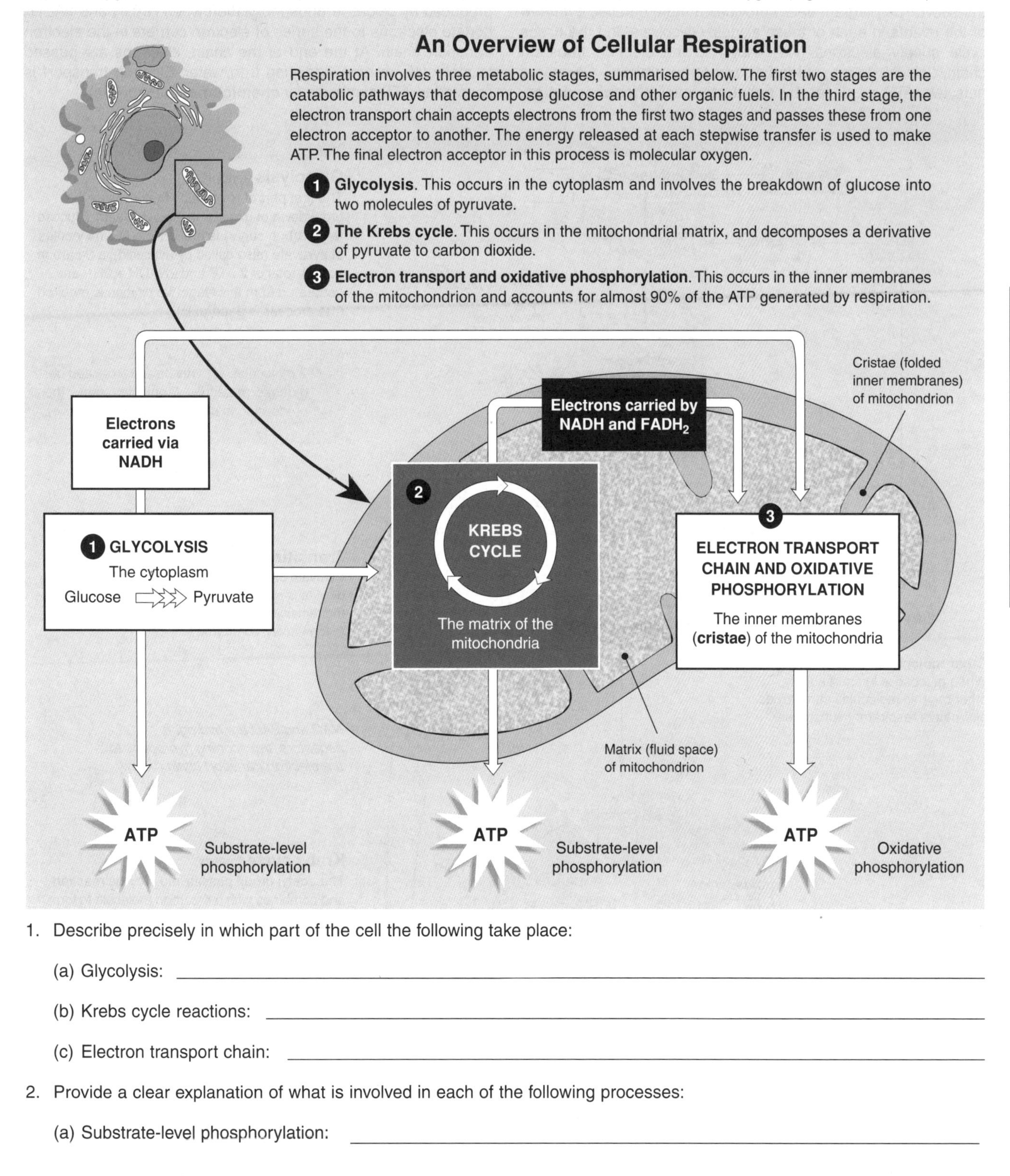

1. Describe precisely in which part of the cell the following take place:

   (a) Glycolysis: ______

   (b) Krebs cycle reactions: ______

   (c) Electron transport chain: ______

2. Provide a clear explanation of what is involved in each of the following processes:

   (a) Substrate-level phosphorylation: ______

   (b) Oxidative phosphorylation: ______

The Biochemistry of Respiration

Cellular respiration is a catabolic, energy yielding pathway. The breakdown of glucose and other organic fuels (such as fats and proteins) to simpler molecules is **exergonic** and releases energy for the synthesis of ATP. As summarised in the previous activity, respiration involves glycolysis, the Krebs cycle, and electron transport. The diagram below provides a more detailed overview of the events in each of these stages. Glycolysis and the Krebs cycle supply electrons (via NADH) to the electron transport chain, which drives **oxidative phosphorylation**. Glycolysis nets two ATP, produced by **substrate-level phosphorylation**.

The conversion of pyruvate (the end product of glycolysis) to **acetyl CoA** links glycolysis to the Krebs cycle. One "turn" of the cycle releases carbon dioxide, forms one ATP by substrate level phosphorylation, and passes electrons to three NAD^+ and one FAD. Most of the ATP generated in cellular respiration is produced by oxidative phosphorylation when NADH and $FADH_2$ donate electrons to the series of electron carriers in the electron transport chain. At the end of the chain, electrons are passed to molecular oxygen, reducing it to water. Electron transport is coupled to ATP synthesis by **chemiosmosis** (opposite).

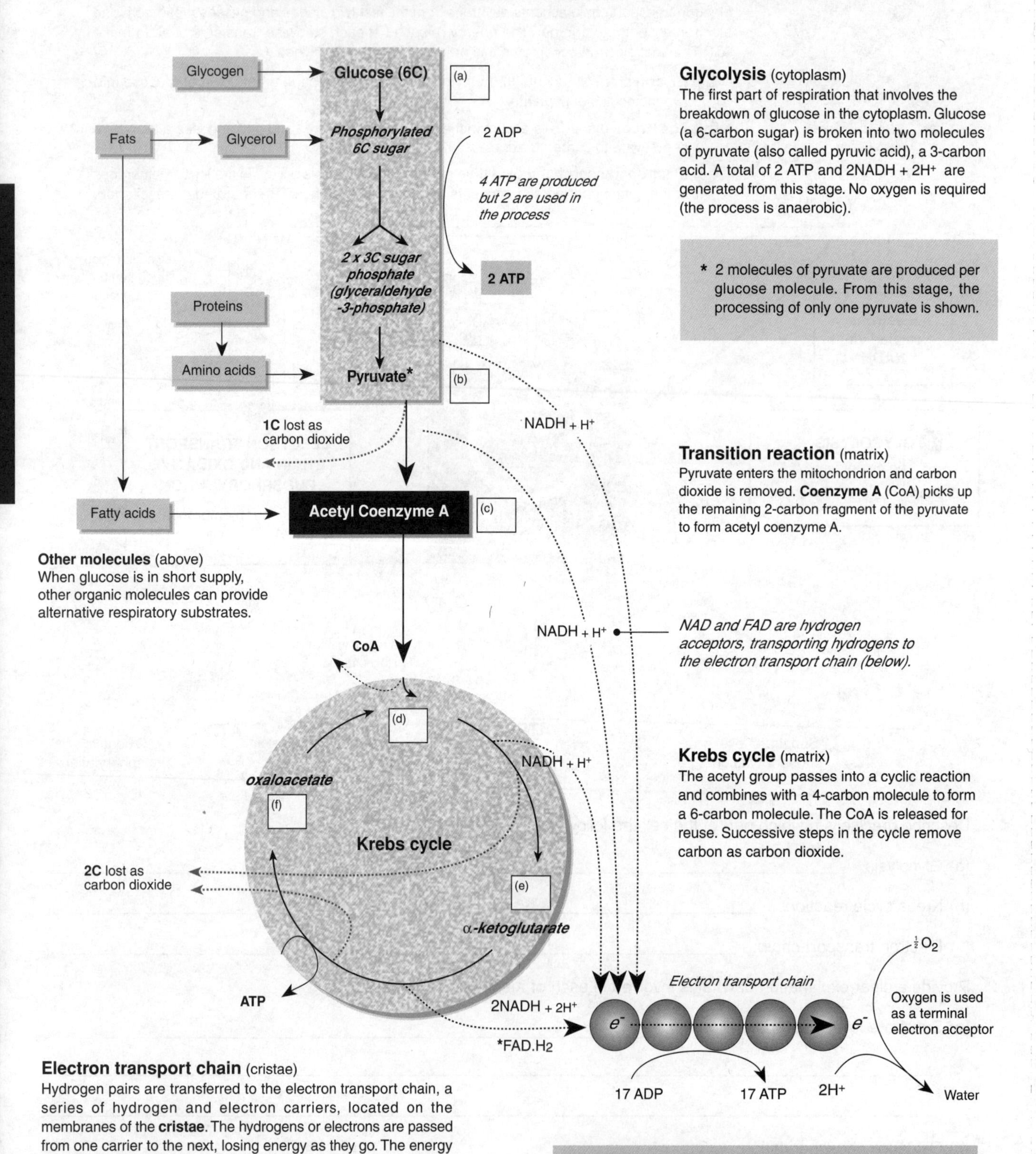

Glycolysis (cytoplasm)
The first part of respiration that involves the breakdown of glucose in the cytoplasm. Glucose (a 6-carbon sugar) is broken into two molecules of pyruvate (also called pyruvic acid), a 3-carbon acid. A total of 2 ATP and 2NADH + $2H^+$ are generated from this stage. No oxygen is required (the process is anaerobic).

* 2 molecules of pyruvate are produced per glucose molecule. From this stage, the processing of only one pyruvate is shown.

Other molecules (above)
When glucose is in short supply, other organic molecules can provide alternative respiratory substrates.

Transition reaction (matrix)
Pyruvate enters the mitochondrion and carbon dioxide is removed. **Coenzyme A** (CoA) picks up the remaining 2-carbon fragment of the pyruvate to form acetyl coenzyme A.

NAD and FAD are hydrogen acceptors, transporting hydrogens to the electron transport chain (below).

Krebs cycle (matrix)
The acetyl group passes into a cyclic reaction and combines with a 4-carbon molecule to form a 6-carbon molecule. The CoA is released for reuse. Successive steps in the cycle remove carbon as carbon dioxide.

Electron transport chain (cristae)
Hydrogen pairs are transferred to the electron transport chain, a series of hydrogen and electron carriers, located on the membranes of the **cristae**. The hydrogens or electrons are passed from one carrier to the next, losing energy as they go. The energy released in this stepwise process is used to produce ATP. Oxygen is the final electron acceptor and is reduced to water.
***Note** FAD enters the electron transport chain at a lower energy level than NAD, and only 2ATP are generated per FAD.H2.

Total ATP yield per glucose
Glycolysis: 2 ATP, *Krebs cycle*: 2 ATP, *Electron transport*: 34 ATP

Chemiosmosis

Chemiosmosis is the process whereby the synthesis of ATP is coupled to electron transport and the movement of protons (H^+ ions). **Electron transport carriers** are arranged over the inner membrane of the mitochondrion and oxidise NADH + H^+ and $FADH_2$. Energy from this process forces protons to move, against their concentration gradient, from the mitochondrial matrix into the space between the two membranes. Eventually the protons flow back into the matrix via ATP synthetase molecules in the membrane. As the protons flow down their concentration gradient, energy is released and ATP is synthesised. Chemiosmotic theory also explains the generation of ATP in the light dependent phase of photosynthesis.

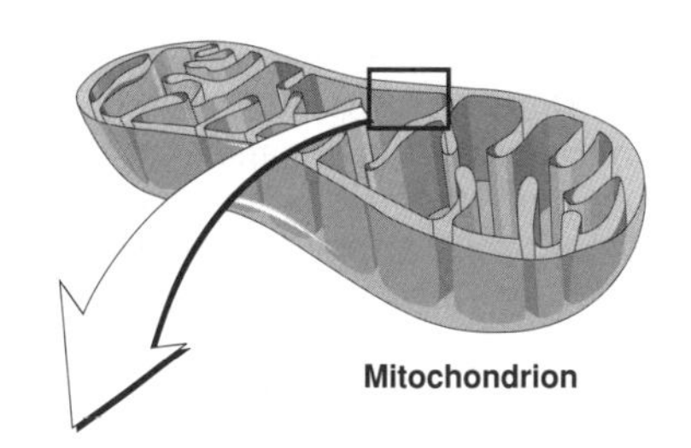

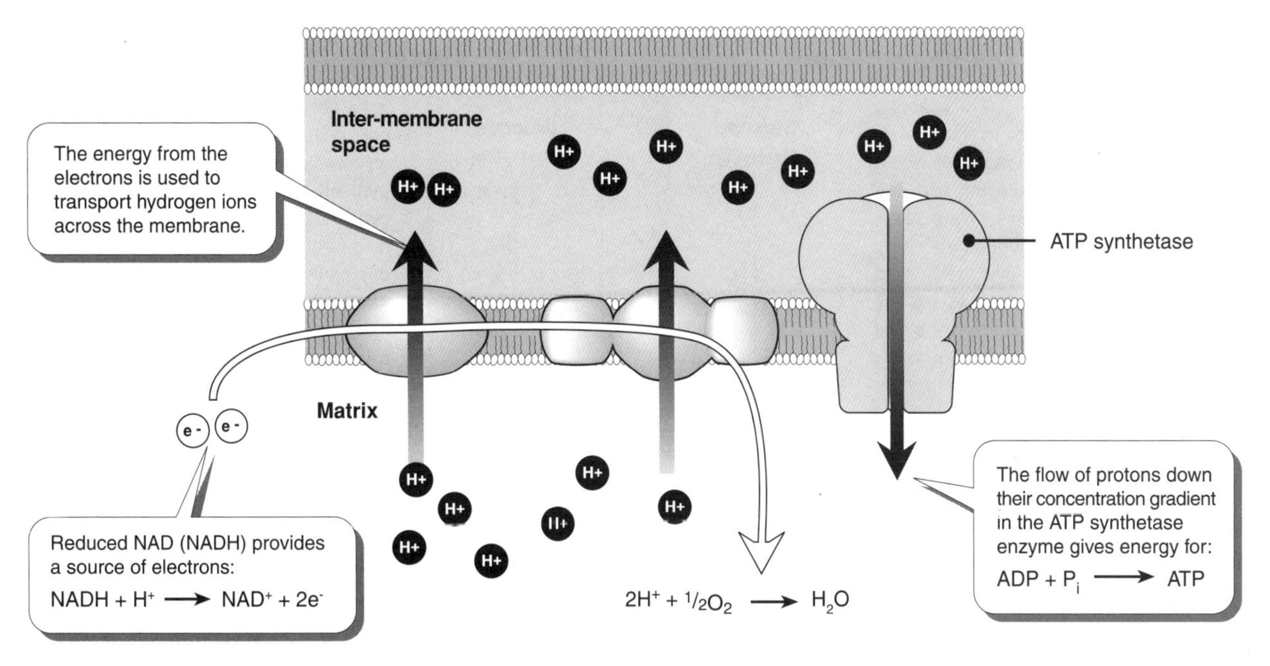

1. On the diagram of cellular respiration (previous page), state the number of carbon atoms in each of the molecules (a) – (f):
2. Determine how many ATP molecules **per molecule of glucose** are generated during the following stages of respiration:

   (a) Glycolysis: __________ (b) Krebs cycle: __________ (c) Electron transport chain: __________ (d) Total: __________

3. Explain what happens to the carbon atoms lost during respiration: ____________________

   ____

4. Describe the role of the following in aerobic cellular respiration:

   (a) Hydrogen atoms: ____________________

   ____

   ____

   (b) Oxygen: ____________________

   ____

   ____

5. (a) Identify the process by which ATP is synthesised in respiration: ____________________

   (b) Briefly summarise this process: ____________________

   ____

   ____

   ____

   ____

   ____

Anaerobic Pathways

All organisms can metabolise glucose anaerobically (without oxygen) using glycolysis in the cytoplasm, but the energy yield from this process is low and few organisms can obtain sufficient energy for their needs this way. In the absence of oxygen, glycolysis soon stops unless there is an alternative acceptor for the electrons produced from the glycolytic pathway. In yeasts and the root cells of higher plants this acceptor is ethanal, and the pathway is called alcoholic fermentation. In the skeletal muscle of mammals, the acceptor is pyruvate itself and the end product is lactic acid. In both cases, the duration of the fermentation is limited by the toxic effects of the organic compound produced. Although fermentation is often used synonymously with anaerobic respiration, they are not the same. Respiration always involves hydrogen ions passing down a chain of carriers to a terminal acceptor, and this does not occur in fermentation. In anaerobic respiration, the terminal H^+ acceptor is a molecule other than oxygen, e.g. Fe^{2+} or nitrate.

Alcoholic Fermentation

In alcoholic fermentation, the H^+ acceptor is ethanal which is reduced to ethanol with the release of CO_2. Yeasts respire aerobically when oxygen is available but can use alcoholic fermentation when it is not. At levels above 12-15%, the ethanol produced by alcoholic fermentation is toxic to the yeast cells and this limits their ability to use this pathway indefinitely. The root cells of plants also use fermentation as a pathway when oxygen is unavailable but the ethanol must be converted back to respiratory intermediates and respired aerobically.

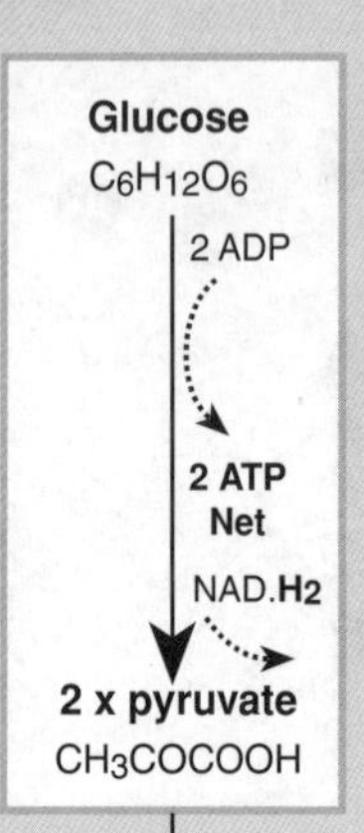

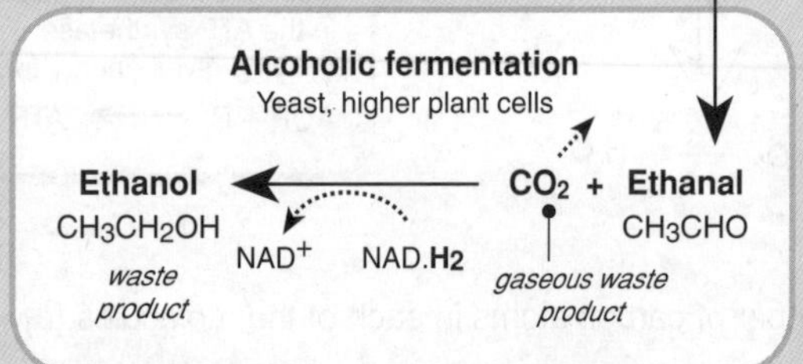

Lactic Acid Fermentation

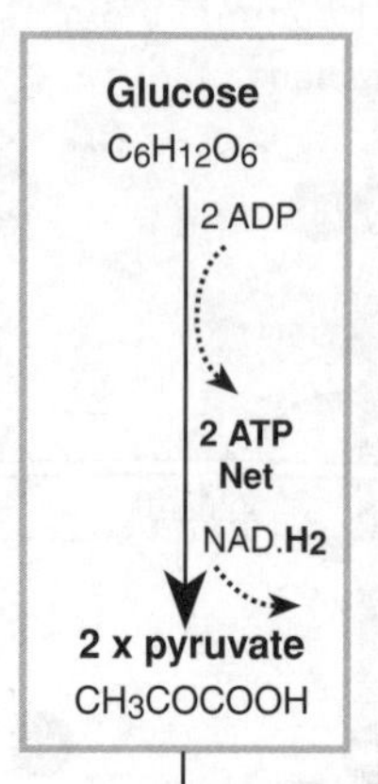

In the absence of oxygen, the skeletal muscle cells of mammals are able to continue using glycolysis for ATP production by reducing pyruvate to lactic acid (the H^+ acceptor is pyruvate itself). This process is called lactic acid fermentation. Lactic acid is toxic and this pathway cannot continue indefinitely. The lactic acid must be removed from the muscle and transported to the liver, where it is converted back to respiratory intermediates and respired aerobically.

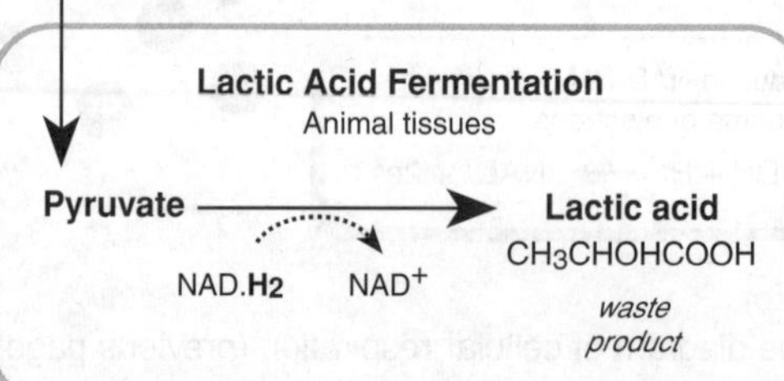

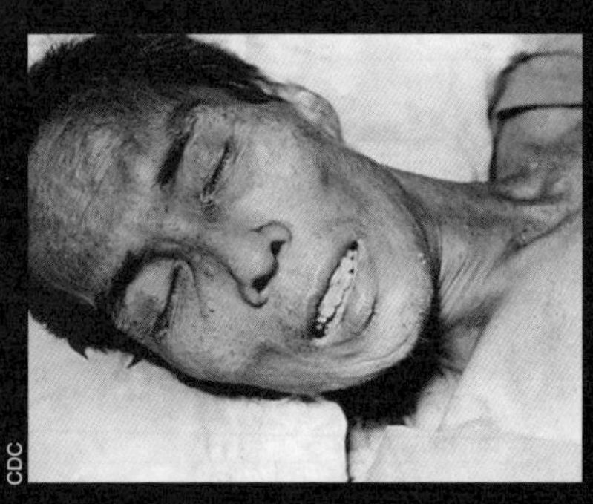

Some organisms respire only in the absence of oxygen and are known as **obligate anaerobes**. Many of these organisms are bacterial pathogens and cause diseases such as tetanus (*above*), gangrene, and botulism.

Vertebrate skeletal muscle is **facultatively anaerobic** because it has the ability to generate ATP for a short time in the absence of oxygen. The energy from this pathway comes from glycolysis and the yield is low.

The products of alcoholic fermentation have been utilised by humans for centuries. The alcohol and carbon dioxide produced from this process form the basis of the brewing and baking industries.

1. Describe the key difference between aerobic respiration and fermentation: ______________________

____

2. (a) Refer to page 30 and determine the efficiency of fermentation compared to aerobic respiration: ________ %

(b) In simple terms, explain why the efficiency of anaerobic pathways is so low: ______________________

____

3. Explain why fermentation cannot go on indefinitely: ______________________

____

____

Leaf Structure

The main function of leaves is to collect the radiant energy from the sun and convert it into a form that can be used by the plant. The sugars produced from photosynthesis have two uses: (1) they can be broken down by respiration to release the stored chemical energy to do cellular work or (2) they can provide the plant with building materials for new tissues (growth and repair). The structure of the leaf is adapted to carry out the job of collecting the sun's energy. To an extent, a leaf is also able to control the amount of carbon dioxide entering and the amount of water leaving the plant. Both the external and internal morphology of leaves are related primarily to habitat, especially availability of light and water. Regardless of their varying forms, foliage leaves comprise epidermal, mesophyll, and vascular tissues. The mesophyll (the packing tissue of the leaf) may be variously arranged according to the particular photosynthetic adaptations of the leaf.

Angiosperm Leaves

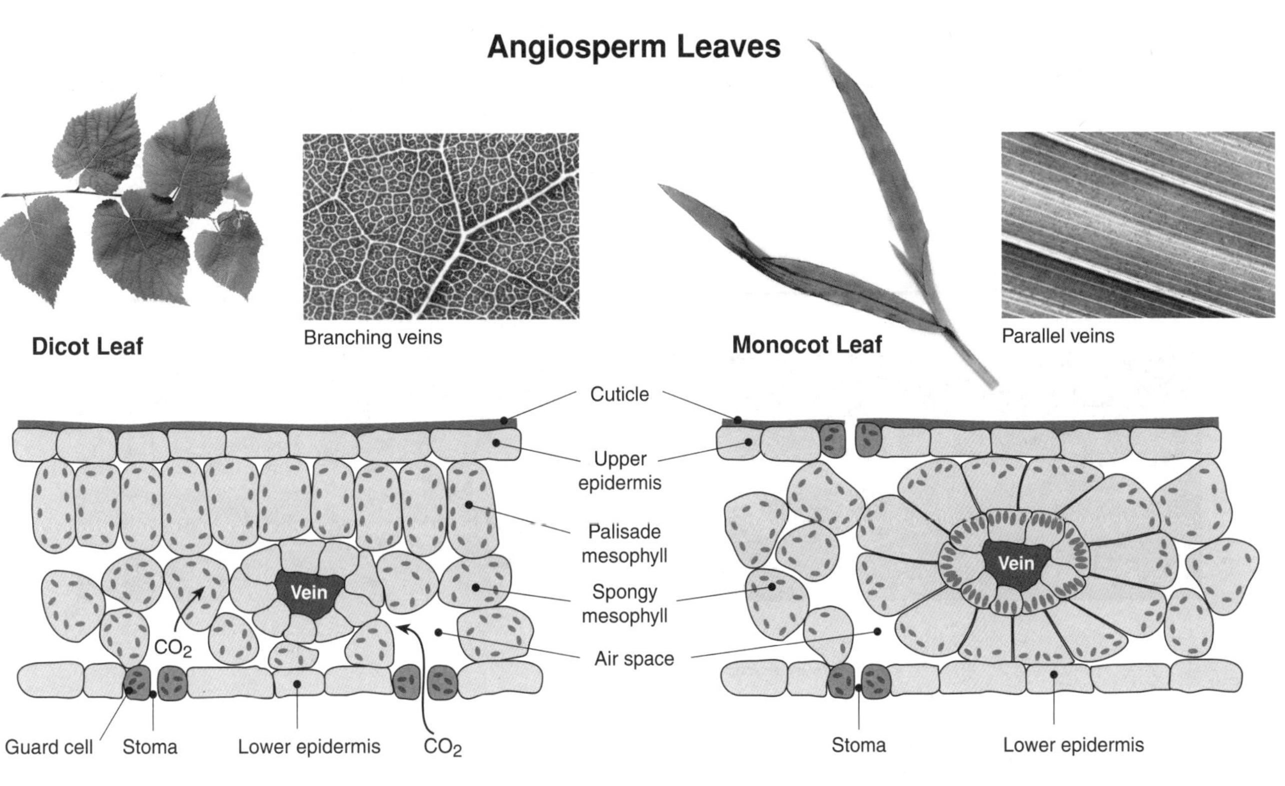

1. The internal arrangement of cells and tissues in typical monocot and dicot leaves are illustrated above. Describe two ways in which the internal leaf structure of the dicot differs from that of the monocot:

   (a) ______________________________

   ____

   (b) ______________________________

   ____

2. (a) Explain the purpose of the waxy cuticle that coats the leaf surface: ______________________________

   ____

   (b) Explain why the leaf epidermis is transparent: ______________________________

   ____

   (c) Explain why leaves are usually broad and flat: ______________________________

   ____

3. (a) Identify the region of a dicot leaf where most of the chloroplasts are found: ______________________________

   (b) Name the important process that occurs in the chloroplasts: ______________________________

4. (a) Explain the purpose of the air spaces in the leaf tissue: ______________________________

   ____

   (b) Describe how gases enter and leave the leaf tissue: ______________________________

   ____

Adaptations for Photosynthesis

In order to photosynthesise, plants must obtain a regular supply of carbon dioxide (CO_2) gas; the raw material for the production of carbohydrate. In green plants, the systems for gas exchange and photosynthesis are linked; without a regular supply of CO_2, photosynthesis ceases. The leaf, as the primary photosynthetic organ, is adapted to maximise light capture and facilitate the entry of CO_2, while minimising water loss. There are various ways in which plant leaves are adapted to do this. The ultimate structure of the leaf reflects the environment of the leaf (sun or shade, terrestrial or aquatic), its resistance to water loss, and the importance of the leaf relative to other parts of the plant that may be photosynthetic, such as the stem.

Sun plant

A **sun leaf**, when exposed to high light intensities, can absorb much of the light available to the cells.

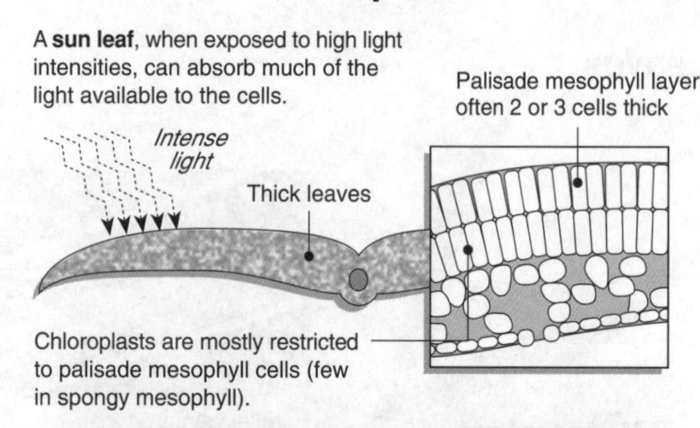

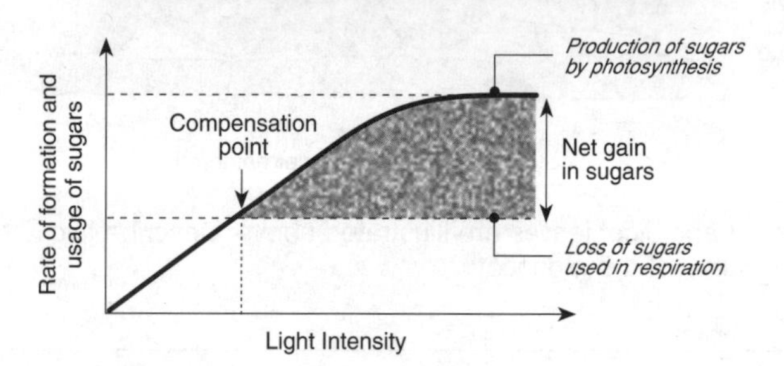

Plants adapted for full sunlight have higher levels of respiration and much higher *compensation points*. **Sun plants** include many weed species found on open ground. They expend much more energy on the construction and maintenance of thicker leaves than do shade plants. The benefit of this investment is that they can absorb the higher light intensities available and grow more quickly.

Shade plant

A **shade leaf** can absorb the light available at lower light intensities. If exposed to high light, most would pass through.

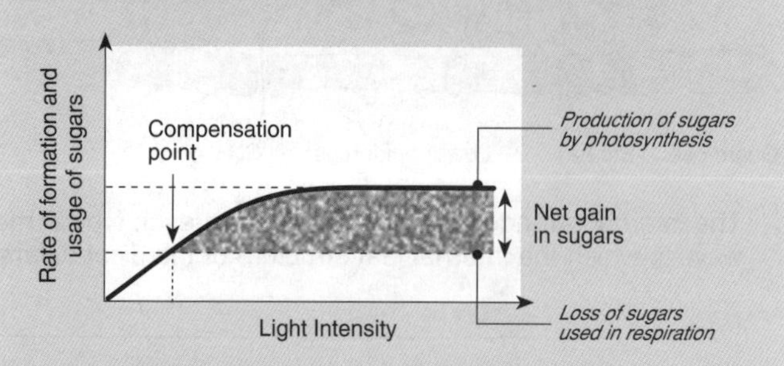

Shade plants typically grow in forested areas, partly shaded by the canopy of larger trees. They have lower rates of respiration than sun plants, mainly because they build thinner leaves. The fewer number of cells need less energy for their production and maintenance. As a result, shade plants reach their *compensation point* at a low light intensity; much sooner than sun plants do.

1. (a) From the diagrams above, determine what is meant by the **compensation point** in terms of sugar production:

   (b) State which type of plant (sun or shade adapted) has the highest level of respiration:

   (c) Explain how the plant compensates for the higher level of respiration:

2. Discuss the adaptations of leaves in **sun** and **shade plants**:

Adaptations for Photosynthesis and Gas Exchange in Plants

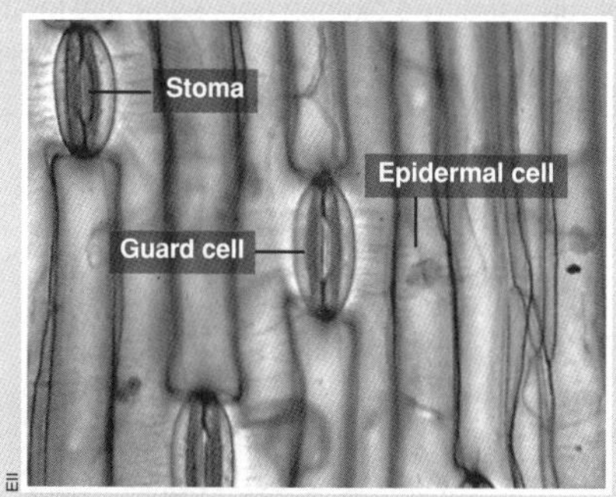

Surface view of stomata on a monocot leaf (grass). The parallel arrangement of stomata is a typical feature of monocot leaves. Grass leaves show properties of **xerophytes**, with several water conserving features (see right).

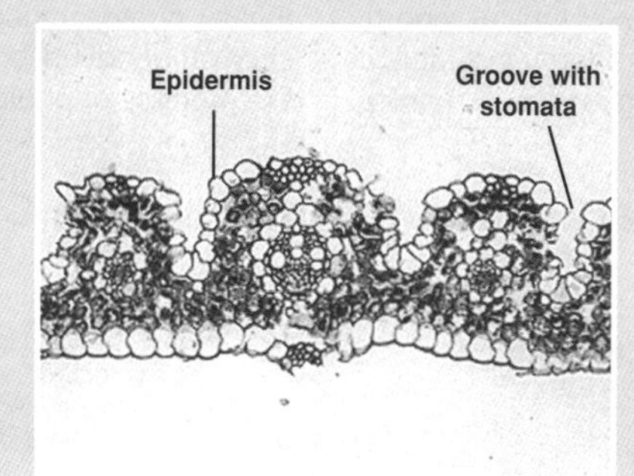

Cross section through a grass leaf showing the stomata housed in grooves. When the leaf begins to dehydrate, it may fold up, closing the grooves and thus preventing or reducing water loss through the stomata.

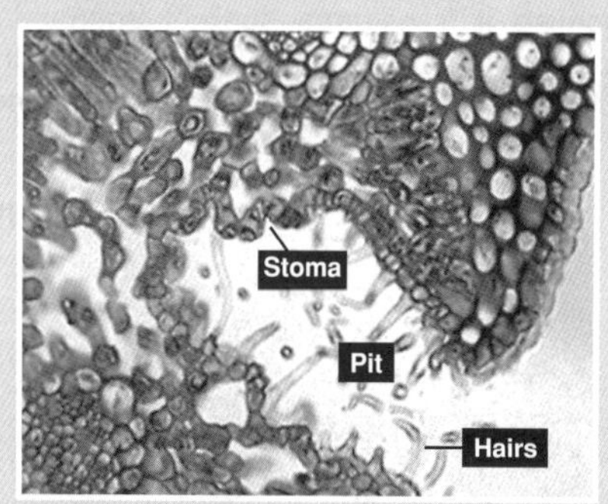

Oleander (above) is a xerophyte that displays many water conserving features. The stomata are found at the bottom of pits on the underside of the leaf. The pits restrict water loss to a greater extent than they reduce CO_2 uptake.

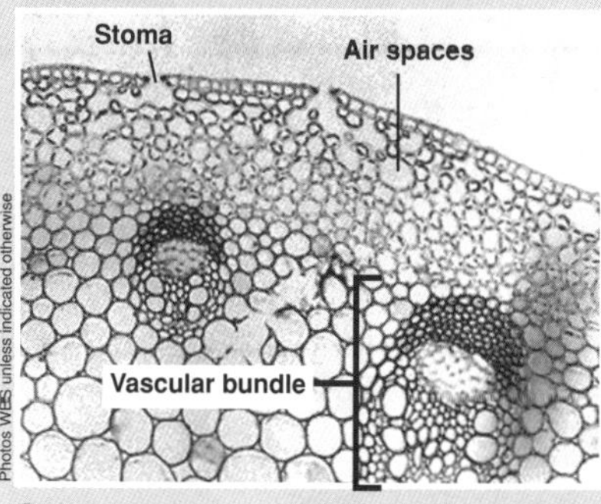

Some plants (e.g. buttercup above) have photosynthetic stems, and CO_2 enters freely into the stem tissue through stomata in the epidermis. The air spaces in the cortex are more typical of leaf mesophyll than stem cortex.

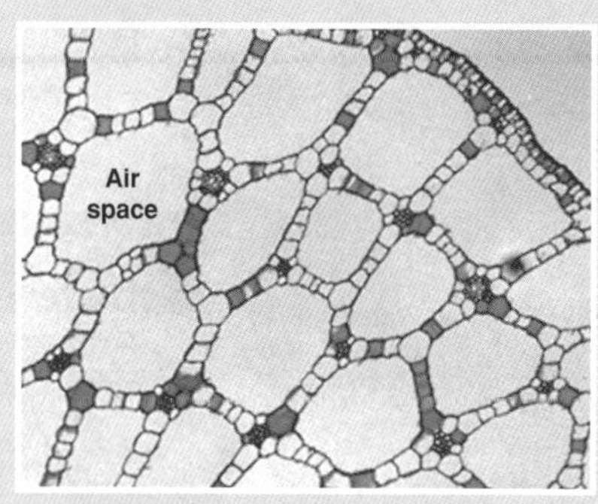

Hydrophytes, such as *Potamogeton*, above, have stems with massive air spaces. The presence of air in the stem means that they remain floating in the zone of light availability and photosynthesis is not compromised.

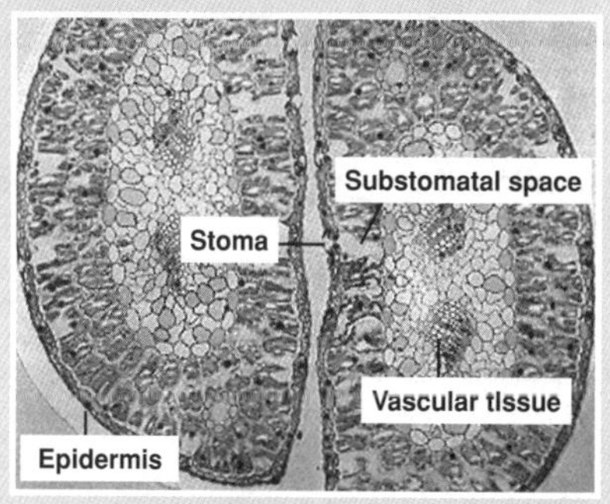

This transverse view of the twin leaves of a two-needle pine shows the sunken stomata and substomatal spaces. This adaptation for arid conditions reduces water loss by creating a region of high humidity around the stoma.

3. Describe two adaptations in plants for reducing water loss while maintaining entry of gas into the leaf:

   (a) ____

   ____

   (b) ____

   ____

4. Describe two adaptations of photosynthetic stems that are not present in non-photosynthetic stems, and explain the reasons for these:

   (a) ____

   ____

   (b) ____

   ____

5. The example of a photosynthetic stem above is from a buttercup, a plant in which the leaves are still the primary organs of photosynthesis.

   (a) Identify an example of the plant where the stem is the **only** photosynthetic organ: ____________________

   (b) Describe the structure of the leaves in your example and suggest a reason for their particular structure:

   ____

   ____

6. Describe one role of the air spaces in the stems of *Potamogeton* related to maintaining photosynthesis:

   ____

Photosynthesis

Photosynthesis is of fundamental importance to living things because it transforms sunlight energy into chemical energy stored in molecules. This becomes part of the energy available in food chains. The molecules that trap the energy in their chemical bonds are also used as building blocks to create other molecules. Finally, photosynthesis releases free oxygen gas, essential for the survival of advanced life forms. Below is a diagram summarising the process of photosynthesis.

Summary of Photosynthesis in a C_3 Plant

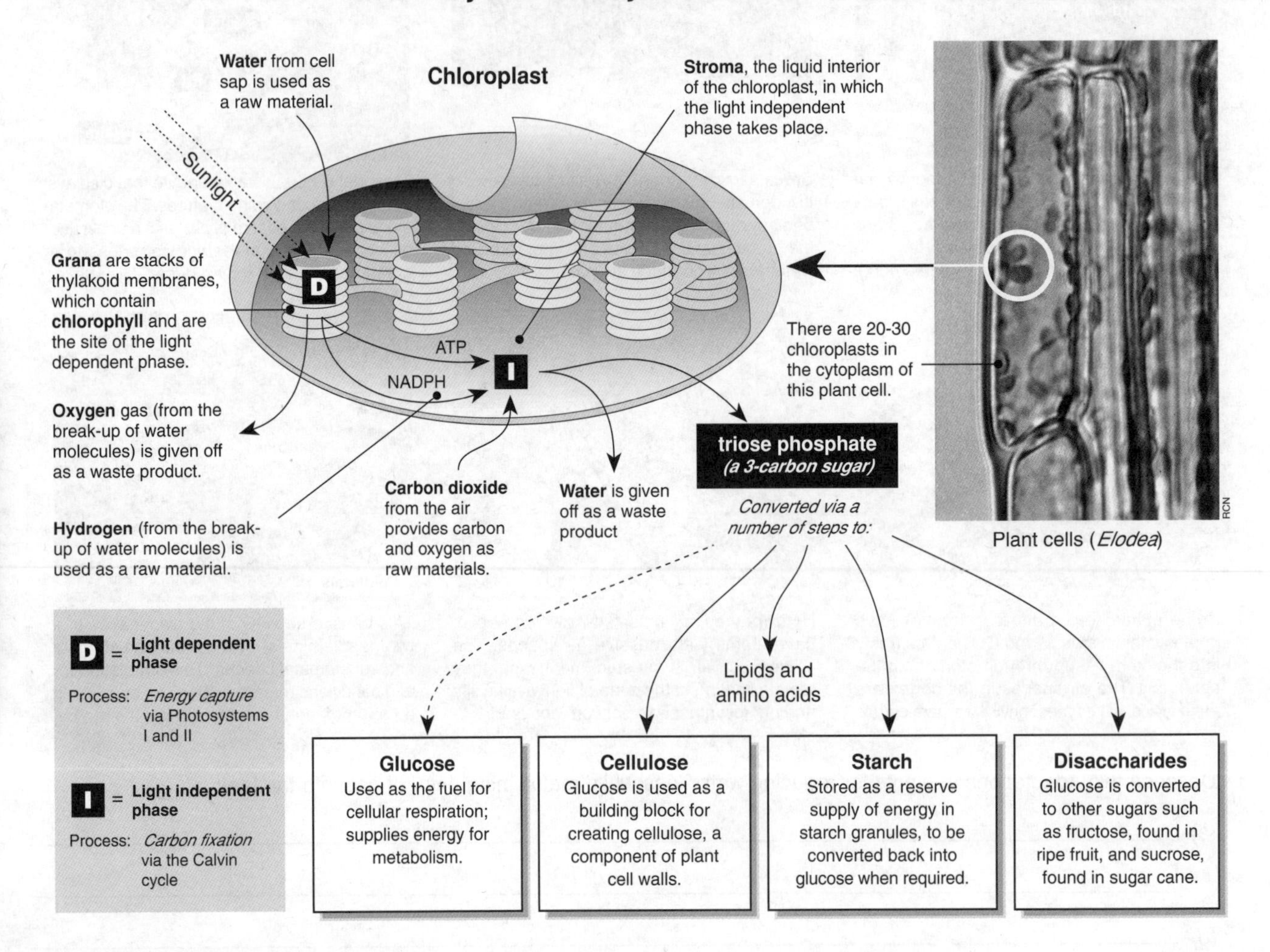

1. Describe the three things of fundamental biological importance provided by photosynthesis:

   (a) ______________________________

   (b) ______________________________

   (c) ______________________________

2. Write the overall chemical equation for photosynthesis using:

   (a) Words: ______________________________

   (b) Chemical symbols: ______________________________

3. Discuss the potential uses for the end products of photosynthesis: ______________________________

   ____

   ____

4. Distinguish between the two different regions of a chloroplast and describe the biochemical processes that occur in each:

   ____

   ____

   ____

Cellular Metabolism

Code: A 2

Pigments and Light Absorption

As light meets matter, it may be reflected, transmitted, or absorbed. Substances that absorb visible light are called **pigments**, and different pigments absorb light of different wavelengths. The ability of a pigment to absorb particular wavelengths of light can be measured with a spectrophotometer. The light absorption vs the wavelength is called the **absorption spectrum** of that pigment. The absorption spectrum of different photosynthetic pigments provides clues to their role in photosynthesis, since light can only perform work if it is absorbed. An **action spectrum** profiles the effectiveness of different wavelength light in fuelling photosynthesis. It is obtained by plotting wavelength against some measure of photosynthetic rate (e.g. CO_2 production). Some features of photosynthetic pigments and their light absorbing properties are outlined below.

The Electromagnetic Spectrum

Light is a form of energy known as electromagnetic radiation. The segment of the electromagnetic spectrum most important to life is the narrow band between about 380 and 750 nanometres (nm). This radiation is known as visible light because it is detected as colours by the human eye (although some other animals, such as insects, can see in the ultraviolet range). It is the visible light that drives photosynthesis.

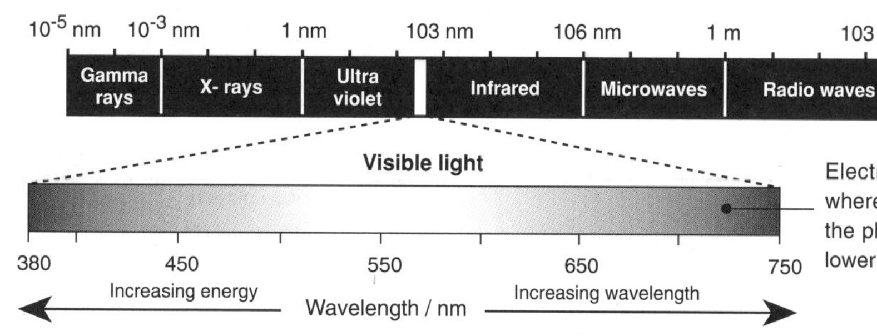

Electromagnetic radiation (EMR) travels in waves, where wavelength provides a guide to the energy of the photons; the greater the wavelength of EMR, the lower the energy of the photons in that radiation.

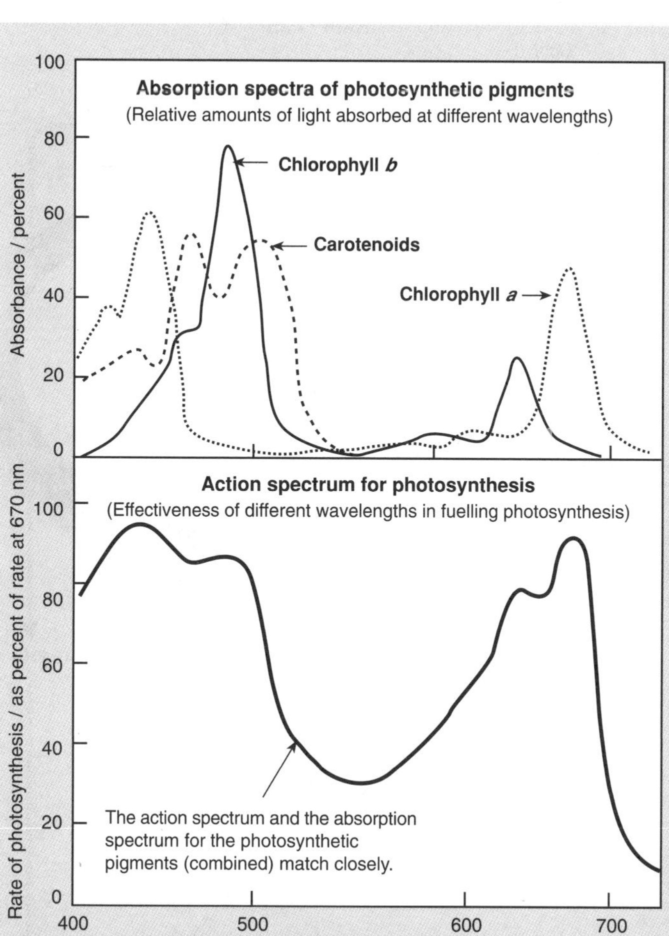

The Photosynthetic Pigments of Plants

The photosynthetic pigments of plants fall into two categories: **chlorophylls** (which absorb red and blue-violet light) and **carotenoids** (which absorb strongly in the blue-violet and appear orange, yellow, or red). The pigments are located on the chloroplast membranes (the thylakoids) and are associated with membrane transport systems.

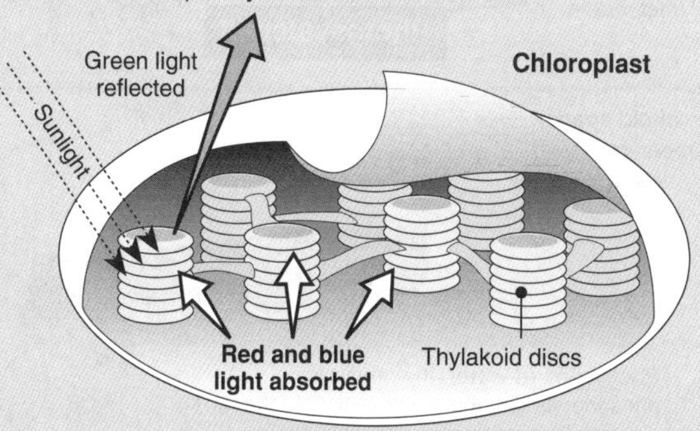

The pigments of chloroplasts in higher plants (above) absorb blue and red light, and the leaves therefore appear green (which is reflected). Each photosynthetic pigment has its own characteristic **absorption spectrum** (left, top graph). Although only chlorophyll *a* can participate directly in the light reactions of photosynthesis, the **accessory pigments** (chlorophyll *b* and carotenoids) can absorb wavelengths of light that chlorophyll *a* cannot. The accessory pigments pass the energy (photons) to chlorophyll *a*, thus broadening the spectrum that can effectively drive photosynthesis.

Left: Graphs comparing absorption spectra of photosynthetic pigments compared with the action spectrum for photosynthesis.

1. Explain what is meant by the absorption spectrum of a pigment: ______________________

____

2. Explain why the action spectrum for photosynthesis does not exactly match the absorption spectrum of chlorophyll *a*.

____

____

The Biochemistry of Photosynthesis

Like cellular respiration, photosynthesis is a redox process, but the electron flow evident in respiration is reversed. In photosynthesis, water is split and electrons are transferred together with hydrogen ions from water to CO_2, reducing it to sugar. The electrons increase in potential energy as they move from water to sugar. The energy to do this is provided by light. Photosynthesis comprises two phases. In the **light dependent phase**, light energy is converted to chemical energy (ATP and reducing power). In the **light independent phase** (or **Calvin cycle**), the chemical energy is used for the synthesis of carbohydrate. The light dependent phase illustrated below shows **non-cyclic phosphorylation**. In **cyclic phosphorylation**, the electrons lost from photosystem II are replaced by those from photosystem I. ATP is generated, but not NADPH.

Light Dependent Phase
(Energy capture)

- This diagram shows **non-cyclic phosphorylation**.
- Photosystem complexes comprise hundreds of pigment molecules, including *chlorophyll a* and *b*.
- **Photosystem II** absorbs light energy to elevate electrons to a moderate energy level.
- **Photosystem I** absorbs light energy to elevate electrons to an even higher level. Its electrons are replaced by electrons from photosystem II.

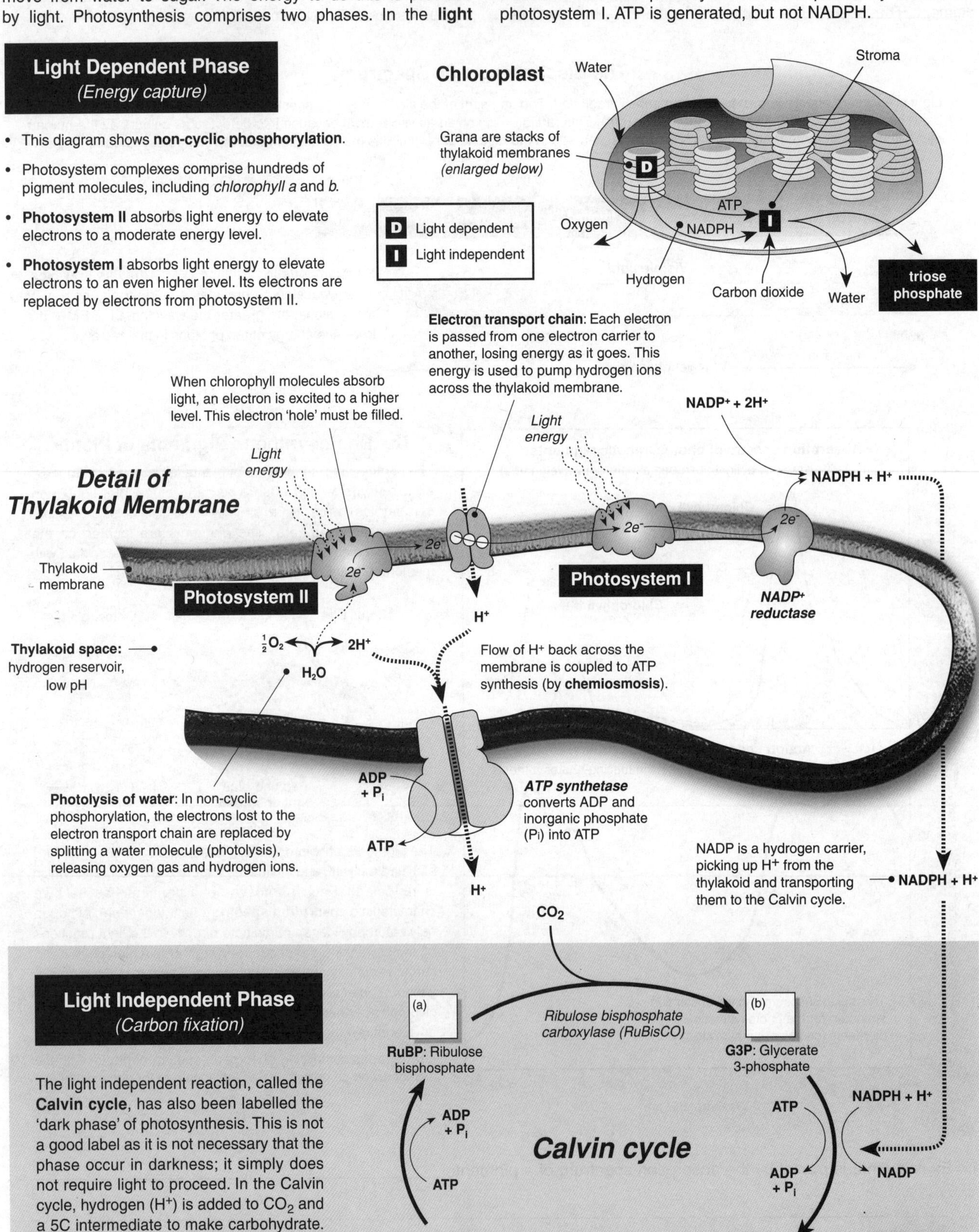

Light Independent Phase
(Carbon fixation)

The light independent reaction, called the **Calvin cycle**, has also been labelled the 'dark phase' of photosynthesis. This is not a good label as it is not necessary that the phase occur in darkness; it simply does not require light to proceed. In the Calvin cycle, hydrogen (H^+) is added to CO_2 and a 5C intermediate to make carbohydrate. The H^+ and ATP are supplied by the light dependent phase above.

1. Describe the role of the carrier molecule **NADP** in photosynthesis:

2. Explain the role of chlorophyll molecules in the process of photosynthesis:

3. On the diagram opposite, write the number of carbon atoms that each molecule has at each stage of the Calvin cycle:

4. Summarise the events in each of the two phases in photosynthesis and identify where each phase occurs:

   (a) **Light dependent phase (D)**:

   (b) **Calvin cycle**:

5. The final product of photosynthesis is triose phosphate. Describe precisely where the carbon, hydrogen and oxygen molecules originate from to make this molecule:

6. Explain how ATP is produced as a result of light striking chlorophyll molecules during the light dependent phase:

7. (a) The diagram of the light dependent phase (opposite) describes **non-cyclic phosphorylation**. Explain what you understand by this term:

   (b) Suggest why this process is also known as non-cyclic **photo**phosphorylation:

   (c) Explain how photophosphorylation differs from the oxidative phosphorylation occurring in cellular respiration:

8. Explain how **cyclic photophosphorylation** differs from non-cyclic photophosphorylation:

Photosynthetic Rate

The rate at which plants can make food (the photosynthetic rate) is dependent on environmental factors, particularly the amount of **light** available, the level of **carbon dioxide** (CO_2) and the **temperature**. The effect of these factors can be tested experimentally by altering one of the factors while holding others constant (a controlled experiment). In reality, a plant is subjected to variations in all three factors at the same time. The interaction of the different factors can also be examined in the same way, as long as only one factor at a time is altered. The results can be expressed in a graph.

Factors Affecting Photosynthetic Rate

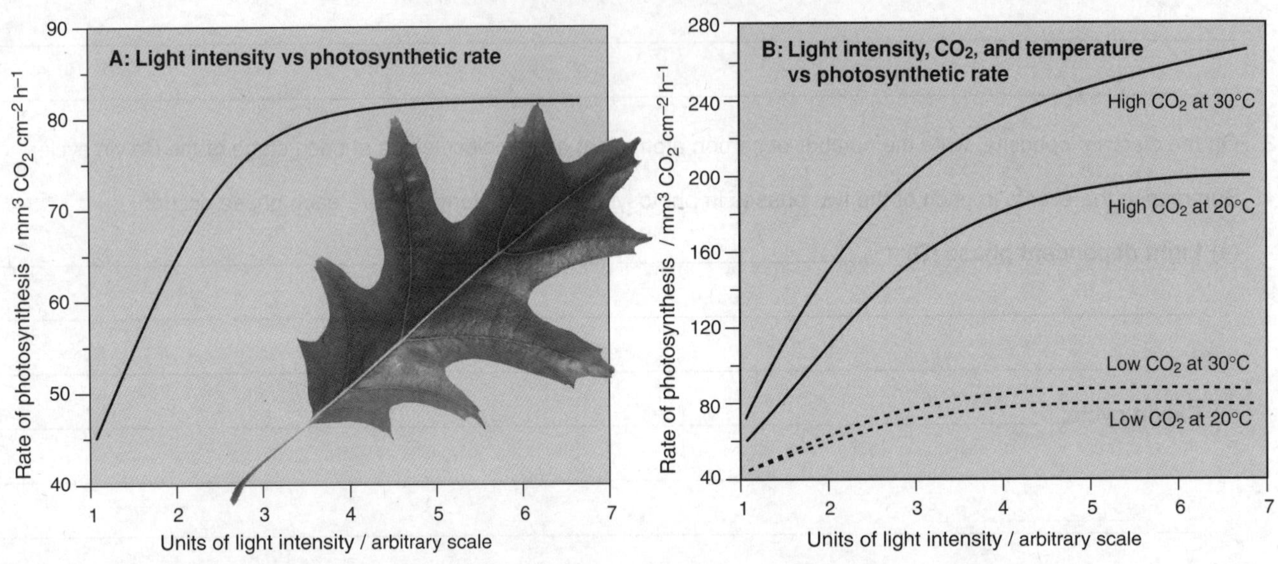

The two graphs above illustrate the effect of different variables on the rate of photosynthesis in cucumber plants. Graph A (above, left) shows the effect of different intensities of light. In this experiment, the level of carbon dioxide available and the temperature were kept constant. Graph B (above, right) shows the effect of different light intensities at two temperatures and two carbon dioxide (CO_2) concentrations. In each of these experiments either the carbon dioxide level or the temperature was raised at each light intensity in turn.

1. (a) Describe the effect of increasing light intensity on the rate of photosynthesis (temperature and CO_2 constant):

   (b) Give a possible explanation for the shape of the curve:

2. (a) Describe the effect of increasing the temperature on the rate of photosynthesis:

   (b) Suggest a reason for this response:

3. Explain why the rate of photosynthesis declines when the CO_2 level is reduced:

4. (a) In the graph above right, explain how the effects of CO_2 level were separated from the effects of temperature:

   (b) State which of the two factors, CO_2 level or temperature, has the greatest effect on photosynthetic rate:

   (c) Explain how you can tell this from the graph:

Plant Nutritional Requirements

Plants require a variety of minerals. **Macronutrients** are required in large quantities, whilst **trace elements** are needed in only very small amounts. Cropping interrupts normal nutrient cycles and contributes to nutrient losses. Nutrients can be replaced by the addition of **fertilisers**: materials that supply nutrients to plants. Other plants form mutualistic associations with mycorrhizal fungi or symbiotic bacteria. These associations aid the plant's nutrition by supplying a nutrient directly or by improving its uptake.

Ion Availability in the Soil

Plants normally obtain minerals from the soil. The availability of ions to plant roots depends on soil texture, since this affects the permeability of the soil to air and water. Mineral ions may be available to the plant in the soil water, adsorbed on to clay particles, or via release from humus and soil weathering. Plant **macronutrients** (e.g. nitrogen, sulphur, and phosphorus) are required in large amounts for building basic constituents like proteins. Trace elements (e.g. manganese, copper, and zinc) are required in smaller amounts. Many are necessary components of, or activators for, enzymes. After being absorbed by epidermal cells mineral ions diffuse down a concentration gradient to the endodermis. From the cytoplasm of the endodermis, the minerals may diffuse or be actively transported to the xylem for transport around the plant.

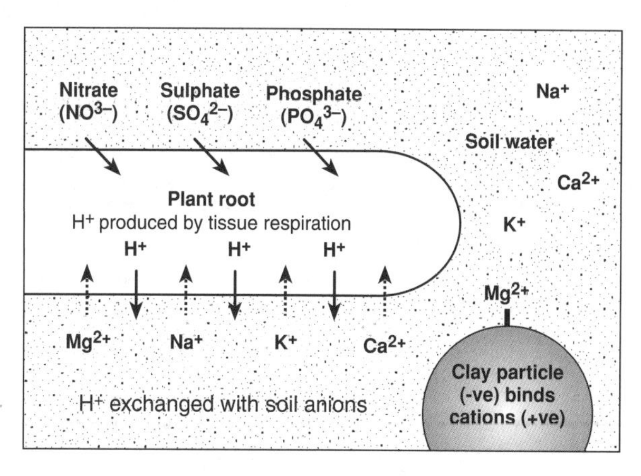

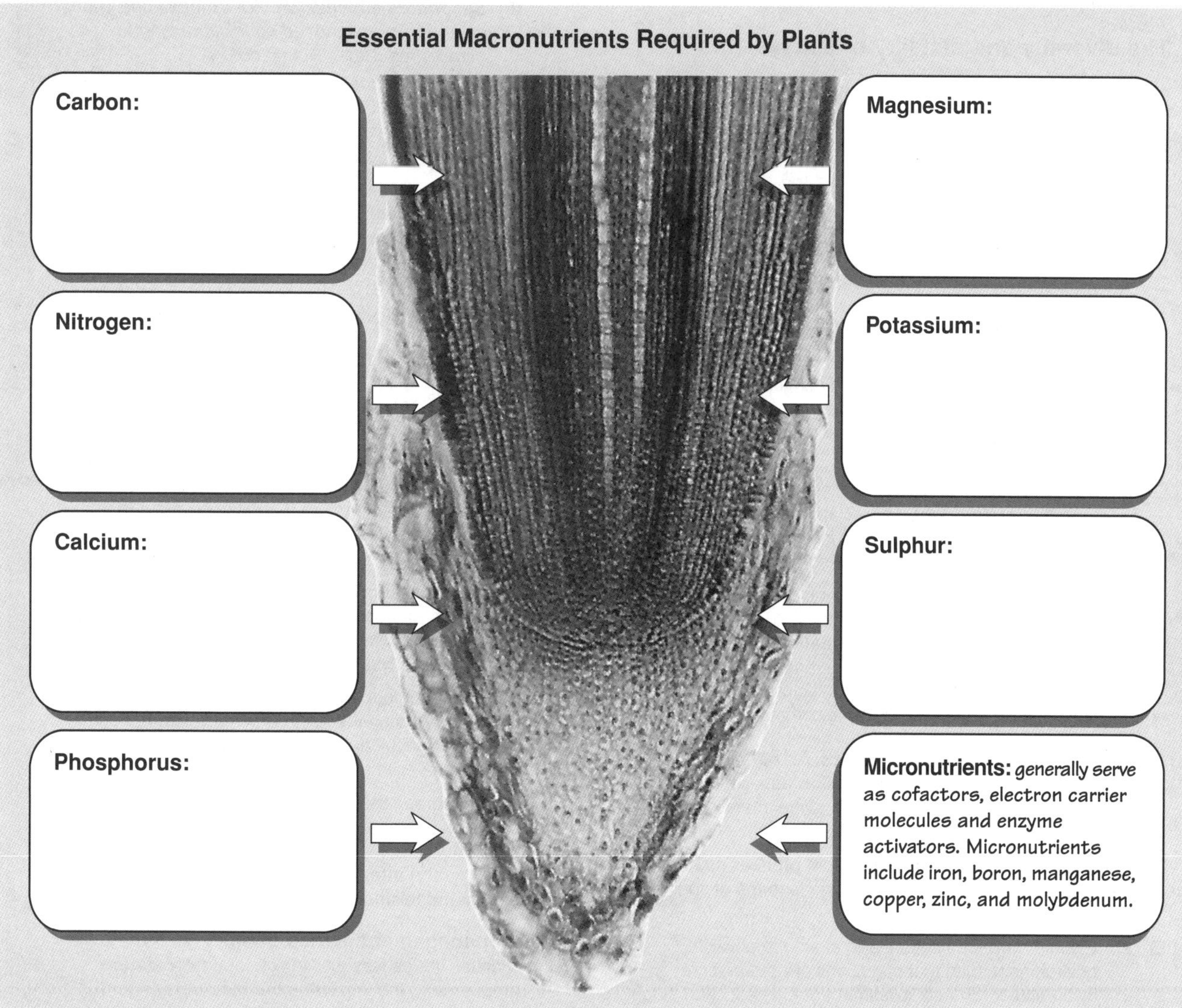

1. Complete the diagram above, outlining the role the stated macronutrients play in plant development.

2. Describe three sources of minerals for plants, apart from artificially applied fertilisers:

(a) ______

(b) ______

(c) ______

Populations and Interactions

AQA-A	AQA-B	CIE	Edexcel	OCR
Complete:	Complete:	Not applicable to the A2 course	Complete:	Complete:
1-5, 7-10, 14, 20-21	*1, 7-10, 13, 15-19, 23, 25-26, 29(c)-(e), 32-34*		*1, 7-12, 15-17, 19, 21, 23, 25-26, 28-29(a)-(d), 30, 33-34*	*1, 7-10, 13, 15-17, 21-24, 26, 28-29(a), (e), 30, 33-34*

Learning Objectives

- ☐ 1. Compile your own glossary from the **KEY WORDS** displayed in **bold type** in the learning objectives below.

The diversity and stability of ecosystems

- ☐ 2. Recognise that the collection of quantitative population data provides the means by which to study ecosystems in a meaningful way (see the topic "*Practical Ecology*").

Ecosystem diversity *(pages 44-45, 72)*

- ☐ 3. Explain what is meant by the **stability** of an ecosystem and identify its components. Explain the relationship between ecosystem stability and diversity.
- ☐ 4. Explain the basis for the following relationships:
    - The usually low stability and low diversity of extreme environments where abiotic factors predominate in determining population distribution and abundance.
    - The usually high stability and high diversity of more moderate environments where biotic factors predominate in determining population distribution and abundance.
- ☐ 5. Demonstrate an ability to calculate and interpret an **index of diversity**. Appreciate the applications and disadvantages of diversity indices in natural ecosystems and in those modified by human activity.
- ☐ 6. With reference to a specific example, describe the main features of a common type of distributional variation within a community, such as **zonation**. Explain how this pattern arises and how it increases the amount of community diversity. Identify the factors determining species distribution within these communities.

Ecological succession *(pages 70-71)*

- ☐ 7. Recognise **ecological succession** as a community pattern in time that is the result of the interaction of species with their environment.
- ☐ 8. Describe primary succession from **pioneer species** to a **climax community**. Identify the features of species typical of each **seral stage**.
- ☐ 9. Explain how each seral stage alters the physical environment such that conditions are made more favourable for the establishment of the next seral stage.
- ☐ 10. Distinguish between **primary** and **secondary succession**, outlining the features of each type. Include reference to the time scale over which these successions take place.
- ☐ 11. Explain how the **climax** vegetation varies according to local, climate, latitude, and altitude.
- ☐ 12. Explain what is meant by a **deflected succession** and describe how such a succession arises. Explain the nature of the **plagioclimax** that develops as a result of a deflected succession.
- ☐ 13. Describe an example of ecological succession:
    - (a) ***Primary succession**: For example, the development of a climax community after volcanic activity or following glacial retreat.*
    - (b) ***Secondary succession**: For example, the change occurring in abandoned agricultural land to established climax woodland.*
    - (c) ***Deflected succession**: As when a particular community composition is maintained through human intervention (e.g. mowing or grazing).*
- ☐ 14. Describe how community diversity changes during the course of a succession (or **sere**). Comment on the stability of the pioneer and climax communities and relate this to the relative importance of abiotic and biotic factors at each stage (also see #4).

The ecological impact of farming

Agricultural ecosystems *(pages 64-67)*

- ☐ 15. Understand that humans manipulate the natural environment to create **agricultural ecosystems** for the production of food and other resources.
- ☐ 16. Explain what is meant by **intensive farming**. Appreciate the implications of intensive farming practices with respect to their impact on the environment and their contribution to world food production.
- ☐ 17. Using examples, discuss potential conflicts of interest between conservation and the need for increased production of resources such as food. Include reference to the use and effects of nitrogen containing fertilisers, and to alternatives to their use (also see #19).
- ☐ 18. Discuss the impact of **monoculture** and the removal of **hedgerows** on the environment. Describe how intensive agricultural ecosystems differ from natural ecosystems with respect to **stability** and **diversity**.
- ☐ 19. Using appropriate contexts, explain how farming practices can enhance **biodiversity** and help to ensure **sustainability**. Include reference to any of: **organic farming** and **intercropping**, **crop rotation** and the use of **organic fertilisers**, and biological pest control.

Land management issues *(pages 68-69)*

- ☐ 20. Discuss the causes and effects of **deforestation**, emphasising the impact on the **biodiversity** and stability of forest ecosystems, and on carbon and nitrogen cycling. Identify regions (globally or locally) where deforestation is a major problem.
- ☐ 21. Appreciate the conflict between the need to **conserve** forests and the rights of humans to obtain a living from the land. Describe methods by which forests may be conserved to ensure the **sustainable provision** of forestry resources. Comment on the feasibility of these.
- ☐ 22. Describe the management of woodland habitats to maintain or increase biodiversity, as illustrated by practices such as **coppicing**.

Features of populations *(pages 46-48, 53)*

- ☐ 23. Distinguish between **population** and **community**, and give an example of each. Explain the term **population density** and distinguish it from population size.
- ☐ 24. Understand that populations are dynamic and that they exhibit attributes not shown by the individuals themselves. Recognise the following attributes of populations and understand their importance to the study of populations: **population density**, **population distribution**, birth rate (**natality**), mean (average) age, death rate (**mortality**), survivorship (age specific survival), migration rate, average brood size, proportion of females breeding, age structure.

  Recognise that these attributes are population specific and represent population patterns.
- ☐ 25. Discuss the distribution patterns (uniform, clumped, random) of organisms within their range. Suggest which factors govern each type of distribution.

Population growth and size *(pages 48-52)*

- ☐ 26. Explain the role of **births**, **deaths**, and **migration** in regulating population growth. Express the relationship in an equation.
- ☐ 27. EXTENSION: Explain the role of a **life tables** and **survivorship curves** in providing information on patterns of population survival and mortality. Describe the features of Type I, II, and III **survivorship curves**.
- ☐ 28. Describe how the trends in population change can be shown in a **population growth curve** of population numbers (Y axis) against time (X axis).
- ☐ 29. Understand the factors that affect final population size, explaining clearly how they operate and providing examples where necessary. Include reference to:
  - *(a)* ***Carrying capacity*** *of the environment.*
  - *(b)* ***Environmental resistance****.*
  - *(c)* ***Density dependent factors****, e.g. intraspecific competition, interspecific competition, predation.*
  - *(d)* ***Density independent factors****, e.g. catastrophic climatic events.*
  - *(e)* ***Limiting factors****, e.g. soil nutrients.*
- ☐ 30. Distinguish between **exponential** and **sigmoidal growth curves**. Create labelled diagrams of these curves, indicating the different phases of growth and the factors regulating population growth at each stage.
- ☐ 31. Recognise patterns of population growth in colonising, stable, declining, and oscillating populations.

Species interactions *(pages 54-63)*

- ☐ 32. Explain the nature of the **interspecific interactions** occurring in communities: **competition**, **mutualism**, **commensalism**, **amensalism**, **allelopathy**, and **exploitation** (**parasitism**, **predation**, **herbivory**).
- ☐ 33. Describing at least one example, explain the possible effects of predator-prey interactions on the **population sizes** of both predator and prey.
- ☐ 34. Describe, and give examples of, **interspecific** and **intraspecific competition**. Explain the effects of inter- or intraspecific competition on the distribution, population size, or niche size of two species.

See the 'Textbook Reference Grid' on pages 8-9 for textbook page references relating to material in this topic.

Supplementary Texts

See pages 4-6 for additional details of these texts:

- Adds, J., *et al.*, 2001. **Genetics, Evolution and Biodiversity**, (NelsonThornes), pp. 40-50, 61-67.
- Allen *et al.*, 2001. **Applied Ecology** (Collins), chpt. 1, 4, and 5 (reading as required).
- Chenn, P., 1999. **Ecology** (John Murray), pp. 1-6, 72-118.
- Reiss, M. & J. Chapman, 2000. **Environmental Biology** (CUP), chpt. 2 (reading as required).

See page 6 for details of publishers of periodicals:

STUDENT'S REFERENCE

Ecosystem diversity and succession

- **Save those Woods!** Biol. Sci. Rev., 12 (2) Nov. 1999, pp. 31-35. *Conservation of woodlands involves understanding succession. This article covers primary and secondary succession and discusses the influence of deflecting factors.*
- **Plant Succession** Biol. Sci. Rev., 14 (2) November 2001, pp. 2-6. *Thorough coverage of primary and secondary succession, including the causes of different types of succession.*
- **Down on the Farm: The Decline in Farmland Birds** Biol. Sci. Rev., 16(4) April 2004, pp. 17-20. *Factors in the decline of bird populations in the UK.*
- **Biodiversity and Ecosystems** Biol. Sci. Rev., 11(4) March 1999, pp. 18-23. *Ecosystem diversity and its relationship to ecosystem stability.*
- **Insecticides and the Conservation of Hedgerows** Biol. Sci. Rev., 16(4) April 2004, pp. 28-31. *Well managed hedgerows can reduce the need for insecticides in adjacent crops by encouraging natural pest control agents.*
- **Birth of an Island** New Scientist, 25 Nov. 1995, pp. 36-39. *The development of an island provides the perfect chance to observe the colonisation of a new and barren environment.*

Species interactions

- **The Other Side of Eden** Biol. Sci. Rev., 15(3) Feb. 2003, pp. 2-7. *An account of the Eden Project; its role in modelling ecosystem dynamics, including the interactions between species, is discussed.*
- **Inside Story** New Scientist, 29 April 2000, pp. 36-39. *Ecological interactions between fungi and plants and animals: what are the benefits?*
- **Symbiosis: Mutual Benefit or Exploitation?** Biol. Sci. Rev., 7(4) March 1995, pp. 8-11. *Symbioses are poorly understood. This article explains them and provides illustrative examples.*
- **The Future of Red Squirrels in Britain** Biol. Sci. Rev., 16(2) Nov. 2003, pp. 8-11. *A further account of the impact of the grey squirrel on Britain's native red squirrel populations.*
- **Reds vs Greys: Squirrel Competition** Biol. Sci. Rev., 10(4) March 1998, pp. 30-31. *The nature of the competition between red and grey squirrels in the UK: an example of competitive exclusion?*
- **The Ecology of Newt Populations** Biol. Sci. Rev., 11(5) May 1999, pp. 5-8. *The ecology of newt populations in Britain (includes factors involved in the regulation of population size).*
- **Logarithms and Life** Biol. Sci. Rev., 13(4) March 2001, pp. 13-15. *The basics of logarithmic growth and its application to real populations.*
- **The Wolf (*Canis lupus*)** Biol. Sci. Rev., 13(5) May 2001, pp. 24-27. *The biology and population ecology of the grey wolf; a species in decline.*
- **Predator-Prey Relationships** Biol. Sci. Rev., 10(5) May 1998, pp. 31-35. *Predator-prey relationships, and the defence strategies of prey.*
- **Fish Predation** Biol. Sci. Rev., 14(1) Sept. 2001, pp. 10-14. *Some fish species in freshwater systems in the UK are important top predators and influence the dynamics of an entire ecosystem.*

TEACHER'S REFERENCE

- **Where Have All the Flowers Gone** Biologist, 48(4) August 2001. *During the war, large areas of Britain's woodlands were cut down and have since regrown. The consequence of this has been a change in the physical conditions and a different flora to that several decades ago.*
- **Life Support** New Scientist, 15 August 1998, pp. 30-34. *What is the true cost of losing species diversity in ecosystems? The importance of biodiversity in the stability of ecosystems and their ability to recover from human impacts.*

See pages 10-11 for details of how to access **Bio Links** from our web site: **www.biozone.co.uk**. From Bio Links, access sites under the topics:

GENERAL BIOLOGY ONLINE RESOURCES > **Online Textbooks and Lecture Notes** • An on-line biology book • Learn.co.uk *... and others*

ECOLOGY: • Ken's bioweb referencing *... and others* > **Ecosystems**: • Bright edges of the world • Desert biome • Freshwater ecosystems • The rocky intertidal zone *... and others* > **Populations and Communities**: • Anemone fishes and their host sea anemones • Bull Shoals Lake 1995 report • Competition • Intraspecific relations: Cooperation and competition • Population ecology • Quantitative population ecology • Species interactions • Death squared

Software and video resources are provided on the Teacher Resource Handbook on CD-ROM

Ecosystem Stability

Ecological theory suggests that all species in an ecosystem contribute in some way to ecosystem function. Therefore, species loss past a certain point is likely to have a detrimental effect on the functioning of the ecosystem and on its ability to resist change (its stability). Although many species still await discovery, we do know that the rate of species extinction is increasing. Scientists estimate that human destruction of natural habitat is driving up to 100 000 species to extinction every year. Every day on Earth 100 - 300 species disappear forever. This has serious implications for the long term stability of many ecosystems.

The Concept of Ecosystem Stability

The stability of an ecosystem refers to its apparently unchanging nature over time. Ecosystem stability has various components, including **inertia** (the ability to resist disturbance) and **resilience** (ability to recover from external disturbances). Ecosystem stability is closely linked to the biodiversity of the system, although it is difficult to predict which factors will stress an ecosystem beyond its range of tolerance. It was once thought that the most stable ecosystems were those with the greatest number of species, since these systems had the greatest number of biotic interactions operating to buffer them against change. This assumption is supported by experimental evidence but there is uncertainty over what level of biodiversity provides an insurance against catastrophe.

Single species crops (monocultures), such as the soy bean crop above left, represent low diversity systems that can be vulnerable to disease, pests, and disturbance. In contrast, natural grasslands (above, right) may appear homogeneous, but contain many species which vary in their predominance seasonally. Although they may be easily disturbed (e.g. by burning) they are very resilient and usually recover quickly.

Tropical rainforests (above, left) represent the highest diversity systems on Earth. Whilst these ecosystems are generally resistant to disturbance, once degraded, (above, right) they have little ability to recover. The biodiversity of ecosystems at low latitudes is generally higher than that at high latitudes, where climates are harsher, niches are broader, and systems may be dependent on a small number of key species.

Community Response to Environmental Change

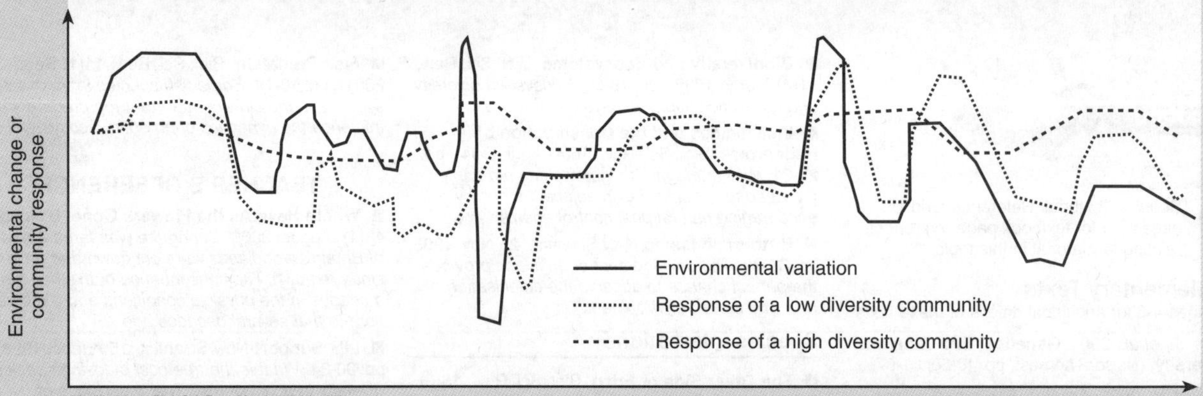

In models of ecosystem function, higher species diversity increases the stability of ecosystem functions such as productivity and nutrient cycling. In the graph above, note how the low diversity system varies more consistently with the environmental variation, whereas the high diversity system is buffered against major fluctuations. In any one ecosystem, some species may be more influential than others in the stability of the system. Such **keystone (key) species** have a disproportionate effect on ecosystem function due to their pivotal role in some ecosystem function such as nutrient recycling or production of plant biomass.

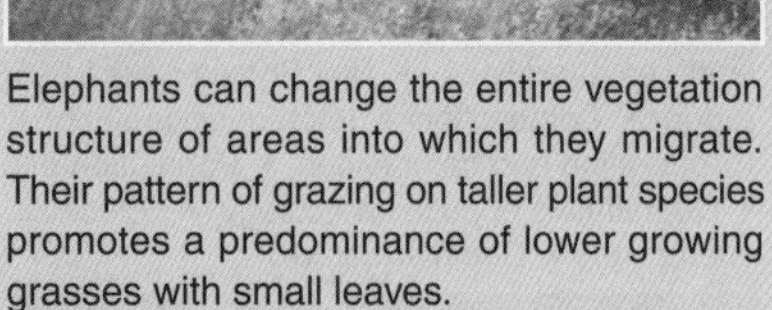

Elephants can change the entire vegetation structure of areas into which they migrate. Their pattern of grazing on taller plant species promotes a predominance of lower growing grasses with small leaves.

Termites are amongst the few larger soil organisms able to break down plant cellulose. They shift large quantities of soil and plant matter and have a profound effect on the rates of nutrient processing in tropical environments.

The starfish *Pisaster* is found along the coasts of North America where it feeds on mussels. If it is removed, the mussels dominate, crowding out most algae and leading to a decrease in the number of herbivore species.

Calculation and use of diversity indices

One of the best ways to determine the health of an ecosystem is to measure the variety of organisms living in it. **Diversity indices** attempt to quantify the degree of diversity and identify indicators for environmental stress or degradation. They are widely used in ecological work, particularly for monitoring ecosystem change or pollution. Most indices of diversity are easy to use and widely applied in ecological studies. Two examples, both of which are derivations of Simpson's index, are described below.

Simpson's Index for finite populations

This diversity index (DI) is a commonly used inversion of Simpson's index, suitable for finite populations.

$$DI = \frac{N(N-1)}{\Sigma n(n-1)}$$

After Smith and Smith as per IOB.

Where:

DI = Diversity index
N = Total number of individuals (of all species) in the sample
n = Number of individuals of each species in the sample

This index ranges between 1 (low diversity) and infinity. The higher the value, the greater the variety of living organisms. It can be difficult to evaluate objectively without reference to some standard ecosystem measure because the values calculated can, in theory, go to infinity.

Complement of Simpson's Index

This diversity index (DI) is the complement of Simpson's original index. It is widely used, although it is based on an infinite population.

$$DI = 1 - \Sigma p_i^2$$

after Krebs: Ecological Methodology 1989

Where:

p_i^2 = N_i/N (the proportion of species i in the community)
N_i = Number of individuals of each species in the sample
N = Total number of individuals (of all species) in the sample

This index ranges between 0 and almost 1. The index is independent of sample distribution and, because of the more limited range of values, is easily interpreted. No single index offers the "best" measure of diversity: they are chosen on their suitability to different situations.

Example of species diversity in a stream

The table below shows the results from a survey of invertebrates living in a stream. Although the species have been identified, this is not necessary in order to calculate diversity as long as the different species can be distinguished from each other. Calculation of the diversity index using Simpson's index for finite populations is as follows:

Species	No. of individuals
A (Common backswimmer)	12
B (Stonefly larva)	7
C (Silver water beetle)	2
D (Caddis fly larva)	6
E (Water spider)	5
Total number of individuals = 32	

$$DI = \frac{32 \times 31}{(12 \times 11) + (7 \times 6) + (2 \times 1) + (6 \times 5) + (5 \times 4)} = \frac{992}{226}$$

DI = 4.39

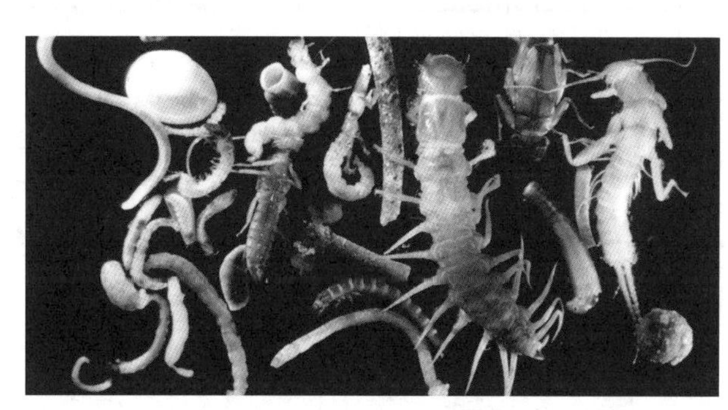

A stream community with a high diversity of macroinvertebrates (above) in contrast to a low diversity stream community (below).

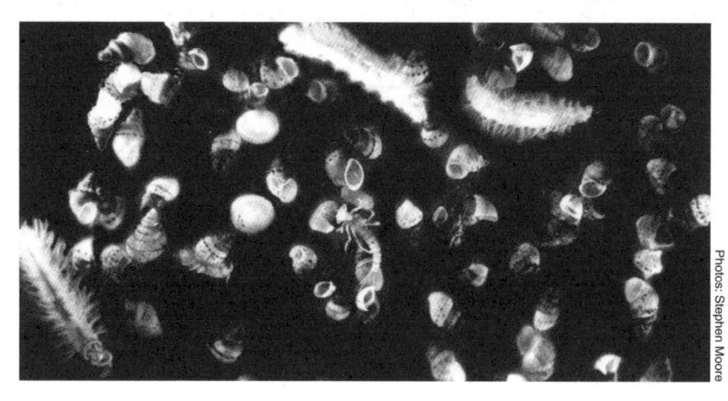

Photos: Stephen Moore

1. Explain what you understand by the term **ecosystem stability**: ______

2. Suggest one probable reason why high biodiversity provides greater ecosystem stability: ______

3. Describe a situation where a species diversity index may provide useful information: ______

4. An area of forest floor was sampled and six invertebrate species were recorded, with counts of 7, 10, 11, 2, 4, and 3 individuals. Using Simpson's index for finite populations, calculate DI for this community:

   (a) DI= ______ DI = ______

   (b) Comment on the diversity of this community: ______

5. Explain why **keystone species** are so important to ecosystem function: ______

Features of Populations

Populations have a number of attributes that may be of interest. Usually, biologists wish to determine **population size** (the total number of organisms in the population). It is also useful to know the **population density** (the number of organisms per unit area). The density of a population is often a reflection of the **carrying capacity** of the environment, i.e. how many organisms a particular environment can support. Populations also have structure; particular ratios of different ages and sexes. These data enable us to determine whether the population is declining or increasing in size. We can also look at the **distribution** of organisms within their environment and so determine what particular aspects of the habitat are favoured over others. One way to retrieve information from populations is to **sample** them. Sampling involves collecting data about features of the population from samples of that population (since populations are usually too large to examine in total). Sampling can be done directly through a number of sampling methods or indirectly (e.g. monitoring calls, looking for droppings or other signs). Some of the population attributes that we can measure or calculate are illustrated on the diagram below.

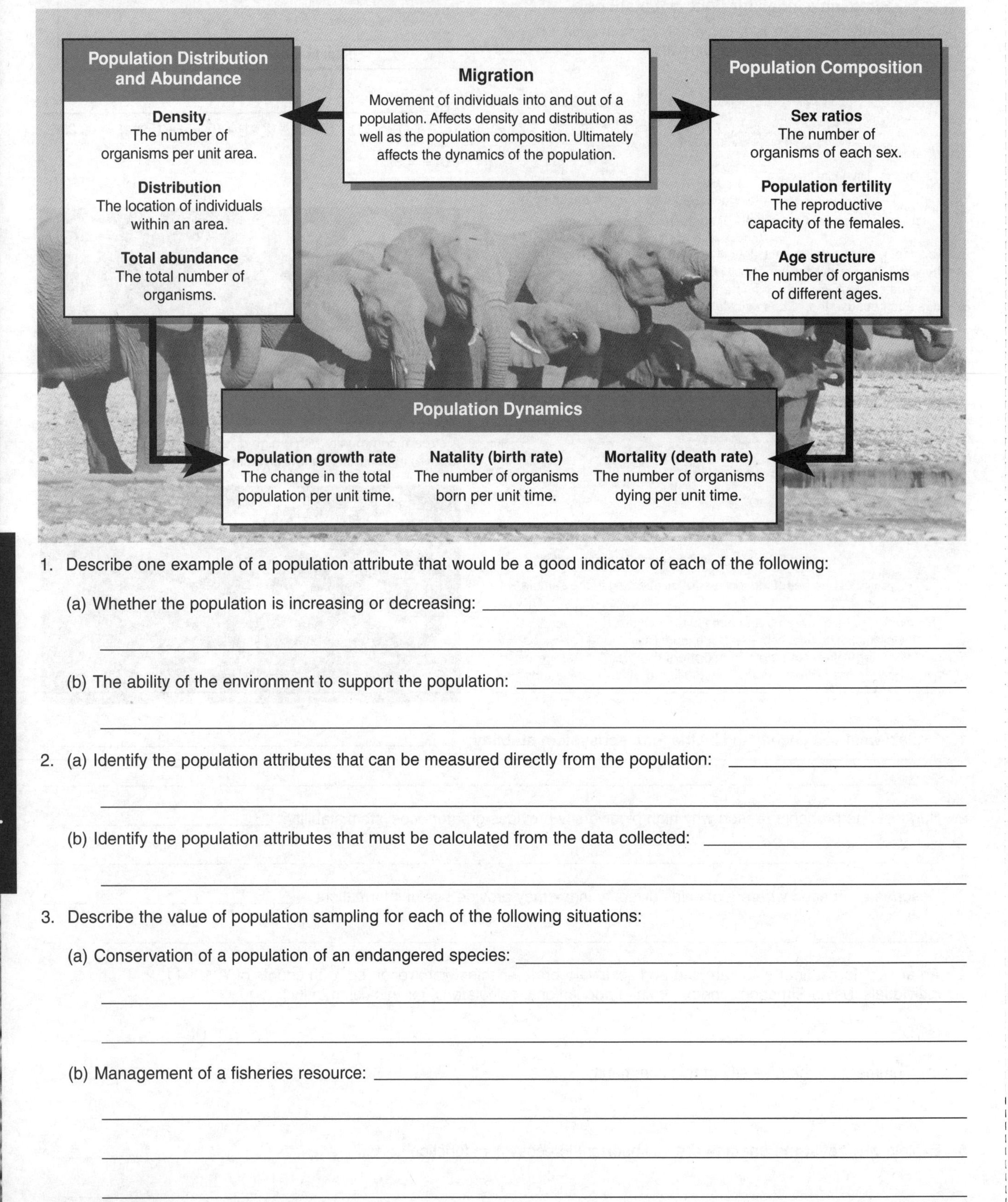

1. Describe one example of a population attribute that would be a good indicator of each of the following:

   (a) Whether the population is increasing or decreasing: ______

   (b) The ability of the environment to support the population: ______

2. (a) Identify the population attributes that can be measured directly from the population: ______

   (b) Identify the population attributes that must be calculated from the data collected: ______

3. Describe the value of population sampling for each of the following situations:

   (a) Conservation of a population of an endangered species: ______

   (b) Management of a fisheries resource: ______

Populations & Interactions

Density and Distribution

Distribution and density are two interrelated properties of populations. Population density is the number of individuals per unit area (for land organisms) or volume (for aquatic organisms). Careful observation and precise mapping can determine the distribution patterns for a species. The three basic distribution patterns are: random, clumped and uniform. In the diagram below, the circles represent individuals of the same species. It can also represent populations of different species.

Low Density

In low density populations, individuals are spaced well apart. There are only a few individuals per unit area or volume (e.g. highly territorial, solitary mammal species).

High Density

In high density populations, individuals are crowded together. There are many individuals per unit area or volume (e.g. colonial organisms, such as many corals).

Tigers are solitary animals, found at low densities.

Termites form well organised, high density colonies.

Random Distribution

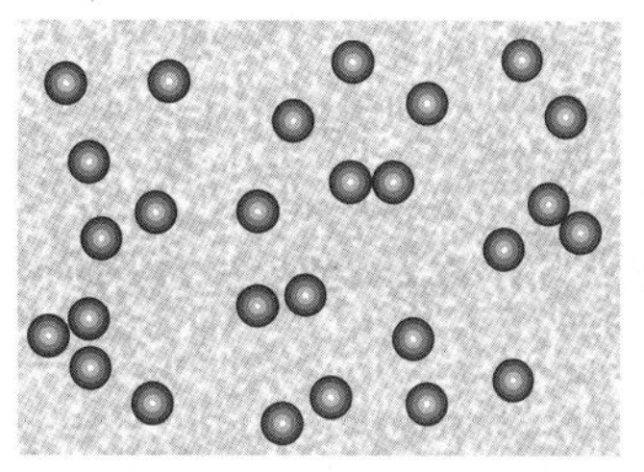

Random distributions occur when the spacing between individuals is irregular. The presence of one individual does not directly affect the location of any other individual. Random distributions are uncommon in animals but are often seen in plants.

Clumped Distribution

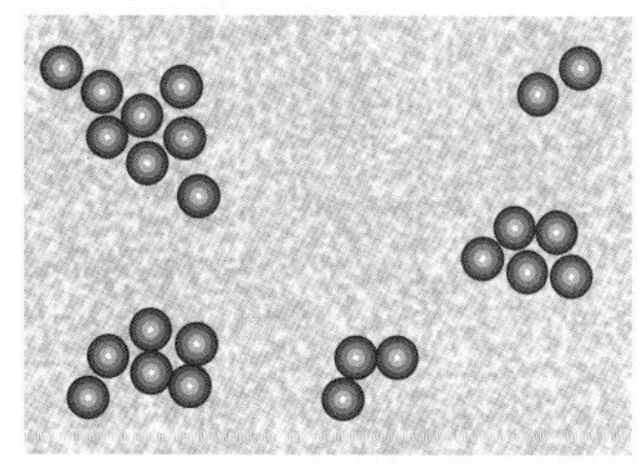

Clumped distributions occur when individuals are grouped in patches (sometimes around a resource). The presence of one individual increases the probability of finding another close by. Such distributions occur in herding and highly social species.

Uniform Distribution

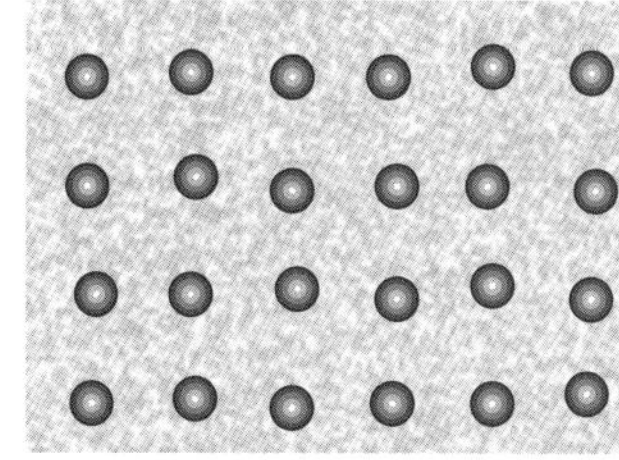

Regular distribution patterns occur when individuals are evenly spaced within the area. The presence of one individual decreases the probability of finding another individual very close by. The penguins illustrated above are also at a high density.

1. Explain why some organisms may exhibit a clumped distribution pattern because of:

   (a) Resources in the environment: ______________________________

   ____

   (b) A group social behaviour: ______________________________

   ____

2. Describe a social behaviour found in some animals that may encourage a uniform distribution: ______________________________

   ____

3. Describe the type of environment that would encourage uniform distribution: ______________________________

   ____

4. Name an organism that exhibits the following type of distribution pattern:

   (a) Clumped: ______________________________

   (b) Random (more or less): ______________________________

   (c) Uniform (more or less): ______________________________

Population Regulation

Very few species show continued exponential growth. Population size is regulated by factors that limit population growth. The diagram below illustrates how population size can be regulated by environmental factors. **Density independent factors** may affect all individuals in a population equally. Some, however, may be better able to adjust to them. **Density dependent factors** have a greater affect when the population density is higher. They become less important when the population density is low.

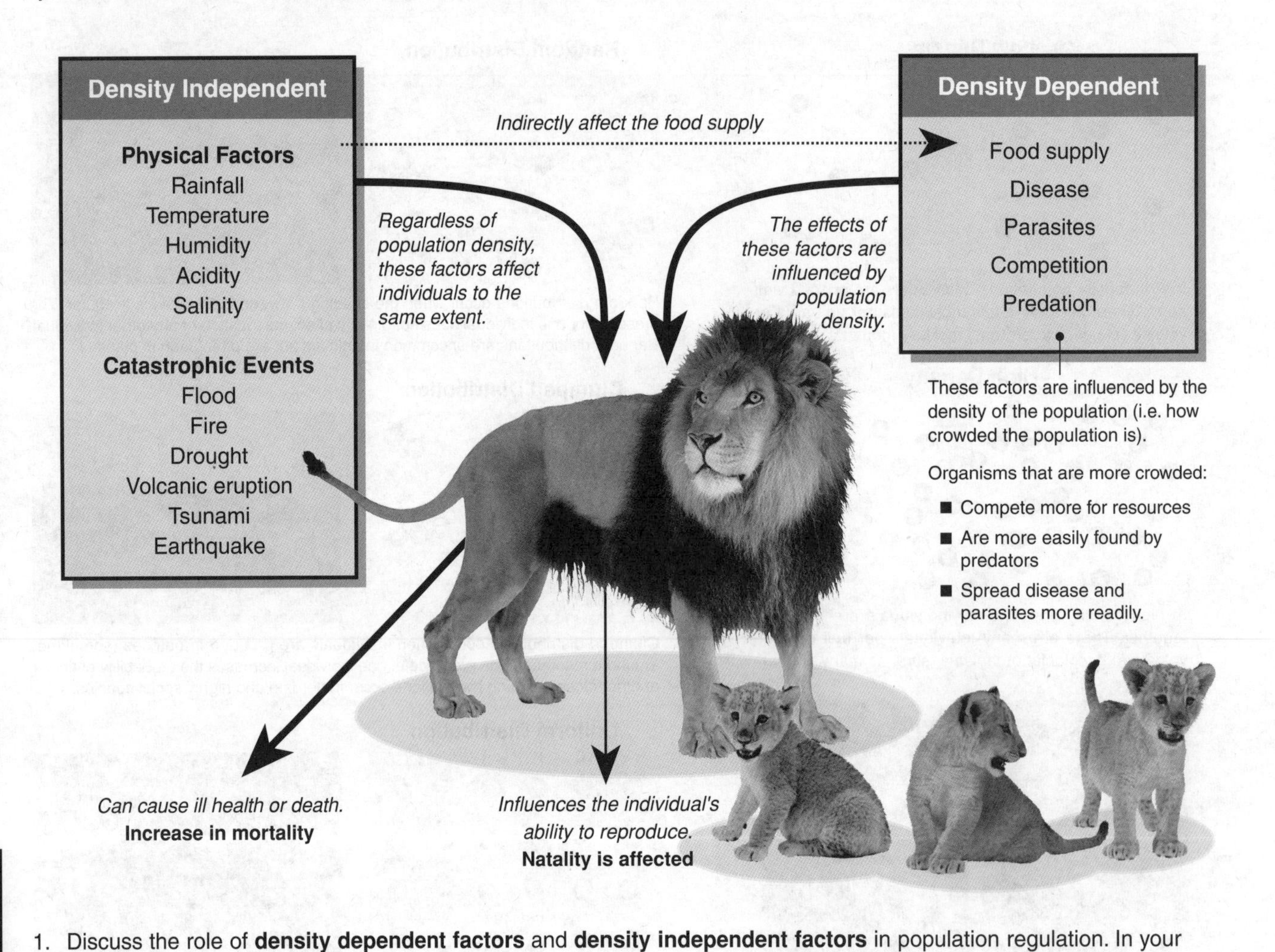

1. Discuss the role of **density dependent factors** and **density independent factors** in population regulation. In your discussion, make it clear that you understand the meaning of each of these terms:

2. Explain how an increase in population density allows disease to have a greater influence in regulating population size:

3. In cooler climates, aphids go through a huge population increase during the summer months. In autumn, population numbers decline steeply. Describe a density dependent and a density independent factor regulating the population:

   (a) Density dependent:

   (b) Density independent:

Population Growth

Organisms do not generally live alone. A **population** is a group of organisms of the same species living together in one geographical area. This area may be difficult to define as populations may comprise widely dispersed individuals that come together only infrequently (e.g. for mating). The number of individuals comprising a population may also fluctuate considerably over time. These changes make populations dynamic: populations gain individuals through births or immigration, and lose individuals through deaths and emigration. For a population in **equilibrium**, these factors balance out and there is no net change in the population abundance. When losses exceed gains, the population declines.

Births, *deaths*, *immigrations* (movements into the population) and *emigrations* (movements out of the population) are events that determine the numbers of individuals in a population. Population growth depends on the number of individuals added to the population from births and immigration, minus the number lost through deaths and emigration. This is expressed as:

Population growth =

Births – Deaths + Immigration – Emigration
(B) (D) (I) (E)

The difference between immigration and emigration gives *net migration*. Ecologists usually measure the **rate** of these events. These rates are influenced by environmental factors and by the characteristics of the organisms themselves. Rates in population studies are commonly expressed in one of two ways:

- Numbers per unit time e.g. 20 150 live births per year.
- Per capita rate (number per head of population) e.g. 122 live births per 1000 individuals per year (12.2%).

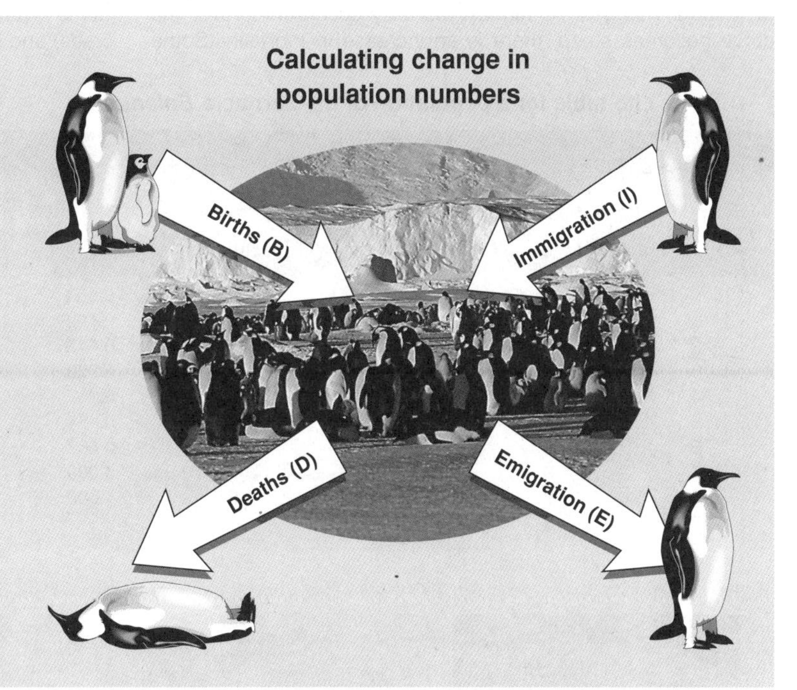

1. Define the following terms used to describe changes in population numbers:

   (a) Death rate (mortality): ________________

   (b) Birth rate (natality): ________________

   (c) Immigration: ________________

   (d) Emigration: ________________

   (e) Net migration rate: ________________

2. Using the terms, B, D, I, and E (above), construct equations to express the following (the first is completed for you):

   (a) A population in equilibrium: B + I = D + E

   (b) A declining population: ________________

   (c) An increasing population: ________________

3. The rate of population change can be expressed as the interaction of all these factors:

   Rate of population change = birth rate – death rate + net migration rate (positive or negative)

   Using the formula above, determine the annual rate of population change for Mexico and the United States in 1972:

	USA	Mexico
Birth rate	1.73%	4.3%
Death rate	0.93%	1.0%
Net migration rate	+0.20%	0.0%

   Rate of population change for USA = ________________

   Rate of population change for Mexico = ________________

4. A population started with a total number of 100 individuals. Over the following year, population data were collected. Calculate birth rates, death rates, net migration rate, and rate of population change for the data below (as percentages):

   (a) Births = 14: Birth rate = ________________

   (b) Net migration = +2: Net migration rate = ________________

   (c) Deaths = 20: Death rate = ________________

   (d) Rate of population change = ________________

   (e) State whether the population is increasing or declining: ________________

Life Tables and Survivorship

The numerical data collected during a population study can be presented as a table of figures called a **life table** or graphically as a **survivorship curve**. These alternative presentations are shown below. Survivorship curves start at 1000 and, as the population ages, the number of survivors progressively declines. The shape of a survivorship curve shows graphically at which life stages the highest mortality occurs. Wherever the curve becomes steep, there is an increase in mortality. Some organisms suffer high losses of early life stages and compensate by producing vast numbers of offspring. Populations with higher survival rates for juveniles usually produce fewer young and have some degree of parental care. Note that many species exhibit a mix of two of the three basic types. Some birds have a high chick mortality (Type III) but adult mortality is fairly constant (Type II). Some invertebrates have high mortality only when moulting (e.g. crabs) and show a stepped curve.

Life table for a population of the barnacle *Balanus*

Age / yr	No. alive at the start of the age interval	Proportion of original no. surviving at the start of the age interval	No. dying during the age interval	Mortality / d
0	142	1.000	80	0.563
1	62	0.437	28	0.452
2	34	0.239	14	0.412
3	20	0.141	5	0.250
4	15	0.106	4	0.267
5	11	0.078	5	0.454
6	6	0.042	4	0.667
7	2	0.014	0	0.000
8	2	0.014	2	1.000
9	0	0.0	_	_

Life tables, such as that shown left, provide a summary of mortality for a population (usually for a group of individuals of the same age). The basic data are just the number of individuals remaining alive at successive sampling times. Life table data can tell us the ages at which most mortality occurs in a population. They can also provide information about life span and population age structure.

Life table data can be presented graphically as a **survivorship curve** (see below). Survivorship curves use a semi-log plot of the number of individuals surviving per thousand in the population against age. They are standardised as the number of survivors per 1000 individuals so that populations of different types can be easily compared.

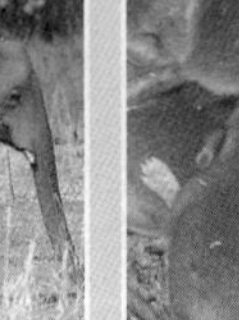

Large mammals: Type I

Rodents: Type II

Hydra: Type II

Barnacles: Type III

Type I survivorship curve
Mortality (death rate) is very low in the infant and juvenile years, and throughout most of adult life. Mortality increases rapidly in old age. **Examples**: humans (in developed countries) and many other large mammals (e.g. big cats, elephants).

Type II survivorship curve
Mortality is relatively constant through all life stages (no one age is more susceptible than another). **Examples**: some invertebrates such as *Hydra*, some birds, some annual plants, some lizards, and many rodents.

Type III survivorship curve
Mortality is very high during early life stages, followed by a very low death rate for the few individuals reaching adulthood. **Examples**: many fish (not mouth brooders) and most marine invertebrates (e.g. oysters, barnacles).

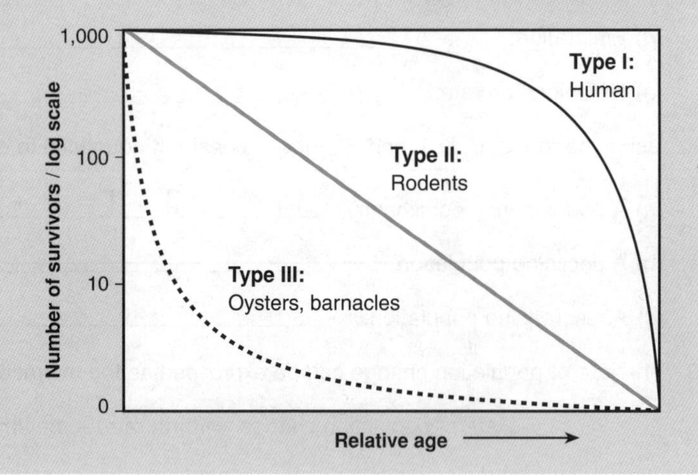

1. Suggest why human populations might not necessarily show a Type I curve: __________

2. Explain how populations with a Type III survivorship compensate for the high mortality during early life stages:

3. Describe the features of a species with a Type I survivorship that aid in high juvenile survival: __________

4. In the *Balanus* example above, state when most of the group die: __________

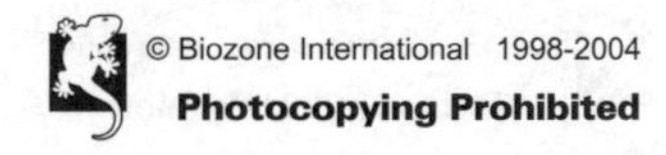

Population Growth Curves

Populations becoming established in a new area for the first time are often termed **colonising populations** (below, left). They may undergo a rapid **exponential** (logarithmic) increase in numbers as there are plenty of resources to allow a high birth rate, while the death rate is often low. Exponential growth produces a J-shaped growth curve that rises steeply as more and more individuals contribute to the population increase. If the resources of the new habitat were endless (inexhaustible) then the population would continue to increase at an **exponential** rate. However, this rarely happens in natural populations. Initially, growth may be exponential (or nearly so), but as the population grows, its increase will slow and it will stabilise at a level that can be supported by the environment (called the carrying capacity or K). This type of growth is called sigmoidal and produces the **logistic growth curve** (below, right). **Established populations** will fluctuate about K, often in a regular way (grey area on the graph below, right). Some species will have populations that vary little from this stable condition, while others may oscillate wildly.

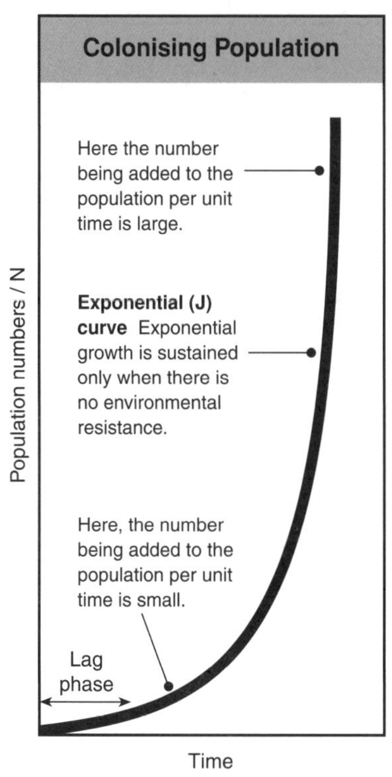

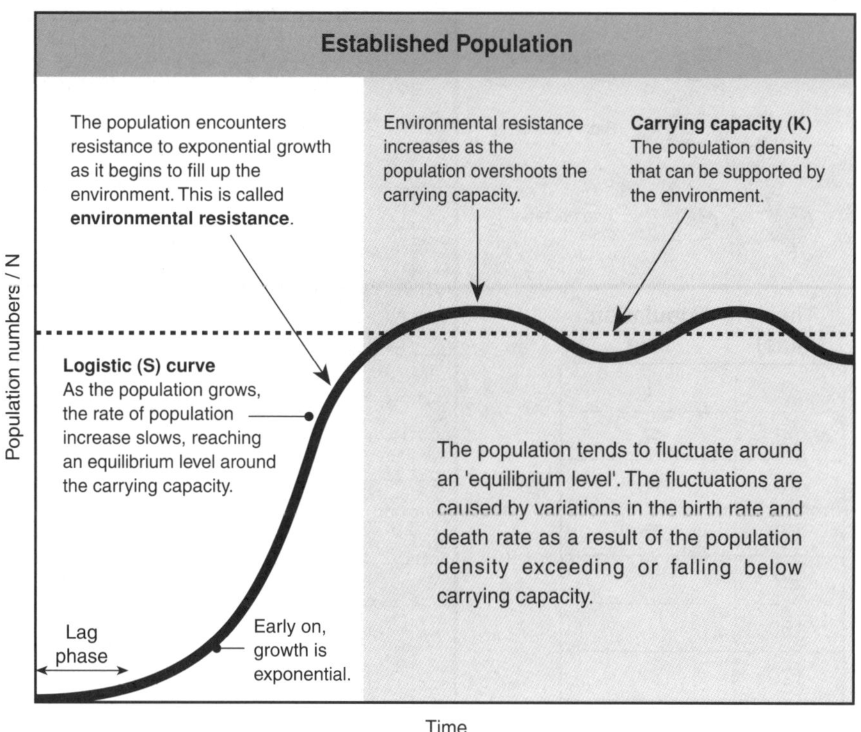

1. Explain why populations tend not to continue to increase exponentially in an environment:

2. Explain what is meant by environmental resistance:

3. (a) Explain what is meant by carrying capacity:

   (b) Explain the importance of **carrying capacity** to the growth and maintenance of population numbers:

4. Species that expand into a new area, such as rabbits when originally introduced to Britain, typically show a period of rapid population growth followed by a slowing of population growth as density dependent factors become more important and the population settles around a level that can be supported by the carrying capacity of the environment.
   (a) Explain why a newly introduced consumer might initially exhibit a period of exponential population growth:

   (b) Describe a likely outcome for a rabbit population after the initial rapid increase had slowed:

5. Describe the effect that high rabbit numbers have had on the appearance and carrying capacity of British open lands:

Growth in a Bacterial Population

Bacteria and protoctistans are able to reproduce by a process called **binary fission**: a simple mitotic cell division that involves a single cell dividing into two. In this activity, you will simulate the growth of a hypothetical bacterial population. Under suitable growing conditions, the bacteria divide every 20 minutes (**doubling time**). Starting with one cell, the population can increase very rapidly. A classic experiment along similar lines was conducted by the biologist Gause with species of *Paramecium*. Gause found that when a single species was grown in a test-tube culture and supplied regularly with fresh medium, its growth continued unchecked for the first few days, before levelling out (as the population reached the test-tube carrying capacity).

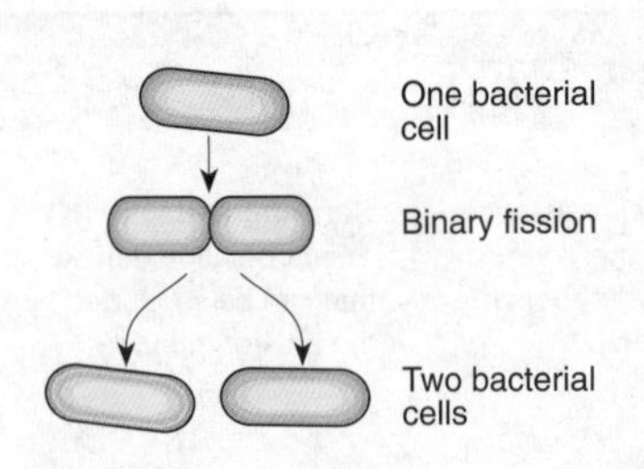

Time (mins)	Population size
0	1
20	2
40	4
60	8
80	
100	
120	
140	
160	
180	
200	
220	
240	
260	
280	
300	
320	
340	
360	

1. Complete the table (above) by doubling the number of bacteria for every 20 minute interval.
2. Graph the results on the graph grid above. Make sure that you choose suitable scales for each axis. Label the axes and mark out (number) the scale for each axis.
3. State how many bacteria were present after: 1 hour: ____________ 3 hours: ____________ 6 hours: ____________
4. Describe the shape of the curve you have plotted: ____________
5. Explain why this hypothetical bacterial population's growth could not go on for ever in the real world: ____________

Population Age Structure

Analysis of the age structure of a population can assist in its management because it can indicate where most of the mortality occurs and whether or not reproductive individuals are being replaced. The age structure of both plant and animal populations can be examined; a common method is through an analysis of size which is often related to age in a predictable way.

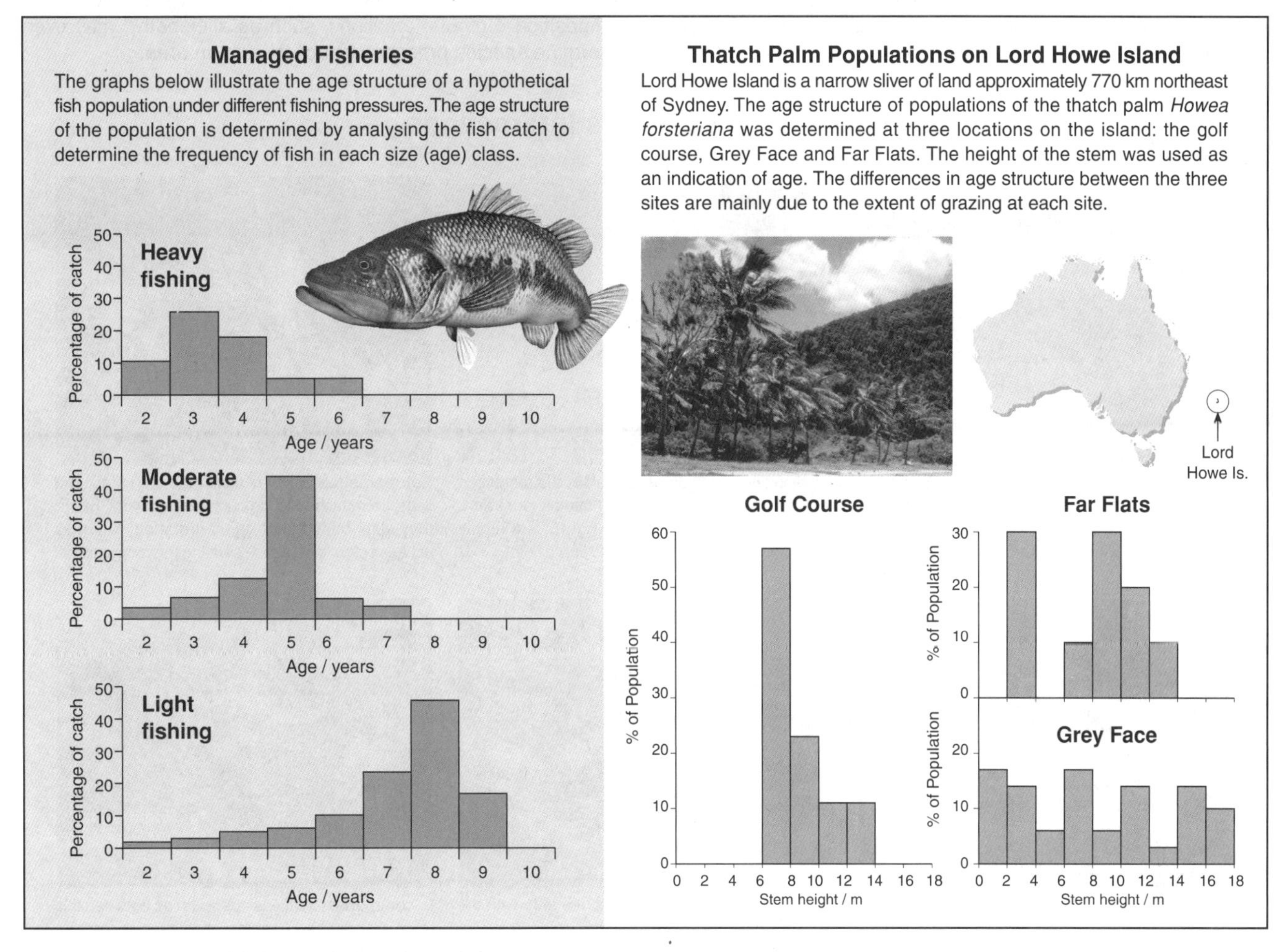

1. For the managed fish population above left:
   (a) Name the general factor that changes the age structure of this fish population: ____________

   (b) Describe how the age structure changes when the fishing pressure increases from light to heavy levels:

   ____

2. State the most common age class for each of the above fish populations with different fishing pressures:

   (a) Heavy: ____________ (b) Moderate: ____________ (c) Light: ____________

3. Determine which of the three sites sampled on Lord Howe Island (above, right), best reflects the age structure of:

   (a) An ungrazed population: ____________

   Reason for your answer: ____________

   (b) A heavily grazed and mown population: ____________

   Reason for your answer: ____________

4. Describe the likely long term prospects for the population at the golf course: ____________

   ____

5. Suggest a potential problem with using size to estimate age: ____________

   ____

6. Give one reason why a knowledge of age structure could be important in managing a resource: ____________

   ____

Species Interactions

No organism exists in isolation. Each takes part in many interactions, both with other organisms and with the non-living components of the environment. Species interactions may involve only occasional or indirect contact (predation or competition) or they may involve close association or **symbiosis**. Symbiosis is a term that encompasses a variety of interactions involving close species contact. There are three types of symbiosis: **parasitism** (a form of exploitation), **mutualism**, and **commensalism**. Species interactions affect population densities and are important in determining community structure and composition. Some interactions, such as allelopathy, may even determine species presence or absence in an area.

Examples of Species Interactions

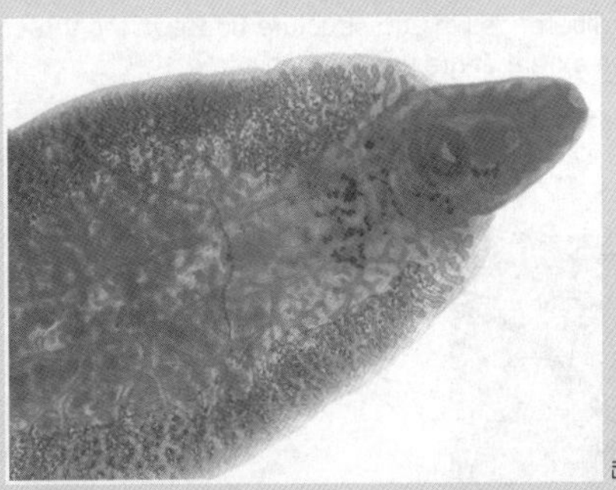

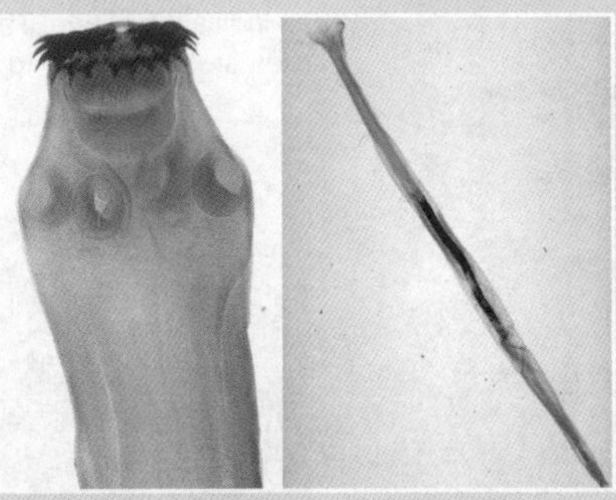

Parasitism is a common exploitative relationship in plants and animals. A parasite exploits the resources of its host (e.g. for food, shelter, warmth) to its own benefit. The host is harmed, but usually not killed. **Endoparasites**, such as liver flukes (left), tapeworms (centre) and nematodes (right)), are highly specialised to live inside their hosts, attached by hooks or suckers to the host's tissues.

Ectoparasites, such as ticks (above), mites, and fleas, live attached to the outside of the host, where they suck body fluids, cause irritation, and may act as vectors for disease causing microorganisms.

Mutualism involves an intimate association between two species that offers advantage to both. **Lichens** (above) are the result of a mutualism between a fungus and an alga (or cyanobacterium).

Termites have a mutualistic relationship with the cellulose digesting bacteria in their guts. A similar mutualistic relationship exists between ruminants and their gut microflora of bacteria and ciliates.

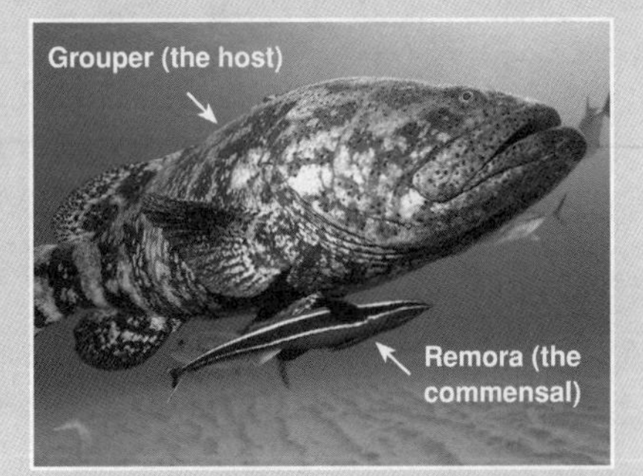

In **commensal** relationships, such as between this large grouper and a remora, two species form an association where one organism, the commensal, benefits and the other is neither harmed or helped.

Many species of decapod crustaceans, such as this anemone shrimp, are commensal with sea anemones. The shrimp gains by being protected from predators by the anemone's tentacles.

Interactions involving **competition** for the same food resources are dominated by the largest, most aggressive species. Here, hyaenas compete for a carcass with vultures and maribou storks.

Predation is an easily identified relationship, as one species kills and eats another (above). Herbivory is similar type of exploitation, except that the plant is usually not killed by the herbivore.

1. Discuss each of the following interspecific relationships, including reference to the species involved, their role in the interaction, and the specific characteristics of the relationship:

   (a) **Mutualism** between ruminant herbivores and their gut microflora: ______________________

(b) **Commensalism** between a shark and a remora: ______________________________

(c) **Parasitism** between a tapeworm and its human host: ______________________________

(d) **Parasitism** between a cat flea and its host: ______________________________

2. Summarise your knowledge of species interactions by completing the following, entering a (+), (–), or (0) for species B, and writing a brief description of each term. Codes: (+): species benefits, (–): species is harmed, (0): species is unaffected.

Interaction	Species A	Species B	Description of relationship
(a) Mutualism	+		
(b) Commensalism	+		
(c) Parasitism	–		
(d) Amensalism	0		
(e) Predation	–		
(f) Competition	–		
(g) Herbivory	+		
(h) Antibiosis	–		

3. For each of the interactions between two species described below, choose the correct term to describe the interaction and assign a +, – or 0 for each species involved in the space supplied. Use the completed table above to help you:

Description	Term	Species A	Species B
(a) A tiny cleaner fish picking decaying food from the teeth of a much larger fish (e.g. grouper).	Mutualism	Cleaner fish +	Grouper +
(b) Ringworm fungus growing on the skin of a young child.		Ringworm	Child
(c) Human effluent containing poisonous substances killing fish in a river downstream of discharge.		Humans	Fish
(d) Humans planting cabbages to eat only to find that the cabbages are being eaten by slugs.		Humans	Slugs
(e) A shrimp that gets food scraps and protection from sea anemones, which seem to gain nothing.		Shrimp	Anemone
(f) Birds follow a herd of antelopes to feed off disturbed insects, antelopes alerted to danger by the birds.		Birds	Antelope

Predator-Prey Strategies

A predator eating its prey is one of the most conspicuous species interactions. In most cases, the predator and prey are different species, though cannibalism is quite common in some species. Predators have acute senses with which to locate and identify prey. Many also have structures such as teeth, claws, and poison to catch and subdue their prey. Animals can avoid being eaten by using passive defences, such as hiding, or active ones, such as rapid escape or aggressive defence.

Predator Avoidance Strategies Among Animals

Batesian mimicry

Harmless prey gain immunity from attack by mimicking harmful animals (called **Batesian mimicry**). The beetle on the right has the same black and yellow colour scheme as the common wasps, which can give a painful sting.

Müllerian mimicry

Many unpalatable species tend to resemble each other; a phenomenon known as **Müllerian mimicry**. The monarch and queen butterflies are both toxic. By looking alike, these mimics present a common image for predators to avoid.

Poisonous

Highly toxic animals, such as arrow poison frogs and lionfish, may advertise this fact with bright colours. The evolution of toxicity may involve kin selection, as the parent may have to die to educate a future predator of its young.

Chemical defence

Some animals can produce offensive smelling chemicals. American skunks squirt a nauseous fluid at attackers.

Hiding

Clown fish

Sea anemone

Animals, such as this clown fish, take refuge in the safety provided by animals with more effective defence.

Detection and escape

Some animals, like this grasshopper, are masters of detecting approaching danger and making a rapid escape.

Camouflage

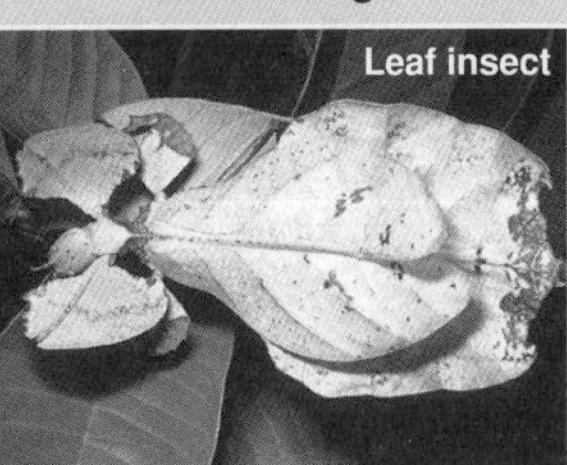

Cryptic shape and coloration allows some animals, such as this leaf insect, to blend into their background.

Startle display

Effective startle displays must come as a surprise. This stick insect will rear up on its hind legs and fan its wings. The frill necked lizard, native to Australia, will rear up suddenly and erect its frill before retreating off at high speed.

Visual deception

Deceptive markings such as large, fake eyes can apparently deceive predators momentarily, allowing the prey to escape. They may also induce the predator to strike at a nonvital end of the prey. Note the fish's real eye is disguised.

Group defence

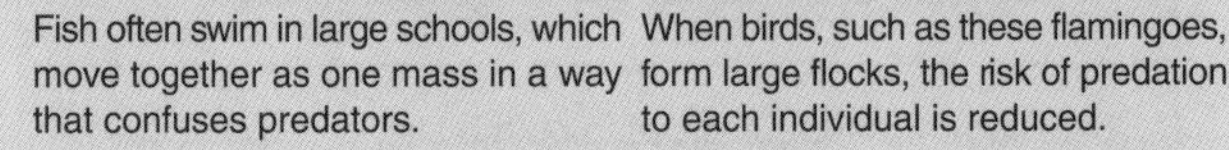

Fish often swim in large schools, which move together as one mass in a way that confuses predators.

When birds, such as these flamingoes, form large flocks, the risk of predation to each individual is reduced.

Armoured defence

Tough outer cases that hamper predators are common in both vertebrate and invertebrate prey. Almost all molluscs have protective shells built around their sensitive parts. Some animals can coil up to protect vulnerable parts.

Offensive weapons

Offensive weapons are essential if prey are to actively fend off an attack by a predator. Many grazing mammals have sharp horns and can repel attacks by predators. Some animal bodies have spikes that offer passive defence.

Prey Capturing Strategies

Concealment

Praying mantis

Some animals camouflage themselves in their surroundings, striking when the prey comes within reach.

Filter feeding

Many marine animals (e.g. barnacles, baleen whales, sponges, manta rays) filter the water to extract tiny plankton.

Lures

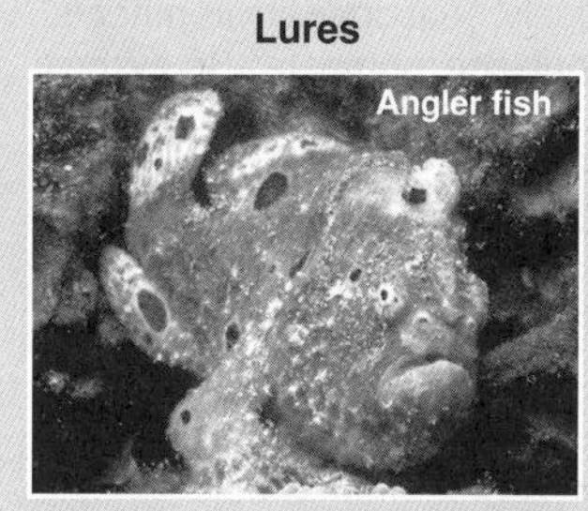

This angler fish, glow worms and a type of spider all use lures to attract prey within striking range.

Traps

Spiders have developed a unique method of trapping their prey. Strong, sticky silk threads trap flying insects.

Tool use

Chimpanzee

Some animals are gifted tool users. Chimpanzees use carefully prepared twigs to extract termites from mounds.

Stealth

The night hunting ability of some poisonous snakes is greatly helped by the presence of infrared senses.

Speed

Some animals, such as cheetahs and some predatory birds, can simply outrun or outfly their prey.

Group attack

Cooperative group behaviour may make prey capture much easier. Pelicans herd fish into 'killing zones'.

1. Explain why poisonous (unpalatable) animals are often brightly coloured so that they are easily seen:

2. Describe the purpose of large, fake eyes on some butterflies and fish:

3. Describe a behaviour typical of a (named) prey species that makes them difficult to detect by a predator:

4. Describe a behaviour of prey that is actively defensive:

5. Describe two behaviours of (named) predators that facilitate prey capture:

   (a)

   (b)

6. Discuss **Batesian and Müllerian mimicry**, explaining the differences between them and any benefits to the species involved directly or indirectly in the relationship:

7. Describe two possible ways in which toxicity could evolve in prey species, given the prey has to be eaten in order for the predator to learn from the experience:

   (a)

   (b)

Predator-Prey Interactions

Some mammals, particularly in highly seasonal environments, exhibit regular cycles in their population numbers. Snowshoe hares in Canada exhibit such a cycle of population fluctuation that has a periodicity of 9–11 years. Populations of lynx in the area show a similar periodicity. Contrary to early suggestions that the lynx controlled the size of the hare population, it is now known that the fluctuations in the hare population are governed by other factors, most probably the availability of palatable grasses. The fluctuations in the lynx numbers however, do appear to be the result of fluctuations in the numbers of hares (their principal food item). This is true of most **vertebrate** predator-prey systems: predators do not usually control prey populations, which tend to be regulated by other factors such as food availability and climatic factors. Most predators have more than one prey species, although one species may be preferred. Characteristically, when one prey species becomes scarce, a predator will "switch" to another available prey item. Where one prey species is the principal food item and there is limited opportunity for prey switching, fluctuations in the prey population may closely govern predator cycles.

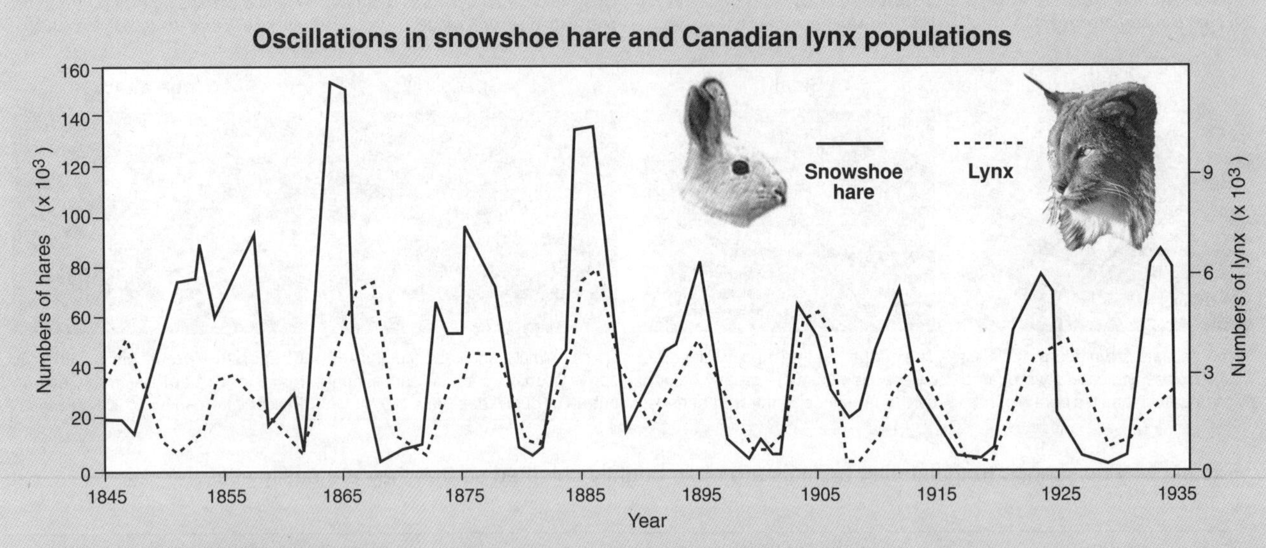

Canadian lynx and snowshoe hare

Regular trapping records of Canadian lynx (left) over a 90 year period revealed a cycle of population increase and decrease that was repeated every 10 years or so. The oscillations in lynx numbers closely matched those of the snowshoe hare (right), their principal prey item. There is little opportunity for prey switching in this system and the lynx are very dependent on the hares for food. Consequently, the oscillations in the two populations have a similar periodicity, with the lynx numbers lagging slightly behind those of the hare.

1. (a) From the graph above, determine the lag time between the population peaks of the hares and the lynx:

   (b) Explain why there is this time lag between the increase in the hare population and the response of the lynx:

2. Suggest why the lynx populations appear to be so dependent on the fluctuations on the hare:

3. (a) In terms of birth and death rates, explain how the availability of palatable food might regulate the numbers of hares:

   (b) Explain how a decline in available palatable food might affect their ability to withstand predation pressure:

Competition and Niche Size

Niche size is affected by competition. The effect on niche size will vary depending on whether the competition is weak, moderate or intense, and whether it is between members of the same species (**intraspecific**) or between different species (**interspecific**). The effects of competition on niche size are outlined in the diagram below.

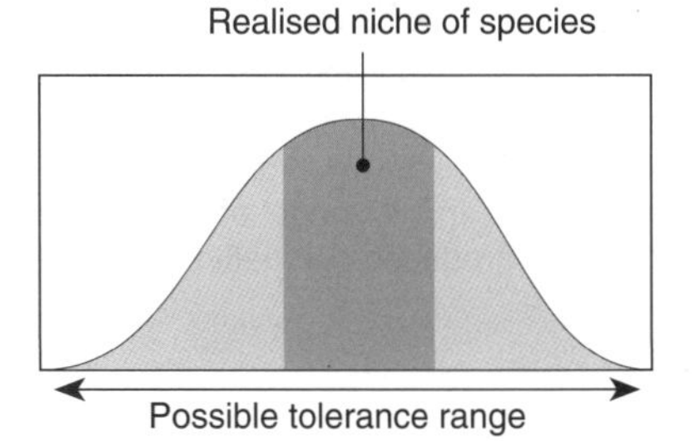

Moderate interspecific competition

The tolerance range represents the potential (**fundamental**) niche a species could exploit. The actual or **realised** niche of a species is narrower than this because of competition. Niches of closely related species may overlap at the extremes, resulting in competition for resources in the zones of overlap.

Narrow niche

Possible tolerance range

Intense interspecific competition

When the competition from one or more closely related species becomes intense, there is selection for a more limited niche. This severe competition prevents a species from exploiting potential resources in the more extreme parts of its tolerance range. As a result, niche breadth decreases (the niche becomes narrower).

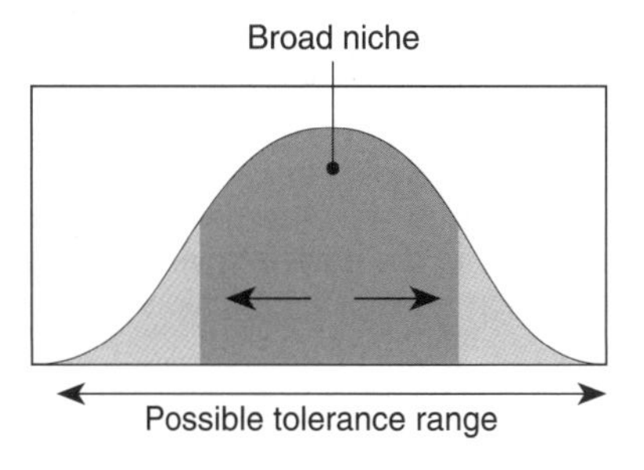

Intense intraspecific competition

Competition is most severe between individuals of the same species, because their resource requirements are usually identical. When intraspecific competition is intense, individuals are forced to exploit resources in the extremes of their tolerance range. This leads to expansion of the realised niche to less preferred areas.

Overlap in resource use between competing species

From the concept of the niche arose the idea that two species with the same niche requirements could not coexist, because they would compete for the same resources, and one would exclude the other. This is known as Gause's "***competitive exclusion principle***". If two species compete for some of the same resources (e.g. food items of a particular size), their resource use curves will overlap. Within the zone of overlap competition between the two species will be intense.

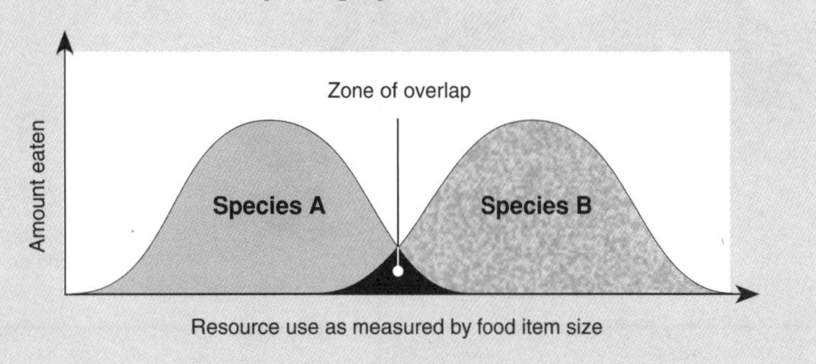

1. Distinguish between interspecific competition and intraspecific competition, and contrast the effect of these two types of competition on niche breadth:

2. Study the diagram above illustrating niche overlap between competing species and then answer the following questions:
   (a) Explain what you would expect to happen if the overlap in resource use of species A and B was to become very large (i.e. they utilised almost the same sized food item):

   (b) Describe how the degree of resource overlap might change seasonally:

   (c) If the zone of overlap between the resource use of the two species increased a little more from what is shown on the diagram, explain what is likely to happen to the breadth of the **realised** niche of each species:

3. Niche breadth can become broader in the presence of intense competition between members of the same species (top diagram). Describe one other reason why niche breadth could be very wide:

Interspecific Competition

In naturally occurring populations, direct competition between different species (**interspecific competition**) is usually less intense than intraspecific competition because coexisting species have evolved slight differences in their realised niches, even though their fundamental niches may overlap (a phenomenon termed **niche differentiation**). However, when two species with very similar niche requirements are brought into direct competition through the introduction of a foreign species, one usually benefits at the expense of the other. The inability of two species with the same described niche to coexist is referred to as the **competitive exclusion principle**. In Britain, introduction of the larger, more aggressive, grey squirrel in 1876 has contributed to a contraction in range of the native red squirrel (below), and on the Scottish coast, this phenomenon has been well documented in barnacle species (opposite). The introduction of ecologically aggressive species is often implicated in the displacement or decline of native species, although there may be more than one contributing factor. Displacement of native species by introduced ones is more likely if the introduced competitor is also adaptable and hardy. It can be difficult to provide evidence of decline in a species as a direct result of competition, but it is often inferred if the range of the native species contracts and that of the introduced competitor shows a corresponding increase.

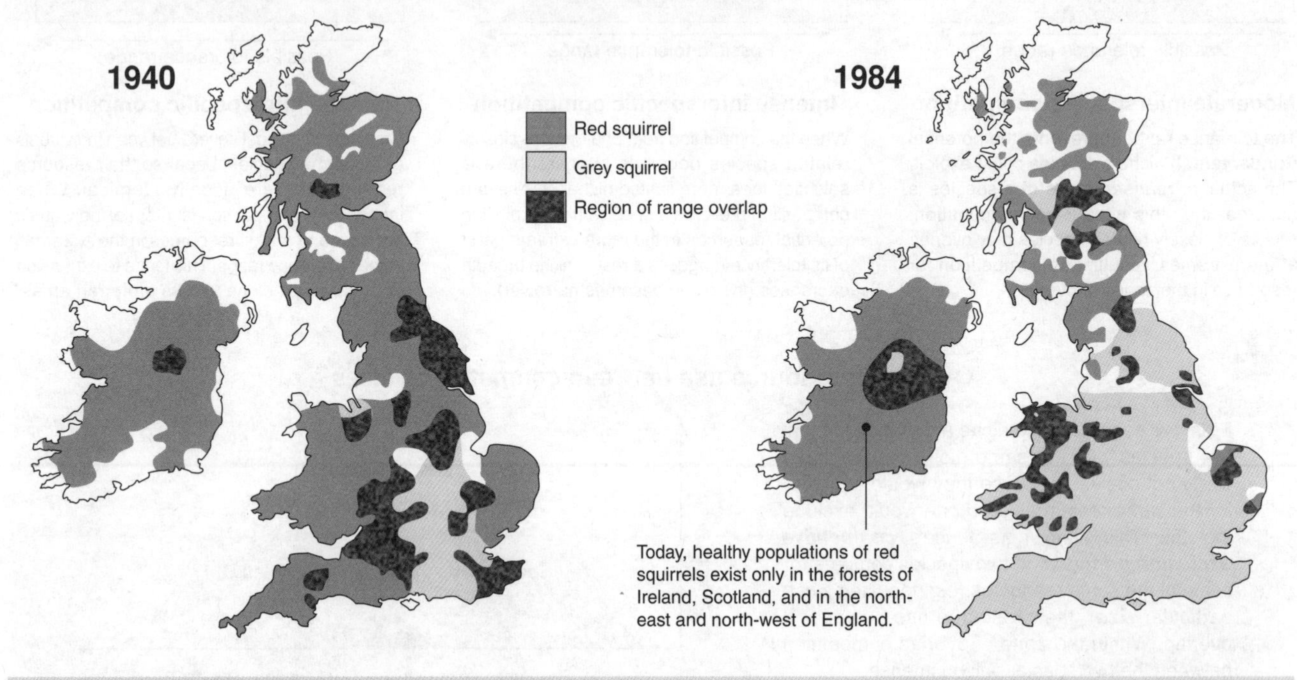

The **European red squirrel**, *Sciurus vulgaris*, was the only squirrel species in Britain until the introduction of the **American grey squirrel**, *Sciurus carolinesis*, in 1876. In 44 years since the 1940 distribution survey (above left), the more adaptable grey squirrel has displaced populations of the native red squirrels over much of the British Isles, particularly in the south (above right). Whereas the red squirrels once occupied both coniferous and broad leafed woodland, they are now almost solely restricted to coniferous forest and are completely absent from much of their former range.

1. Outline the evidence to support the view that the red-grey squirrel distributions in Britain are an example of the competitive exclusion principle:

2. Some biologists believe that competition with grey squirrels is only one of the factors contributing to the decline in the red squirrels in Britain. Explain the evidence from the 1984 distribution map that might support this view:

Competitive Exclusion in Barnacles

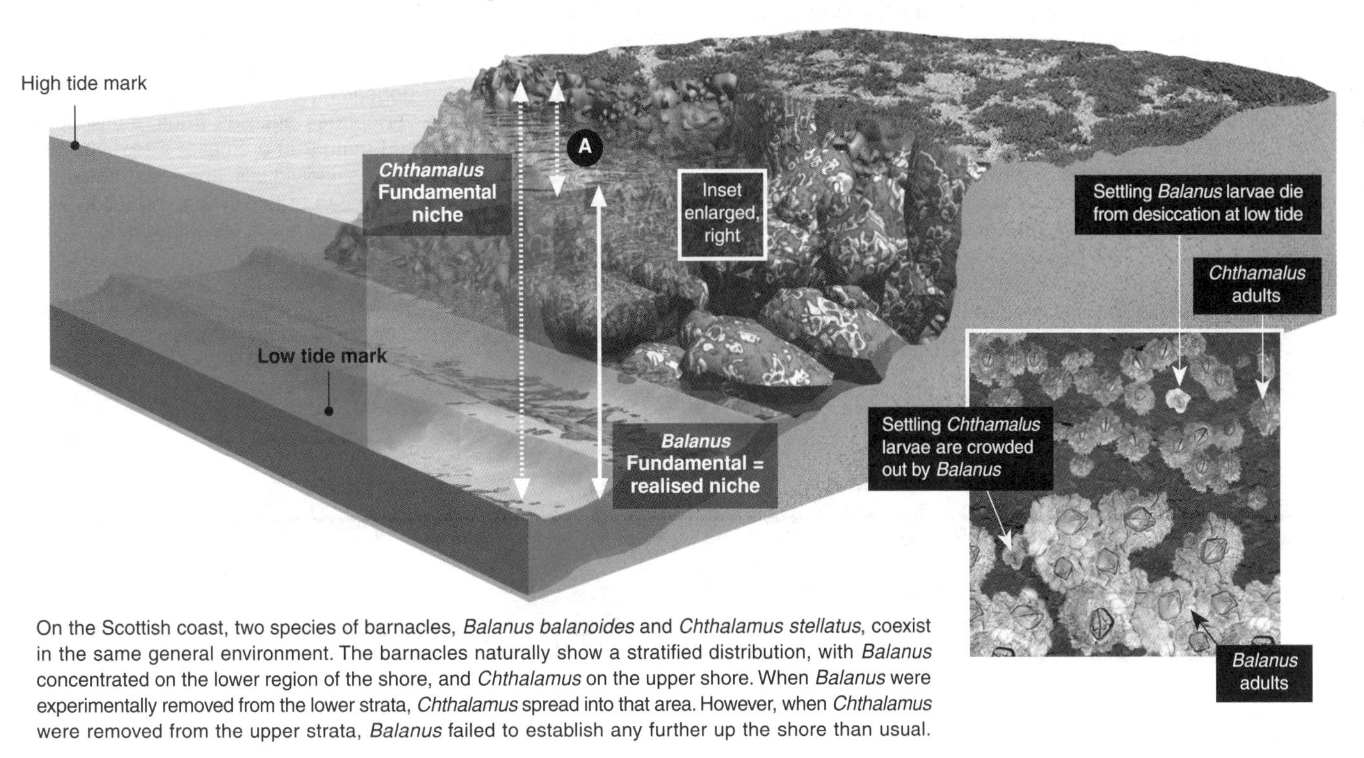

On the Scottish coast, two species of barnacles, *Balanus balanoides* and *Chthalamus stellatus*, coexist in the same general environment. The barnacles naturally show a stratified distribution, with *Balanus* concentrated on the lower region of the shore, and *Chthalamus* on the upper shore. When *Balanus* were experimentally removed from the lower strata, *Chthalamus* spread into that area. However, when *Chthalamus* were removed from the upper strata, *Balanus* failed to establish any further up the shore than usual.

3. The ability of red and grey squirrels to coexist appears to depend on the diversity of habitat type and availability of food sources (reds appear to be more successful in regions of coniferous forest). Suggest why careful habitat management is thought to offer the best hope for the long term survival of red squirrel populations in Britain:

4. Suggest other conservation methods that might aid the survival of viable red squirrel populations:

5. (a) In the example of the barnacles (above), describe what is represented by the zone labelled with the arrow A:

   (b) Outline the evidence for the barnacle distribution being the result of competitive exclusion:

6. Describe two aspects of the biology of a named introduced species that have helped its success as an invading competitor:

   Species:

   (a)

   (b)

Intraspecific Competition

Some of the most intense competition occurs between individuals of the same species (**intraspecific competition**). Most populations have the capacity to grow rapidly, but their numbers cannot increase indefinitely because environmental resources are finite. Every ecosystem has a **carrying capacity** (K), defined as the number of individuals in a population that the environment can support. Intraspecific competition for resources increases with increasing population size and, at carrying capacity, it reduces the per capita growth rate to zero. When the demand for a particular resource (e.g. food, water, nesting sites, nutrients, or light) exceeds supply, that resource becomes a **limiting factor**. Populations respond to resource limitation by reducing their population growth rate (e.g. through lower birth rates or higher mortality). The response of individuals to limited resources varies depending on the organism. In many invertebrates and some vertebrates such as frogs, individuals reduce their growth rate and mature at a smaller size. In some vertebrates, territoriality spaces individuals apart so that only those with adequate resources can breed. When resources are very limited, the number of available territories will decline.

Intraspecific Competition

Scramble competition in caterpillars

Direct competition for available food between members of the same species is called **scramble competition**. In some situations where scramble competition is intense, none of the competitors gets enough food to survive.

Contest competition in wolves

In some cases, competition is limited by hierarchies existing within a social group. Dominant individuals receive adequate food, but individuals low in the hierarchy must **contest** the remaining resources and may miss out.

Display of a male anole

Intraspecific competition may be for mates or breeding sites, as well as for food. In anole lizards (above), males have a bright red throat pouch and use much of their energy displaying to compete with other males for available mates.

Competition Between Tadpoles of *Rana tigrina*

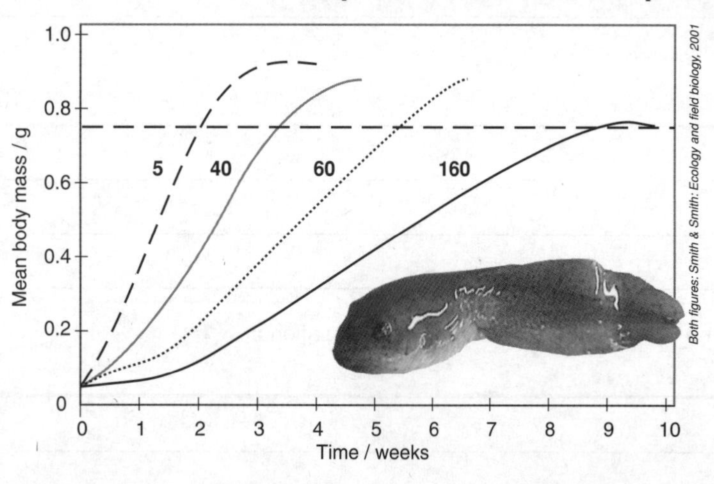

Food shortage reduces both individual growth rate and survival, and population growth. In some organisms, where there is a metamorphosis or a series of moults before adulthood (e.g. frogs, crustacean zooplankton, and butterflies), individuals may die before they mature.

The graph (left) shows how the growth rate of tadpoles (*Rana tigrina*) declines as the density increases from 5 to 160 individuals (in the same sized space).

- At high densities, tadpoles grow more slowly, taking longer to reach the minimum size for metamorphosis (0.75 g), and decreasing their chances of successfully metamorphosing from tadpoles into frogs.
- Tadpoles held at lower densities grow faster, to a larger size, metamorphosing at an average size of 0.889 g.
- In some species, such as frogs and butterflies, the adults and juveniles reduce the intensity of intraspecific competition by exploiting different food resources.

1. Using an example, predict the likely effects of **intraspecific competition** on each of the following:

   (a) Individual growth rate: ______

   (b) Population growth rate: ______

   (c) Final population size: ______

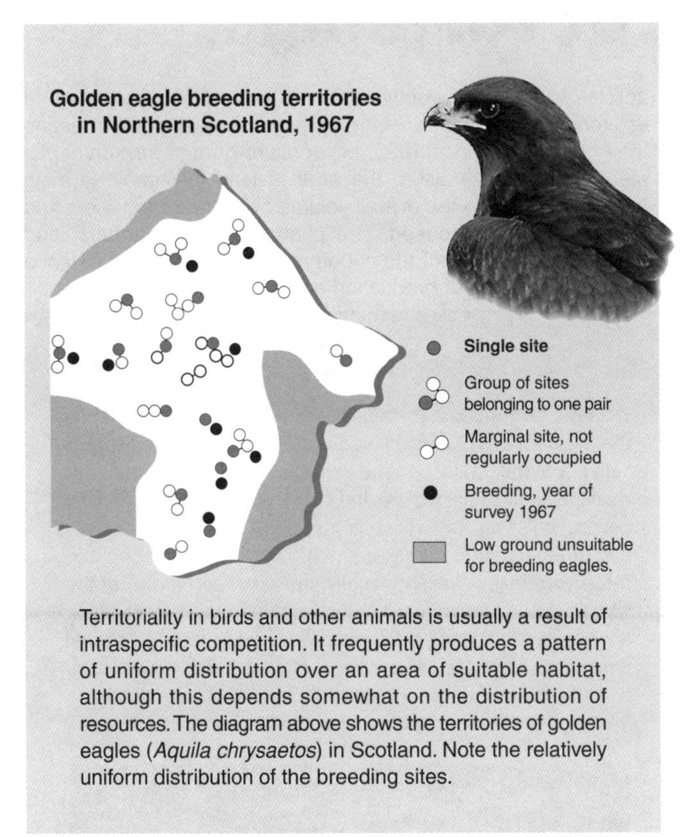

Territoriality in birds and other animals is usually a result of intraspecific competition. It frequently produces a pattern of uniform distribution over an area of suitable habitat, although this depends somewhat on the distribution of resources. The diagram above shows the territories of golden eagles (*Aquila chrysaetos*) in Scotland. Note the relatively uniform distribution of the breeding sites.

Territoriality in Great Tits (*Parus major*)

Six breeding pairs of great tits were removed from an oak woodland (below, left). Within three days, four new pairs had moved into the unoccupied areas (*below, right*) and some residents had expanded their territories. The new birds moved in from territories in hedgerows, considered to be suboptimal habitat. This type of territorial behaviour limits the density of breeding animals in areas of optimal habitat.

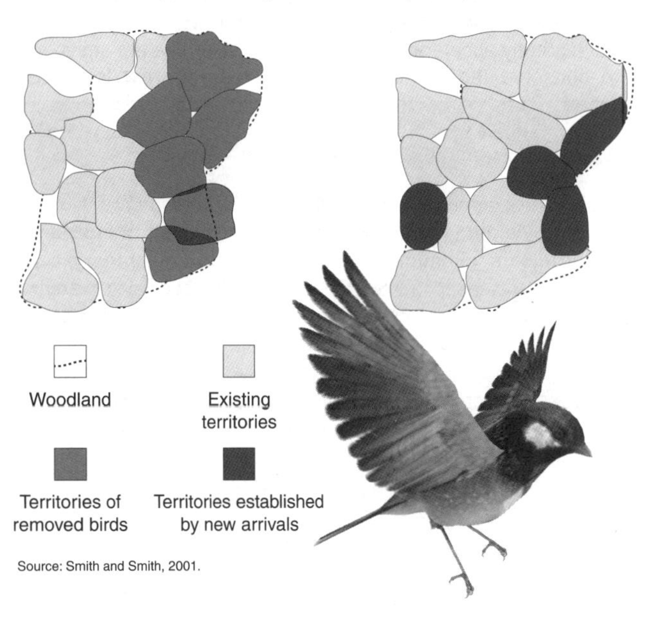

Source: Smith and Smith, 2001.

2. In the tank experiment with *Rana* (opposite), the tadpoles were contained in a fixed volume with a set amount of food:

   (a) Describe how *Rana* tadpoles respond to resource limitation: ______

   (b) Categorise the effect on the tadpoles as density-dependent / density-independent (delete one).

   (c) Comment on how much the results of this experiment are likely to represent what happens in a natural population:

3. Identify two ways in which animals can reduce the intensity of intraspecific competition:

   (a) ______

   (b) ______

4. (a) Suggest why carrying capacity of an ecosystem might decline: ______

   (b) Predict how a decline in carrying capacity might affect final population size: ______

5. Using appropriate examples, discuss the role of territoriality in reducing intraspecific competition:

The Impact of Farming

The English countryside has been shaped by many hundreds of years of agriculture. The landscape has changed as farming practices evolved through critical stages. Farming has always had an impact on Britain's rich biodiversity (generally in a negative manner). Modern farming practices, such as increasing mechanisation and the move away from mixed farming operations, have greatly accelerated this decline. In recent years active steps to conserve the countryside, such as **hedgerow legislation**, policies to increase woodland cover, and schemes to promote environmentally sensitive farming practices are slowly meeting their objectives. Since 1990, expenditure on agri-environmental measures has increased, the area of land in organic farming has increased, and the overall volume of inorganic fertilisers and pesticides has decreased. The challenge facing farmers, and those concerned about the countryside, is to achieve a balance between the goals of production and conservation.

Intensive Farming

Intensive farming techniques flourished after World War II. Using **high-yielding hybrid cultivars** and large inputs of **inorganic fertilisers**, **chemical pesticides**, and **farm machinery**, crop yields increased to 3 or 4 times those produced using the more extensive (low-input) methods of 5 decades ago. Large areas planted in monocultures (single crops) are typical. Irrigation and fertiliser programmes are often extensive to allow for the planting of several crops per season. Given adequate irrigation and continued fertiliser inputs, yields from intensive farming are high. Over time, these yields decline as soils are eroded or cannot recover from repeated cropping.

Intensive agriculture relies on the heavy use of irrigation, inorganic fertilisers (produced using fossil fuels), pesticides, and farm machinery. Such farms may specialise in a single crop for many years.

Impact on the environment

- Pesticide use is escalating yet pesticide effectiveness is decreasing. This causes a reduction in species diversity, particularly among the invertebrates.
- Mammals and birds may be affected by **bioaccumulation** of pesticides in the food chain and loss of food sources as invertebrate species diminish.
- Fertiliser use is increasing, resulting in a continued decline in soil and water quality.
- More fertiliser leaches from the soil and enters groundwater as a pollutant, relative to organic farming practices.
- Large fields lacking hedgerows create an impoverished habitat and cause the isolation of remaining wooded areas.
- A monoculture regime leads to reduced biodiversity.

Sustainable Agricultural Practices

Organic farming is a sustainable form of agriculture based on the avoidance of chemicals and applied *inorganic* fertilisers. It relies on mixed (crop and livestock) farming and crop management, combined with the use of environmentally friendly pest controls (e.g., biological controls and flaming), and livestock and green manures. Organic farming uses **crop rotation** and **intercropping**, in which two or more crops are grown at the same time on the same plot, often maturing at different times. If well cultivated, these plots can provide food, fuel, and natural pest control and fertilisers on a sustainable basis. Yields are typically lower than on intensive farms, but the produce can fetch high prices, and pest control and fertiliser costs are reduced.

Some traditional farms in the UK use low-input agricultural practices similar to those used in modern organic farming. However, many small farming units find it difficult to remain economically viable.

Impact on the environment

- Pesticides do not persist in the environment nor accumulate in the food chain.
- Produce is pesticide free and produced in a sustainable way.
- Alternative pest control measures, such as using natural predators and pheromone traps, reduce the dependence on pesticides.
- The retention of hedgerows increases habitat diversity and produces corridors for animal movement between forested areas.
- Crop rotation (alternation of various crops, including legumes) prevents pests and disease species building up to high levels.
- Conservation tillage (ploughing crop residues into the topsoil) as part of the crop rotation cycle improves soil structure.

1. Discuss the conflict of interest between the need for high agricultural production and the need for habitat conservation:

The Hedgerow Issue

A particularly significant factor of landscape change in recent years has been the amalgamation of fields and the removal of traditional hedgerows. Many traditional, mixed farms (right), which required hedgerows to contain livestock, have been converted to arable farms, and fields have become larger to accommodate modern machinery. In Britain, this conversion has resulted in the loss of thousands of kilometres of hedgerows each year.

Hedgerows are ecologically important because they increase the diversity of wildlife by:

- Providing food and habitats for birds and other animals.
- Acting as corridors, along which animals can move.
- Providing habitat for predators of pest species.

2. From an environmental perspective, outline two advantages of using hedgerows as a form of farm fencing:

   (a) ___

   (b) ___

3. From the perspective of the farmer, outline two disadvantages of using hedgerows:

   (a) ___

   (b) ___

4. Discuss the differences between intensive farming and organic farming, identifying the specific features of each:

___

___

___

___

___

___

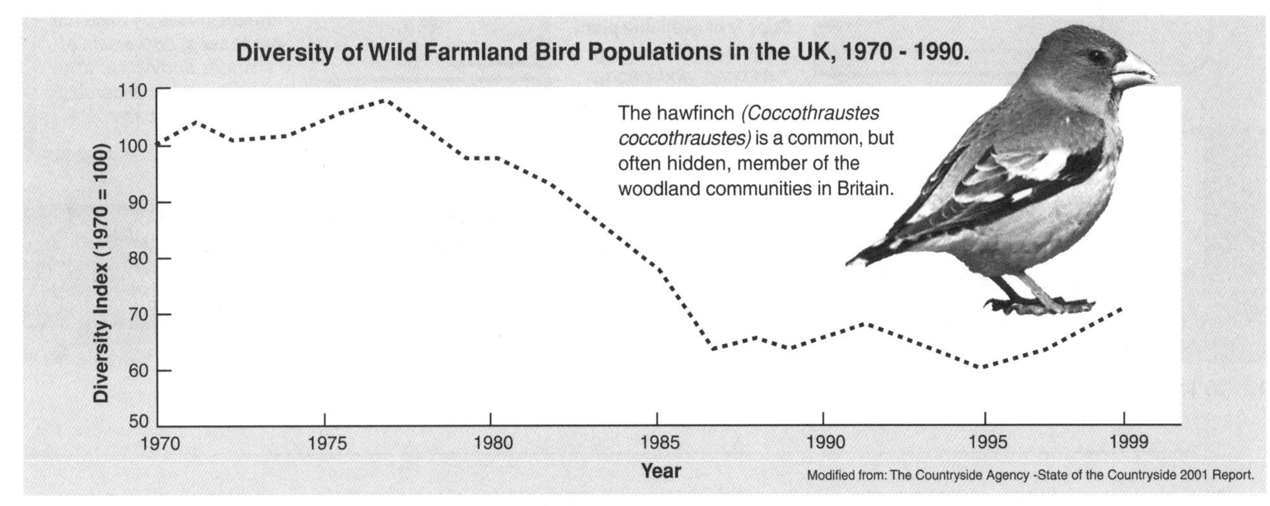

Populations of wild farmland bird species, because of their wide distribution and position near the top of the food chain, provide good indicators of the state of other wildlife species and of environmental health in general. Over the last 25 years, there has been a marked net decline in the diversity of farmland bird populations. However, since 1986, diversity has ceased to decline further and, in recent years, has actually showed an increase.

5. Suggest two possible reasons for this decline:

   (a) ___

   (b) ___

6. List three initiatives local and national government have implemented in an attempt to reverse this decline:

___

Fertilisers and Land Use

A variety of minerals are required by plants. **Macronutrients** are required in large quantities, whilst **trace elements** are needed in only very small amounts. Harvesting interrupts the normal recycling of nutrients and contributes to nutrient losses. These nutrients can be replaced by the addition of fertilisers: materials that supply nutrients to plants. Some plants form symbiotic associations that aid in their nutrition. Legumes in particular (e.g. peas, beans, clover), are well known for their association with nitrogen-fixing bacteria that share their rich source of fixed nitrogen with their host plants. Legumes have been extensively used to maintain pasture productivity for grazing livestock, as well as in crop rotation to rejuvenate nitrogen-depleted soils. Historically, crop rotation was an important method to manage soil exhaustion on farms in the United Kingdom. Such systems have become relatively obsolete as demand for production during the 1980s and early 1990s increased. The availability of low cost artificial fertilisers and pesticides (**agrichemicals**) have made continuous single-crop farms economically viable.

The supply of nutrients to plants (in soil) depends on **soil fertility**. This refers to the condition of the soil relative to the amount and availability to plants of elements required for growth. Soil fertility depends both on the supply of chemical plant nutrients (e.g. nitrogen, phosphorus, and potassium), and on the supply of moisture and oxygen required to provide the appropriate environment for mineral uptake. Nutrients lost through harvesting, erosion, and denitrification can be replenished through the addition of organic and inorganic fertilisers, the decomposition of organic matter, and nitrogen fixation.

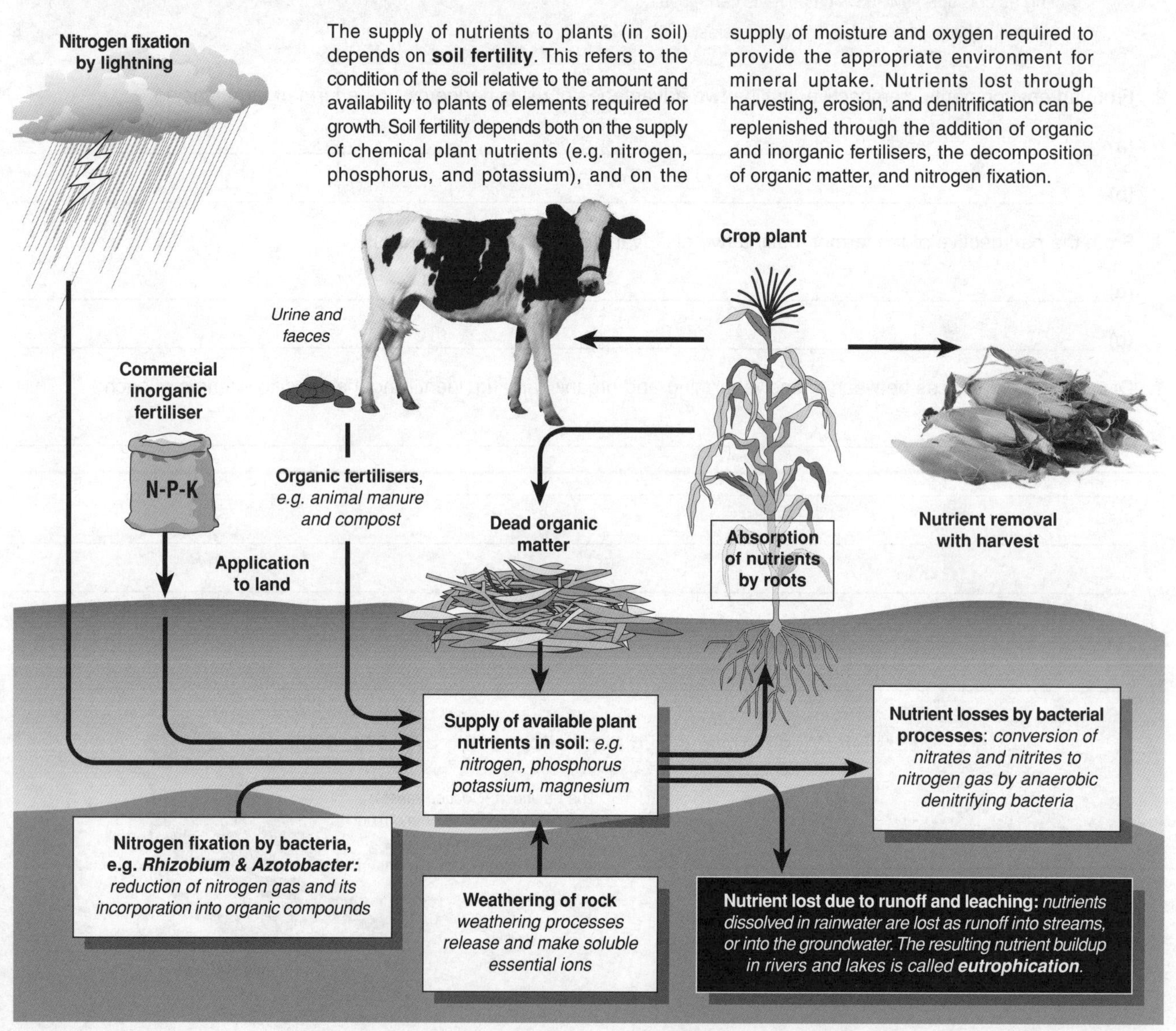

1. (a) Identify three sources of minerals for plants, apart from artificially applied fertilisers:

   (b) Suggest why these sources alone provide inadequate soil nutrients for plant growth in most cropping situations:

2. (a) Distinguish between an organic fertiliser and an inorganic fertiliser:

Crop harvesting (above) removes nutrients from the land. If left unreplenished, the soil becomes nutrient deficient. The addition of organic fertilisers (**animal manure**, **green manure**, and **compost**) and inorganic fertilisers (chemically synthesised) restores the soil's fertility. The production of inorganic fertilisers consumes considerable amounts of fossil fuel (e.g. urea may be made from natural gas). In contrast, the use of organic fertilisers, reduces waste and has the added benefit of improving soil structure.

Tilling (above) prepares the land for crop sowing. Sometimes, the crop residue from the previous crop is left on the ground and mixed into the surface layers of the soil. This practice, called **conservation tillage**, improves soil structure through aeration and returns nutrients to the soil. It is best suited to crop rotations where the crop residue involved changes seasonally. In continuous cropping systems, conservation tillage leaves the same type of residue in the soil all year round, and this may harbour pests and disease.

Crop rotation is an agricultural practice in which different crops are cultivated in succession on the same ground in successive years. Its purpose is to maintain soil fertility and reduce the adverse effects of pests. Legumes, such as peas, beans, clover, vetch, and soybean (above) are important in the rotation as they restore nitrogen to the soil. In the UK, other crops that may be included in a typical rotation are wheat, barley, and root crops.

Animal manure, seen here being spread by a muck spreader, improves soil structure, adds organic nitrogen, and stimulates beneficial soil bacteria and fungi. Sustainable farming practices, such as crop rotation and the application of animal manures, have substantially declined in recent years as they do not accommodate the shift to large-scale farming practices and the move away from mixed crop and livestock farming.

(b) Identify which of these two types of fertiliser provides nutrients most readily to plants and why: ______________________

____

____

3. (a) Predict the likely consequence of continued harvesting without replenishment of soil nutrients: ______________________

____

(b) Describe ways in which this consequence could be avoided: ______________________

____

4. Discuss the benefits and drawbacks of **crop rotation** and **conservation tillage** as agricultural management tools:

____

____

____

____

5. Explain how overuse of inorganic and organic fertilisers may contribute to excessive **leaching** and **eutrophication**:

____

____

____

____

Deforestation

Tropical rainforests prevail in places where the climate is very moist throughout the year (200 to 450 cm of rainfall per year). Almost half of the world's rainforests are in just three countries: **Indonesia** in Southeast Asia, **Brazil** in South America, and **Zaire** in Africa. Much of the world's biodiversity resides in rainforests. Destruction of the forests will contribute towards global warming through a large reduction in photosynthesis. In the Amazon, 75% of deforestation has occurred within 50 km of Brazil's roads. Many potential drugs could still be discovered in rainforest plants, and loss of species through deforestation may mean they will never be found. Rainforests can provide economically sustainable crops (rubber, coffee, nuts, fruits, and oils) for local people.

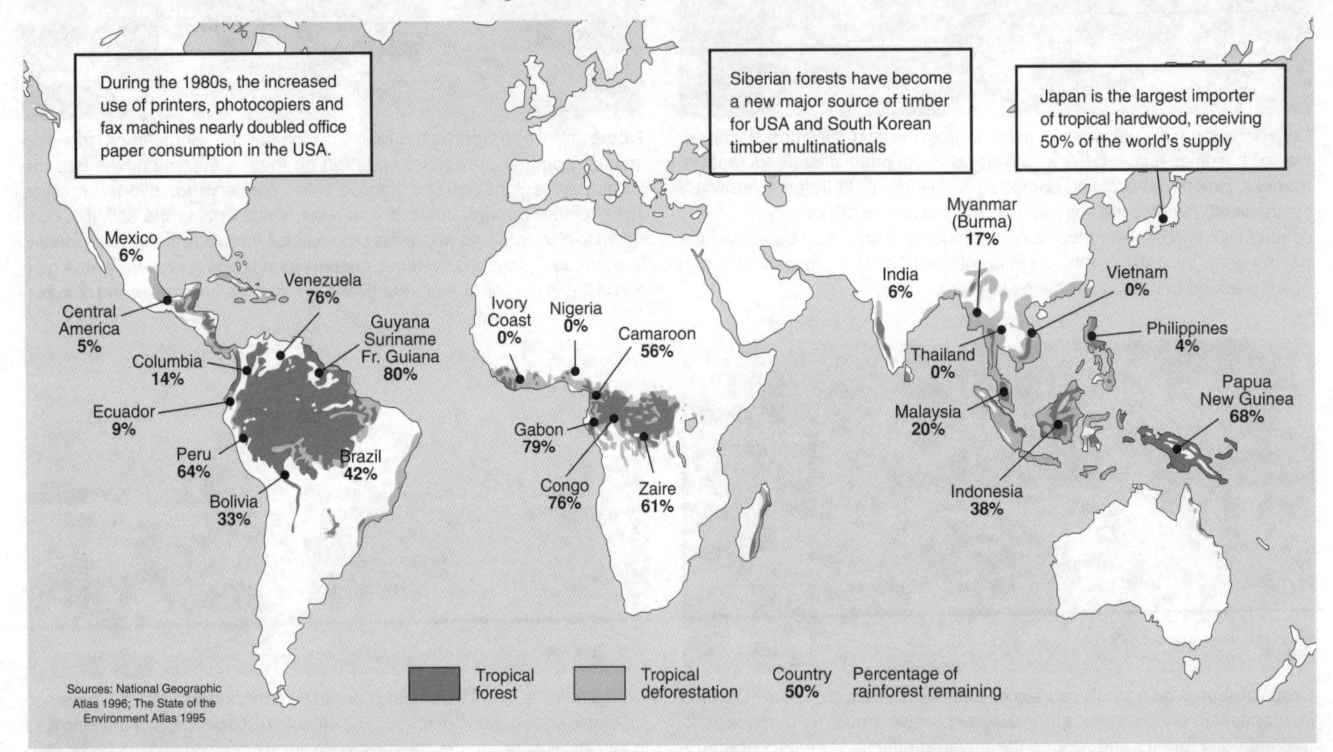

The felling of rainforest trees is taking place at an alarming rate as world demand for tropical hardwoods increases and land is cleared for the establishment of agriculture. The resulting farms and plantations often have shortlived productivity.

Huge forest fires have devastated large amounts of tropical rainforest in Indonesia and Brazil in 1997/98. The fires in Indonesia were started by people attempting to clear the forest areas for farming in a year of particularly low rainfall.

The building of new road networks into regions with tropical rainforests causes considerable environmental damage. In areas with very high rainfall there is an increased risk of erosion and loss of topsoil.

1. Describe three reasons why tropical rainforests should be conserved:

   (a) ______________________________

   (b) ______________________________

   (c) ______________________________

2. Identify the three main human activities that cause tropical deforestation and discuss their detrimental effects:

   ____

   ____

   ____

Managing Woodlands

The United Kingdom was once largely covered with woodland, but over many centuries it was cleared to meet the needs of a growing population. Currently only one tenth of the land area of the UK is covered by forest or other wooded areas. Although the percentage cover is still low, it has doubled since World War I as a result of major **afforestation programmes**. Currently, more than 80% of all forests are available for wood supply, with the remainder protected for conservation reasons. In the UK, commercial forests and woods are mainly composed of conifers and other softwood trees. With careful management it is possible to make use of the forest resource in a sustainable manner without destroying the particular features of the ecosystem.

Clear Cutting

A section of a mature forest is selected (based on tree height, girth, or species), and all the trees are removed. During this process the understorey is destroyed. A new forest of economically desirable trees may be planted. In plantation forests, the trees are generally of a single species and may even be clones. Clear cutting is a very productive and economical method of managing a forest, however it is also the most damaging to the natural environment. In plantation forests this may not be of concern, but clear cutting of old growth forests causes enormous ecological damage.

Selective Logging

A mature forest is examined, and trees are selected for removal based on height, girth, or species. These trees are felled individually and directed to fall in such a way as to minimise the damage to the surrounding younger trees. The forest is managed in such a way as to ensure continual regeneration of young seedlings and provide a balance of tree ages that mirrors the natural age structure.

Coppicing

Coppicing is the ancient practice of harvesting wood for weaving, thatching, firewood, or for making charcoal. A selection of deciduous trees is **coppiced** (cut close to the ground), leaving stumps known as **stools**. Most deciduous trees have the ability to regrow from their bases when they are cut down. However their growth form changes from a single stem to producing many stems. It is these stems that provide the wood for harvesting in the future.

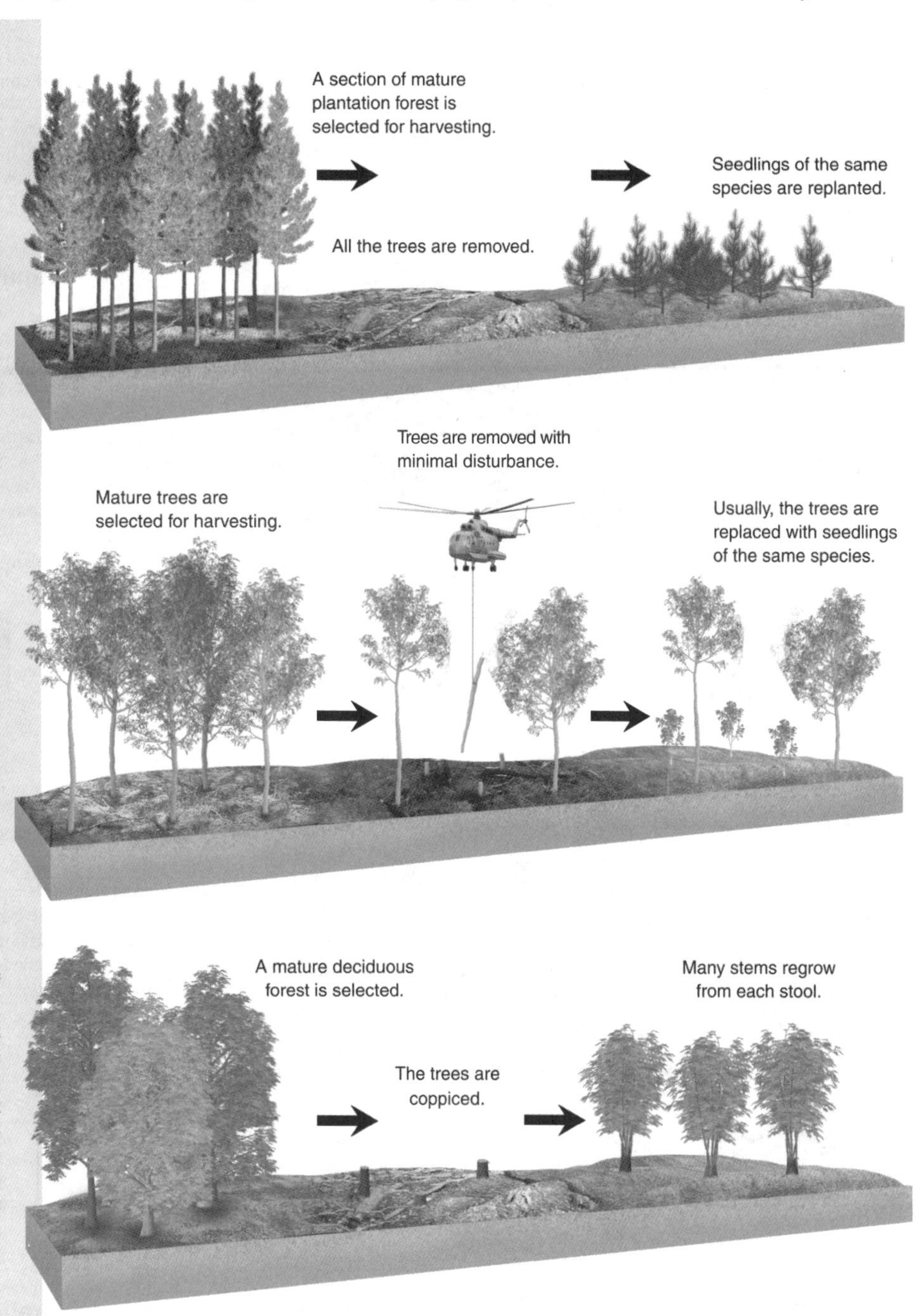

1. Using the terms *high*, *medium*, *low,* and *no change,* complete the following table.

Method of logging	Volume of logs harvested	Habitat destruction	Risk of soil erosion	Damage to surrounding trees	Biodiversity of regenerating forest
Coppicing					
Selective logging					
Clear cutting					

2. Suggest which logging method best balances the aims of productivity and conservation: ______________________

Ecological Succession

Ecological succession is the process by which communities in a particular area change over time. Succession takes place as a result of complex interactions of biotic and abiotic factors. Early communities modify the physical environment causing it to change. This in turn alters the biotic community, which further alters the physical environment and so on. Each successive community makes the environment more favourable for the establishment of new species. A succession (or **sere**) proceeds in **seral stages**, until the formation of a **climax community**, which is stable until further disturbance. Early successional communities are characterised by a low species diversity, a simple structure, and broad niches. In contrast, climax communities are complex, with a large number of species interactions, narrow niches, and high species diversity.

Composition of the community changes with time

Past community → **Present community** → **Future community**

Some species in the **past community** were outcompeted, and/or did not tolerate altered abiotic conditions.

The **present community** modifies such abiotic factors as:
- Light intensity
- Light quality
- Wind speed
- Wind direction
- Air temperature
- Soil water
- Soil composition
- Humidity

Changing conditions in the **present community** will allow new species to become established. These will make up the **future community**.

Primary Succession

TG

Primary succession refers to colonisation of regions where there is no preexisting community. Examples include regions where the previous community has been extinguished by a volcanic eruption (such as the Indonesian island of Krakatau (Krakatoa), which erupted in 1883), newly formed glacial moraines, or newly formed volcanic islands (as when Surtsey appeared off Iceland in 1963). The sequence of colonisation described below is typical of a Northern hemisphere lithosere; a succession on bare rock. This sequence is not necessarily the same as that occurring on another substrate, such as volcanic ash, which allows the earlier establishment of grasses.

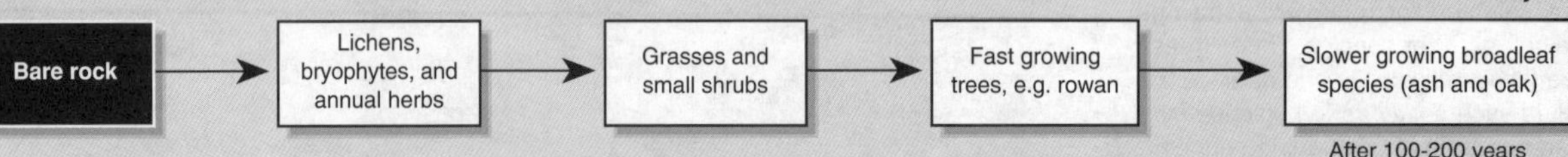

Secondary Succession in Cleared Land

(150+ years for mature woodland to develop again)

Pioneer community: grasses invade

Climax woodland

A secondary succession takes place after a land clearance (e.g. from fire or landslide). Such events do not involve loss of the soil and so tend to be more rapid than primary succession, although the time scale depends on the species involved and on climatic and edaphic (soil) factors. Humans may deflect the natural course of succession (e.g. by mowing) and the climax community that results will differ from the natural community. A climax community arising from a **deflected succession** is called a **plagioclimax**.

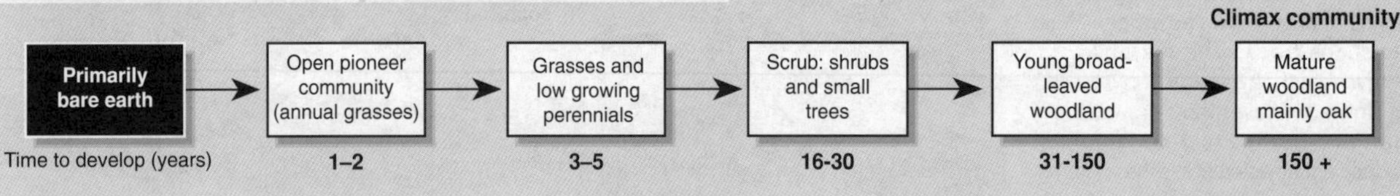

Gap Regeneration Cycle in a Rainforest

(500-700 years)

BH

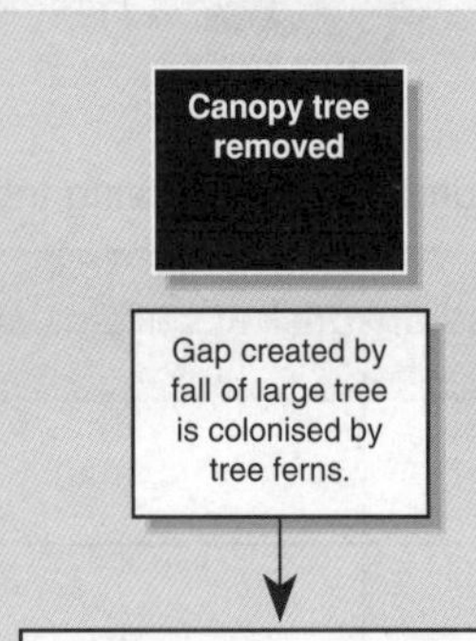

Large canopy trees have a profound effect on the make-up of a rainforest community, reducing light penetration and impeding the growth of saplings. When a large tree falls, it opens a hole in the canopy that lets in sunlight. Saplings then compete to fill the gap. The photograph on the left shows a large canopy tree in temperate rainforest that has recently fallen, leaving a gap in the canopy through which light can penetrate.

Growth of subcanopy trees suppresses tree ferns. Seedlings of canopy trees grow beneath the subcanopy.

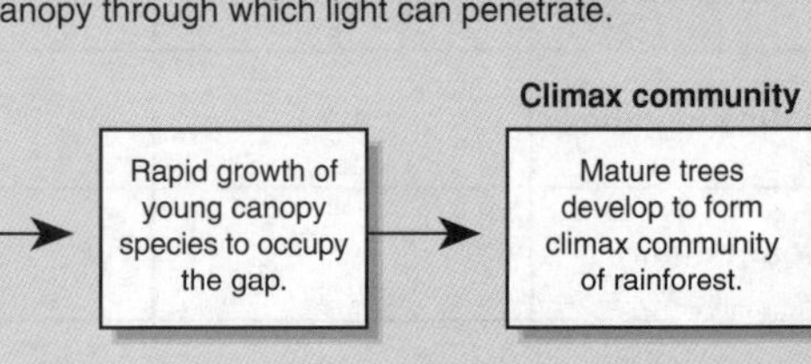

Populations & Interactions

Wetland areas present a special case of ecological succession. Wetlands are constantly changing as plant invasion of open water leads to siltation and infilling. This process is accelerated by **eutrophication**. In well drained areas, pasture or **heath** may develop as a result of succession from freshwater to dry land. When the soil conditions remain non-acid and poorly drained, a swamp will eventually develop into a seasonally dry **fen**. In special circumstances (see below) an acid **peat bog** may develop. The domes of peat that develop produce a hummocky landscape with a unique biota. Wetland peat ecosystems may take more than 5000 years to form but are easily destroyed by excavation and lowering of the water table.

Wetland Succession

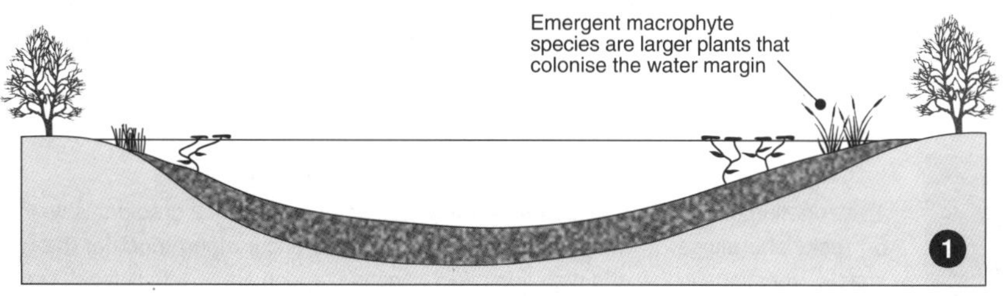
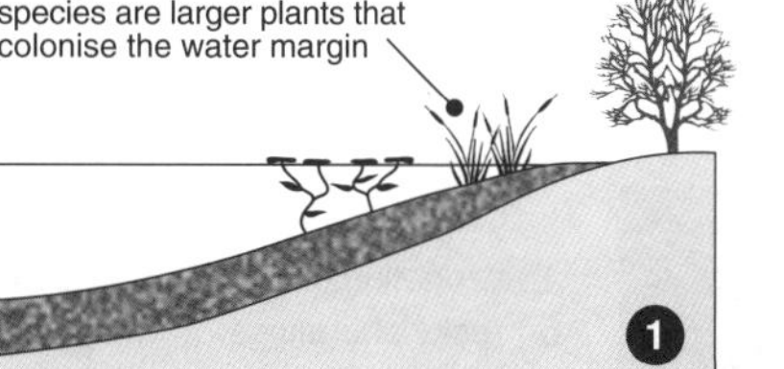

An open body of water, with time, becomes silted up and is invaded by aquatic plants. Emergent macrophyte species colonise the accumulating sediments, driving floating plants towards the remaining deeper water.

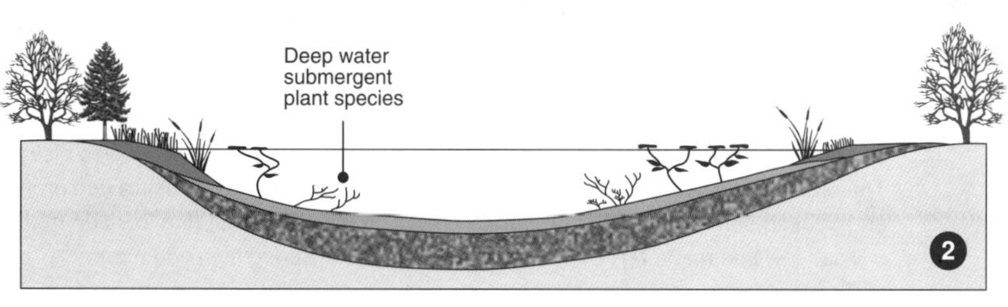

The increasing density of rooted emergent, submerged, and floating macrophytes encourages further sedimentation by slowing water flows and adding organic matter to the accumulating silt.

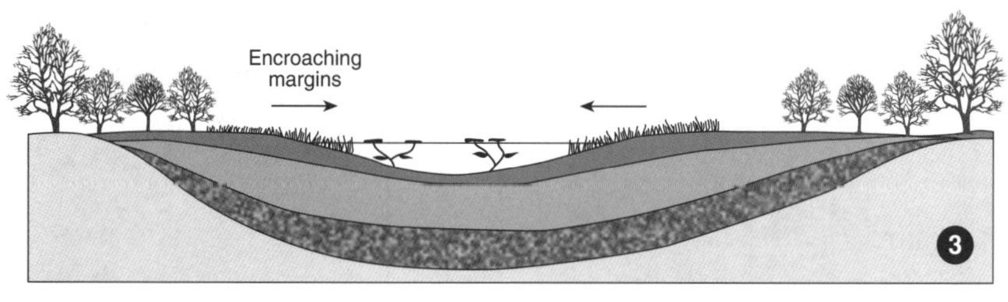

The resulting **swamp** is characterised by dense growths of emergent macrophytes and permanent (although not necessarily deep) standing water. As sediment continues to accumulate the swamp surface may, in some cases, dry off in summer.

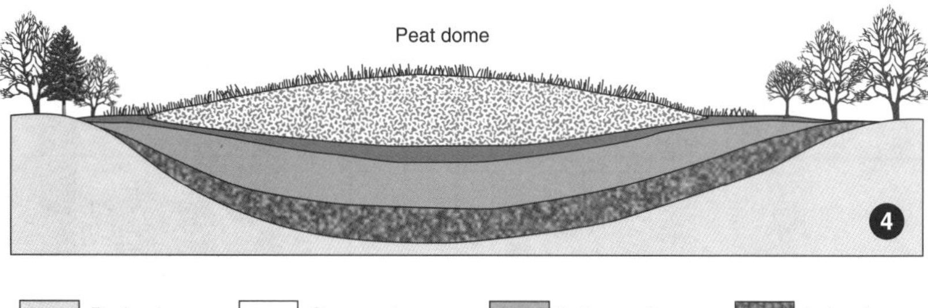

In colder climates, low evaporation rates, high rainfall, and invasion by *Sphagnum* moss leads to development of a peat **bog**; a low pH, nutrient poor environment where acid-tolerant plants such as sundew replace swamp species.

1. (a) Explain what is meant by primary succession: ______________________________

   (b) Explain how secondary succession differs from primary succession: ______________________________

2 (a) Explain what is meant by a deflected succession: ______________________________

   (b) Explain the role that deflected succession has had in shaping much of Britain's managed woodland:

   ____

3. (a) In bog development, suggest how *Sphagnum* alters the environment to favour the establishment of other bog species:

   ____

   (b) Identify the two most important abiotic factors in determining whether or not a wetland will develop into bog:

   ____

Shoreline Zonation Patterns

Zonation refers to the division of an ecosystem into distinct zones that experience similar abiotic conditions. In a more global sense, differences in latitude and altitude create distinctive zones of vegetation type, or **biomes**. Ecosystem zonation is particularly clear on a rocky seashore, where assemblages of different species form a banding pattern approximately parallel to the waterline. This effect is marked in temperate regions where the prevailing weather comes from the same general direction. Exposed shores show the clearest zonation. On sheltered rocky shores there is considerable species overlap and it is only on the upper shore that distinct zones are evident. Rocky shores exist where wave action prevents the deposition of much sediment. The rock forms a stable platform for the secure attachment of organisms such as large seaweeds and barnacles. Sandy shores are less stable than rocky shores and the organisms found there are adapted to the more mobile substrate (see below).

Rocky shore at Sleahead, Ireland.

Seashore Zonation Patterns

The zonation of species distribution according to an environmental gradient is well shown on rocky shorelines. In Britain, exposed rocky shores occur along much of the western coasts. Variations in low and high tide affect zonation, and in areas with little tidal variation, zonation is restricted. High on the shore, some organisms may be submerged only at spring high tide. Low on the shore, others may be exposed only at spring low tide. There is a gradation in extent of exposure and the physical conditions associated with this. Zonation patterns generally reflect the vertical movement of seawater. Sheer rocks can show marked zonation as a result of tidal changes with little or no horizontal shift in species distribution. The profiles below left, show zonation patterns on an exposed rocky shore (left profile) with an exposed sandy shore for comparison (right profile). **SLT** = Spring low tide mark, **MLT** = Mean low tide mark, **MHT** = Mean tide mark, **SHT** = Spring high tide mark.

Key to species

1 Lichen: sea ivory
2 Small periwinkle *Littorina neritoides*
3 Lichen *Verrucaria maura*
4 Rough periwinkle *Littorina saxatilis*
5 Common limpet *Patella vulgaris*
6 Laver *Porphyra*
7 Spiral wrack *Fucus spiralis*
8 Australian barnacle
9 Common mussel *Mytilus edulis*
10 Common whelk *Buccinum undatum*
11 Grey topshell *Gibbula cineraria*
12 Carrageen (Irish moss) *Chondrus crispus*
13 Thongweed *Himanthalia elongata*
14 Toothed wrack *Fucus serratus*
15 Dabberlocks *Alaria esculenta*
16 Common sandhopper
17 Sandhopper *Bathyporeia pelagica*
18 Common cockle *Cerastoderma edule*
19 Lugworm *Arenicola marina*
20 Sting winkle *Ocinebra erinacea*
21 Common necklace shell *Natica alderi*
22 Rayed trough shell *Mactra corallina*
23 Sand mason worm *Lanice conchilega*
24 Sea anemone *Halcampa*
25 Pod razor shell *Ensis siliqua*
26 Sea potato *Echinocardium* (a heart urchin)

Note: Where several species are indicated within a single zonal band, they occupy the entire zone, not just the position where their number appears.

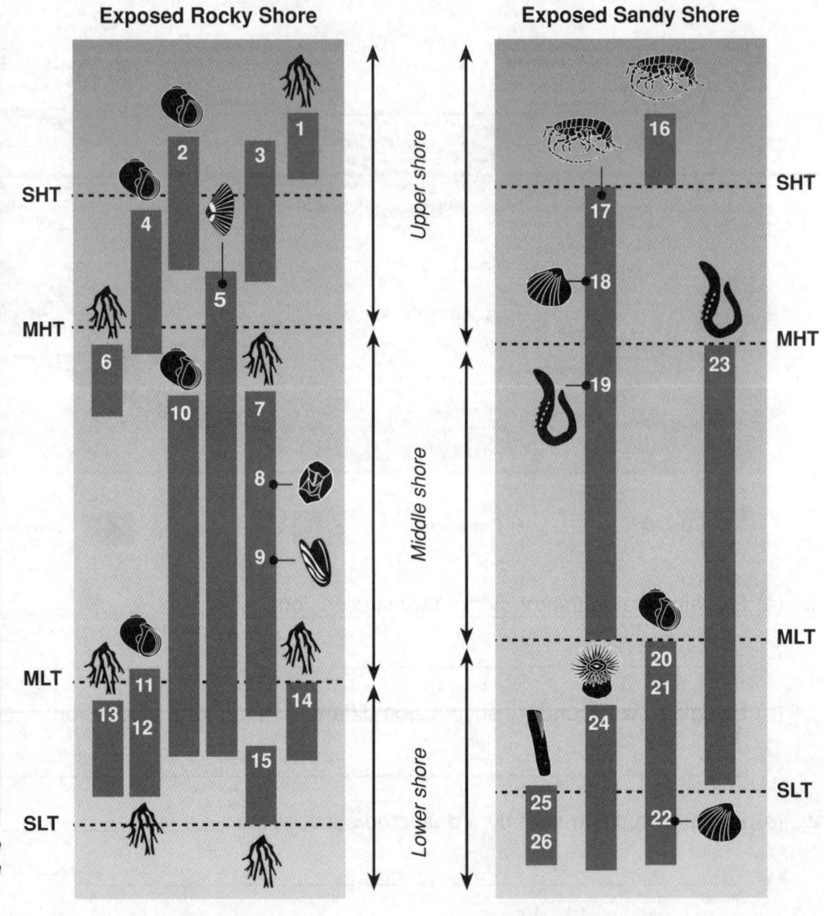

1. (a) Suggest why the time of exposure above water is a major factor controlling species distribution on a rocky shore:

   (b) Identify two other abiotic factors that might influence species distribution on a rocky shore:

   (c) Identify two biotic factors that might influence species distribution on a rocky shore:

2. Describe the zonation pattern on a rocky shore:

Practical Ecology

AQA-A	AQA-B	CIE	Edexcel	OCR
Complete: *1-4, 6 (b)-(d), 10* *Extension: 5, 7, 9, 11-12*	Complete: *1-4, 6, 9-12* *Extension: 5, 7*	Complete: *1-12 as required*	Complete: *1-4, 6, 8, 10-12* *Extension: 5, 7, 9*	Complete: *1-6 (b), (c), 10* *Extension: 7, 9*

Learning Objectives

☐ 1. Compile your own glossary from the **KEY WORDS** displayed in **bold type** in the learning objectives below.

Sampling populations *(pages 74-75, 78-93)*

☐ 2. Ecosystems comprise different groups, or **populations**, of species. Identify the sort of information that is gained from population studies (e.g. **abundance**, **density**, **age structure**, **distribution**).

☐ 3. Explain what is meant by **sampling** as it relates to population studies. Explain why we sample populations and describe the advantages and drawbacks involved.

☐ 4. A field study should enable you to test a hypothesis about a certain aspect of a population. You should provide a clear outline of your study with attention to: *the methods you will use to collect* ***data****, the type of data you are collecting (relevance to the hypothesis), the number of recordings you will make, the size of your sampling unit (e.g. quadrat size) and the number of* ***samples*** *you will take, the assumptions that you are making in the investigation, and appropriate* ***controls***.

☐ 5. Explain how and why sample size affects the accuracy of population estimates. Explain how you would decide on a suitable sample size. Discuss the compromise between sampling accuracy and sampling effort.

☐ 6. Describe the following techniques used to study aspects of populations (e.g. **distribution**, **abundance**, **density**). Identify the advantages and limitations of each method with respect to sampling time, cost, and the suitability to the organism and specific habitat type:
(a) **Direct counts**
(b) **Frame** and/or **point quadrats**
(c) **Belt** and/or **line transects**
(d) **Mark and recapture** and the **Lincoln index**
(e) **Netting** and **trapping**

☐ 7. Recognise the value to population studies of **radio-tracking** and **indirect methods** of sampling such as counting nests, and recording calls and droppings.

☐ 8. Describe **qualitative methods** for investigating the distribution of organisms in specific habitats (e.g. terrestrial, freshwater, marine littoral).

☐ 9. Describe the methods used to ensure **random sampling**, and appreciate why this is important.

☐ 10. Recognise the appropriate ways in which different types of data may be recorded, analysed, and presented. Demonstrate an ability to apply statistical tests, as appropriate, to data collected from populations. Recognise that the type of data collected and the design of the field study will determine how the data can be analysed.

Measuring abiotic factors *(pages 76-77)*

☐ 11. Describe methods to measure abiotic factors in a habitat. Include reference to the following (as appropriate): pH, light, temperature, dissolved oxygen, current speed, total dissolved solids, and conductivity.

☐ 12. Appreciate the influence of abiotic factors on the distribution and abundance of organisms in a habitat.

See the 'Textbook Reference Grid' on pages 8-9 for textbook page references relating to material in this topic.

Supplementary Texts

See pages 4-6 for additional details of these texts:

■ Adds, J. *et al.*, 1999. **Tools, Techniques and Assessment in Biology** (NelsonThornes), chpt. 6.

■ Adds, J., *et al.*, 2001. **Genetics, Evolution and Biodiversity**, (NelsonThornes), pp. 36-40, 46, 59-60.

■ Allen *et al.*, 2001. **Applied Ecology** (Collins), pp. 6-14 (includes diversity indices).

■ Chenn, P., 1999. **Ecology** (John Murray), chpt. 9.

■ Jones, A., *et al.*, 1994. **Practical Skills in Biology** (Addison-Wesley), as required.

■ Reiss, M. & J. Chapman, 2000. **Environmental Biology** (CUP), chpt. 7.

See page 6 for details of publishers of periodicals:

STUDENT'S REFERENCE

■ **Ecological Projects** Biol. Sci. Rev., 8(5) May 1996, pp. 24-26. *Planning and carrying out a field-based project: from choosing the topic to data analysis and reporting.*

■ **Fieldwork - Sampling Animals** Biol. Sci. Rev., 10(4) March 1998, pp. 23-25. *Excellent article covering the appropriate methodology for collecting different types of animals in the field. Includes a synopsis of the mark and recapture technique.*

■ **Fieldwork Sampling - Plants** Biol. Sci. Rev., 10(5) May 1998, pp. 6-8. *Excellent article covering the methodology for sampling plant communities (line and belt transects and quadrat techniques).*

■ **Bird Ringing** Biol. Sci. Rev., 14(3) Feb. 2002, pp. 14-19. *The practical investigation of populations of these highly mobile organisms. Includes discussion of mark and recapture methods, ringing techniques, and diversity indices.*

■ **British Butterflies in Decline** Biol. Sci. Rev., 14(4) April 2002, pp. 10-13. *Documented changes in the distribution of British butterfly species. This account includes a description of the techniques used to monitor changes in butterfly numbers in different regions of the UK. How this information is used to indicate patterns of population abundance in relation to habitat change is also discussed.*

TEACHER'S REFERENCE

■ **Ecology Fieldwork in 16 to 19 Biology** SSR, 84(307) December 2002, pp. 87. *Fieldwork is recognised as an important element of education in biology, yet the anecdotal evidence suggests that it is declining. This article examines the fieldwork opportunities provided to 16-19 students and suggests how these could be enhanced and firmly established for all students studying biology.*

Software and video resources are provided on the Teacher Resource Handbook on CD-ROM

Designing your Field Study

The following provides an example and some ideas for designing a field study. It provides a framework which can be modified for most simple comparative field investigations. For reasons of space, the full methodology is not included.

Observation

A student read that a particular species of pill millipede (left) is extremely abundant in forest leaf litter, but a search in the litter of a conifer-dominated woodland near his home revealed only very low numbers of this millipede species.

Hypothesis

This millipede species is adapted to a niche in the leaf litter of oak woodlands and is abundant there. However, it is rare in the litter of coniferous woodland. The **null hypothesis** is that there is no difference between the abundance of this millipede species in oak and coniferous woodland litter.

Oak or coniferous woodland

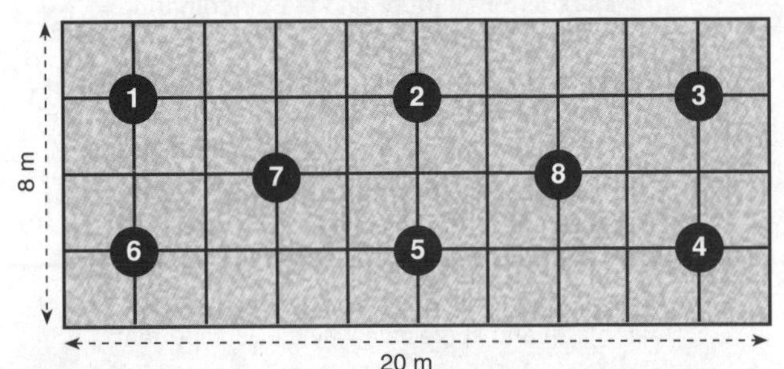

Sampling sites numbered 1-8 at evenly spaced intervals on a 2 x 2 m grid within an area of 20 m x 8 m.

Sampling equipment: leaf litter light trap

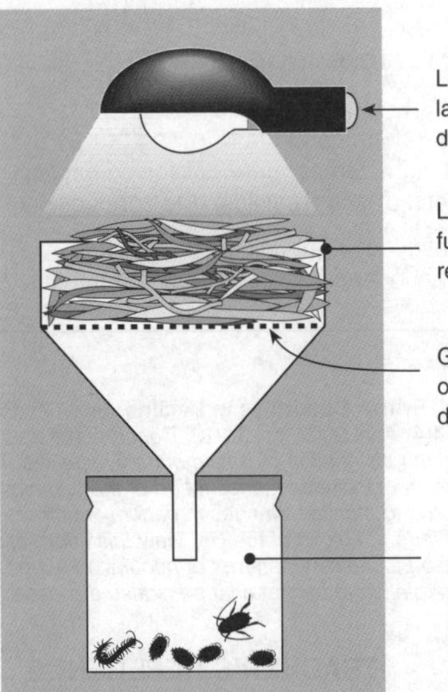

Notes on collection and analysis

- Mean millipede abundance was calculated from the counts from the eight sites. The difference in abundance at the sites was tested using a Student's *t* test.
- After counting and analysis of the samples, all the collected invertebrates were returned to the sites.

Sampling Programme

A sampling programme was designed to test the prediction that the millipedes are more abundant in the leaf litter of oak woodlands than in coniferous woodlands.

Equipment and Procedure

Sites: For each of the two woodland types, an area 20 x 8 m was chosen and marked out in 2 x 2 m grids. Eight sampling sites were selected, evenly spaced along the grid as shown.

- The general area for the study chosen was selected on the basis of the large amounts of leaf litter present.
- Eight sites were chosen as the largest number feasible to collect and analyse in the time available.
- The two woodlands were sampled on sequential days.

Capture of millipedes: At each site, a 0.4 x 0.4 m quadrat was placed on the forest floor and the leaf litter within the quadrat was collected. Millipedes and other leaf litter invertebrates were captured using a simple gauze lined funnel containing the leaf litter from within the quadrat. A lamp was positioned over each funnel for two hours and the invertebrates in the litter moved down and were trapped in the collecting jar.

- After two hours each jar was labelled with the site number and returned to the lab for analysis.
- The litter in each funnel was bagged, labelled with the site number and returned to the lab for weighing.
- The number of millipedes at each site was recorded.
- The numbers of other invertebrates (classified into major taxa) were also noted for reference.

Assumptions

- The areas chosen in each woodland were representative of the woodland types in terms of millipede abundance.
- Eight sites were sufficient to adequately sample the millipede populations in each forest.
- A quadrat size of 0.4 x 0.4 m contained enough leaf litter to adequately sample the millipedes at each site.
- The millipedes were not preyed on by any of the other invertebrates captured in the collecting jar.
- All the invertebrates within the quadrat were captured.
- Millipedes moving away from the light are effectively captured by the funnel apparatus and cannot escape.
- Two hours was long enough for the millipedes to move down through the litter and fall into the trap.

  Note that these last two assumptions could be tested by examining the bagged leaf litter for millipedes after returning to the lab.

A note about sample size

When designing a field study, the size of your sampling unit (e.g. quadrat size) and the sample size (the number of samples you will take) should be major considerations. There are various ways to determine the best quadrat size. Usually, these involve increasing the quadrat size until you stop finding new species. For simple field studies, the number of samples you take (the sample size or *n* value) will be determined largely by the resources and time that you have available to collect and analyse your data. It is usually best to take as many samples as you can, as this helps to account for any natural variability present and will give you greater confidence in your data. For a summary of these aspects of study design as well as coverage of collecting methods see: *Jones, A. et al. (1998) Practical Skills in Biology* (or the earlier 1994 edition).

1. Explain the importance of recognising any assumptions that you are making in your study:

2. Describe how you would test whether the quadrat size of 0.4 x 0.4 m was adequate to effectively sample the millipedes:

3. Suggest why the litter was bagged, returned to the lab and then weighed properly for the analysis:

4. Suggest why the numbers of other invertebrates were also recorded even though it was only millipede abundance that was being investigated:

YOUR CHECKLIST FOR FIELD STUDY DESIGN

The following provides a checklist for a field study. Check off the points when you are confident that you have satisfied the requirements in each case:

1. **Preliminary:**
   - ☐ (a) Makes a hypothesis based on observation(s).
   - ☐ (b) The hypothesis (and its predictions) are testable using the resources you have available (the study is feasible).
   - ☐ (c) The organism you have chosen is suitable for the study and you have considered the ethics involved.
2. **Assumptions and site selection:**
   - ☐ (a) You are aware of any assumptions that you are making in your study.
   - ☐ (b) You have identified aspects of your field design that could present problems (such as time of year, biological rhythms of your test organism, difficulty in identifying suitable habitats etc.).
   - ☐ (c) The study sites you have selected have the features necessary in order for you to answer the questions you have asked in your hypothesis.
3. **Data collection:**
   - ☐ (a) You are happy with the way in which you are going to take your measurements or samples.
   - ☐ (b) You have considered the size of your sampling unit and the number of samples you are going to take (and tested for these if necessary).
   - ☐ (c) You have given consideration to how you will analyse the data you collect and made sure that your study design allows you to answer the questions you wish to answer

Monitoring Physical Factors

Most ecological studies require us to measure the physical factors (parameters) in the environment that may influence the abundance and distribution of organisms. In recent years there have been substantial advances in the development of portable, light-weight meters and dataloggers. These enable easy collection and storage of data in the field.

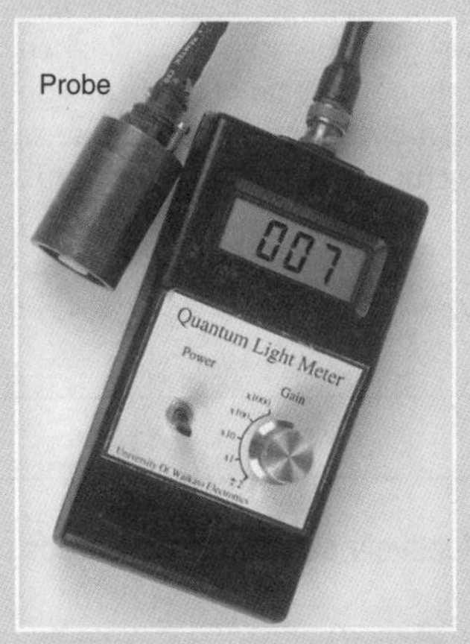

Quantum light meter: Measures light intensity levels. It is not capable of measuring light quality (wavelength).

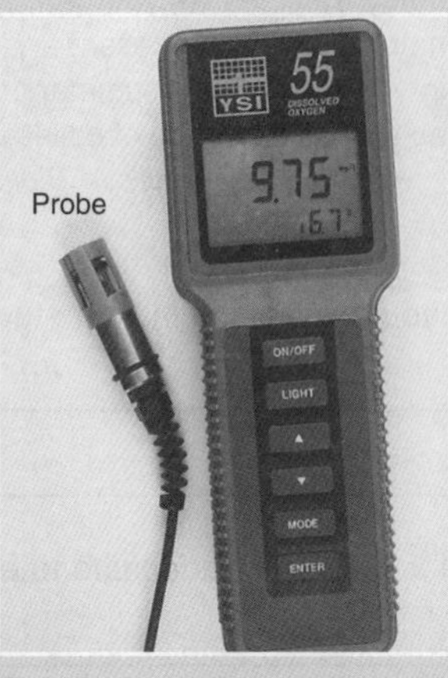

Dissolved oxygen meter: Measures the amount of oxygen dissolved in water (expressed as mgl^{-1}).

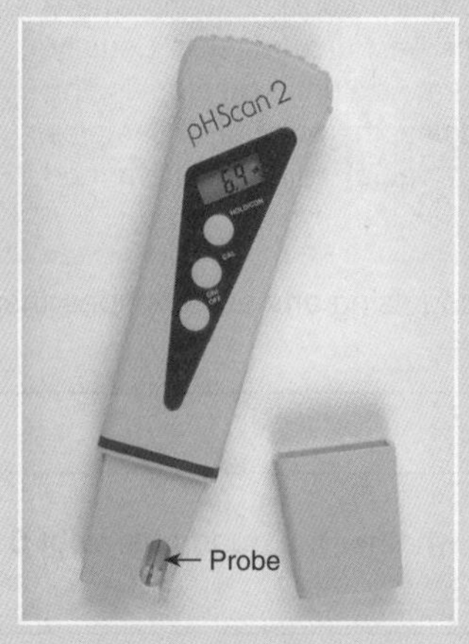

pH meter: Measures the acidity of water or soil, if it is first dissolved in pure water (pH scale 0 to 14).

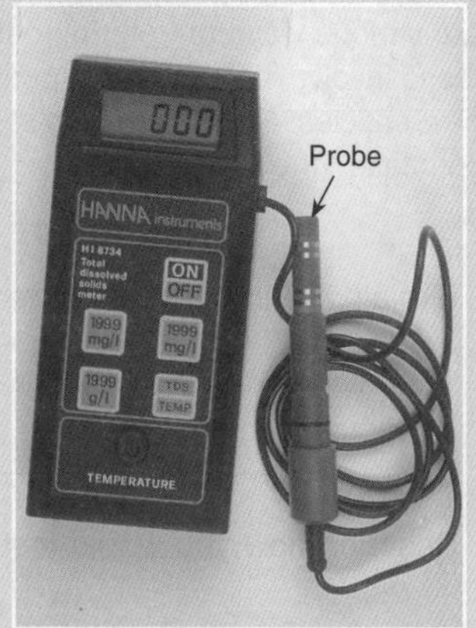

Total dissolved solids (TDS) meter: Measures content of dissolved solids (as ions) in water in mgl^{-1}.

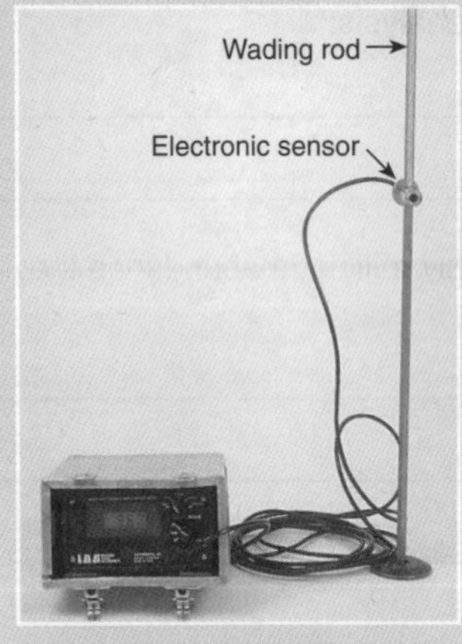

Current meter: The electronic sensor is positioned at set depths in a stream or river on the calibrated wading rod as current readings are taken.

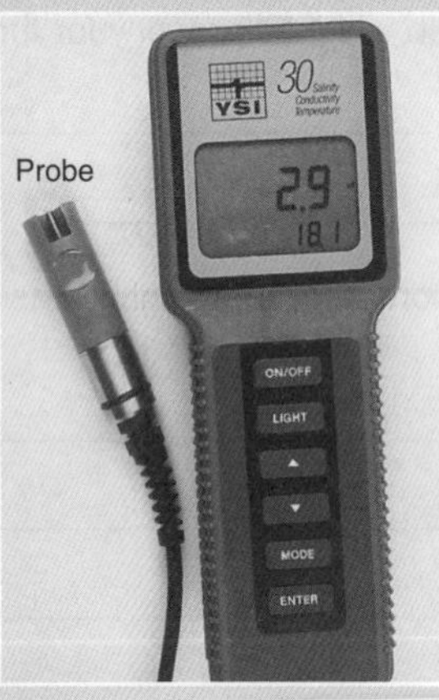

Multipu
func
sali
si

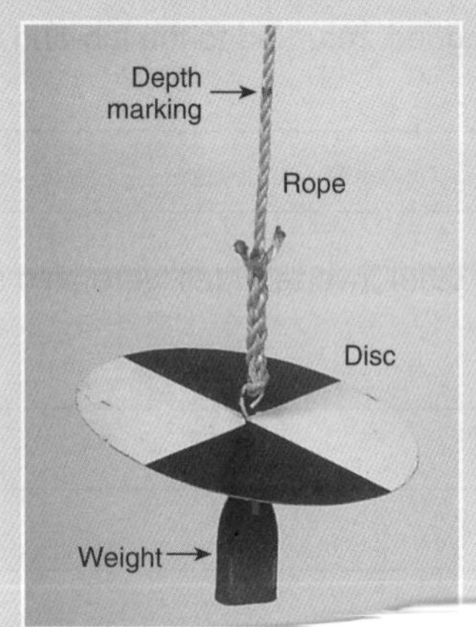

for nutrients, oxygen and pH.

hoto JDG. Others, Campus photography University of Waikato

Dataloggers and Environmental Sensors

Dataloggers are electronic instruments that record measurements over time. They are equipped with a microprocessor, data storage facility, and sensor. Different sensors are employed to measure a range of variables in water (photos A and B) or air (photos C and D), as well as make physiological measurements. The datalogger is connected to a computer, and software is used to set the limits of operation (e.g. the sampling interval) and initiate the logger. The logger is then disconnected and used remotely to record and store data. When reconnected to the computer, the data are downloaded, viewed, and plotted. Dataloggers, such as those pictured here from PASCO, are being increasingly used in professional and school research. They make data collection quick and accurate, and they enable prompt data analysis.

Dataloggers are now widely used to monitor conditions in aquatic environments. Different variables such as pH, temperature, conductivity, and dissolved oxygen can be measured by changing the sensor attached to the logger.

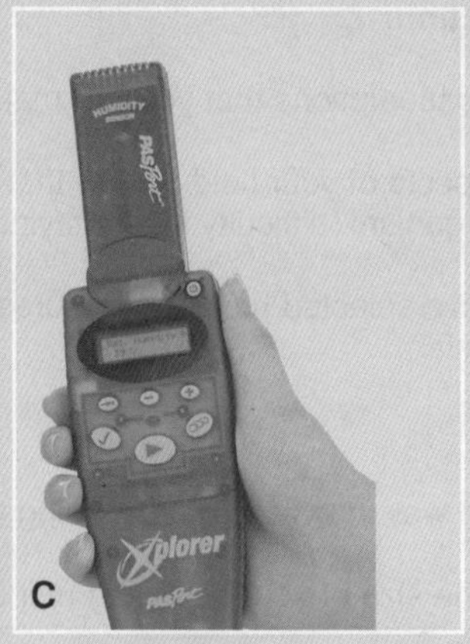

Datalogger photos courtesy of PASCO

Dataloggers fitted with sensors are portable and easy to use in a wide range of terrestrial environments. They are used to measure variables such as air temperature and pressure, relative humidity, light, and carbon dioxide gas.

1. The table below lists a variety of ecosystems and physical factors that may influence the abundance and distribution of organisms within them. For each ecosystem, identify which environmental measurements would be most useful in a study of the ecosystems listed (the first one has been completed for you):

Ecosystem	Temperature	Wind velocity	Light intensity	pH	Dissolved oxygen	Specific ions	Humidity
Freshwater stream	✓		✓	✓	✓	✓	
Polluted stream							
Ocean waters							
Estuarine mudflat							
Woodland leaf litter							
Open field							
Small pond (diurnal changes)							
Soil							
Peat bog							

2. The physical factors of an exposed rocky shore and a sheltered estuarine mudflat differ markedly. For each of the facto listed below, briefly describe how they may differ (if at all):

Environmental parameter	Exposed rocky coastline	Estuarine mudflat
Severity of wave action		
Light intensity and quality		
Salinity/ conductivity		
Temperature change (diurnal)		
Substrate/ sediment type		
Oxygen concentration		
Exposure time to air (tide out)		

Sampling Populations

Information about the populations of rare organisms in isolated populations may, in some instances, be collected by direct measure (direct counts and measurements of all the individuals in the population). However, in most cases, populations are too large to be examined directly and they must be sampled in a way that still provides information about them. Most practical exercises in population ecology involve the collection or census of living organisms, with a view to identifying the species and quantifying their abundance and other population features of interest. Sampling techniques must be appropriate to the community being studied and the information you wish to obtain. Some of the common strategies used in ecological sampling, and the situations for which they are best suited, are outlined in the table below. It provides an overview of points to consider when choosing a sampling regime. One must always consider the time and equipment available, the organisms involved, and the impact of the sampling method on the environment. For example, if the organisms involved are very mobile, sampling frames are not appropriate. If it is important not to disturb the organisms, observation alone must be used to gain information.

Method	Equipment and procedure	Information provided and considerations for use
Point sampling 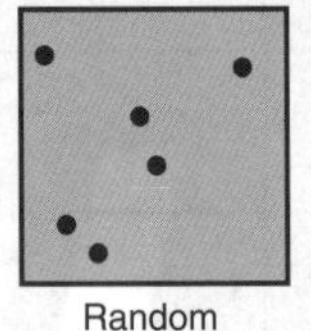 Random 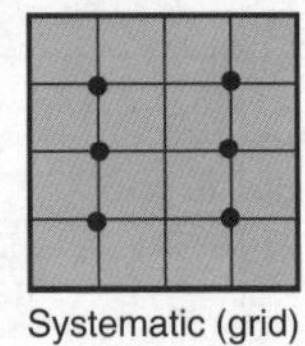 Systematic (grid)	Individual points are chosen on a map (using a grid reference or random numbers applied to a map grid) and the organisms are sampled at those points. Mobile organisms may be sampled using traps or nets.	**Useful for**: Determining species abundance and community composition. If samples are large enough, population characteristics (e.g. age structure, reproductive parameters) can be determined. **Considerations**: Time efficient. Suitable for most organisms. Depending on methods used, disturbance to the environment can be minimised. Species occurring in low abundance may be missed.
Transect sampling 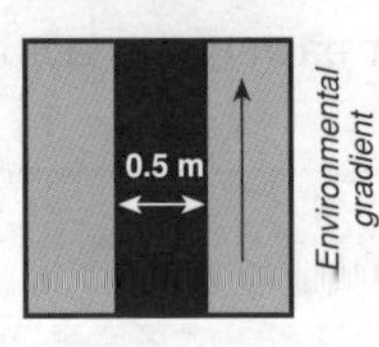 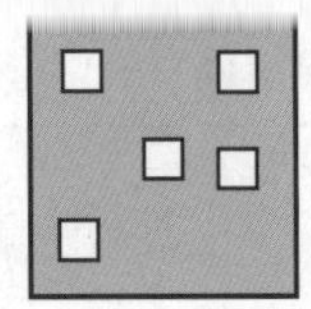 	Lines are drawn across a map and organisms occurring along the line are sampled. **Line transects**: Tape or rope marks the line. The species occurring on the line are recorded (all along the line or, more usually, at regular intervals). Lines can be chosen randomly (left) or may follow an environmental gradient. **Belt transects**: A measured strip is located across the study area to highlight any transitions. Quadrats are used to sample the plants and animals at regular intervals along the belt. Plants and immobile animals are easily recorded. Mobile or cryptic animals need to be trapped or recorded using appropriate methods.	**Useful for**: Well suited to determining changes in community composition along an environmental gradient. When placed randomly, they provide a quick measure of species occurrence. **Considerations for line transects**: Time efficient. Most suitable for plants and immobile or easily caught animals. Disturbance to the environment can be minimised. Species occurring in low abundance may be missed. **Considerations for belt transects**: Time consuming to do well. Most suitable for plants and immobile or easily caught animals. Good chance of recording most or all species. Efforts should be made to minimise disturbance to the environment.
Quadrat sampling 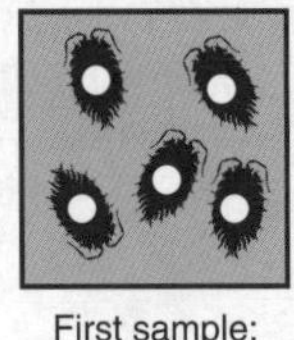	Sampling units or quadrats are placed randomly or in a grid pattern on the sample area. The occurrence of organisms in these squares is noted. Plants and slow moving animals are easily recorded. Rapidly moving or cryptic animals need to be trapped or recorded using appropriate methods.	**Useful for**: Well suited to determining community composition and features of population abundance: species density, frequency of occurrence, percentage cover, and biomass (if harvested). **Considerations**: Time consuming to do well. Most suitable for plants and immobile or easily caught animals. Quadrat size must be appropriate for the organisms being sampled and the information required. Some disturbance if organisms are removed.
Mark and recapture (capture-recapture)  First sample: marked Second sample: proportion recaptured	Animals are captured, marked, and then released. After a suitable time period, the population is resampled. The number of marked animals recaptured in a second sample is recorded as a proportion of the total.	**Useful for**: Determining total population density for highly mobile species in a certain area (e.g. butterflies). Movements of individuals in the population can be tracked (especially when used in conjunction with electronic tracking devices). **Considerations**: Time consuming to do well. Not suitable for immobile species. Population should have a finite boundary. Period between samplings must allow for redistribution of marked animals in the population. Marking should present little disturbance and should not affect behaviour.

1. Briefly explain why we **sample** populations: ______________________

____

2. Suggest what sampling technique would be appropriate for determining the following:

   (a) The percentage cover of a plant species in pasture: ______________________

   (b) The density and age structure of a plankton population: ______________________

   (c) Change in community composition from low to high altitude on a mountain: ______________________

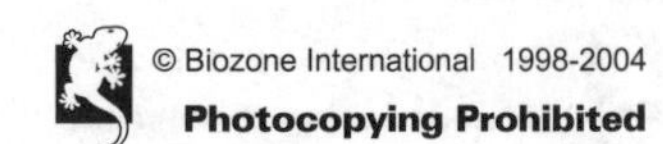

Quadrat Sampling

Quadrat sampling is a method by which organisms in a certain proportion (sample) of the habitat are counted directly. As with all sampling methods, it is used to estimate population parameters when the organisms present are too numerous to count in total. It can be used to estimate population **abundance** (number), **density**, **frequency of occurrence**, and **distribution**. Quadrats may be used without a transect when studying a relatively uniform habitat. In this case, the quadrat positions are chosen randomly using a random number table.

The general procedure is to count all the individuals (or estimate their percentage cover) in a number of quadrats of known size and to use this information to work out the abundance or percentage cover value for the whole area. The number of quadrats used and their size should be appropriate to the type of organism involved (e.g. grass vs tree).

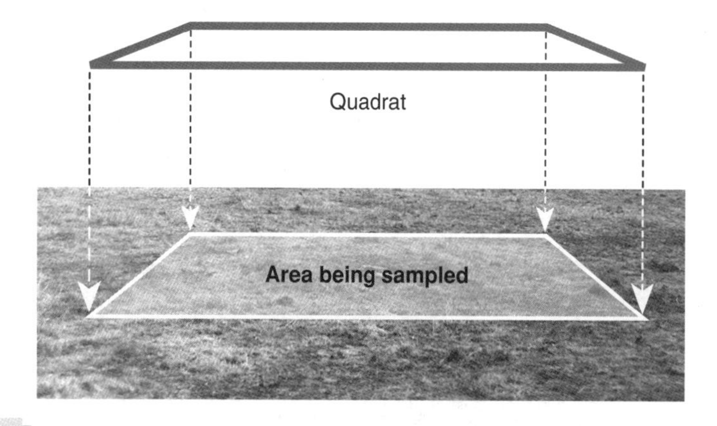

$$\text{Estimated average density} = \frac{\text{Total number of individuals counted}}{\text{Number of quadrats} \times \text{area of each quadrat}}$$

Guidelines for Quadrat Use:

1. The **area of each quadrat** must be known exactly and ideally quadrats should be the same shape. The quadrat does not have to be square (it may be rectangular, hexagonal etc.).
2. **Enough quadrat samples** must be taken to provide results that are representative of the total population.
3. The **population of each quadrat** must be known exactly. Species must be distinguishable from each other, even if they have to be identified at a later date. It has to be decided beforehand what the count procedure will be and how organisms over the quadrat boundary will be counted.
4. The size of the quadrat should be appropriate to the organisms and habitat, e.g. a large size quadrat for trees.
5. The quadrats must be **representative of the whole area**. This is usually achieved by **random sampling** (right).

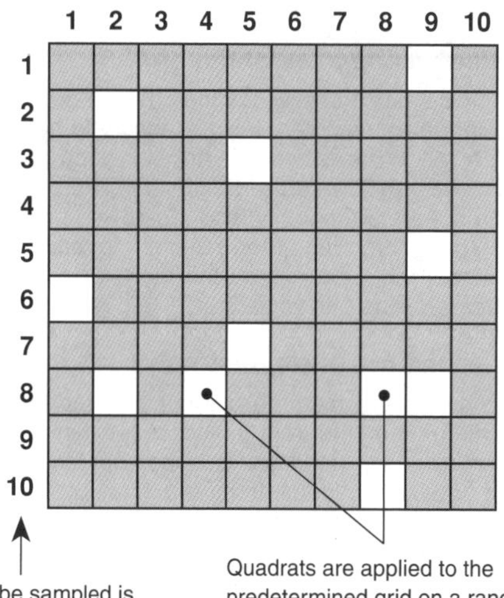

The area to be sampled is divided up into a grid pattern with indexed coordinates

Quadrats are applied to the predetermined grid on a random basis. This can be achieved by using a random number table.

Sampling a centipede population

A researcher by the name of Lloyd (1967) carried out a sampling of centipedes in Wytham Woods, near Oxford in England. A total of 37 hexagon–shaped quadrats were used, with a diameter of 30 cm (see diagram on right). These were arranged in a pattern so that they were all touching each other. Use the data in the diagram to answer the following questions.

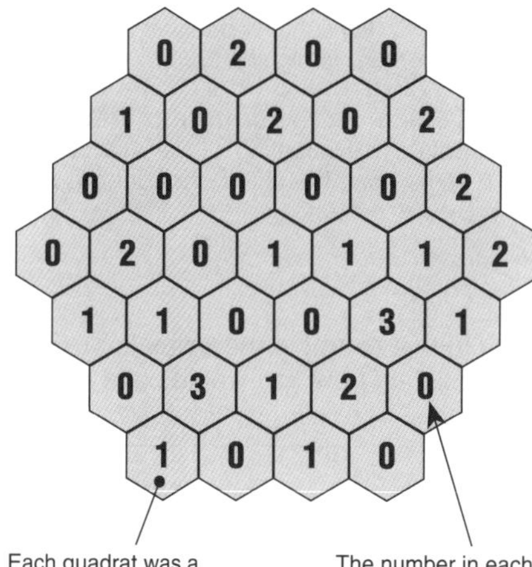

Each quadrat was a hexagon with a diameter of 30 cm and an area of 0.08 square metres.

The number in each hexagon indicates how many centipedes were caught in that quadrat.

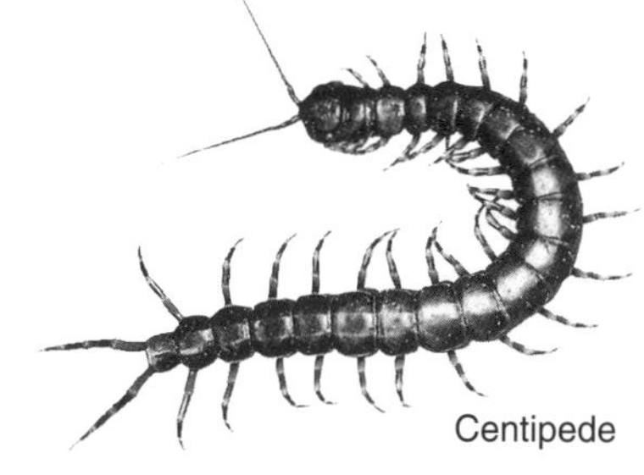

1. Determine the average number of centipedes captured per quadrat:

____

2. Calculate the estimated average density of centipedes per square metre (remember that each quadrat is 0.08 square metres in area):

____

3. Looking at the data for individual quadrats, describe in general terms the distribution of the centipedes in the sample area:

____

____

4. Describe one factor that might account for the distribution pattern:

____

____

Practical Ecology

Sampling a Leaf Litter Population

The diagram on the facing page represents an area of leaf litter from a forest floor with a resident population of organisms. The distribution of four animal species as well as the arrangement of leaf litter is illustrated. Leaf litter comprises leaves and debris that have dropped off trees to form a layer of detritus. This exercise is designed to practice the steps required in planning and carrying out a sampling of a natural population. It is desirable, but not essential, that students work in groups of 2–4.

1. Decide on the sampling method

For the purpose of this exercise, it has been decided that the populations to be investigated are too large to be counted directly and a quadrat sampling method is to be used to estimate the average density of the four animal species as well as that of the leaf litter.

2. Mark out a grid pattern

Use a ruler to mark out 3 cm intervals along each side of the sampling area (area of quadrat = 0.03 x 0.03 m). **Draw lines** between these marks to create a 6 x 6 grid pattern (total area = 0.18 x 0.18 m). This will provide a total of 36 quadrats that can be investigated.

3. Number the axes of the grid

Only a small proportion of the possible quadrat positions are going to be sampled. It is necessary to select the quadrats in a random manner. It is not sufficient to simply guess or choose your own on a 'gut feeling'. The best way to choose the quadrats randomly is to create a numbering system for the grid pattern and then select the quadrats from a random number table. Starting at the *top left hand corner*, **number the columns** and **rows** from 1 to 6 on each axis.

4. Choose quadrats randomly

To select the required number of quadrats randomly, use random numbers from a random number table. The random numbers are used as an index to the grid coordinates. Choose 6 quadrats from the total of 36 using table of random numbers provided for you at the bottom of the facing page. Make a note of which column of random numbers you choose. Each member of your group should choose a different set of random numbers (i.e. different column: A–D) so that you can compare the effectiveness of the sampling method.

Column of random numbers chosen: ______

NOTE: Highlight the boundary of each selected quadrat with coloured pen/highlighter.

5. Decide on the counting criteria

Before the counting of the individuals for each species is carried out, the criteria for counting need to be established.

There may be some problems here. You must decide before sampling begins as to what to do about individuals that are only partly inside the quadrat. Possible answers include:

(a) Only counting individuals if they are completely inside the quadrat.

(b) Only counting individuals that have a clearly defined part of their body inside the quadrat (such as the head).

(c) Allowing for 'half individuals' in the data (e.g. 3.5 snails).

(d) Counting an individual that is inside the quadrat by half or more as one complete individual.

Discuss the merits and problems of the suggestions above with other members of the class (or group). You may even have counting criteria of your own. Think about other factors that could cause problems with your counting.

6. Carry out the sampling

Carefully examine each selected quadrat and **count the number of individuals** of each species present. Record your data in the spaces provided on the facing page.

7. Calculate the population density

Use the combined data TOTALS for the sampled quadrats to estimate the average density for each species by using the formula:

$$\textbf{Density} = \frac{\textbf{Total number in all quadrats sampled}}{\textbf{Number of quadrats sampled X area of a quadrat}}$$

Remember that a total of 6 quadrats are sampled and each has an area of 0.0009 m^2. The density should be expressed as the number of individuals *per square metre (no.* m^{-2}*)*.

Woodlouse: ______ False scorpion: ______

Centipede: ______ Leaf: ______

Springtail: ______

8. (a) In this example the animals are not moving. Describe the problems associated with sampling moving organisms. Explain how you would cope with sampling these same animals if they were really alive and very active:

(b) Carry out a direct count of all four animal species and the leaf litter for the whole sample area (all 36 quadrats). Apply the data from your direct count to the equation given in (7) above to calculate the actual population density (remember that the number of quadrats in this case = 36):

Woodlouse: ______ Centipede: ______ False scorpion: ______ Springtail: ______ Leaf: ______

Compare your estimated population density to the actual population density for each species:

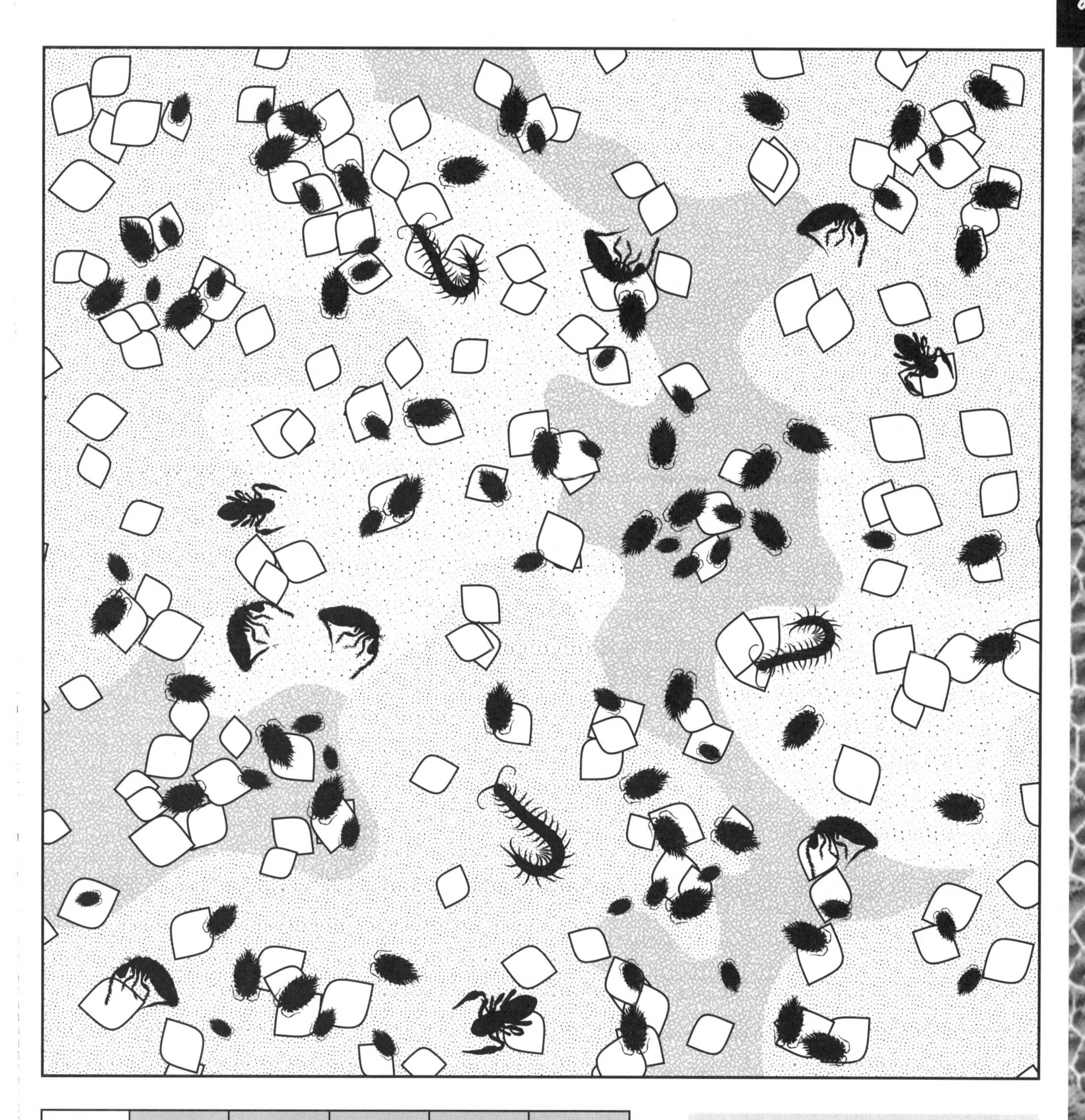

Coordinates for each quadrat	Woodlouse	Centipede	False scorpion	Springtail	Leaf
1:					
2:					
3:					
4:					
5:					
6:					
TOTAL					

Table of random numbers

A	B	C	D
2 2	3 1	6 2	2 2
3 2	1 5	6 3	4 3
3 1	5 6	3 6	6 4
4 6	3 6	1 3	4 5
4 3	4 2	4 5	3 5
5 6	1 4	3 1	1 4

The table above has been adapted from a table of random numbers from a statistics book. Use this table to select quadrats randomly from the grid above. Choose one of the columns (A to D) and use the numbers in that column as an index to the grid. The first digit refers to the row number and the second digit refers to the column number. To locate each of the 6 quadrats, find where the row and column intersect, as shown below:

Example: **5 2** refers to the 5th row and the 2nd column

Transect Sampling

A **transect** is a line placed across a community of organisms. Transects are usually carried out to provide information on the **distribution** of species in the community. This is of particular value in situations where environmental factors change over the sampled distance. This change is called an **environmental gradient** (e.g. up a mountain or across a seashore). The usual practice for small transects is to stretch a string between two markers. The string is marked off in measured distance intervals, and the species at each marked point are noted. The sampling points along the transect may also be used for the siting of quadrats, so that changes in density and community composition can be recorded. Belt transects are essentially a form of continuous quadrat sampling. They provide more information on community composition but can be difficult to carry out. Some transects provide information on the vertical, as well as horizontal, distribution of species (e.g. tree canopies in a forest).

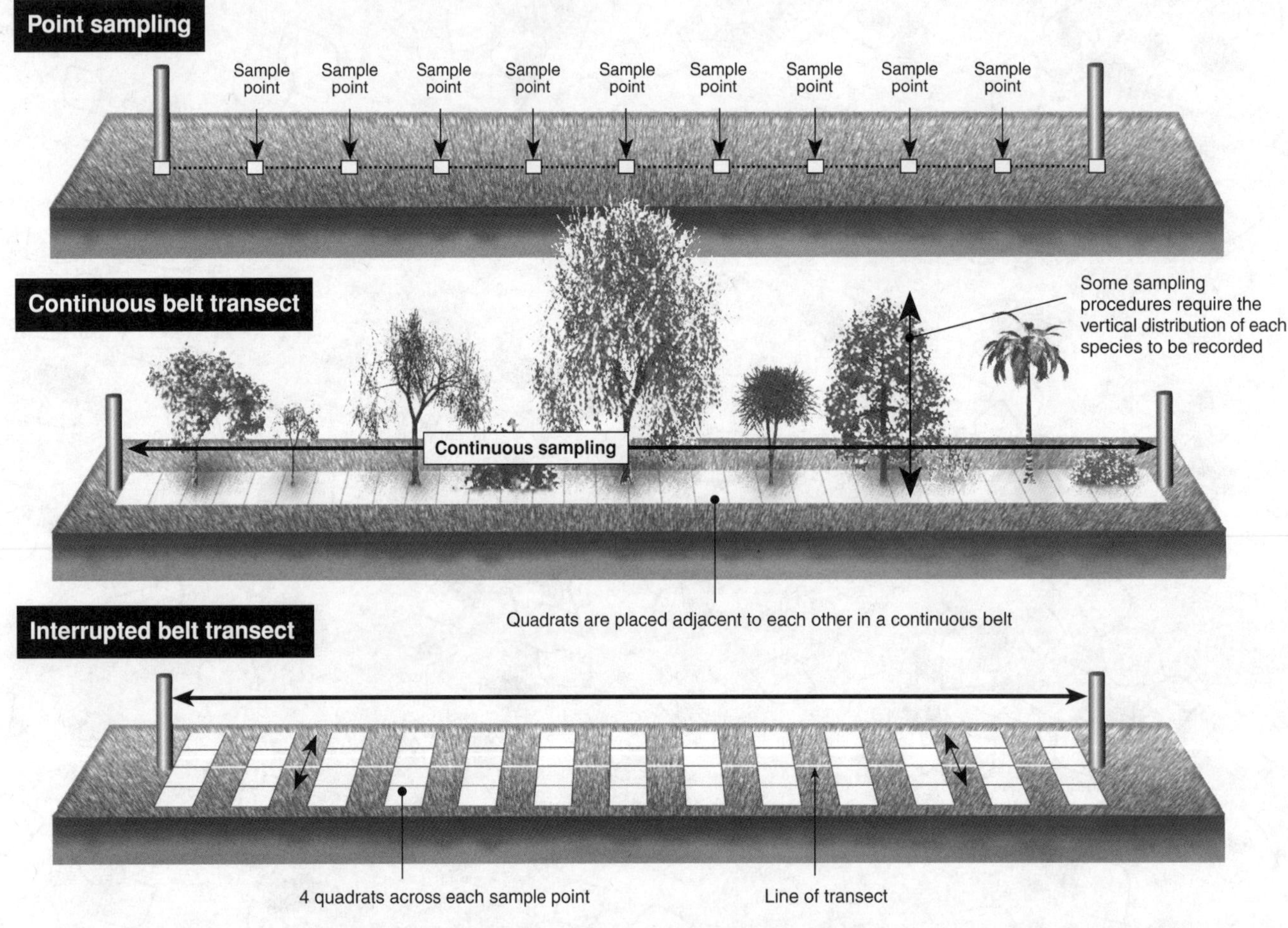

1. Belt transect sampling uses quadrats placed along a line at marked intervals. In contrast, point sampling transects record only the species that are touched or covered by the line at the marked points.

   (a) Describe one disadvantage of belt transects: ______

   (b) Explain why line transects may give an unrealistic sample of the community in question: ______

   (c) Explain how belt transects overcome this problem: ______

   (d) Describe a situation where the use of transects to sample the community would be inappropriate: ______

2. Explain how you could test whether or not a transect sampling interval was sufficient to accurately sample a community: ______

Practical Ecology

Code: DA 2

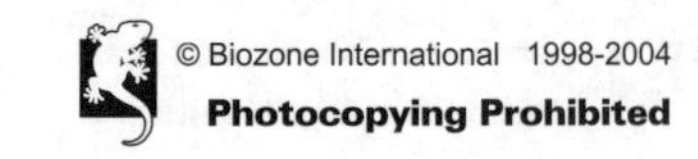

Kite graphs are an ideal way in which to present distributional data from a belt transect (e.g. abundance or percentage cover along an environmental gradient. Usually, they involve plots for more than one species. This makes them good for highlighting probable differences in habitat preference between species. Kite graphs may also be used to show changes in distribution with time (e.g. with daily or seasonal cycles).

3. The data on the right were collected from a rocky shore field trip. Periwinkles from four common species of the genus *Littorina* were sampled in a continuous belt transect from the low water mark, to a height of 10 m above that level. The number of each of the four species in a 1 m^2 quadrat was recorded.

   Plot a **kite graph** of the data for all four species on the grid below. Do not forget to include a scale so that the number at each point on the kite can be calculated.

Field data notebook

Numbers of periwinkles (4 common species) showing vertical distribution on a rocky shore

Height above low water (m)	Periwinkle species: *L. littorea*	*L. saxatalis*	*L. neritoides*	*L. littoralis*
0-1	0	0	0	0
1-2	1	0	0	3
2-3	3	0	0	17
3-4	9	3	0	12
4-5	15	12	0	1
5-6	5	24	0	0
6-7	2	9	2	0
7-8	0	2	11	0
8-9	0	0	47	0
9-10	0	0	59	0

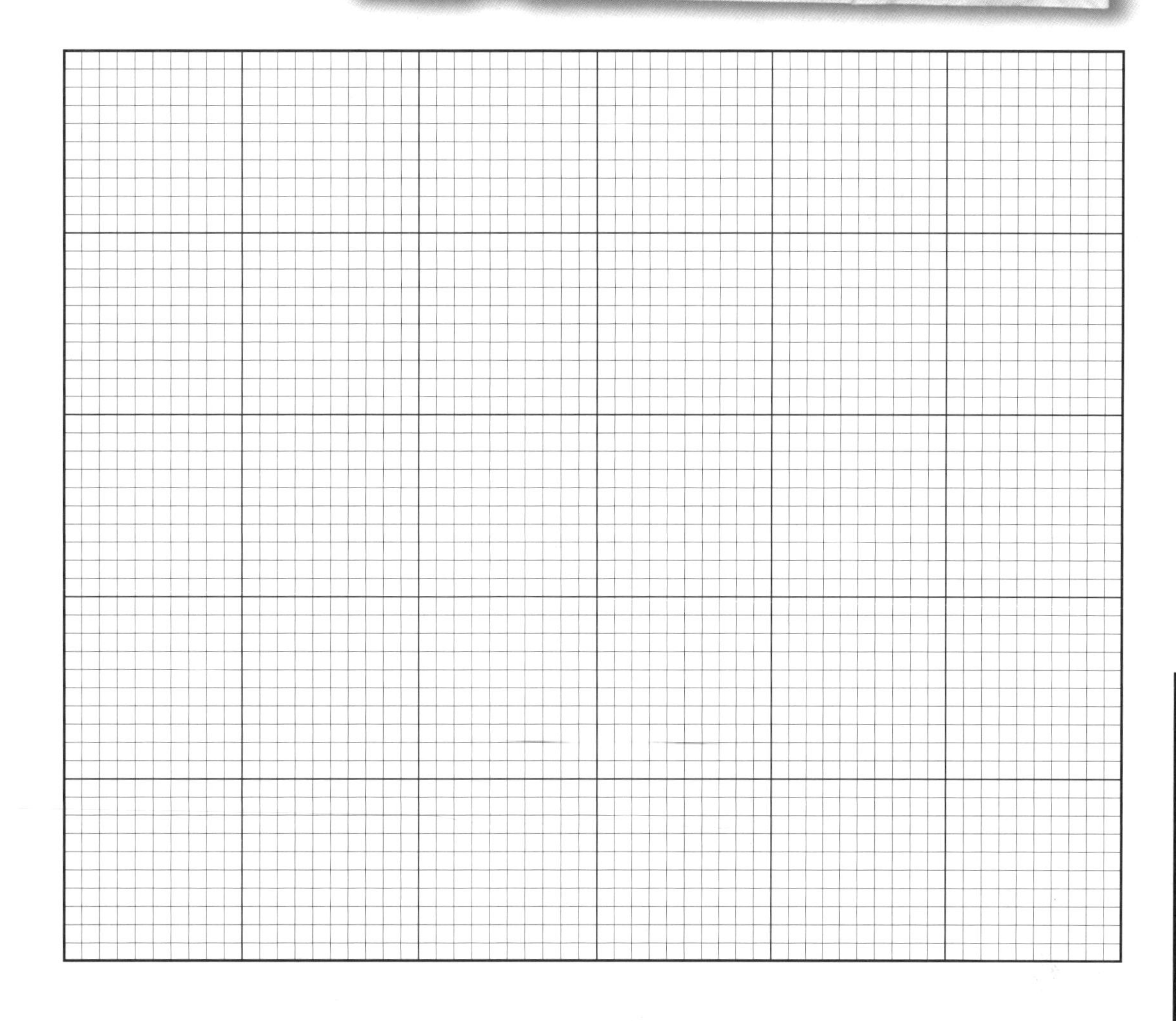

Sampling Animal Populations

Unlike plants, most animals are highly mobile and present special challenges in terms of sampling them **quantitatively** to estimate their distribution and abundance. The equipment available for sampling animals ranges from various types of nets and traps, illustrated on this page, to the more complex electronic devices illustrated opposite.

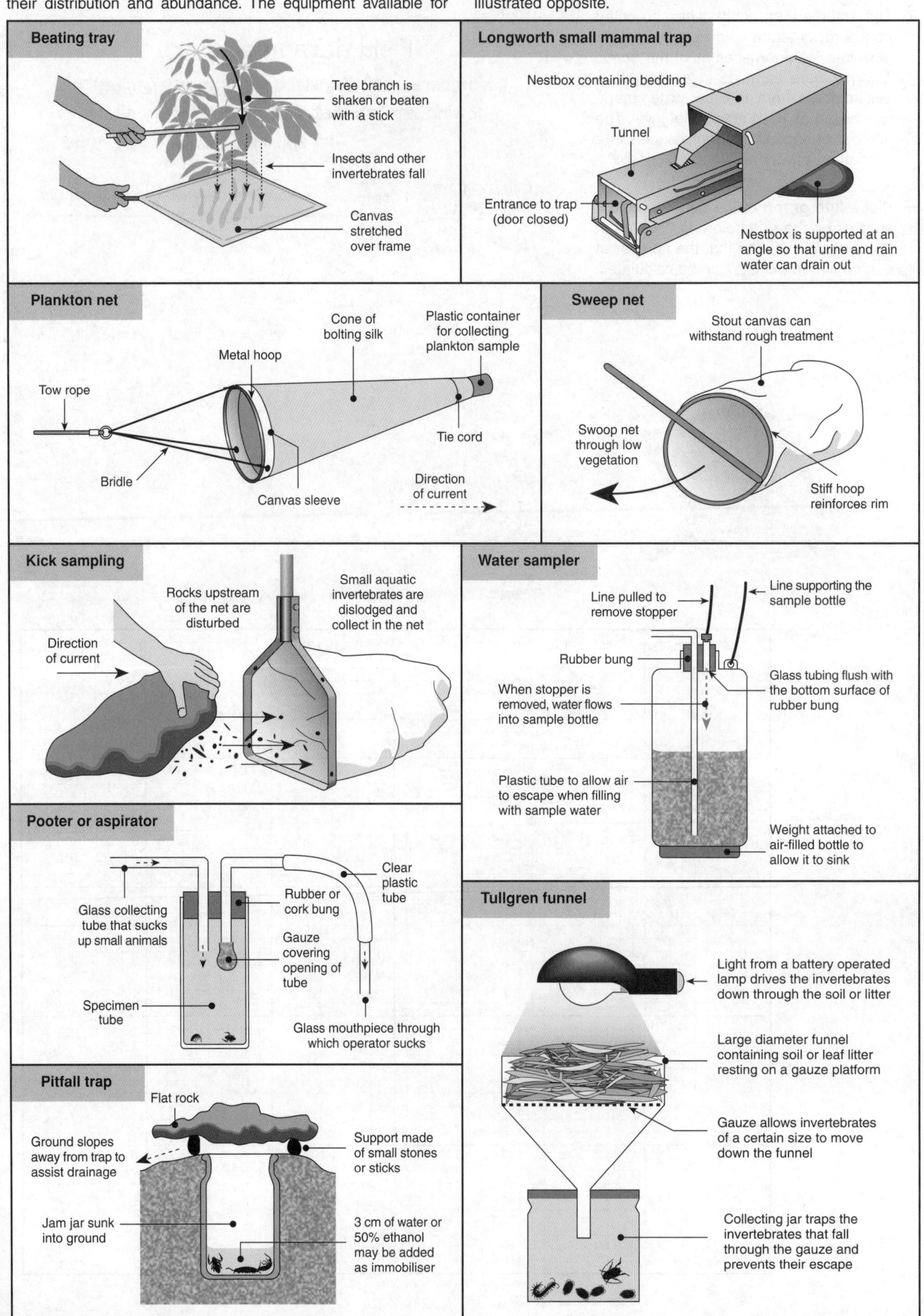

Electro-fishing a stream (Sweden)

Electro-fishing: An effective, but expensive method of sampling larger stream animals (e.g. fish). Wearing a portable battery backpack, the operator walks upstream holding the anode probe and a net. The electrical circuit created by the anode and the stream bed stuns the animals, which are netted and placed in a bucket to recover. After analysis (measurement, species, weights) the animals are released.

Adelie penguins fitted with transmitters

Radio-tracking: A relatively non-invasive method of examining many features of animal populations, including movement, distribution, and habitat use. A small transmitter with an antenna (arrowed) is attached to the animal. The transmitter emits a pulsed signal which is picked up by a receiver. In difficult terrain, a tracking antenna can be used in conjunction with the receiver to accurately fix an animal's position.

Electronic bat detector showing frequency dial

Electronic detection devices: To sample nocturnal, highly mobile species such as bats, electronic devices, such as the bat detector illustrated above, can be used to estimate population density. In this case, the detector is tuned to the particular frequency of the hunting clicks emitted by the bat species of interest. The number of calls recorded per unit time can be used to estimate numbers within a certain area.

1. Describe what each the following types of sampling equipment is used for in a sampling context:

   (a) Kick sampling technique: Provides a semi-quantitative sample of substrate-dwelling stream invertebrates

   (b) Beating tray:

   (c) Longworth small mammal trap:

   (d) Plankton net:

   (e) Sweep net:

   (f) Water sampler:

   (g) Pooter or aspirator:

   (h) Tullgren funnel:

   (i) Pitfall trap:

2. Suggest why pitfall traps are not recommended for estimates of population density:

3. (a) Suggest what influence mesh size might have on the sampling efficiency of a plankton net:

   (b) Explain how this would affect your choice of mesh size when sampling animals in a pond:

Mark and Recapture Sampling

The mark and recapture method of estimating population size is used in the study of animal populations where individuals are highly mobile. It is of no value where animals do not move or move very little. The number of animals caught in each sample must be large enough to be valid. The technique is outlined in the diagram below.

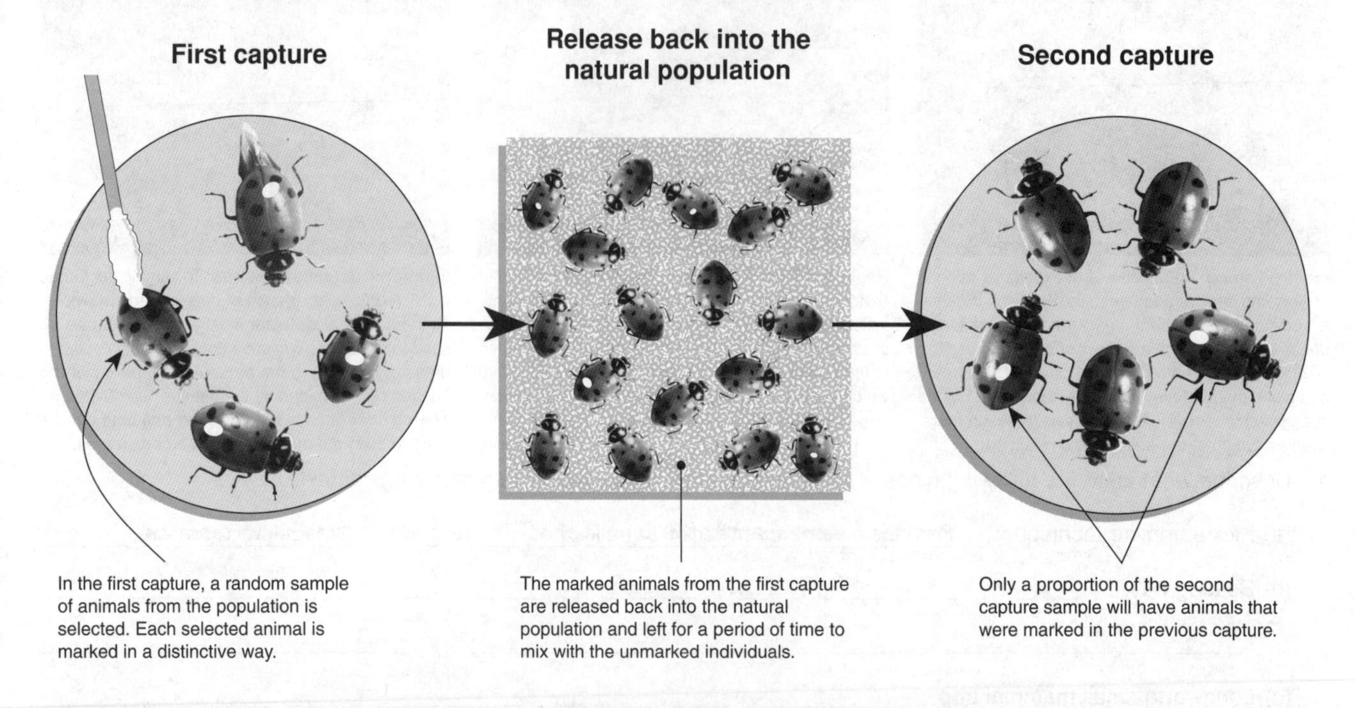

In the first capture, a random sample of animals from the population is selected. Each selected animal is marked in a distinctive way.

The marked animals from the first capture are released back into the natural population and left for a period of time to mix with the unmarked individuals.

Only a proportion of the second capture sample will have animals that were marked in the previous capture.

The Lincoln Index

$$\text{Total population} = \frac{\text{No. of animals in 1st sample (all marked)} \times \text{Total no. of animals in 2nd sample}}{\text{Number of marked animals in the second sample (recaptured)}}$$

The mark and recapture technique comprises a number of simple steps:

1. The population is sampled by capturing as many of the individuals as possible and practical.
2. Each animal is marked in a way to distinguish it from unmarked animals (unique mark for each individual not required).
3. Return the animals to their habitat and leave them for a long enough period for complete mixing with the rest of the population to take place.
4. Take another sample of the population (this does not need to be the same sample size as the first sample, but it does have to be large enough to be valid).
5. Determine the numbers of marked to unmarked animals in this second sample. Use the equation above to estimate the size of the overall population.

1. For this exercise you will need several boxes of matches and a pen. Work in a group of 2-3 students to 'sample' the population of matches in the full box by using the mark and recapture method. Each match will represent one animal.

   (a) Take out 10 matches from the box and mark them on 4 sides with a pen so that you will be able to recognise them from the other unmarked matches later.
   (b) Return the marked matches to the box and shake the box to mix the matches.
   (c) Take a sample of 20 matches from the same box and record the number of marked matches and unmarked matches.
   (d) Determine the total population size by using the equation above.
   (e) Repeat the sampling 4 more times (steps b-d above) and record your results:

	Sample 1	Sample 2	Sample 3	Sample 4	Sample 5
Estimated population					

   (f) Count the actual number of matches in the matchbox : ______________________

   (g) Compare the actual number to your estimates. By how much does it differ?: ______________________

Practical Ecology

2. In 1919 a researcher by the name of Dahl wanted to estimate the number of trout in a Norwegian lake. The trout were subject to fishing so it was important to know how big the population was in order to manage the fish stock. He captured and marked 109 trout in his first sample. A few days later, he caught 177 trout in his second sample, of which 57 were marked. Use the **Lincoln index** (on the previous page) to estimate the total population size:

   Size of 1st sample: ______________________

   Size of 2nd sample: ______________________

   No. marked in 2nd sample: ______________________

   Estimated total population: ______________________

3. Describe some of the problems with the mark and recapture method if the second sampling is:

   (a) Left too long a time before being repeated: ______________________

   (b) Too soon after the first sampling: ______________________

4. Describe two important assumptions being made in this method of sampling that would cause the method to fail if they were not true:

   (a) ______________________

   (b) ______________________

5. Some types of animal would be **unsuitable** for this method of population estimation (i.e. the method would not work).

   (a) Name an animal for which this method of sampling would not be effective: ______________________

   (b) Give a reason for your answer above: ______________________

6. Name three methods for marking animals for mark and recapture sampling. Take into account the possibility of animals shedding their skin, or being difficult to get close to again:

   (a) ______________________

   (b) ______________________

   (c) ______________________

7. Scientists in Australia are involved in a computerised tagging programme for Southern bluefin tuna (a species widely distributed in the southern oceans). Describe the type of information obtainable through this tagging programme:

   ____

Student's t Test Exercise

Data from two flour beetle populations are provided below. The numbers of beetles in each of ten samples were counted. The experimenter wanted to test if the densities of the two populations were significantly different. The exercise below involves manual computation to determine a *t* value. Follow the steps to complete the test. You can also use a spreadsheet programme such as *Microsoft Excel* to do the computations (opposite) or perform the entire analysis for you (see the Teacher Resource Handbook).

1. (a) Complete the calculations to perform the *t* test for these two populations. Some calculations are provided for you.

x (counts)		$x - \bar{x}$ (deviation from the mean)		$(x - \bar{x})^2$ (deviation from mean)^2	
Popn A	Popn B	Popn A	Popn B	Popn A	Popn B
465	310	9.3	–10.6	86.5	112.4
475	310	19.3	–10.6	372.5	112.4
415	290				
480	355				
436	350				
435	335				
445	295				
460	315				
471	316				
475	330				
$n_A = 10$	$n_B = 10$			$\Sigma(x - \bar{x})^2$	$\Sigma(x - \bar{x})^2$

The number of samples in each data set

The sum of each column is called the sum of squares

(b) The variance for population A: $s^2_A =$

The variance for population B: $s^2_B =$

(c) The difference between the population means

$(\bar{x}_A - \bar{x}_B) =$

(d) t (calculated) =

(e) Determine degrees of freedom (d.f.)

d.f. $(n_A + n_B - 2) =$

(f) $P =$

t (critical value) =

(g) Your decision is:

Step 1: Summary statistics

Tabulate the data as shown in the first 2 columns of the table (left). Calculate the mean and give the *n* value for each data set. Compute the standard deviation if you wish.

Popn A $\bar{x}_A = 455.7$, $n_A = 10$, $s_A = 21.76$

Popn B $\bar{x}_B = 320.6$, $n_B = 10$, $s_B = 21.64$

Step 2: State your null hypothesis

Step 3: Decide if your test is one or two-tailed

Calculating the t value

Step 4a: Calculate sums of squares

Complete the computations outlined in the table left. The sum of each of the final two columns (left) is called the sum of squares.

Step 4b: Calculate the variances

Calculate the variance (s^2) for each set of data. This is the sum of squares divided by $n - 1$ (number of samples in each data set – 1). In this case the *n* values are the same, but they need not be.

$$s^2_A = \frac{\Sigma(x - \bar{x})^2_{(A)}}{n_A - 1} \qquad s^2_B = \frac{\Sigma(x - \bar{x})^2_{(B)}}{n_B - 1}$$

Step 4c: Difference between means

Calculate the *actual* difference between the means $(\bar{x}_A - \bar{x}_B)$

Step 4d: Calculate t

Calculate the *t* value. Ask for assistance if you find interpreting the lower part of the equation difficult

$$t = \frac{(\bar{x}_A - \bar{x}_B)}{\sqrt{\frac{s^2_A}{n_A} + \frac{s^2_B}{n_B}}}$$

Step 4e: Determine degrees of freedom

Degrees of freedom (d.f.) are defined by the number of samples (e.g. counts) taken: d.f. = $n_A + n_B - 2$ where n_A and n_B are the number of counts in each of populations A and B.

Step 5: Consult the t table

Consult the *t*-tables (opposite page) for the critical *t* value at the appropriate degrees of freedom and the acceptable probability level (e.g. $P = 0.05$).

Step 5a: Make your decision

Make your decision whether or not to reject H_0. If your *t* value is large enough you may be able to reject H_0 at a lower *P* value (e.g. 0.001), increasing your confidence in the alternative hypothesis.

2. The previous example (manual calculation for two beetle populations) is outlined below in a spreadsheet (created in *Microsoft Excel*). The spreadsheet has been shown in a special mode with the formulae displayed. Normally, when using a spreadsheet, the calculated values will appear as the calculation is completed (entered) and a formula is visible only when you click into an individual cell. When setting up a spreadsheet, you can arrange your calculating cells wherever you wish. What is important is that you accurately identify the cells being used for each calculation. Also provided below is a summary of the spreadsheet notations used and a table of critical values of *t* at different levels of *P*. Note that, for brevity, only some probability values have been shown. To be significant at the appropriate level of probability, calculated values must be greater than those in the table for the appropriate degrees of freedom.

   (a) Using the data in question 1, set up a spreadsheet as indicated below to calculate *t*. Save your spreadsheet. Print it out and staple the print-out into your manual.

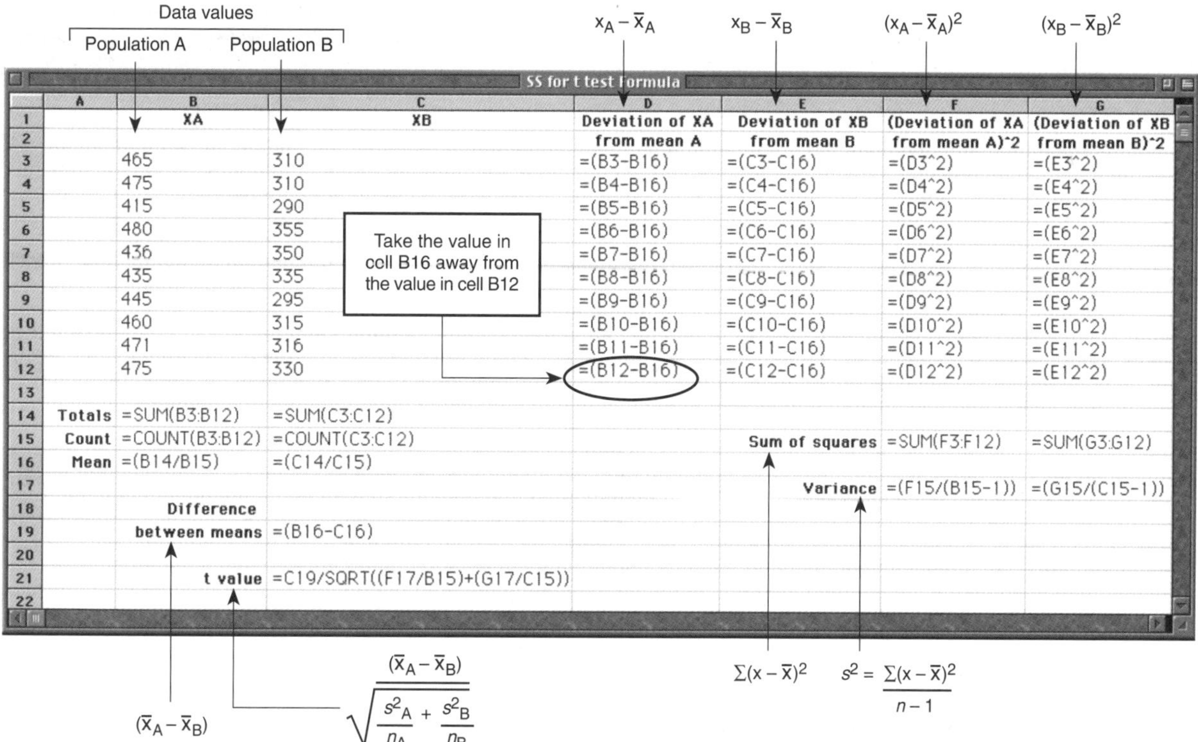

	A	B	C	D	E	F	G
1		XA	XB	Deviation of XA	Deviation of XB	(Deviation of XA	(Deviation of XB
2				from mean A	from mean B	from mean A)^2	from mean B)^2
3		465	310	=(B3-B16)	=(C3-C16)	=(D3^2)	=(E3^2)
4		475	310	=(B4-B16)	=(C4-C16)	=(D4^2)	=(E4^2)
5		415	290	=(B5-B16)	=(C5-C16)	=(D5^2)	=(E5^2)
6		480	355	=(B6-B16)	=(C6-C16)	=(D6^2)	=(E6^2)
7		436	350	=(B7-B16)	=(C7-C16)	=(D7^2)	=(E7^2)
8		435	335	=(B8-B16)	=(C8-C16)	=(D8^2)	=(E8^2)
9		445	295	=(B9-B16)	=(C9-C16)	=(D9^2)	=(E9^2)
10		460	315	=(B10-B16)	=(C10-C16)	=(D10^2)	=(E10^2)
11		471	316	=(B11-B16)	=(C11-C16)	=(D11^2)	=(E11^2)
12		475	330	=(B12-B16)	=(C12-C16)	=(D12^2)	=(E12^2)
13							
14	Totals	=SUM(B3:B12)	=SUM(C3:C12)				
15	Count	=COUNT(B3:B12)	=COUNT(C3:C12)		Sum of squares	=SUM(F3:F12)	=SUM(G3:G12)
16	Mean	=(B14/B15)	=(C14/C15)				
17					Variance	=(F15/(B15-1))	=(G15/(C15-1))
18		Difference					
19		between means	=(B16-C16)				
20							
21		t value	=C19/SQRT((F17/B15)+(G17/C15))				
22							

Notation	Meaning
Columns and rows	Columns are denoted A, B, C ... at the top of the spreadsheet, rows are 1, 2, 3, ... on the left. Using this notation a cell can be located e.g. C3
=	An "equals" sign *before* other entries in a cell denotes a formula.
( )	Parentheses are used to group together terms for a single calculation. This is important for larger calculations (see cell C21 above)
C3:C12	Cell locations are separated by a colon. C3:C12 means "every cell between and including C3 and C12"
SUM	Denotes that what follows is added up. =SUM(C3:C12) means "add up the values in cells C3 down to C12"
COUNT	Denotes that the number of values is counted =COUNT(C3:C12) means "count up the number of values in cells C3 down to C12"
SQRT	Denotes "take the square root of what follows"
^2	Denotes an exponent e.g. x^2 means that value x is squared.

Above is a table explaining some of the spreadsheet notations used for the calculation of the *t* value for the exercise on the previous page. It is not meant to be an exhaustive list for all spreadsheet work, but it should help you to become familiar with some of the terms and how they are used. This list applies to *Microsoft Excel*. Different spreadsheets may use different notations. These will be described in the spreadsheet manual.

Table of critical values of *t* at different levels of *P*.

Degrees of freedom	Level of Probability		
	0.05	0.01	0.001
1	12.71	63.66	636.6
2	4.303	9.925	31.60
3	3.182	5.841	12.92
4	2.776	4.604	8.610
5	2.571	4.032	6.869
6	2.447	3.707	5.959
7	2.365	3.499	5.408
8	2.306	3.355	5.041
9	2.262	3.250	4.781
10	2.228	3.169	4.587
11	2.201	3.106	4.437
12	2.179	3.055	4.318
13	2.160	3.012	4.221
14	2.145	2.977	4.140
15	2.131	2.947	4.073
16	2.120	2.921	4.015
17	2.110	2.898	3.965
18	2.101	2.878	3.922
19	2.093	2.861	3.883
20	2.086	2.845	3.850

   (b) Save your spreadsheet under a different name and enter the following new data values for population B: **425, 478, 428, 465, 439, 475, 469, 445, 421, 438**. Notice that, as you enter the new values, the calculations are updated over the entire spreadsheet. Re-run the *t*-test using the new *t* value. State your decision for the two populations now:

   New *t* value: ____________________ Decision on null hypothesis (delete one): Accept/Reject H_0

Using Chi-Squared in Ecology

The following exercise illustrates the use of chi-squared (χ^2) in ecological studies of habitat preference. In the first example, it is used for determining if the flat periwinkle *(Littorina littoralis)* shows significant preference for any of the four species of seaweeds with which it is found. Using quadrats, the numbers of periwinkles associated with each seaweed species were recorded. The data from this investigation are provided for you in Table 1. In the second example, the results of an investigation into habitat preference in woodlice (also called pillbugs, sowbugs, or slaters) are presented for analysis (Table 2).

1. (a) State your null hypothesis for this investigation (H_0):

____

____

____

(b) State the alternative hypothesis (H_A): ______________________

____

____

Table 1: Number of periwinkles associated with different seaweed species

Seaweed species	Number of periwinkles
Spiral wrack	9
Bladder wrack	28
Toothed wrack	19
Knotted wrack	64

2. Use the chi-squared (χ^2) test to determine if the differences observed between the samples are significant or if they can be attributed to chance alone. The table of critical values of χ^2 at different *P* values is available in *Advanced biology AS*", on the Teacher Resource Handbook on CD-ROM, or in any standard biostatistics book.

(a) Enter the observed values (no. of periwinkles) and complete the table to calculate the χ^2 value:

Category	O	E	O — E	$(O - E)^2$	$\frac{(O-E)^2}{E}$
Spiral wrack					
Bladder wrack					
Toothed wrack					
Knotted wrack					
	Σ				Σ

(b) Calculate χ^2 value using the equation:

$$\chi^2 = \sum \frac{(O-E)^2}{E} \qquad \chi^2 = \_\_\_\_\_\_$$

(c) Calculate the degrees of freedom: _______

(d) Using the appendix, state the *P* value corresponding to your calculated χ^2 value:

____

(e) State whether or not you reject your null hypothesis:

reject H_0 / do not reject H_0 *(circle one)*

3. Students carried out an investigation into habitat preference in woodlice. In particular, they were wanting to know if the woodlice preferred a humid atmosphere to a dry one, as this may play a part in their choice of habitat. They designed a simple investigation to test this idea. The woodlice were randomly placed into a choice chamber for 5 minutes where they could choose between dry and humid conditions (atmosphere). The investigation consisted of five trials with ten woodlice used in each trial. Their results are shown on Table 2 (right):

Table 2: Number of woodlice in each habitat

	Atmosphere	
Trial	Dry	Humid
1	2	8
2	3	7
3	4	6
4	1	9
5	5	5

(a) State the null and alternative hypotheses (H_0 and H_A):

____

____

Use a separate piece of paper to calculate the chi-squared value and summarise your answers below:

(b) Calculate the χ^2 value: ____________

(c) Calculate the degrees of freedom and state the *P* value corresponding to your calculated χ^2 value: ____________

(d) State whether or not you reject your null hypothesis (H_0): reject H_0 / do not reject H_0 *(circle one)*

Practical Ecology

Code: DA 3

Comparing More Than Two Groups

The Student's *t* test is limited to comparing two groups of data. To compare more than two groups at once you need to use a test that is appropriate to this aim. One such test, appropriate for normally distributed data, is an analysis of variance (**ANOVA**), as described in the next activity. A good place to start with any such an analysis is to plot your data, together with some measure of the reliability of the statistic you calculate as an indicator of the true population parameter, For normally distributed data, this is likely to be the mean and the 95% confidence interval (95% CI). See *Advanced Biology AS* for an introduction to this statistic. In the example described below, students recorded the survival of weevil larvae on five different pasture types and calculated descriptive statistics for the data. After you have worked through the analysis, you should be able to enter your own data and calculate descriptive statistics using *Microsoft Excel*. The plot and full analysis of these data are presented in the next activity.

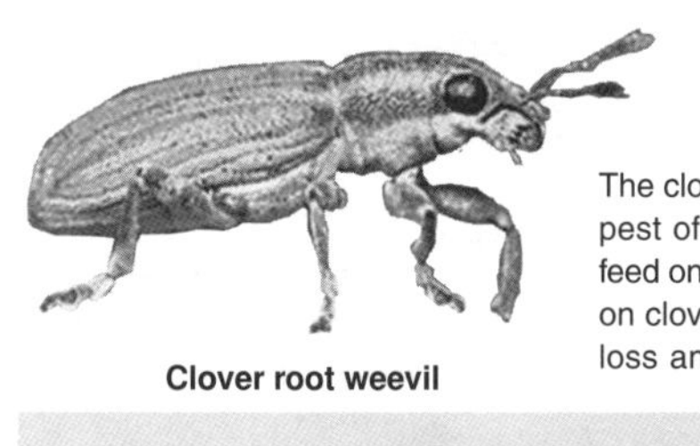

Clover root weevil

The clover root weevil (*Sitona lepidus*) is a pest of white clover pastures. The adults feed on clover leaves, while the larvae feed on clover nodules and roots, causing root loss and a reduction in nitrogen fixation.

50 weevils	50 weevils	50 weevils	50 weevils	50 weevils
Rye grass $n = 6$	Fescue $n = 6$	White clover $n = 6$	Red clover $n = 6$	Chicory $n = 6$

Comparing the Means of More Than Two Experimental Groups

Research has indicated that different pastures have different susceptibility to infestation by a pest insect, the clover root weevil (left). Armed with this knowledge, two students decided to investigate the effect of pasture type on the survival of clover root weevils. The students chose five pasture types, and recorded the number of weevil larvae (from a total of 50) surviving in each pasture type after a period of 14 days. Six experimental pots were set up for each pasture type (n = 6). Their results and the first part of their analysis (calculating the descriptive statistics) are presented in this activity.

1 Calculating Descriptive Statistics

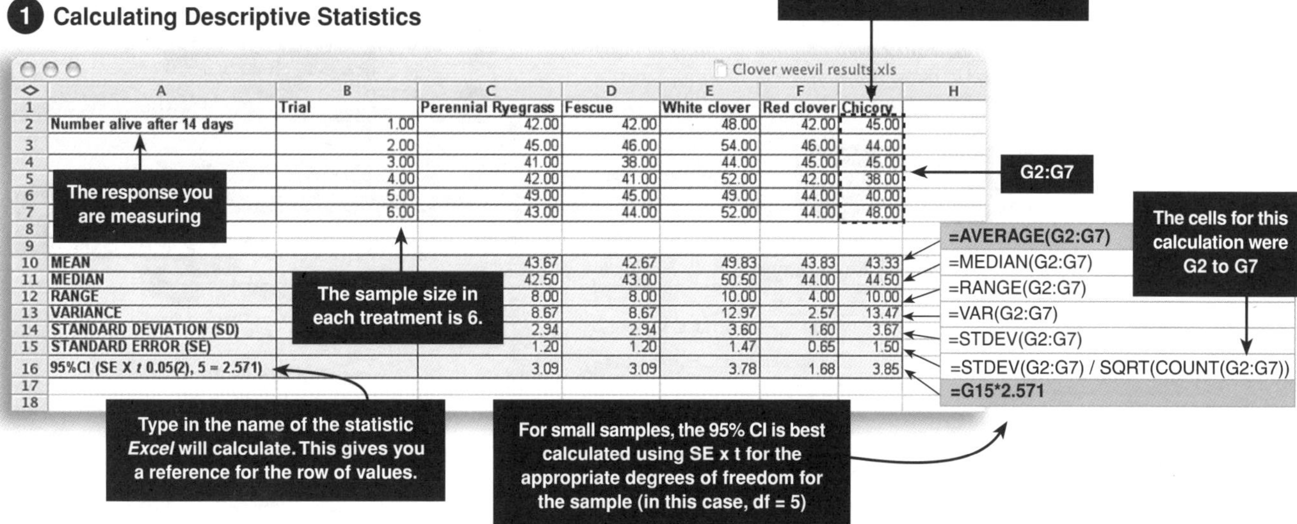

	A	B	C	D	E	F	G	H
1		Trial	Perennial Ryegrass	Fescue	White clover	Red clover	Chicory	
2	Number alive after 14 days	1.00	42.00	42.00	48.00	42.00	45.00	
3		2.00	45.00	46.00	54.00	46.00	44.00	
4		3.00	41.00	38.00	44.00	45.00	45.00	
5		4.00	42.00	41.00	52.00	42.00	38.00	
6		5.00	49.00	45.00	49.00	44.00	40.00	
7		6.00	43.00	44.00	52.00	44.00	48.00	
8								
9								
10	MEAN		43.67	42.67	49.83	43.83	43.33	=AVERAGE(G2:G7)
11	MEDIAN		42.50	43.00	50.50	44.00	44.50	=MEDIAN(G2:G7)
12	RANGE		8.00	8.00	10.00	4.00	10.00	=RANGE(G2:G7)
13	VARIANCE		8.67	8.67	12.97	2.57	13.47	=VAR(G2:G7)
14	STANDARD DEVIATION (SD)		2.94	2.94	3.60	1.60	3.67	=STDEV(G2:G7)
15	STANDARD ERROR (SE)		1.20	1.20	1.47	0.65	1.50	=STDEV(G2:G7) / SQRT(COUNT(G2:G7))
16	95%CI (SE X t 0.05(2), 5 = 2.571)		3.09	3.09	3.78	1.68	3.85	=G15*2.571
17								
18								

	J	K	L	M	N
	nial Ryegrass	Fescue	White clover	Red clover	Chicory
Mean	43.67	42.67	49.83	43.83	43.33
95% CI	3.09	3.09	3.78	1.68	3.85

Enter the values for means and 95% CI in columns under the categories.

2 To plot a graph in *Excel*, you will need to enter the values for the data you want to plot in a format that *Excel* can use. To do this, enter the values in columns under each category (left). Each column will have two entries: mean and 95% CI. In this case, we want to plot the mean survival of weevils in different swards and add the 95% CI as error bars. You are now ready to plot the data set.

1. Identify the number of treatments in this experimental design: ______________________

2. Identify the independent variable for this experiment: ______________________

3. Identify the dependent variable for this experiment: ______________________

4. Identify the type of graph you would use to plot this data set and explain your choice: ______________________

____

5. Explain what the 95% confidence interval tells you about the means for each treatment: ______________________

____

____

Analysis of Variance

Analysis of variance (or **ANOVA**) could be considered beyond the scope of the statistical analyses you would do at school. However, using *Excel* (previous activity and below), it is not difficult and is an appropriate test for a number of situations in experimental biology. It is described below for the data on weevil survival described in the previous activity. A plot of the data is also described as an aid to interpreting the statistical analysis. An ANOVA may also be appropriate when the independent variable is continuous, but in these instances a regression is likely to be a more suitable analysis (see *Advanced Biology AS*). After you have worked through this activity you should be able to:

- Use *Excel* to plot data appropriately with error bars.
- Interpret the graphically presented data and reach tentative conclusions about the findings of the experiment.
- Support your conclusions with an ANOVA, in *Microsoft Excel*, to test the significance of differences in the sample data.

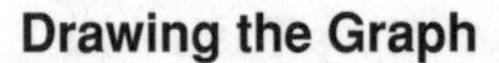

Drawing the Graph

Recall the design of this experiment. The independent variable is categorical, so the correct graph type is a column chart. To plot the graph, select the row of mean values (including column headings) from the small table of results you constructed (screen, right).

1. From the menu bar choose: **Insert** > **Chart** > **Column**. This is **STEP 1** in the Chart Wizard. Click **Next**.
2. At **STEP 2** in Chart Wizard: The data range you have selected (the source data for the chart) will appear in the "Data range" window. Click **Next**.
3. At **STEP 3** in Chart Wizard: You have the option to add a title, labels for your X and Y axes, turn off gridlines, and add (or remove) a legend (a legend is useful when you have two or more columns, such as males and females, for each treatment). When you have added all the information you want, click **Next**.
4. At **STEP 4** in Chart Wizard: Specify the chart location. It should appear in the active sheet by default. Click on the chart and move it to reveal the data.

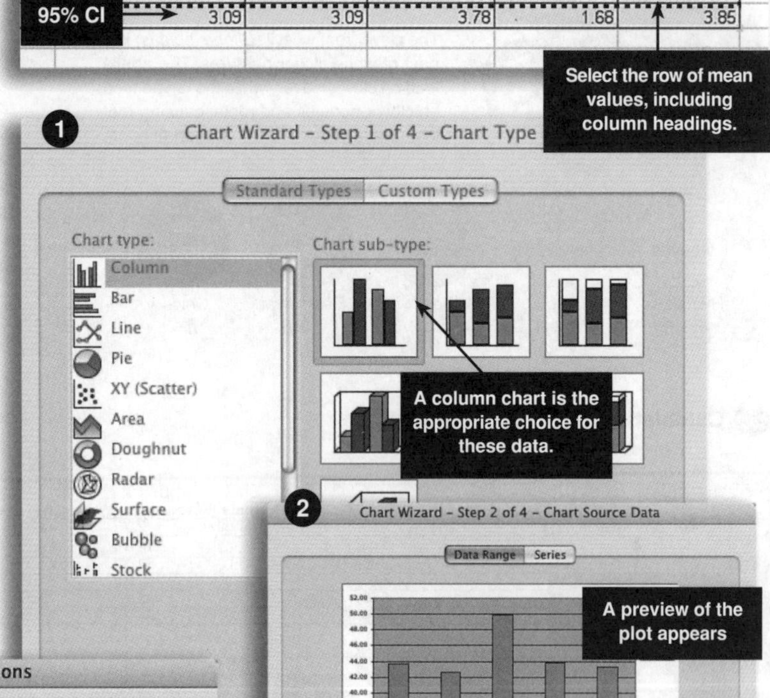

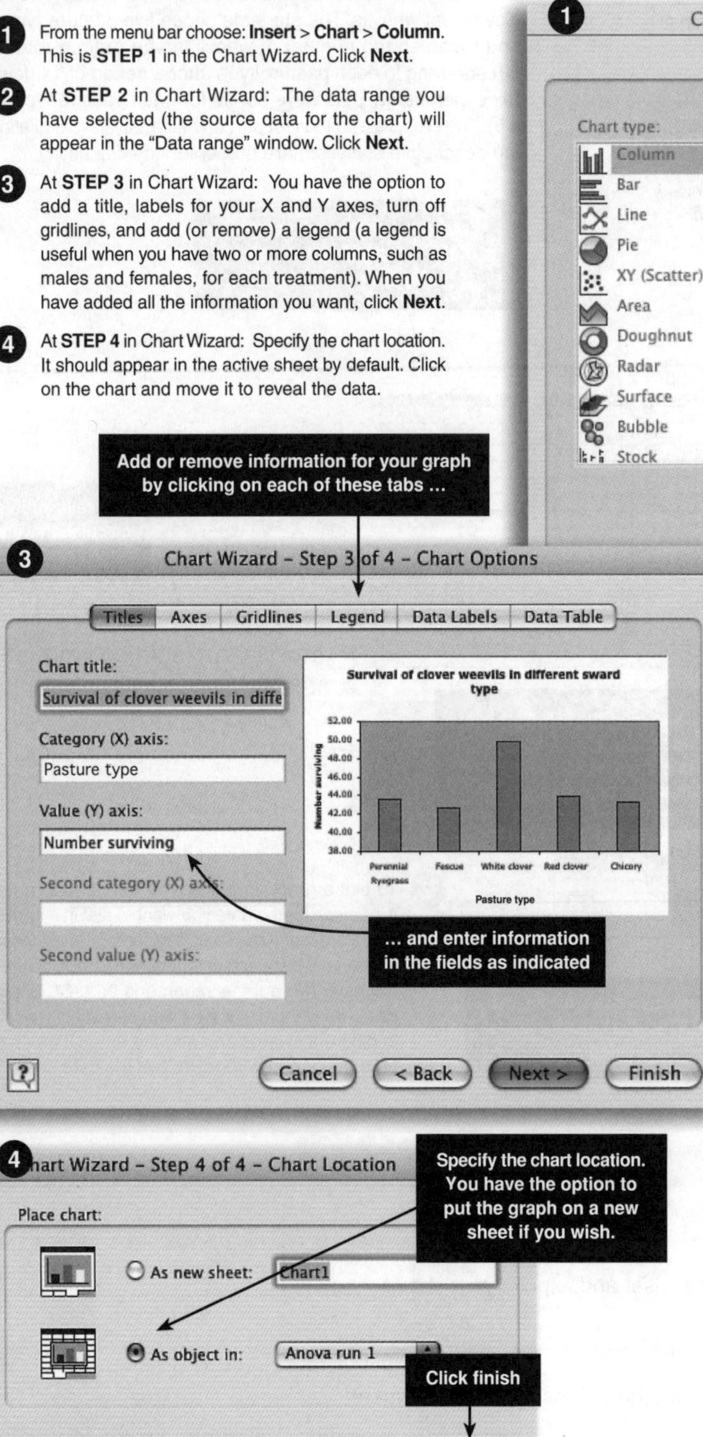

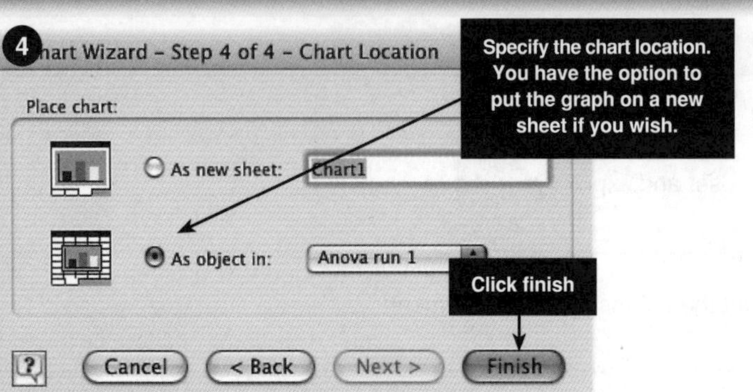

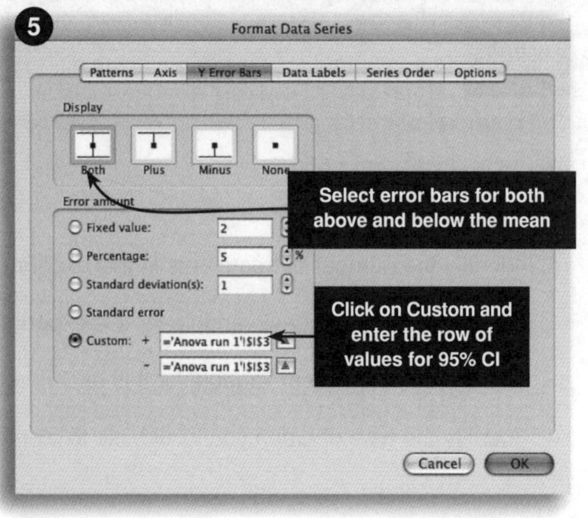

5. **Adding error bars:** A chart will appear. **Right click** (Ctrl-click on Mac) on any part of any column and choose **Format data series** (below). To add error bars, select the **Y error bars** tab, and click on the symbol that shows **Display both**. Click on **Custom**, and use the data selection window to select the row of 95% CI data. Click OK.

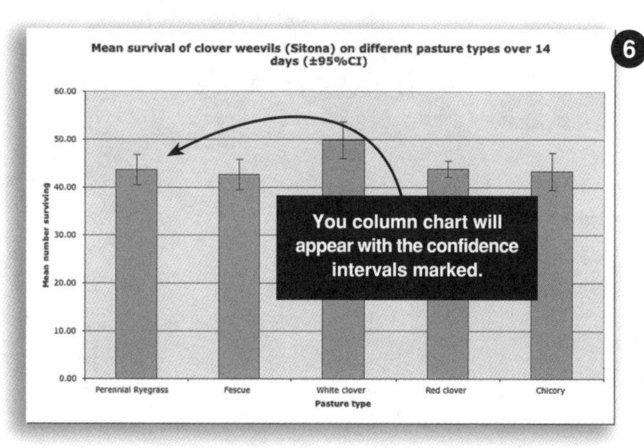

6 After you have added error bars using the "Format data series" option, your chart will plot with error bars. These error bars represent the 95% confidence intervals for the mean of each treatment. The completed plot gives you a visual representation of your data and an immediate idea as to how confident you can be about the differences between the five treatments. You are now ready to run your analysis of variance.

About ANOVA

ANOVA is a test for normally distributed data, which can determine whether or not there is a real (statistically significant) difference between two or more sample means (an ANOVA for two means is a *t* test). ANOVA does this by accounting for the amount of variability (in the measured variable) within a group and comparing it to the amount of variability between the groups (treatments). It is an ideal test when you are looking for the effect of a particular treatment on some kind of biological response, e.g. plant growth with different fertilisers. ANOVA is a very useful test for comparing multiple samples because, like the *t* test, it does not require large sample sizes.

Running the ANOVA

ANOVA is part of the *Excel* "Data Analysis Toolpak", which is part of normal *Excel*, but is not always installed. If there is no Data Analysis option on the Tools menu, you need to run *Excel* from the original installation disk to install the "Analysis Toolpak add-in".

From the Tools menu select Data Analysis and **ANOVA single factor**. This brings up the ANOVA dialogue box (numbered points 1 and 2). Here we are concerned with the effect of one variable on another, so the ANOVA is single factor (also called one way ANOVA).

Click in the **Output Range** box and click on a free cell on the worksheet which will become the left cell of the results table. Click OK.

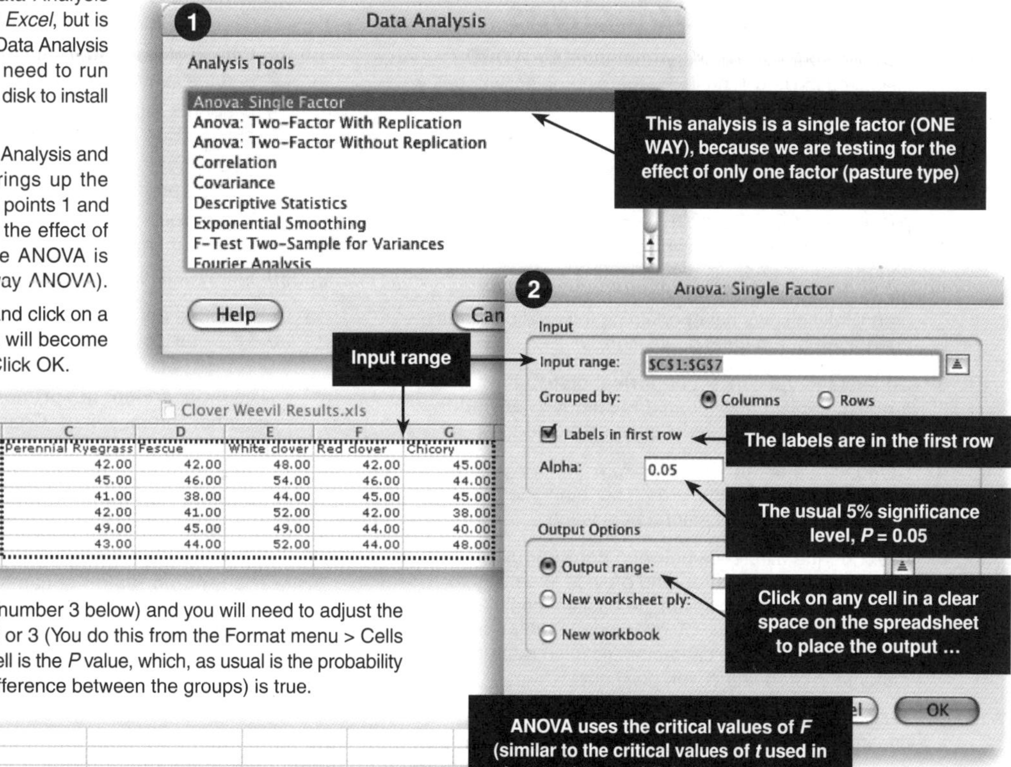

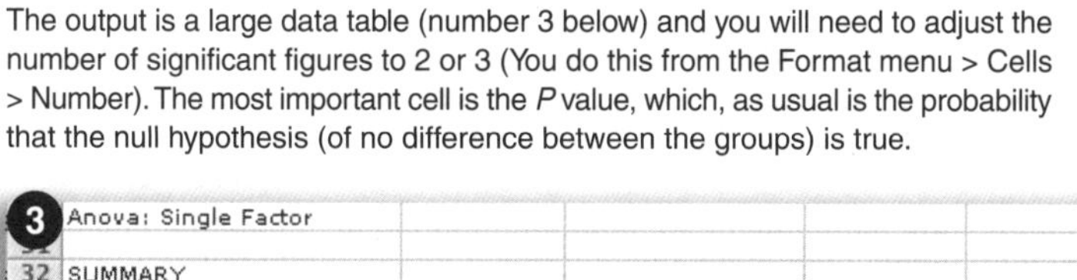

Clover Weevil Results.xls

A	B	C	D	E	F	G
	Trial	Perennial Ryegrass	Fescue	White clover	Red clover	Chicory
Number alive after 14 days	1.00	42.00	42.00	48.00	42.00	45.00
	2.00	45.00	46.00	54.00	46.00	44.00
	3.00	41.00	38.00	44.00	45.00	45.00
	4.00	42.00	41.00	52.00	42.00	38.00
	5.00	49.00	45.00	49.00	44.00	40.00
	6.00	43.00	44.00	52.00	44.00	48.00

The output is a large data table (number 3 below) and you will need to adjust the number of significant figures to 2 or 3 (You do this from the Format menu > Cells > Number). The most important cell is the *P* value, which, as usual is the probability that the null hypothesis (of no difference between the groups) is true.

3 Anova: Single Factor

32	SUMMARY						
33	*Groups*	*Count*	*Sum*	*Average*	*Variance*		
34	Perennial Ryegrass	6.000	262.000	43.667	8.667		
35	Fescue	6.000	256.000	42.667	8.667		
36	White clover	6.000	299.000	49.833	12.967		
37	Red clover	6.000	263.000	43.833	2.567		
38	Chicory	6.000	260.000	43.333	13.467		
39							
40							
41	ANOVA						
42	*Source of Variation*	*SS*	*df*	*MS*	*F*	*P-value*	*F crit*
43	Between Groups	205.000	4.000	51.250	5.531	0.002	2.759
44	Within Groups	231.667	25.000	9.267			
45							
46	Total	436.667	29.000				

ANOVA uses the critical values of *F* (similar to the critical values of *t* used in the *t* test) to establish the significance of the difference. However, it is the *P* value that is important to you.

ANOVA looks at the variation within the groups and compares it to the variation between groups.

$P < 0.05$; at least one of the groups is significantly different from the others. ANOVA cannot tell you which one.

Important note

ANOVA can tell you if there is a significant difference between at least one of your groups (or treatments) and the others, but it cannot tell you which one it is. To find this out, you need a more complicated analysis, called a multiple range test. However, if you have plotted your data with confidence intervals, it is likely that you will be able to find this out from your graph.

1. The ANOVA cannot tell us which one (or more) of the groups is significantly different from the others. Explain how you could you use your graphical analysis to answer this question:

____

____

____

Sources of Variation

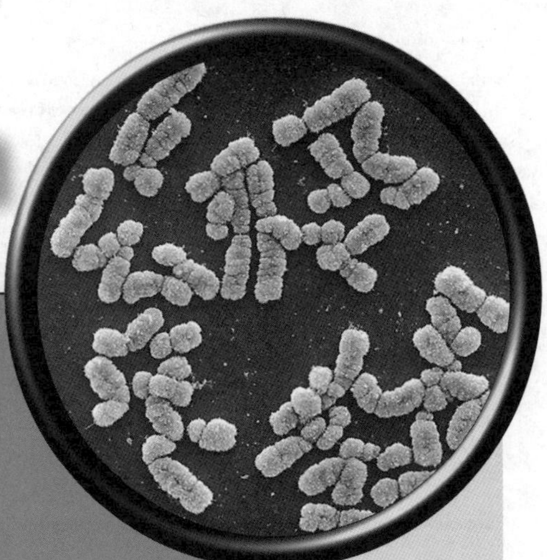

AQA-A	AQA-B	CIE	Edexcel	OCR
Complete: *1-5, 7-13, 17, 23-24 (a)-(b)* Extension: *6, 14-16*	Complete: *1, 5-12 (a)-(c), 13, 17* Extension: *2-4, 14-16*	Complete: *1, 7-13, 23-25, 27*	Complete: *1, 5-13, 17-19, 23-25* Extension: *5, 16, 21, 24*	Complete: *1, 7-13, 23-25, 27*

Learning Objectives

☐ 1. Compile your own glossary from the **KEY WORDS** displayed in **bold type** in the learning objectives below.

Variation

Continuous vs discontinuous variation *(pages 98-99, 138-139)*

☐ 2. Recognise the need for **random sampling** and the importance of **chance** in contributing to differences between samples. Demonstrate an ability to collect and graphically display data pertaining to variation in the phenotype of a natural population.

☐ 3. Understand the concept of **normal distribution** about a mean. Explain how mean and **standard deviation** are used to measure variation in a sample.

☐ 4. Demonstrate an ability to *calculate* and *interpret* **standard error** of sample data pertaining to variation in the phenotype of a natural population.

☐ 5. Describe the difference between **continuous** and **discontinuous variation** and provide examples to illustrate each type. Explain the basis of continuous and discontinuous variation by reference to the number of genes controlling the characteristic (also refer to the material in "*Inheritance*" on polygeny).

☐ 6. Collect data pertaining to human traits that show continuous variation, e.g. foot size, hand span, height. From the same set of subjects, collect data pertaining to human traits that show discontinuous variation, e.g. ear lobe shape or mid digit hair. Plot these sets of data separately to illustrate the differences in distribution.

Genotype and environment *(pages 100-101)*

☐ 7. Define: **gene-environment interaction**. Recognise that both genotype and environment contribute to phenotypic variation. Understand that this relationship can be expressed mathematically as: $V_P = V_G + V_E$.

☐ 8. Use examples to describe how the environment may affect the phenotype.

Meiosis, linkage, and recombination

Meiosis *(pages 96, 102-104)*

☐ 9. Appreciate the general significance of meiosis in generating genetic variation. Identify how meiosis creates new allele combinations in the gametes.

☐ 10. Define: **homologous chromosomes** (=homologues), **genome**, **chromatid**, **centromere**. Describe how chromosome numbers can vary between somatic cells (**diploid** 2N) and gamete cells (**haploid** 1N).

☐ 11. Recognise that meiosis, like mitosis, involves DNA replication during interphase in the parent cell, but that this is followed by *two* cycles of nuclear division.

☐ 12. Using simple diagrams, summarise the principal events associated with meiosis, to include:
(a) Pairing of **homologous chromosomes** (synapsis) and the formation of **bivalents**.
(b) **Chiasma** formation and exchange between **chromatids** in the first, (**reduction**) **division**.
(c) Separation of chromatids in the second division and the production of **haploid cells**.
(d) The associated behaviour of the **nuclear envelope**, plasma membrane, and **centrioles**.
(e) Identification of the names of the main stages.

☐ 13. Describe the behaviour of homologous chromosomes (and their associated alleles) during meiosis and fertilisation, with reference to:
- The **independent assortment** of maternal and paternal chromosomes. Chance governs which pole each chromosome of a bivalent moves to, ensuring random combinations of non-homologous chromosomes in the haploid nuclei.
- The **recombination** of segments of maternal and paternal homologous chromosomes in **crossing over**.
- The **random fusion** of gametes during fertilisation.

Linkage and recombination *(pages 96, 103-106)*

☐ 14. With respect to inheritance, explain what is meant by the terms: **linkage** and **linkage group**. Explain the consequences of linkage to the inheritance of alleles.

☐ 15. Recall that **recombination** refers to the exchange of alleles between homologous chromosomes as a result of **crossing over**. Explain the consequences of recombination to the inheritance of alleles.

☐ 16. Explain the effect of linkage and recombination on the phenotypic ratios from dihybrid crosses.

Mutations as a source of variation *(pages 96, 107-109, 112, also see page 167)*

☐ 17. Define **mutation**. Understand the significance of mutations as the ultimate source of all new alleles.

☐ 18. Define: **spontaneous mutation**. Recognise that each species has its own frequency of naturally occurring mutations (called the **background mutation rate**).

☐ 19. Define the terms: **mutagen**, **induced mutation**. Identify the environmental factors that can cause mutations. Describe the effects of chemical mutagens and radiation on DNA and the rate of mutation.

☐ 20. Mutations may occur in different types of cell. The location of a mutation can have a different significance in terms of producing heritable change. Clearly define: **germ line**, **somatic mutation**, **gametic mutation**.

☐ 21. Recognise that mutations may have different survival value in different environments. Using examples, explain how mutations may be **harmful**, **beneficial**, or **neutral** (silent) in their effect on the organism. Explain the evolutionary importance of **neutral mutations**.

☐ 22. In more detail than above, identify the genetic basis of a beneficial mutation, as illustrated by **antibiotic resistance** in bacteria.

Types of mutations

Gene mutations *(pages 110-111, 113-115)*

☐ 23. Explain what is meant by a **gene mutation** and provide examples. Distinguish between gene mutations involving change in a single nucleotide (often called **point mutations**), and those involving changes to a triplet (e.g. triplet deletion or triplet repeat). Explain why gene mutations have the most evolutionary potential.

☐ 24. Understand point mutations as illustrated by **base**:
(a) **Deletion**
(b) **Substitution**
(c) **Insertion**

Understand what is meant by the term **frame shift**, explain how a frame shift can arise and explain the consequences of this type of error.

☐ 25. Describe the effect of a point mutation on the resulting amino acid sequence and on the phenotype, as illustrated by the **sickle cell mutation** in humans.

☐ 26. Recognise other disorders that arise as a result of gene mutations: **β-thalassaemia**, **cystic fibrosis**, and **Huntington disease**. Describe the genetic basis of one or more of these diseases.

☐ 27. Outline the implications of the **Human Genome Project** for the detection and diagnosis of genetic disorders. Suggest how our deeper knowledge of genetic inheritance and the basis of genetic disorders will assist in genetic counselling in the future.

See the 'Textbook Reference Grid' on pages 8-9 for textbook page references relating to material in this topic.

Supplementary Texts

See pages 4-6 for additional details of these texts:

■ Adds, J., *et al.*, 2001. **Genetics, Evolution and Biodiversity**, (NelsonThornes), pp. 72-73, 88-99.

■ Clegg, C.J., 1999. **Genetics and Evolution**, (John Murray), pp. 19-22, 34-39.

■ Jones, N., *et al.*, 2001. **Essentials of Genetics**, 1-4, 17-25, 61-75, 113-121, 157-188, 257.

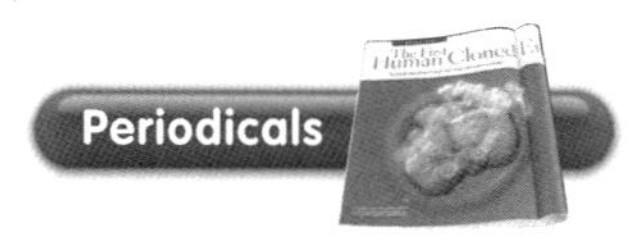

See page 6 for details of publishers of periodicals:

STUDENT'S REFERENCE

■ **What is a Gene?** Biol. Sci. Rev., 15(2) Nov. 2002, pp. 9-11. *A good synopsis of genes and their role in heredity, mutations, and transcriptional control of gene expression.*

■ **Spermatogenesis** Biol. Sci. Rev., 15(4), April 2003, pp. 10-14. *The process and control of sperm production. This account includes a discussion of the cell division processes (both meiotic and mitotic) involved in the production of viable sperm.*

■ **Mechanisms of Meiosis** Biol. Sci. Rev., 15(4), April 2003, pp. 20-24. *A clear and thorough account of the events and mechanisms of meiosis. There is a discussion of what happens when meiosis goes wrong, as well as an examination of synapsis and crossing over.*

■ **How do Mutations Lead to Evolution?** New Scientist, 14 June 2003, pp. 32-39, 48-51. *An account of the five most common points of discussion regarding evolution and the mechanisms by which it occurs. This article looks at how mutations to the promoter regions of DNA, which control gene expression, may account for the evolution of morphology and behaviour.*

■ **Secrets of The Gene** National Geographic, 196(4) October 1999, pp. 42-75. *The nature of genes and the location of some human mutations.*

■ **Radiation and Risk** New Scientist, 18 March 2000 (Inside Science). *An excellent account of the biological effects of radiation.*

■ **Globins, Genes and Globinopathies** Biol. Sci. Rev., 9(4) March 1997, pp. 2-5. *The structure of globins, and inborn errors of globin synthesis and assembly (sickle cell and thalassaemias).*

■ **Fragile X Syndrome** Biol. Sci. Rev., 10(3) Jan. 1998, pp. 12-14. *One in 1500 boys has fragile X syndrome. This article outlines the nature of the syndrome and the search for the culprit gene.*

■ **Antibiotic Resistance** Biol. Sci. Rev., 12(2) November 1999, pp. 28-30. *The genetic basis of antibiotic resistance in bacteria.*

■ **Genes, the Genome, and Disease** New Scientist, 17 Feb. 2001, (Inside Science). *Understanding the human genome: producing genome maps, the role of introns in gene regulation, and the future of genomic research.*

■ **Bioinformatics** Biol. Sci. Rev., 15(3), February 2003, pp. 13-15 and 15(4) April 2003, pp. 2-6. *Two accounts of the burgeoning field of bioinformatics: what it is and what it offers biology and medicine.*

TEACHER'S REFERENCE

■ **DNA 50** SSR, 84(308), March 2003, pp. 17-80. *A special issue celebrating 50 years since the discovery of DNA. There are various articles examining the practical and theoretical aspects of teaching molecular genetics and inheritance.*

■ **Radiation Roulette** New Scientist, 11 October 1997, pp. 36-40. *Ionising radiation damages DNA and causes chromosomal defects. Sometimes the damage is not immediately apparent.*

■ **The Enigma of Huntington's Disease** Scientific American, December 2002, pp. 60-95. *An account of the nature and cause of this inherited disorder. it includes a diagram illustrating the molecular basis of the disease.*

■ **Cystic Fibrosis** Scientific American, December 1995, pp. 36-43. *The basis of the cystic fibrosis mutation: the nature of the mutation and how it brings about the symptoms of the disease.*

■ **Weapons of Mass Disruption** Scientific American, Nov. 2002, pp. 58-63. *A look at the risks of the 'weapons of mass destruction' and, if used, what their effect might be on mutation rates.*

■ **X-Rated Brains** New Scientist, 25 May 2002, pp. 26-30. *Mutations in certain X-linked genes may be associated with mental impairment. Selection for normal versions of these genes may have been important in the evolution of human intelligence.*

■ **Day of the Mutators** New Scientist, 14 Feb. 1998, pp. 38-42. *The role of mutations in bacterial evolution: bacteria may direct their own evolution through the accumulation of favourable mutations.*

■ **Proteomics** Biologist 49(2) April 2002. *To really understand cellular functions, it is the proteome you need to analyse. What is involved in this?*

See pages 10-11 for details of how to access **Bio Links** from our web site: **www.biozone.co.uk**. From Bio Links, access sites under the topics:

GENERAL BIOLOGY ONLINE RESOURCES > **Online Textbooks and Lecture Notes**: • An on-line biology book • Actionbioscience.org • Biology-Online.org • Bursary topics • The Biology Project ... *and others* > **General Online Biology Resources**: • Access excellence • Biology I interactive tutorials ... *and others* > **Glossaries**: • Genetics glossary ... *and others*

CELL BIOLOGY AND BIOCHEMISTRY: • Cell & molecular biology online • MIT biology hypertextbook • Molecular biology web book > **Cell Division**: • Cell division: Meiosis and sexual reproduction • Comparison of mitosis and meiosis

GENETICS > **Molecular Genetics**: • Beginner's guide to molecular biology ... *and others* > **Inheritance**: • Online Mendelian inheritance in man • Patterns of inheritance > **Mutations and Genetic Disorders**: • Mutant fruit flies • Mutations • National PKU news • PKU fact sheet • Cystic fibrosis • Mutations causing cystic fibrosis • Joint center for sickle cell and thalassaemic disorders • Sickle cell disease • Sickle cell information centre ... *and others*

Software and video resources are provided on the Teacher Resource Handbook on CD-ROM

Sources of Genetic Variation

The genetic variability between individuals is what makes us all different from each other. Brothers and sisters may look similar to each other but there are always significant differences between them (unless they happen to be identical twins). The differences between close relatives is due mostly to a **shuffling** of the existing genetic material into new combinations. In addition to this is the new variation that originates from the **mutation** of existing genes. While most mutations are harmful, there are a significant number that are thought to be 'silent' and do not appear to have any effect on the individual. On rare occasions, a mutation may even prove to be beneficial. Mutations create new **alleles** and form an important part of the evolutionary process.

Mutations

Gene mutations; chromosome mutations

Mutations are the source of all **new** genetic information. Existing genes are modified by base substitutions and deletions, causing the formation of new alleles.

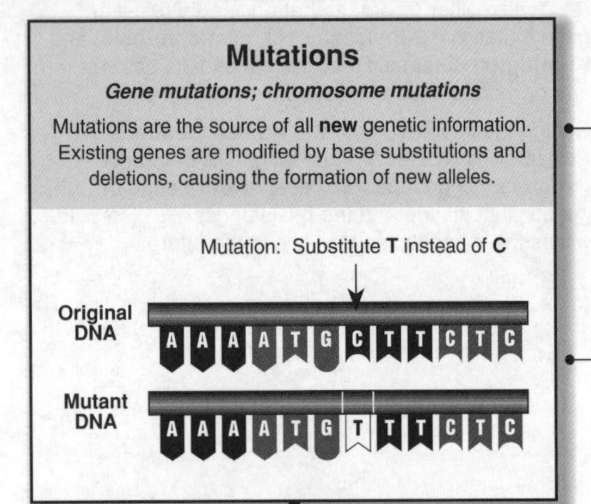

Gene Mutations

Mutations may cause alterations in the genetic instructions coded in the DNA of chromosomes. Most mutations are harmful, some are neutral (no effective change), while a very few may provide some improvement on the earlier version of the gene. Mutations may be accumulated (inherited) over many generations.

Chromosome Mutations

Pieces of chromosome may be rearranged during meiosis. Pieces may be turned upside-down, duplicated, moved from one chromosome to another or lost altogether. Most instances are harmful, but occasionally they may be beneficial.

Sexual Reproduction

Independent assortment; crossing over and recombination; mate selection

Sexual reproduction provides a rearrangement and shuffling of the genetic material into new combinations.

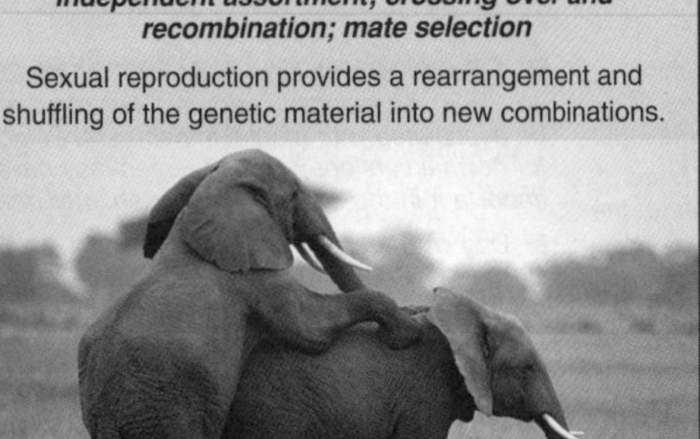

Independent Assortment

Genes are carried on chromosomes, 23 pairs in the case of humans. Each chromosome pair is sorted independently of the other pairs during meiosis. This random shuffling produces a huge variety of gametes from a single individual (parent).

Recombination

Pieces of chromosome are often exchanged with a chromosome's homologue (its paired chromosome with equivalent genes). This increases shuffling of allele combinations.

Mate Selection

Variation is further enhanced by the choice of mate to produce offspring. Different combinations of genes will come together in the offspring, depending on which two parents mate together.

Genotype

Determines the **genetic potential** of an individual

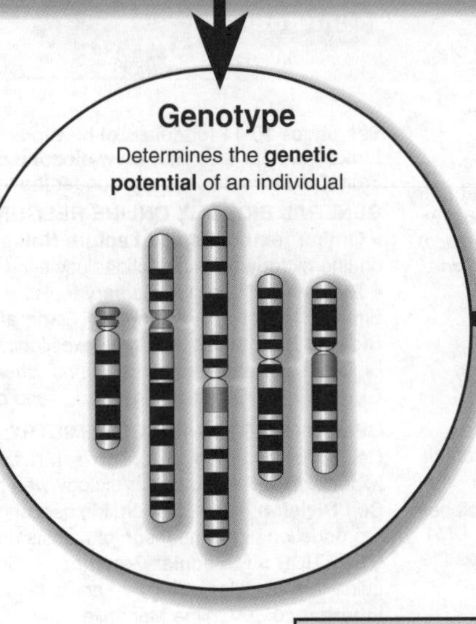

Dominant, recessive, codominant and multiple allele systems, as well as interactions between genes, combine in their effects.

Phenotype

The phenotype expressed in an individual is the result of all the factors listed on this page. The genetic instructions for creating the individual may be modified along the way, or at least modified by environmental influences.

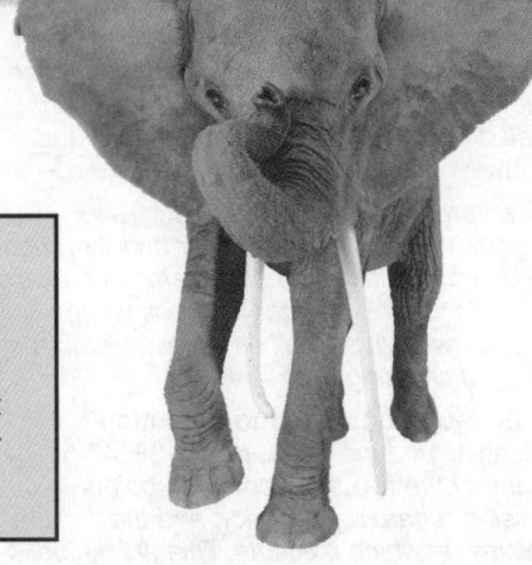

Environmental Factors

Environmental factors may influence the expression of the genotype. These factors may include physical factors such as temperature, light intensity, presence of groundwater, diet or nutrients, wind exposure, and pH. The presence of other organisms may also affect the expression of the genotype.

1. Describe three ways in which sexual reproduction can provide genetic variation in individuals:

   (a) ______

   (b) ______

   (c) ______

2. Explain how the environment of a particular genotype can affect the phenotype:

3. Describe three common ways by which humans can, by choice, alter their phenotype:

   (a) ______

   (b) ______

   (c) ______

4. Explain why siblings (brothers and sisters) have a family similarity but are not identical (unless they happen to be identical twins):

5. (a) Explain what is meant by a neutral (silent) mutation: ______

   (b) Discuss how neutral mutations can be important in the evolution of populations: ______

Investigating Human Variation

An estimated 80 000 or so genes determine all human characteristics (traits). Some human traits are determined by a single gene (see examples below). Single gene traits show **discontinuous variation** in a population; individuals show only one of a limited number of phenotypes (usually two or three). Other traits, such as skin colour, are polygenic (controlled by more than one gene) and show **continuous variation** in a population, with a spread of phenotypes across a normal distribution range. It is possible to classify a small part of your own genotype, and that of your classmates, for the six traits below. Investigations of continuous variation in human genotypes, arising as a result of polygeny, is covered in the next topic, "Inheritance".

Trait: Eye colour

Dominant

Phenotype: Brown, green, hazel or grey

Recessive

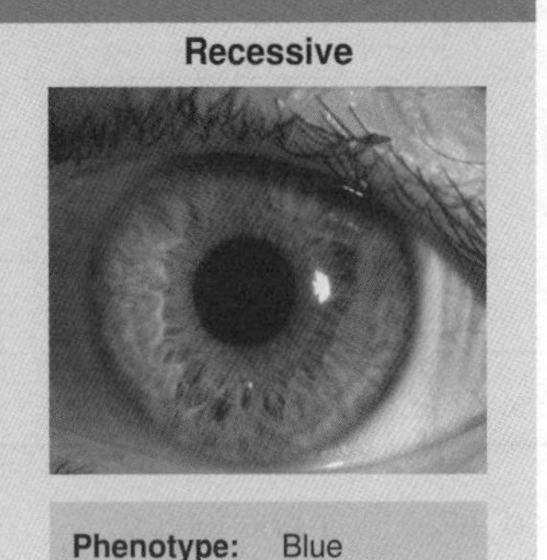

Phenotype: Blue

Trait: Handedness

Dominant

Phenotype: Right-handed

Recessive

Phenotype: Left-handed

Trait: Tongue roll

Dominant

Phenotype: Can roll tongue
Allele: R

Recessive

Phenotype: Cannot roll tongue
Allele: r

The ability to roll the tongue into a U-shape (viewed from the front) is controlled by a dominant allele. There are rare instances where a person can roll it in the opposite direction (to form an n-shape).

Trait: Middle digit hair

Dominant

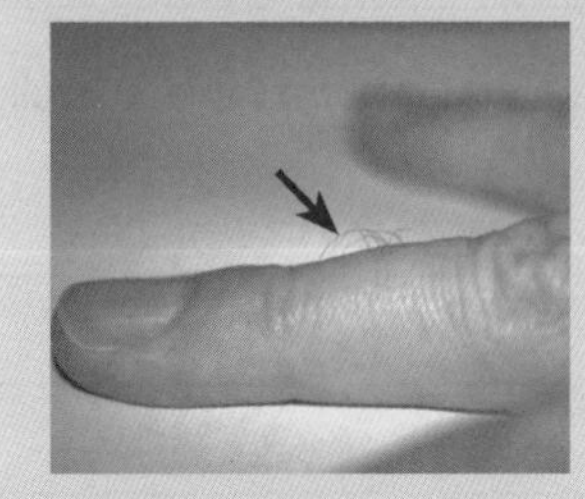

Phenotype: Hair on middle segment
Allele: M

Recessive

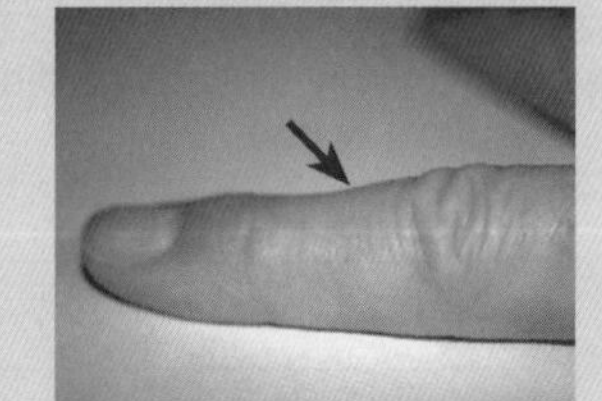

Phenotype: No hair on mid segment
Allele: m

Some people have a dominant allele that causes hair to grow on the middle segment of their fingers. It may not be present on all fingers, and in some cases may be very fine and hard to see.

Trait: Ear lobe shape

Dominant

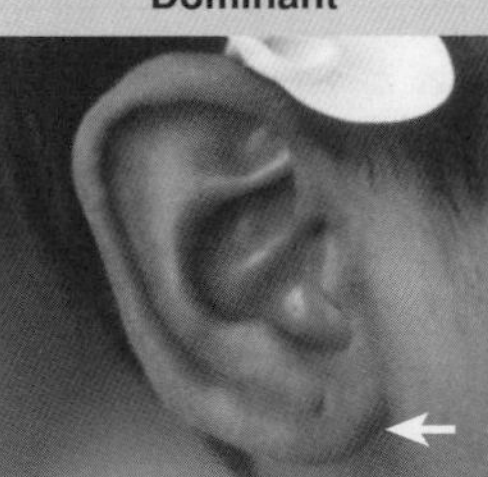

Phenotype: Lobes free
Allele: F

Recessive

Phenotype: Lobes attached
Allele: f

In people with only the recessive allele (homozygous recessive), ear lobes are attached to the side of the face. The presence of a dominant allele causes the ear lobe to hang freely.

Trait: Thumb hyperextension

Dominant

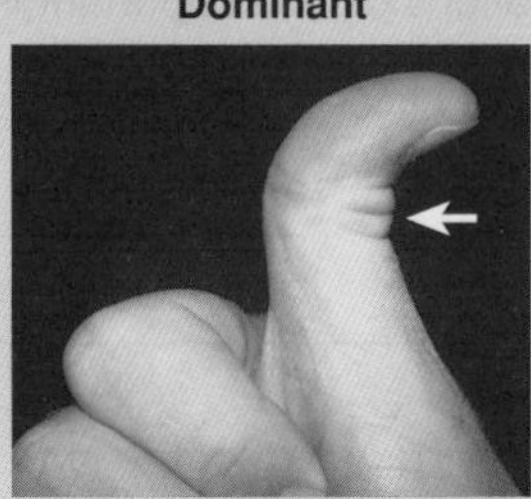

Phenotype: 'Hitchhiker's thumb'
Allele: H

Recessive

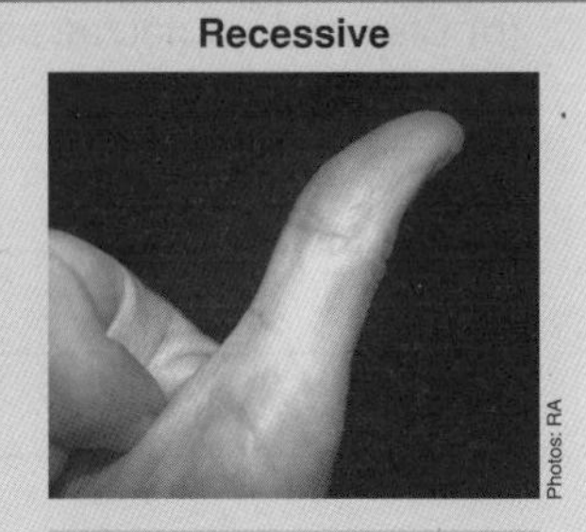

Phenotype: Normal thumb
Allele: h

Photos: RA

There is a gene that controls the trait known as 'hitchhiker's thumb' which is technically termed distal hyperextensibility. People with the dominant phenotype are able to curve their thumb backwards without assistance, so that it forms an arc shape.

Your Genotype Profile

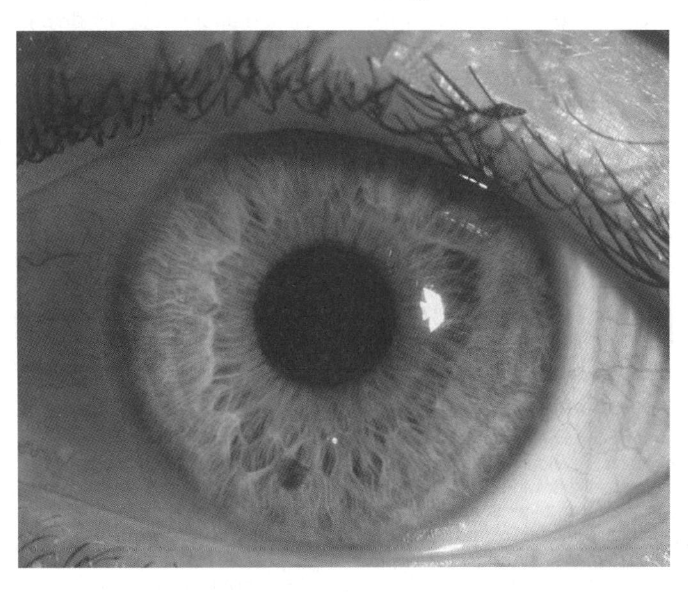

Use the descriptions and the symbols on the previous page to determine your own genotype. In situations where you exhibit the dominant form of the trait, it may be helpful to study the features of your family to determine whether you are homozygous dominant or heterozygous. If you do not know whether you are heterozygous for a given trait, assume you are.

Your traits:	Thumb	Ear lobes	Eye colour	Middle digit hair	Handedness	Tongue roll
Phenotype:						
Genotype:						

1. Enter the details of your own genotype in the table above. The row: 'Phenotype' requires that you write down the version of the trait that is expressed in you (e.g. blue eyes). Each genotype should contain two alleles.
2. Use a piece of paper and cut out 12 squares. Write the symbols for your alleles listed in the table above (each of the two alleles on two separate squares for the six traits) and write your initials on the back.
3. Move about the class, shaking hands with other class members to simulate mating (this interaction does not have to be with a member of the opposite sex).
4. Proceed to determine the possible genotypes and phenotypes for your offspring with this other person by:
   (a) Selecting each of the six characters in turn
   (b) Where a genotype for a person is known to be homozygous (dominant or recessive) t' erson will simply place down one of the piece aper with their allele for that gene. If they ar erozygous for this trait, toss a coin to deter which gets 'donated' with heads being th inant allele and tails being the recessive.
   (c) The partner places their using the same method as in (b) abov etermine their contribution to this trait.
   (d) Write down the resulting genotype in the table below and determine the phenotype for that trait.
   (e) Proceed on to the next trait.
5. Try another mating with a different partner or the same partner and see if you end up with a child of the same phenotype.

Child 1	Thumb	Ear lobes	Eye colour	Middle digit hair	Handedness	Tongue roll
Phenotype:						
Genotype:						

Child 2	Thumb	Ear lobes	Eye colour	Middle digit hair	Handedness	Tongue roll
Phenotype:						
Genotype:						

Gene-Environment Interactions

External environmental factors can modify the phenotype coded by genes. This can occur both in the development of the embryo and later in life. Even identical twins, which are essentially clones, have minor differences in their appearance due to factors such as diet. Environmental factors that affect the phenotype of plants and animals include nutrients or diet, temperature, and the presence of other organisms.

The Effect of Temperature

The sex of some animals is determined by the temperature at which they were incubated during their embryonic development. Examples include turtles, crocodiles, and the American alligator. In some species, high incubation temperatures produce males and low temperatures produce females. In other species, the opposite is true. The advantages of temperature regulated sex determination may arise through prevention of inbreeding (since all siblings will tend to be of the same sex).

The Effect of Other Organisms

The presence of other individuals of the same species may control the determination of sex in other individuals of the group. Some fish species, including some in the wrasse family (e.g. *Coris sandageri,* above), show such a change in phenotype. The fish live in groups consisting of a single male with attendant females and juveniles. In the presence of a male, all juvenile fish of this species grow into females. When the male dies, the dominant female will undergo physiological changes to become a male for the group. The male has distinctive vertical bands behind the gills. The female is pale in colour and has very faint markings.

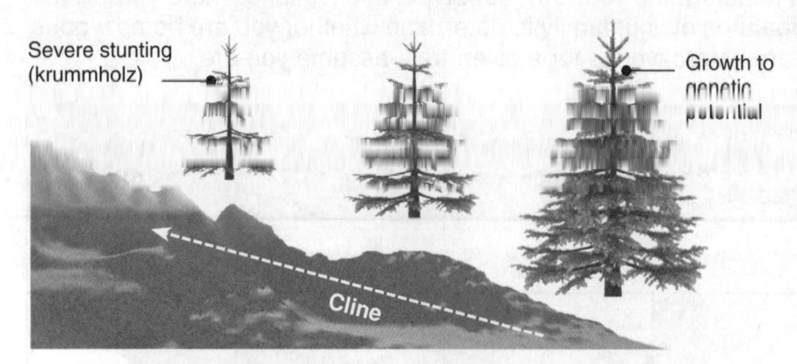

The Effect of Altitude

Increasing altitude can stunt the phenotype of plants with the same genotype. In some conifers, e.g. Engelmann spruce (*Picea engelmannii*), plants at low altitude grow to their full genetic potential, but become progressively more stunted as elevation increases, forming krummholz (gnarled bushy growth forms) at the highest, most severe sites. This situation, where there is a continuous, or nearly continuous, gradation in a phenotypic character within a species, associated with a change in an environmental variable, is called a **cline**.

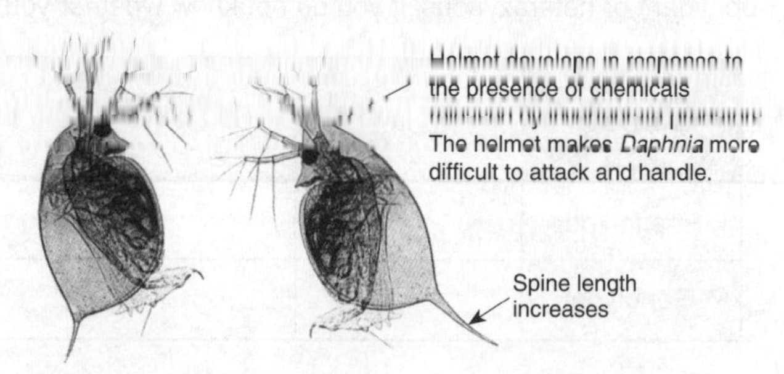

Non-helmeted form **Helmeted form with long tail spine**

Some organisms respond to the presence of other, potentially harmful, organisms by changing their morphology or body shape. Invertebrates such as *Daphnia* will grow a large helmet when a predatory midge larva (*Chaoborus*) is present. Such responses are usually mediated through the action of chemicals produced by the predator (or competitor), and are common in plants as well as animals.

1. Giving appropriate examples, distinguish clearly between **genotype** and **phenotype**:

2. Identify some of the physical factors associated with altitude that could affect plant phenotype:

3. The hydrangea is a plant that exhibits a change in the colour of its flowers according to the condition of the soil. Identify the physical factor that causes hydrangea flowers to be blue or pink. If you can, find out how this effect is exerted:

4. Colour pointing in some breeds of cats such as the siamese, involves the activity of a temperature sensitive enzyme that produces the pigment melanin. Explain why the darker patches of fur are found only on the face, paws and tail:

At conception (the formation of the zygote), an organism possesses a genetic potential to grow into an adult form with certain characteristics. The exact form it takes is determined largely by the genes in its chromosomes, but it is also strongly influenced by a vast range of environmental factors acting upon it. These factors may subject an organism to stresses and may limit its growth to something less than it is capable of (e.g. plants that are grown at high altitude or in very exposed locations will often have stunted growth). Changes in the phenotype due solely to environmental factors are not inherited. Traumatic events such as the loss of a limb on a tree, or the removal of the tail in a young mammal (e.g. lambs, pups), does not affect the phenotype of the next generation (trees do not grow with limbs missing, and not one lamb has been born without a tail).

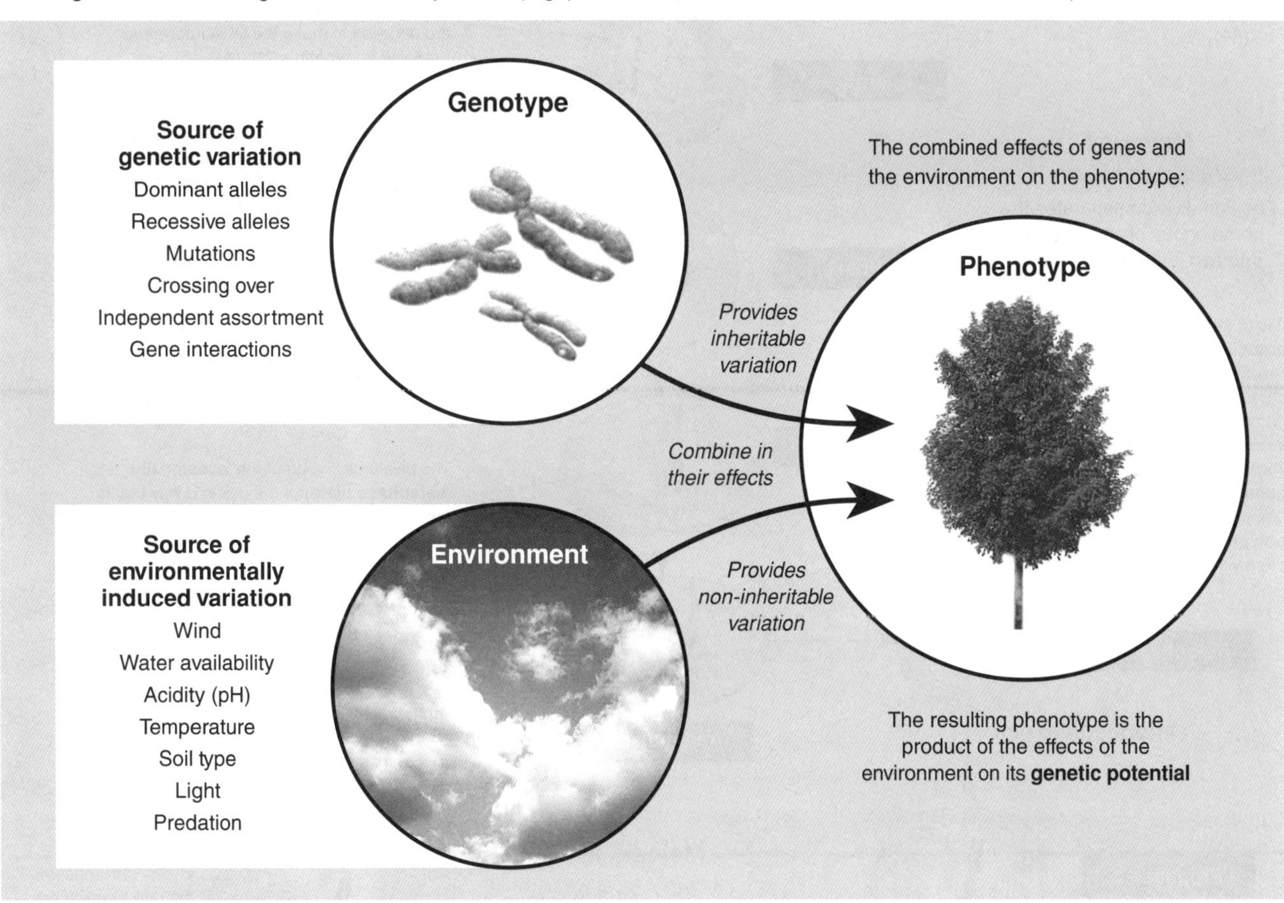

5. There has been much amusement over the size of record-breaking vegetables, such as enormous pumpkins, produced for competitions. Explain how you could improve the chance that a vegetable would reach its maximum genetic potential:

6. (a) Explain what is meant by a **cline**:

   (b) On a windswept portion of a coast, two different species of plant (species A and species B) were found growing together. Both had a low growing (prostrate) phenotype. One of each plant type was transferred to a greenhouse where "ideal" conditions were provided to allow maximum growth. In this controlled environment, species B continued to grow in its original prostrate form, but species A changed its growing pattern and became erect in form. Identify the **cause** of the prostrate phenotype in each of the coastal grown plant species and explain your answer:

   Plant species A:

   Plant species B:

   (c) Identify which of these species (A or B) would be most likely to exhibit clinal variation:

Meiosis

The process of **meiosis** is a special type of cell division concerned with producing sex cells (gametes) for the purpose of sexual reproduction. This cell division occurs in the sex organs of plants and animals. If genetic mistakes (**gene** and **chromosome mutations**) occur here, they will be passed on to the offspring (they will be inherited).

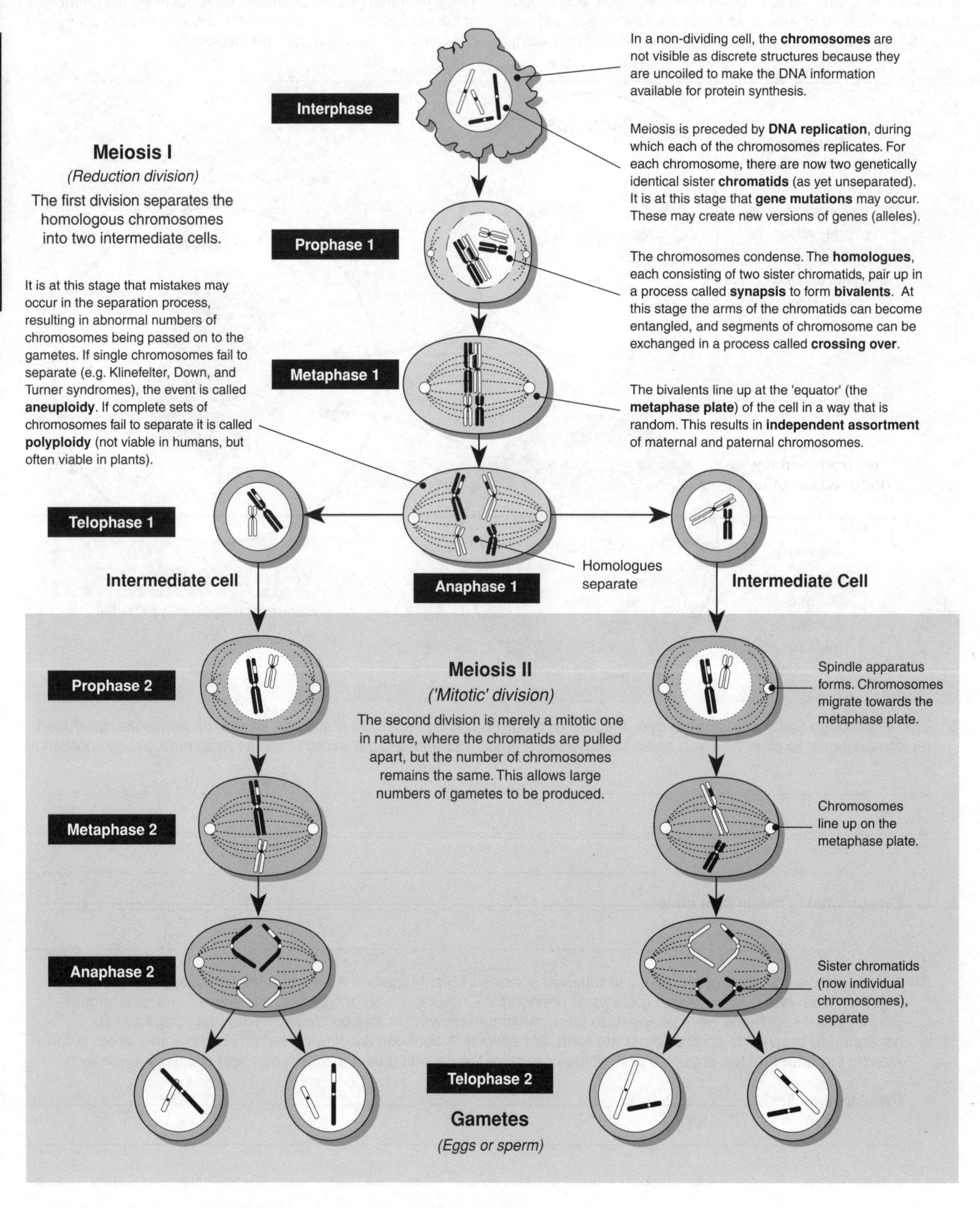

1. Describe the behaviour of the chromosomes in the first division of meiosis: ______________________________

____

2. Describe the behaviour of the chromosomes in the second division of meiosis: ______________________________

____

Crossing Over

Crossing over refers to the mutual exchange of pieces of chromosome and involves the swapping of whole groups of genes between the **homologous** chromosomes. This process can occur only during the first division of **meiosis**. Errors in crossing over can result in **chromosome mutations** (see the activity on this topic), which can be very damaging to development. Crossing over can upset expected frequencies of offspring in dihybrid crosses. The frequency of crossing over (COV) for different genes (as followed by inherited, observable traits) can be used to determine the relative positions of genes on a chromosome and provide a **genetic map**. There has been a recent suggestion that crossing over may be necessary to ensure accurate cell division.

Pairing of Homologous Chromosomes

Every somatic cell has a pair of each type of chromosome in its nucleus. These chromosome pairs, one from each parent, are called **homologous** pairs or **homologues**. In prophase of the first division of **meiosis**, the homologues pair up to form **bivalents** in a process called **synapsis**. This allows the chromatids of the homologous chromosomes to come in very close contact.

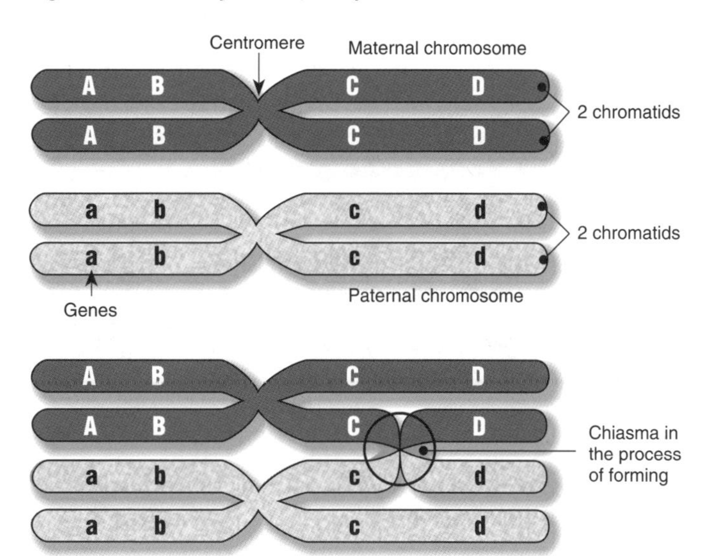

Chiasma Formation and Crossing Over

The pairing of the homologues allows **chiasmata** to form between the chromatids of homologous chromosomes. These are places where the chromatids become criss-crossed and the chromosomes exchange segments. In the diagram, the chiasma are in the process of forming and the exchange of pieces of chromosome have not yet taken place. Every point where the chromatids have crossed is a **chiasma**.

Separation

New combinations of genes arise from crossing over, resulting in what is called **recombination**. When the homologues separate at anaphase of meiosis I, each of the chromosomes pictured will have new genetic material (mixed types) that will be passed into the gametes soon to be formed. This process of recombination is an important source of variation for the gene pool of a population.

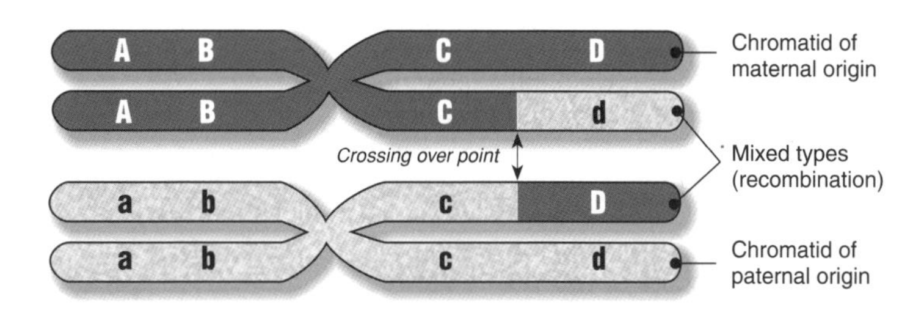

Gamete Formation

Once the final division of meiosis is complete, the two chromatids that made up each replicated chromosome become separated and are now referred to as chromosomes. Because chromatid segments were exchanged, **four** chromosomes that are quite different (genetically) are produced. If no crossing over had occurred, there would have been only two types (two copies of each). Each of these chromsomes will end up in a different gamete (sperm or egg).

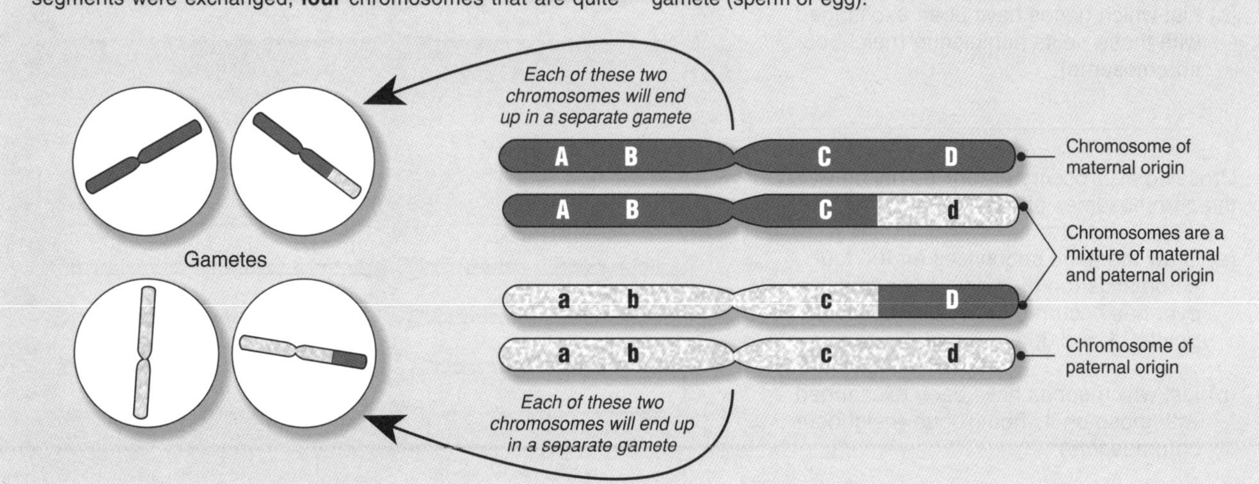

1. Briefly explain how the process of crossing over is going to alter the genotype of gametes: ____________________

____

2. Describe the importance of crossing over in the process of evolution: ____________________

____

Crossing Over Problems

The diagram below shows a pair of homologous chromosomes about to undergo chiasma formation during the first division of meiosis. There are known crossover points along the length of the chromatids (same on all four chromatids shown in the diagram). In the prepared spaces below, draw the gene sequences after crossing over has occurred on three unrelated and separate occasions (it would be useful to use different coloured pens to represent the genes from the two different chromosomes). See the diagrams on the previous page as a guide.

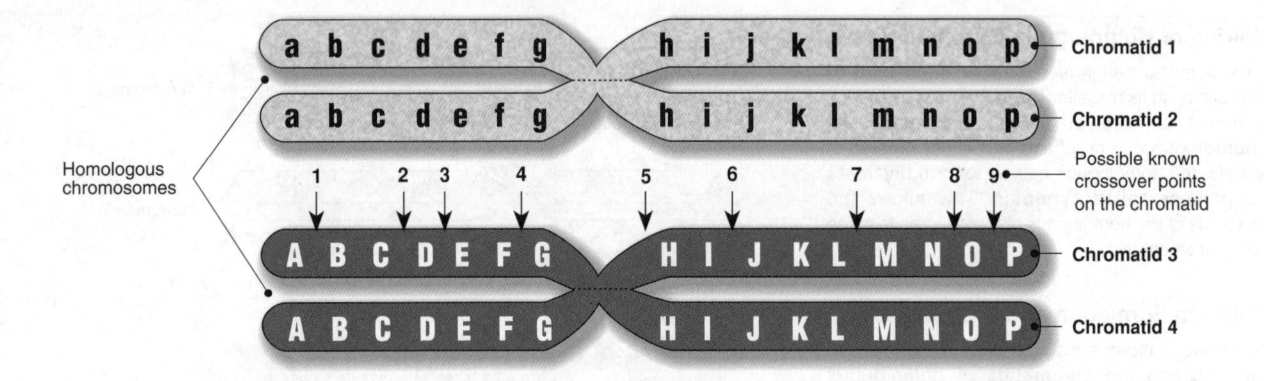

1. Crossing over occurs at a **single** point between the chromosomes above.

   (a) Draw the gene sequences for the four chromatids (on the right), after crossing over has occurred at crossover point: **2**

   (b) List which genes have been exchanged with those on its homologue (neighbour chromosome):

   ____

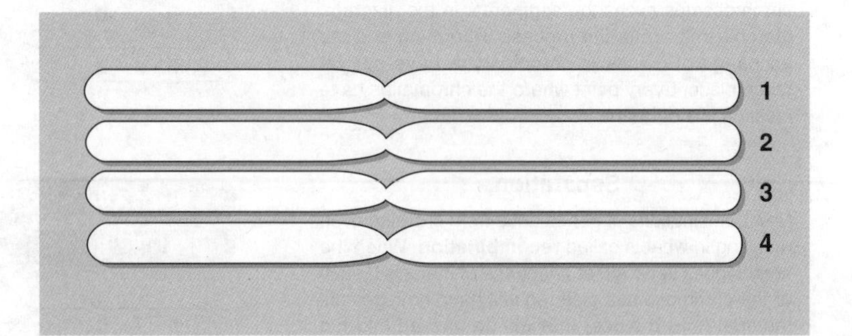

2. Crossing over occurs at **two** points between the chromosomes above.

   (a) Draw the gene sequences for the four chromatids (on the right), after crossing over has occurred between crossover points: **6** and **7**.

   (b) List which genes have been exchanged with those on its homologue (neighbour chromosome):

   ____

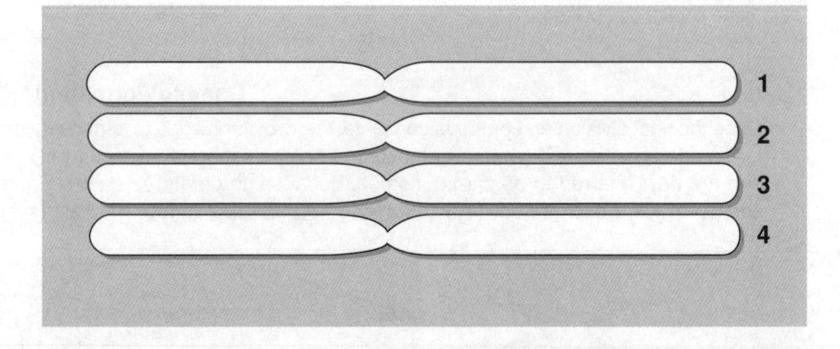

3. Crossing over occurs at **four** points between the chromosomes above.

   (a) Draw the gene sequences for the four chromatids (on the right), after crossing over has occurred between crossover points: **1** and **3**, and **5** and **7**.

   (b) List which genes have been exchanged with those on its homologue (neighbour chromosome):

   ____

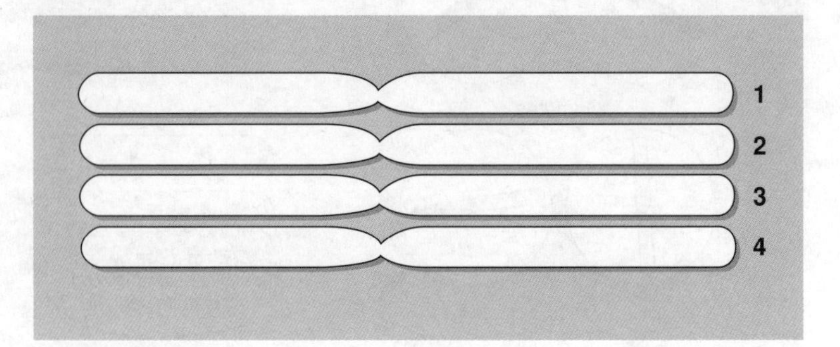

4. Explain the genetic significance of **crossing over**: ______________________

____

____

____

Linkage

Linkage refers to genes that are located on the same chromosome. Linked genes tend to be inherited together and fewer genetic combinations of their alleles are possible. Linkage reduces the variety of offspring that can be produced (contrast this with recombination). In genetic crosses, linkage is indicated when a greater proportion of the progeny resulting from a cross are of the parental type (than would be expected if the alleles were assorting independently). If the genes in question had been on separate chromosomes, there would have been more genetic variation in the gametes and therefore in the offspring. Note that in the example below, wild type alleles are dominant and are denoted by an upper case symbol of the mutant phenotype (Cu or Eb). This symbology used for *Drosophila* departs from the convention of using the dominant gene to provide the symbol. This is necessary because there are many mutant alternative phenotypes to the wild type. Alternatively, the wild type is sometimes denoted with a raised plus sign e.g. cu^+cu^+ and all symbols are in lower case.

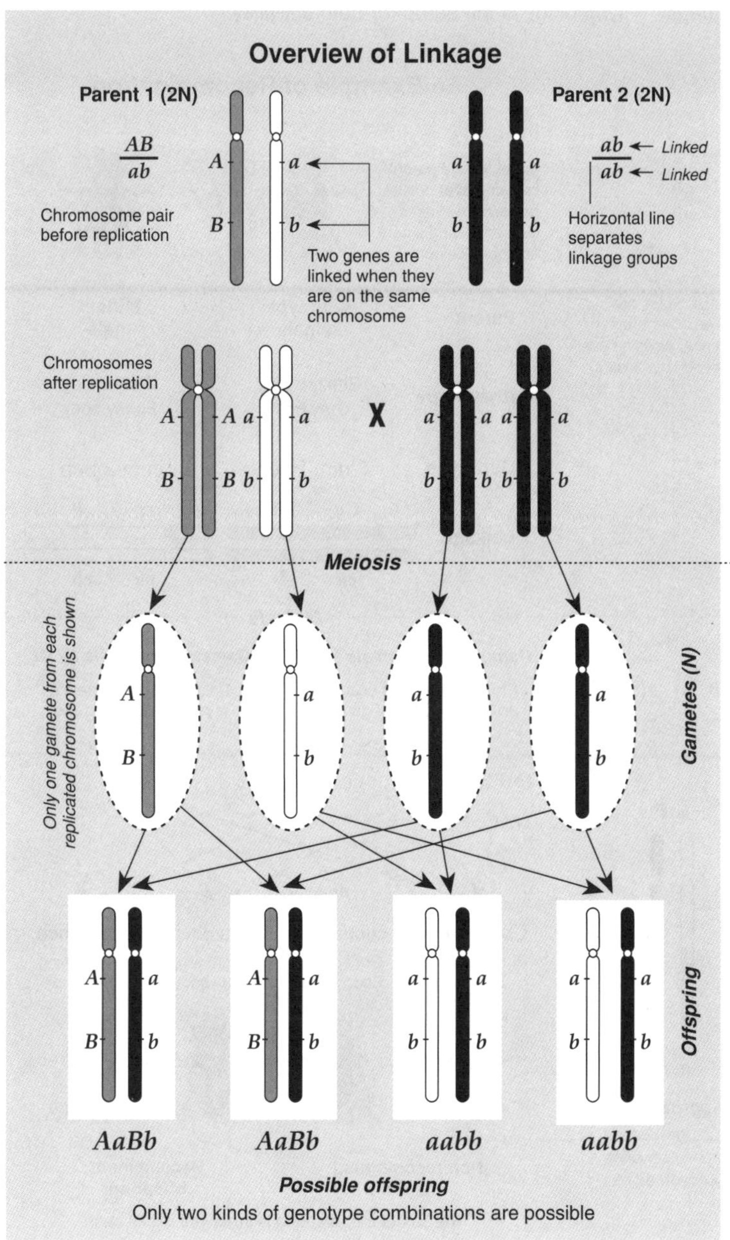

Possible offspring
Only two kinds of genotype combinations are possible

An Example of Linked Genes in *Drosophila*

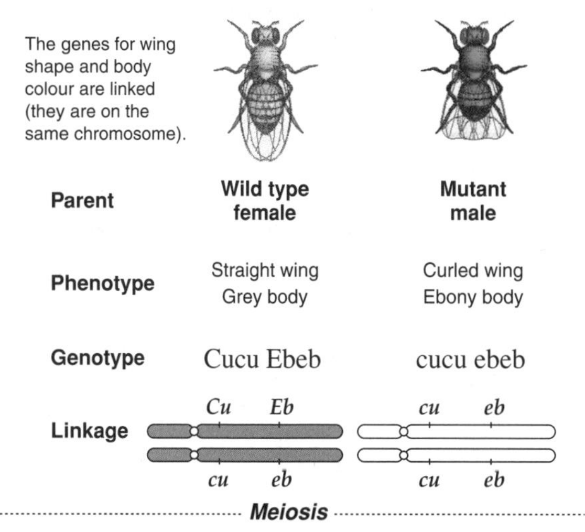

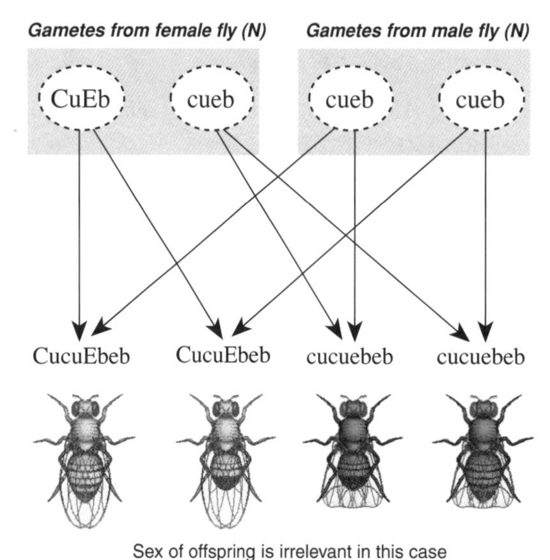

Sex of offspring is irrelevant in this case

Contact **Newbyte Educational Software** for details of their superb *Drosophila Genetics* software package which includes coverage of linkage and recombination. *Drosophila* images © Newbyte Educational Software.

1. Describe the effect of **linkage** on the inheritance of genes: ______________________

____

2. (a) List the possible genotypes in the offspring (above, left) if genes A and B had been on **separate chromosomes**:

____

(b) If the female *Drosophila* had been homozygous for the dominant wild type alleles (CuCu EbEb), state:

The genotype(s) of the F_1: ______________ The phenotype(s) of the F_1: ______________

3. Explain how linkage decreases the amount of genetic variation in the offspring: ______________________

____

Recombination

Genetic recombination refers to the exchange of alleles between homologous chromosomes as a result of **crossing over**. The alleles of parental linkage groups separate and new associations of alleles are formed in the gametes. Offspring formed from these gametes show new combinations of characteristics and are known as **recombinants** (offspring with genotypes unlike either parent). The proportion of recombinants in the offspring can be used to calculate the frequency of recombination (crossover value). These values are fairly constant for any given pair of alleles and can be used to produce gene maps indicating the relative positions of genes on a chromosome. In contrast to linkage, recombination increases genetic variation. Recombination between the alleles of parental linkage groups is indicated by the appearance of recombinants in the offspring, although not in the numbers that would be expected had the alleles been on separate chromosomes (independent assortment). The example below uses the same genotypes as the previous activity, *Linkage*, but in this case crossing over occurs between the alleles in a linkage group in one parent. The symbology is the same for both activities.

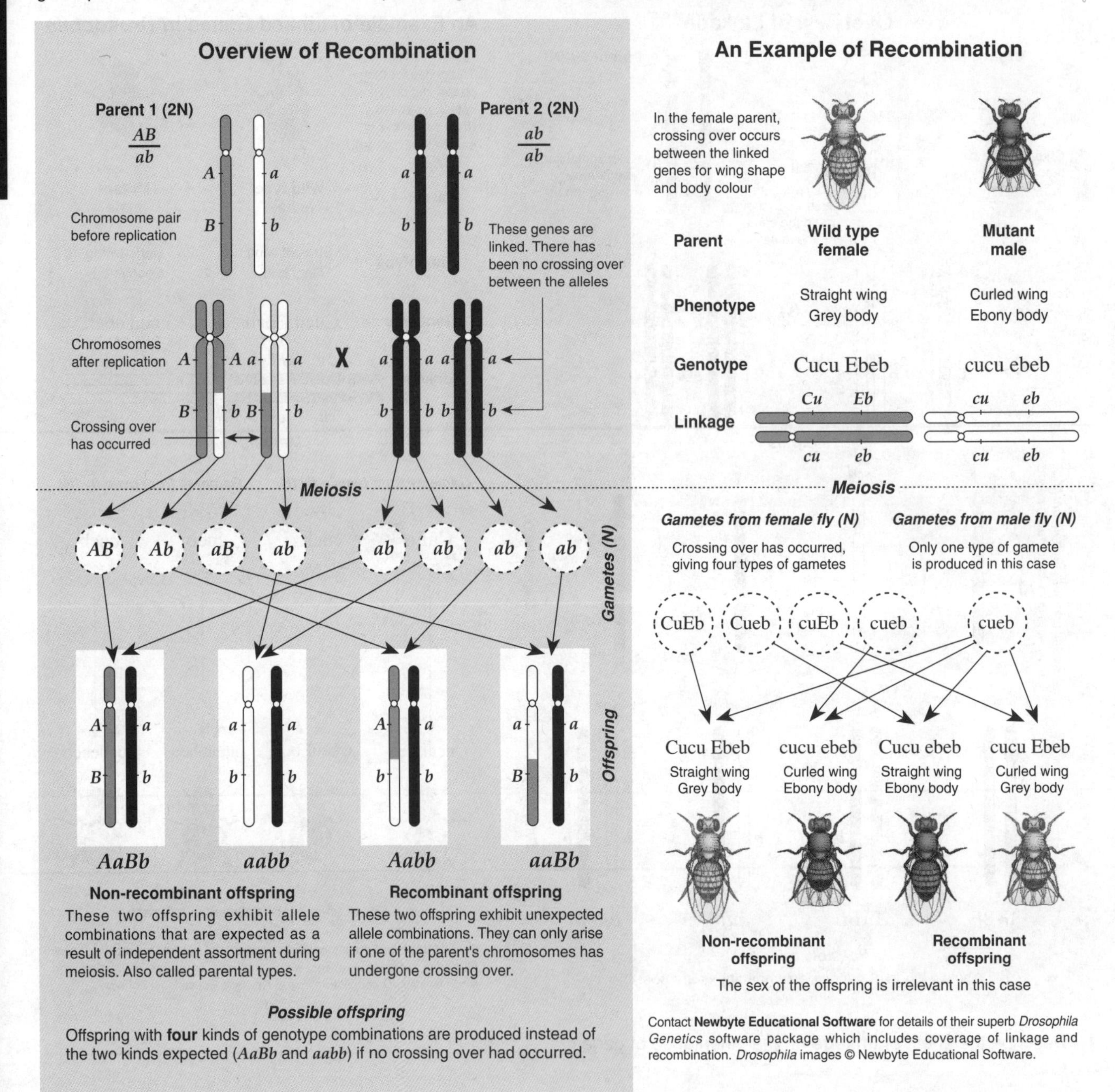

Contact **Newbyte Educational Software** for details of their superb *Drosophila Genetics* software package which includes coverage of linkage and recombination. *Drosophila* images © Newbyte Educational Software.

1. Describe the effect of **recombination** on the inheritance of genes: ______

2. Explain how recombination increases the amount of genetic variation in offspring: ______

3. Explain why it is not possible to have a recombination frequency of greater than 50% (half recombinant progeny): ______

Harmful Mutations

There are many well-documented examples of mutations that cause harmful effects. Examples are the mutations giving rise to **cystic fibrosis** (CF) and **sickle cell disease**. The sickle cell mutation involves a change to only one base in the DNA sequence, whereas the CF mutation involves the loss of a single triplet (three nucleotides). The malformed proteins that result from these mutations cannot carry out their normal biological functions. **Albinism** is caused by a mutation in the gene that produces an enzyme in the metabolic pathway to produce melanin. It occurs in a large number of animals e.g the alligator, right. Albinos are uncommon in the wild because they tend to be more vulnerable to predation and damaging UV radiation.

Beneficial Mutations

Tolerance to high cholesterol levels in humans: In the small village of **Limone**, about 40 villagers have extraordinarily high levels of blood cholesterol, with no apparent harmful effects on their coronary arteries. The village has a population of 980 inhabitants and was, until recently, largely isolated from the rest of the world. The villagers possess a mutation that alters the protein produced by just **one amino acid**. This improved protein is ten times more effective at mopping up excess cholesterol. No matter how much excess cholesterol is taken in by eating, it can always be disposed of. All carriers of the mutation are related and descended from one couple who arrived in Limone in 1636. Generally, the people of Limone live longer and show a high resistance to heart disease.

High blood cholesterol and dietary fat are implicated in the formation of plaques in the coronary arteries and in the development of cardiovascular disease.

Italy
Limone
Lake Garda
Brescia
Verona
CDC

Neutral Mutations

Some mutations are neither harmful nor beneficial to the organism in which they occur. However, they may be very important in an evolutionary sense. A mutation may have no 'adaptive value' when it occurs but this may not be the case in the future. Neutral or **silent** mutations are virtually impossible to detect because they have no observable effect. The example below shows how a change to the DNA sequence can be silenced if no change to the amino acid sequence occurs.

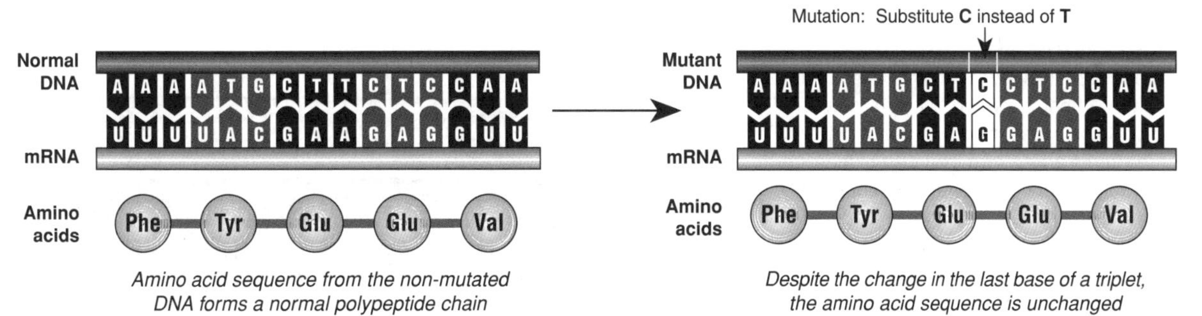

Amino acid sequence from the non-mutated DNA forms a normal polypeptide chain

Despite the change in the last base of a triplet, the amino acid sequence is unchanged

5. (a) Explain the difference between neutral (silent), beneficial, and harmful mutations:

(b) State which of these mutations is the most common and suggest a reason why this would be the case:

6. Explain how the mutation that 40 of the villagers of Limone possess is beneficial under current environmental conditions:

Gene Mutations

Gene mutations are small, localised changes in the structure of a DNA strand. These mutations may involve change in a single nucleotide (these are often called **point mutations**), or changes to a triplet (e.g. triplet deletion or triplet repeat). If one amino acid in a protein is wrong, the biological function of the entire protein can be disrupted. Not all mutations may result in altered proteins. Because of the degeneracy of the genetic code, a substitution of the 3rd base in a codon may code for the same amino acid. The diagrams below and opposite show how various point mutations can occur. These alterations in the DNA are at the **nucleotide** level where individual **codons** are affected. Alteration of the precise nucleotide sequence of a coded gene in turn alters the mRNA transcribed from the mutated DNA and may affect the polypeptide chain that it is designed to create.

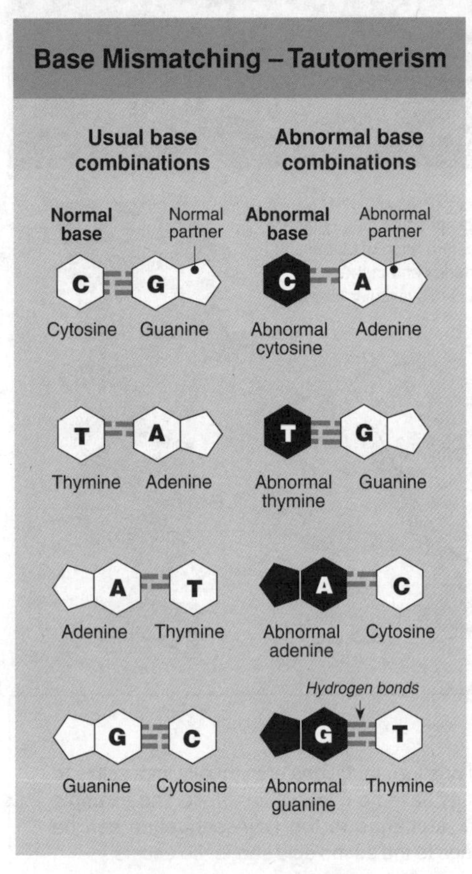

Base Mismatching

Watson and Crick proposed a theory of how base mismatching could occur. The diagram on the left shows suggested changes in bases and the resulting mismatch of complementary bases. On rare occasions some bases may have altered hydrogen-bond positions. As a result, during DNA replication, such abnormal bases pair with incorrect complementary bases. This gives rise to mutations in DNA molecules, which in turn are expressed as altered forms of mRNA and often altered proteins. NOTE: The abnormal bases on the right hand side of the diagram have a different arrangement of hydrogen bonds than normal.

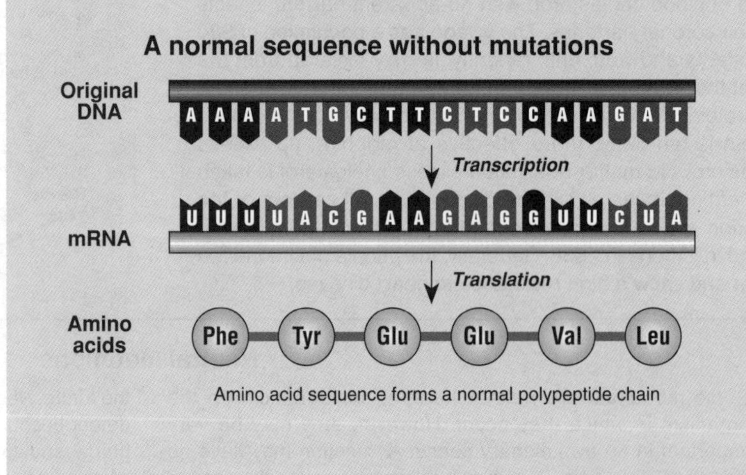

Missense substitution

A single base is substituted for another base which may result in a codon that codes for a different amino acid. Some substitutions, however, may still code for the same amino acid, because of the high degree of degeneracy in the genetic code (i.e., many amino acids have 4 or 6 codons coding for them). In the illustrated example, placing a T where a C should have been, results in the amino acid **lysine** appearing where **glutamic** acid should be. This could affect how this protein functions.

Nonsense substitution

Some amino acids can be coded for by 4 or 6 different codons and are therefore less affected by substitutions. In the example illustrated, a single base substitution in the first nucleotide of the third codon has a dramatic effect on the nature of the polypeptide chain it is coding for. The codon no longer codes for an amino acid, but instead is an instruction for the termination of the translation process of protein synthesis. This results in a very short polypeptide chain that is likely to have little or no function since the **STOP** codon is introduced near the **START** codon.

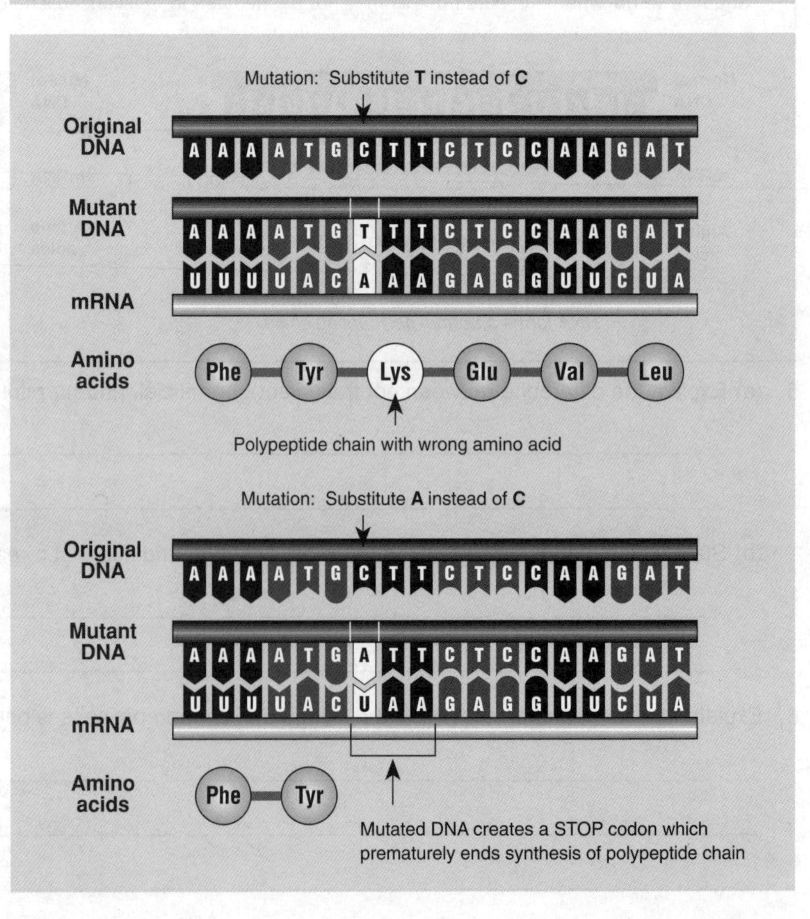

Reading frame shift by insertion

A major upset can occur when a single extra base is inserted into the DNA sequence. This has the effect of displacing all the other bases along one position and thereby creating a whole new sequence of codons. Such mutations are almost always likely to lead to a non-functional protein, but this does depend on the distance of the insertion or deletion from the START codon (i.e. the closer the insertion is to the START codon, the more the protein will be affected).

NOTE: could also lead to nonsense

Reading frame shift by deletion

In the same way that an insertion of an extra base into the DNA sequence has a large scale damaging effect, a deletion may also cause a frame shift. Again the result is usually a polypeptide chain of doubtful biological activity.

NOTE: could also lead to nonsense

Partial reading frame shift

Both an insertion and a deletion of bases within a gene can cause a frame shift effect where each codon no longer has the correct triplet of three bases. In this example, three codons have been affected, along with the amino acids they code for. The error is limited to the codons including and between the insertion and deletion. There is no biological activity if the amino acids altered are important to the functioning of the resulting protein.

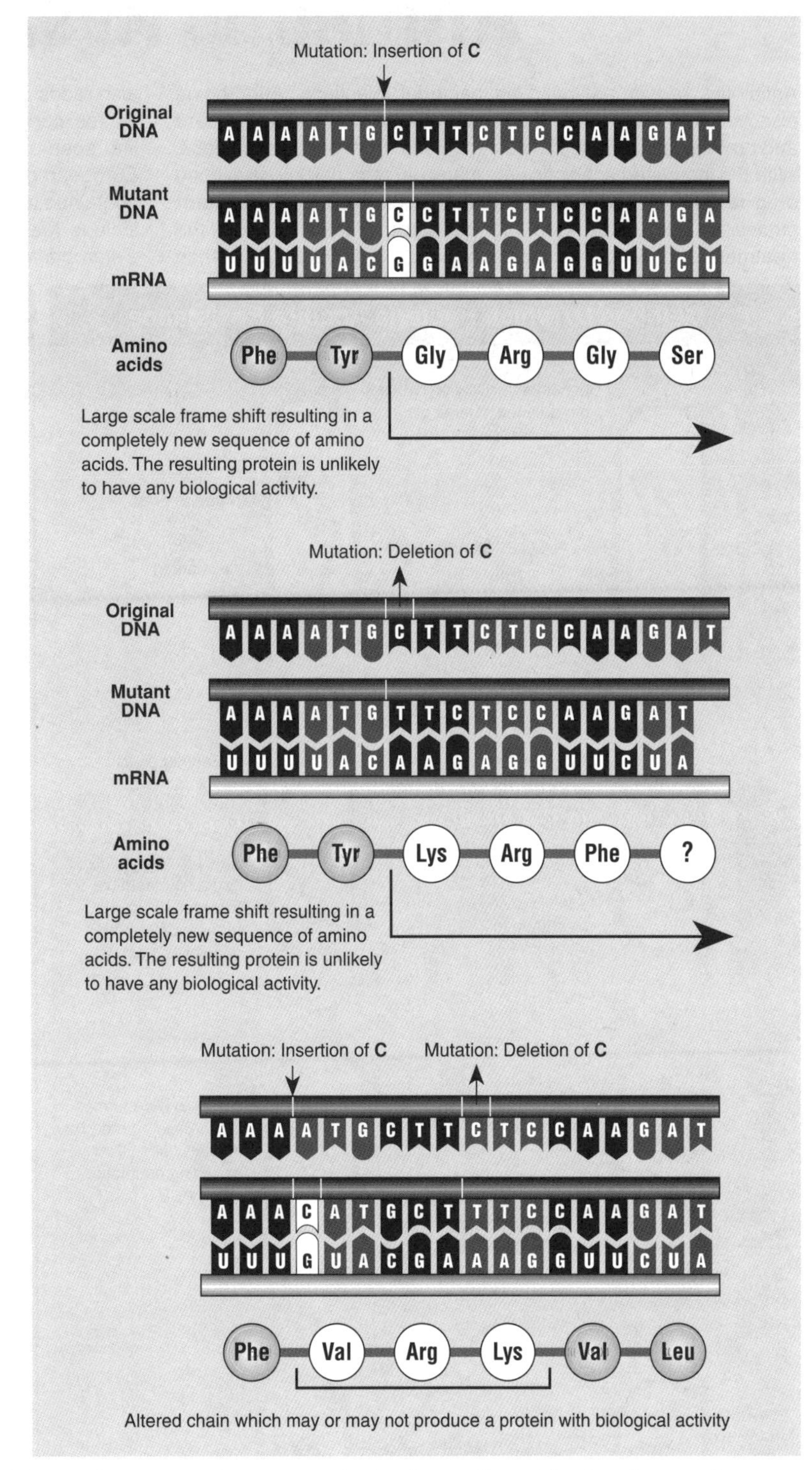

Altered chain which may or may not produce a protein with biological activity

Mutations involving change in only one nucleotide may have no observable effect on the phenotype of the organism; the subtle changes in the DNA sequence may still produce a chain of identical amino acids in the protein, or at least produce a protein that is unaffected by the change. Because of the degeneracy of the genetic code, many mutations of this type are unlikely to cause any change in the biological activity of the protein (there are exceptions, e.g. sickle cell disease).

1. Explain what is meant by a **reading frame shift**: ____________________

____

2. Not all point mutations have the same effect on the organism, some are more disruptive than others.

   (a) Identify which type of point mutations are the most damaging to an organism: ____________________

   (b) Explain why they are the most disruptive: ____________________

   ____

3. Explain why **biological activity** of a protein might be affected by a reading frame shift: ____________________

____

Antibiotic Resistance

Antibiotics are drugs that fight bacterial infections. After being discovered in the 1940s, they rapidly transformed medical care and dramatically reduced illness and death from bacterial disease. With the increased antibiotic use, many bacteria quickly developed drug resistance. The increasing number of **multi-drug resistant** strains is particularly worrying; resistant infections inhibit the treatment of patients and increase mortality. Antibiotic resistance also adds considerably to the costs of treating disease and, as resistance spreads, new drugs have an increasingly limited life span during which they are effective. A survey in five European countries showed that around 40% of hospital samples contained at least one antibiotic resistant strain. Resistant bacteria include *Klebsiella*, *Enterococcus*, *E. coli*, *Staphylococcus aureus*, *Enterobacter*, *Pseudomonas*, and *Mycobacterium tuberculosis.*

Methods by which Bacteria Acquire Resistance

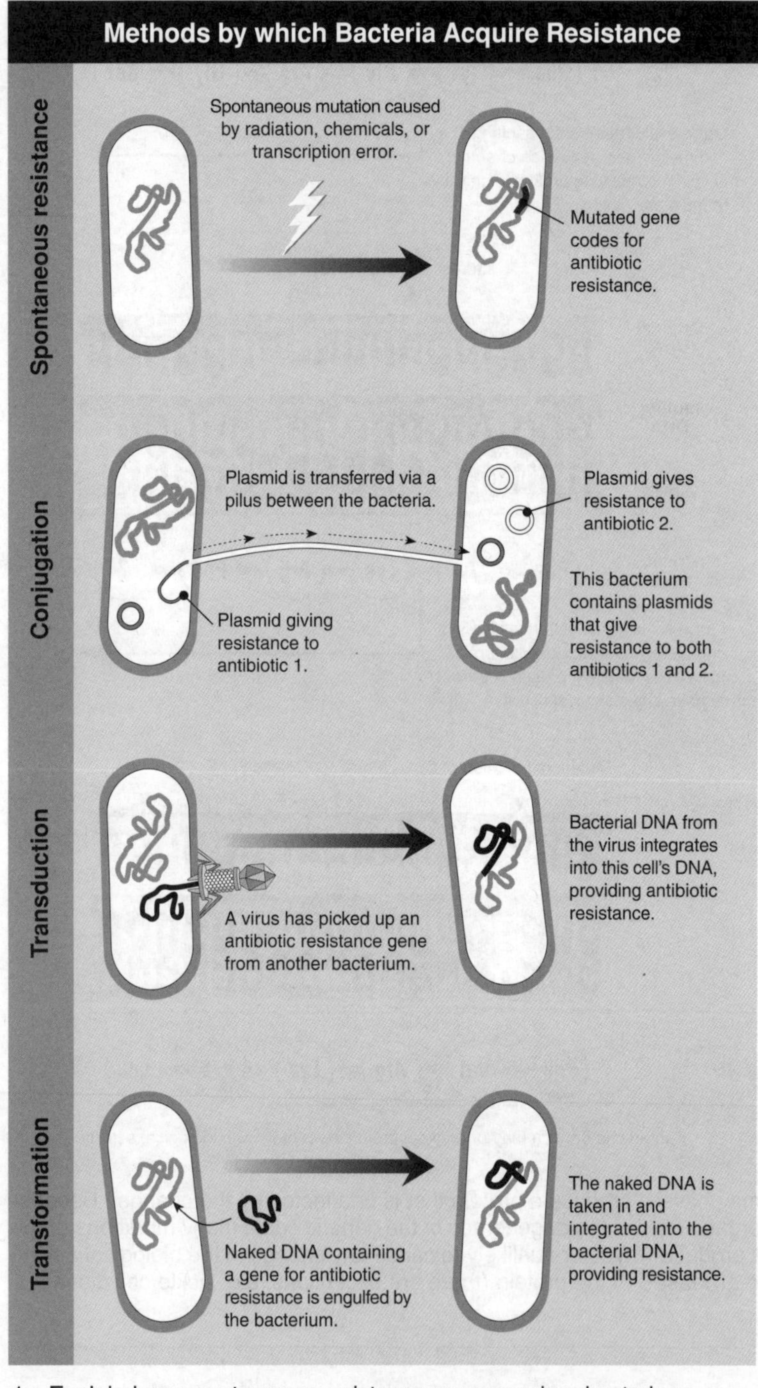

Mechanisms of Resistance

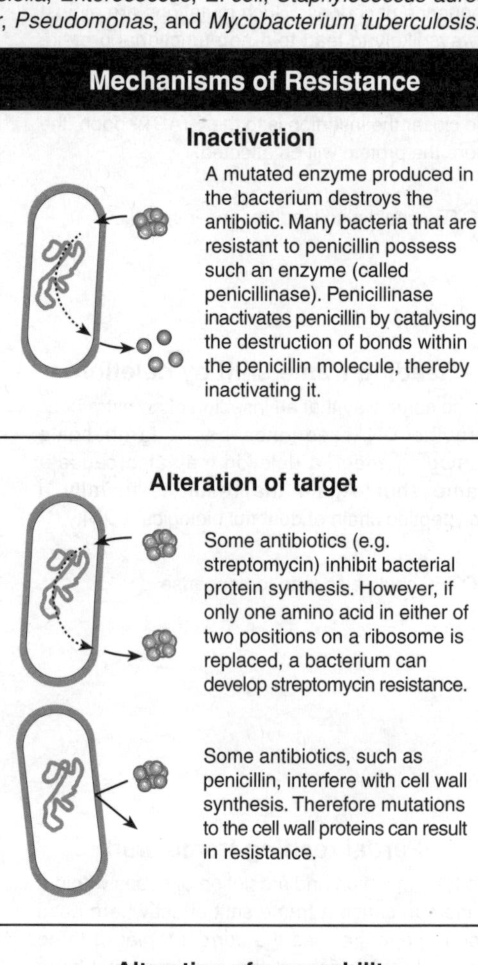

Alteration of permeability

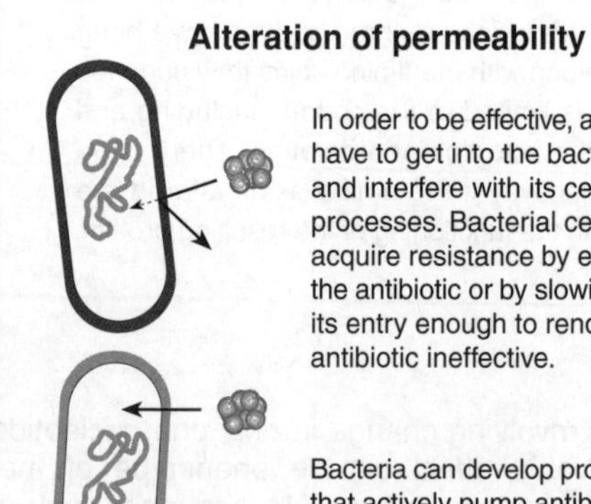

In order to be effective, antibiotics have to get into the bacterial cell and interfere with its cellular processes. Bacterial cells can acquire resistance by excluding the antibiotic or by slowing down its entry enough to render the antibiotic ineffective.

Bacteria can develop proteins that actively pump antibiotics out of their cell faster than the antibiotics can enter.

1. Explain how spontaneous resistance can occur in a bacterium:

2. Explain how the misuse of antibiotics by patients can lead to the development of antibiotic resistant bacteria:

Cystic Fibrosis Mutation

Cystic fibrosis is an inherited disorder caused by a faulty gene which in turn codes for a faulty **CFTR protein**. This is the most common lethal genetic mutation of caucasians with 5% of the population thought to be carriers (heterozygous). The DNA sequence below is part of the transcribing sequence for the **normal** cystic fibrosis gene.

Normal CFTR protein *(1480 amino acids)*
Correctly controls chloride ion balance in the cell

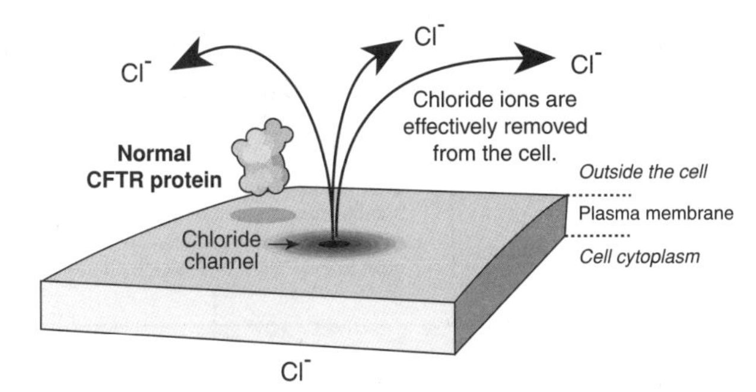

Abnormal CFTR protein *(1479 amino acids)*
Does not control chloride ion balance in the cell

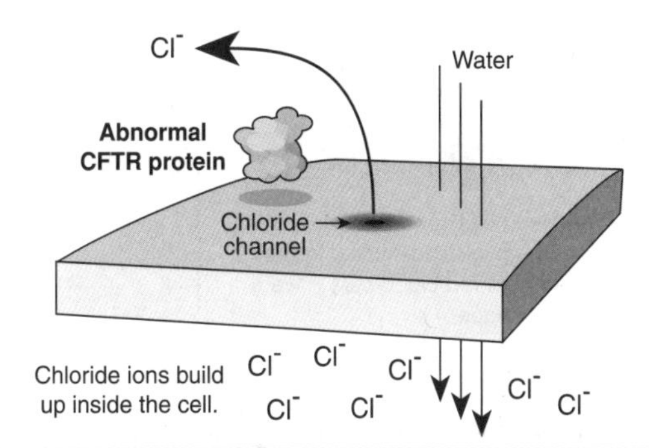

The CFTR gene on chromosome 7

The CFTR gene is located on chromosome 7. Cystic fibrosis results from a deletion mutation where 3 nucleotides are lost. This in turn causes the loss of a single **amino acid** (the 508th in a total chain of 1480) from an important protein called CFTR. This protein normally regulates the chloride channels in cell membranes, but the mutant form fails to achieve this adequately. The portion of the DNA containing the mutation site is shown below:

p
q
CFTR gene

The CFTR protein consists of 1480 amino acids

CFTR protein

The mutant form of the CFTR (Cystic Fibrosis Transmembrane Conductance Regulator) protein does not work properly and excessive amounts of chloride ions remain in the cell. This in turn leads to water from the tissue fluid outside the cell entering the cell. This accounts for the symptoms of this genetic disease, where mucus-secreting glands, particularly in the lungs and pancreas, become fibrous and produce abnormally thick mucus.

Base 1630

DNA: C C G T G G T A A T T T C T T T T A T A G T A G A A A C C A C C A

This triplet codes for the 500th amino acid

The 508th triplet is lost (not present) in the mutant form

1. Identify how many of the following are exhibited in the DNA sequence above:

   (a) Bases: ______________ (b) Triplets: ______________ (c) Amino acids coded for: ______________

2. Write the mRNA sequence for the transcribing DNA strand above:

   ____

3. Use a mRNA-amino acid table to determine the amino acid sequence coded by the mRNA (in question 2 above) for the fragment of the normal protein we are studying here (consult a textbook or *Advanced Biology AS*):

   ____

4. The mutation that causes cystic fibrosis has nucleotide bases missing at positions 1654-1656.

   (a) Rewrite the transcribing DNA sequence from the above diagram, but without the 508th triplet shown (white):

   ____

   (b) State what kind of mutation this is: ______________________________

5. Write the mRNA sequence for the mutant DNA strand above:

   ____

6. Use a mRNA-amino acid table to determine the amino acid sequence coded by the mRNA (in question 5 above) for the fragment of the mutant protein we are studying here (consult a textbook or *Advanced Biology AS*):

   ____

7. Identify the amino acid that has been removed from the protein by this mutation: ______________________________

The Sickle Cell Mutation

Sickle cell disease (formerly called sickle cell anaemia) is an inherited disorder caused by a gene mutation which codes for a faulty beta (β) chain haemoglobin (Hb) protein. This in turn causes the red blood cells to deform causing a whole range of medical problems. The DNA sequence below is the beginning of the transcribing sequence for the **normal** β-chain Hb molecule.

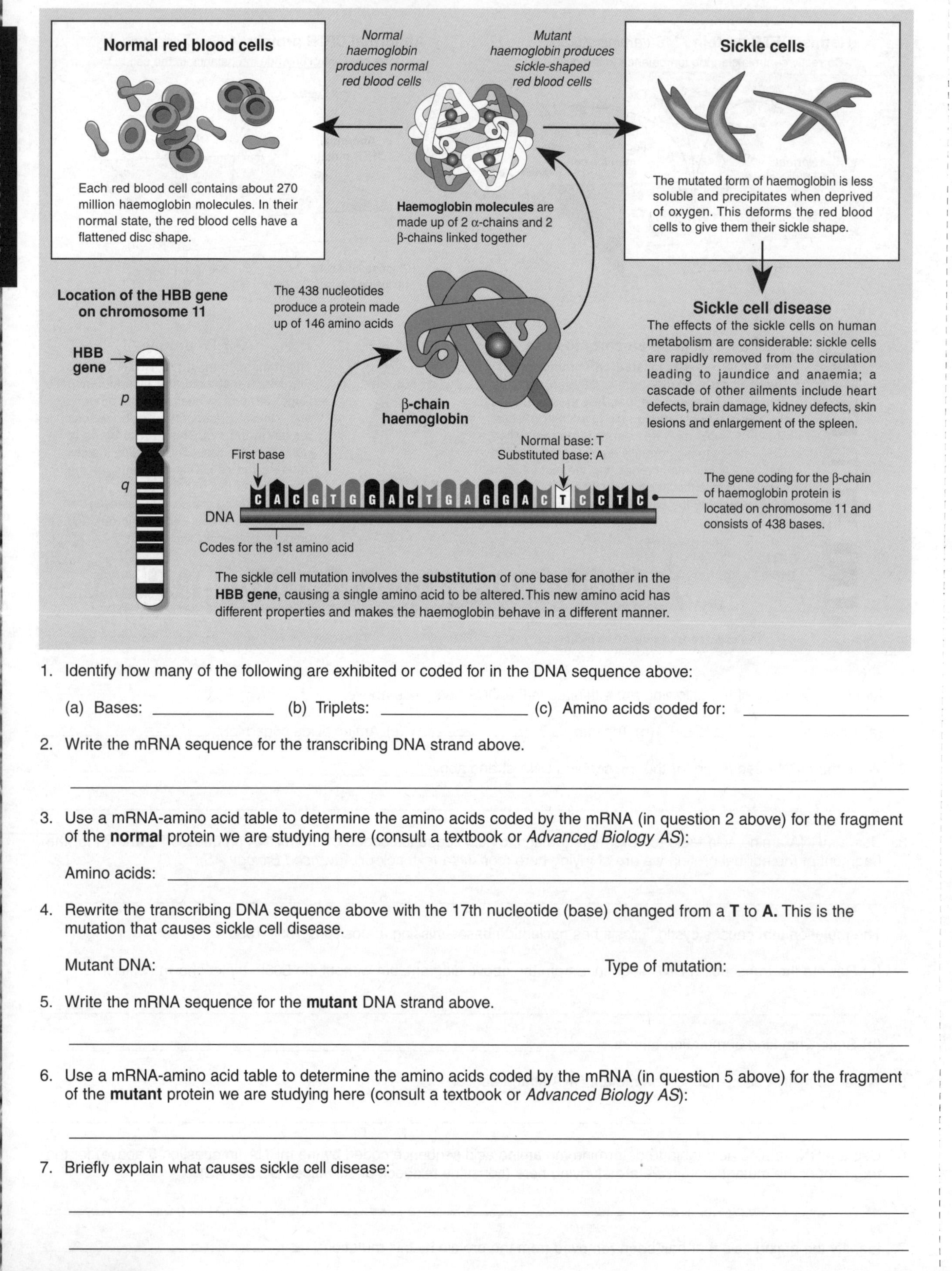

1. Identify how many of the following are exhibited or coded for in the DNA sequence above:

   (a) Bases: ______________ (b) Triplets: ________________ (c) Amino acids coded for: ________________

2. Write the mRNA sequence for the transcribing DNA strand above.

   ____

3. Use a mRNA-amino acid table to determine the amino acids coded by the mRNA (in question 2 above) for the fragment of the **normal** protein we are studying here (consult a textbook or *Advanced Biology AS*):

   Amino acids: ____

4. Rewrite the transcribing DNA sequence above with the 17th nucleotide (base) changed from a **T** to **A.** This is the mutation that causes sickle cell disease.

   Mutant DNA: ______________________________________ Type of mutation: ________________

5. Write the mRNA sequence for the **mutant** DNA strand above.

   ____

6. Use a mRNA-amino acid table to determine the amino acids coded by the mRNA (in question 5 above) for the fragment of the **mutant** protein we are studying here (consult a textbook or *Advanced Biology AS*):

   ____

7. Briefly explain what causes sickle cell disease: ______________________________________

   ____

   ____

Inherited Metabolic Disorders

Humans have more than 6000 physiological diseases attributed to mutations in single genes and over one hundred syndromes known to be caused by chromosomal abnormality. The number of genetic disorders identified increases every year. Rapid progress of the Human Genome Project is enabling the identification of the genetic basis of these disorders. This will facilitate the development of new drug therapies and gene therapies. Four genetic disorders are summarised below.

Sickle Cell Disease

Synonym: Sickle cell anaemia

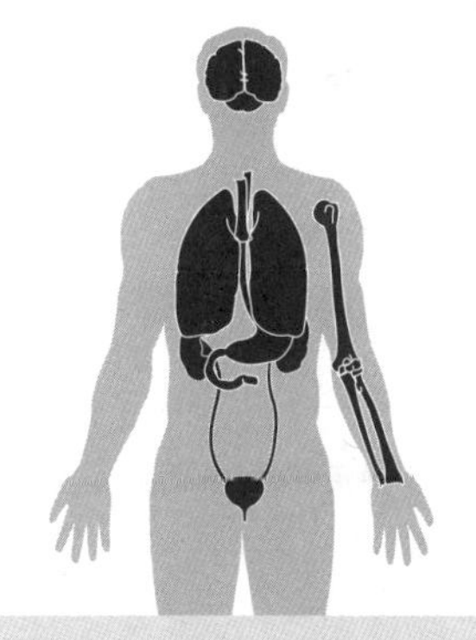

Incidence: Occurs most commonly in people of African ancestry.
West Africans: 1% (10-45% carriers)
West Indians: 0.5%

Gene type: Autosomal recessive mutation which results in the substitution of a single nucleotide in the HBB gene that codes for the beta chain of haemoglobin.

Gene location: Chromosome 11

Symptoms: Include the following: pain, ranging from mild to severe, in the chest, joints, back, or abdomen; swollen hands and feet; jaundice; repeated infections, particularly pneumonia or meningitis; kidney failure; gallstones (at an early age); strokes (at an early age), anaemia.

Treatment and outlook: Patients are given folic acid. Acute episodes may require oxygen therapy, intravenous infusions of fluid, and antibiotic drugs. Experimental therapies include bone marrow transplants and gene therapy.

ß-Thalassaemia

Synonyms: Cooley anaemia, Mediterranean anaemia

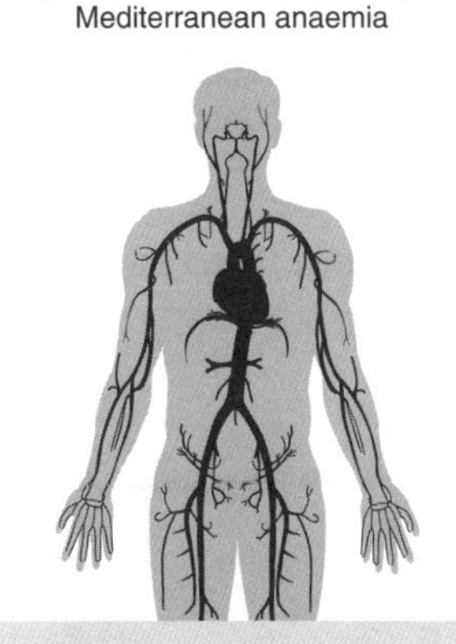

Incidence: Most common type of thalassaemia affecting 1% of some populations. More common in Asia, Middle East and Mediterranean.

Gene type: Autosomal recessive mutation of the HBB gene coding for the haemoglobin beta chain. It may arise through a gene deletion or a nucleotide deletion or insertion.

Gene location: Chromosome 11

Symptoms: The result of haemoglobin with few or no beta chains, causes a severe anaemia during the first few years of life. People with this condition are tired and pale because not enough oxygen reaches the cells.

Treatment and outlook: Patients require frequent blood transfusions which causes iron build-up in their heart, liver, and other organs. Bone marrow transplants and gene therapy hold promise and are probable future treatments.

Cystic Fibrosis

Synonyms: Mucoviscidosis, CF

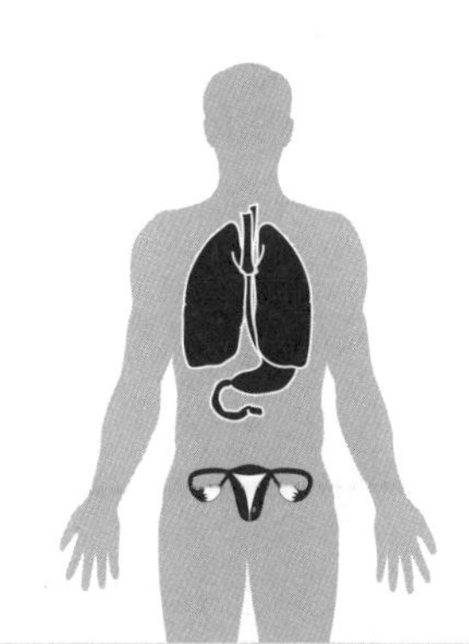

Incidence: Varies with populations:
Northern Ireland: 1 in 1800
Asians in England: 1 in 10 000
Caucasians: 1 in 20-28 are carriers

Gene type: Autosomal recessive. Over 500 different recessive mutations (deletions, missense, nonsense, terminator codon) of the CFTR gene have been identified.

Gene location: Chromosome 7

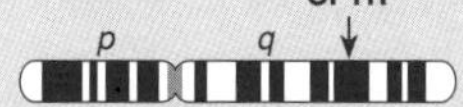

Symptoms: Disruption of glands: the *pancreas*; *intestinal glands*; *biliary tree* (biliary cirrhosis); *bronchial glands* (chronic lung infections); and *sweat glands* (high salt content of which becomes depleted in a hot environment); *infertility* occurs in males/females.

Treatment and outlook: Conventional: chest physiotherapy, a modified diet, and the use of TOBI antibiotic to control lung infections. Outlook: Gene transfer therapy inserting normal CFTR gene using adenovirus vectors and liposomes.

Huntington Disease

Synonyms: Huntington's chorea, HD (abbreviated)

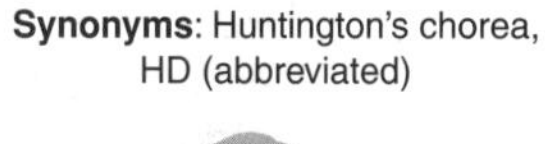

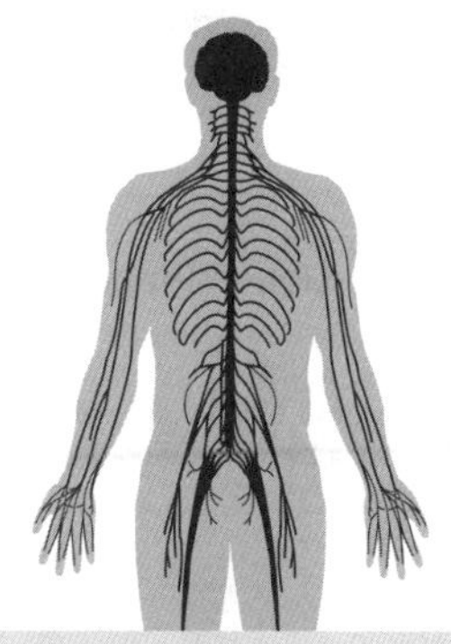

Incidence: An uncommon genetic disease present in 1 in 20 000. The UK has about 8000 sufferers and some 48 000 at risk.

Gene type: An autosomal dominant mutation of the HD gene (IT15) caused by an increase in the length (36-125) of a CAG repeat region (normal range is 11-30 repeats).

Gene location: Chromosome 4

Symptoms: Mutant gene forms defective protein: **Huntingtin**. Progressive, selective *nerve cell death* associated with chorea (jerky, involuntary movements), *psychiatric disorders,* and *dementia* (memory loss, disorientation, impaired ability to reason, and personality changes).

Treatment and outlook: Surgical treatment may be possible. Research is underway to discover drugs that interfere with *Huntingtin* protein. Genetic counselling coupled with genetic screening of embryos may be developed into the future.

1. For each of the genetic disorder below, indicate the following:

(a) Sickle cell disease: Gene name: HBB Chromosome: 11 Mutation type: Substitution

(b) ß-thalassaemia: Gene name: ______ Chromosome: ____ Mutation type: ______

(c) Cystic fibrosis: Gene name: ______ Chromosome: ____ Mutation type: ______

(d) Huntington disease: Gene name: ______ Chromosome: ____ Mutation type: ______

2. Explain the cause of the symptoms for people suffering from β-thalassaemia: ______

3. Suggest a reason for the differences in the country-specific incidence rates for some genetic disorders: ______

Implications of the HGP

The **Human Genome Project** (HGP) is a publicly funded venture involving many different organisations throughout the world. In 1998, Celera Genomics in the USA began a competing project, as a commercial venture, in a race to be the first to determine the human genome sequence. In 2000, both organisations reached the first draft stage, and the entire genome is now available as a high quality (golden standard) sequence. In addition to determining the order of bases in the human genome, genes are being identified, sequenced, and mapped (their specific chromosomal location identified). The next challenge is to assign functions to the identified genes. By identifying and studying the protein products of genes (a field known as **proteomics**), scientists can develop a better understanding of genetic disorders. Long term benefits of the HGP are both medical and non-medical (opposite). Many biotechnology companies have taken out patents on gene sequences. This practice is controversial because it restricts the use of the sequence information to the patent holders. Other genome sequencing projects have arisen as a result of the initiative to sequence the human one. A controversial project to map the differences between racial and ethnic groups is the **Human Genome Diversity Project** (HGDP). It aims to understand the degree of diversity amongst individuals in the human species. It is still in its planning stages, seeking the best way to achieve its goals.

Gene Mapping

This process involves determining the precise position of a gene on a chromosome. Once the position is known, it can be shown on a diagram.

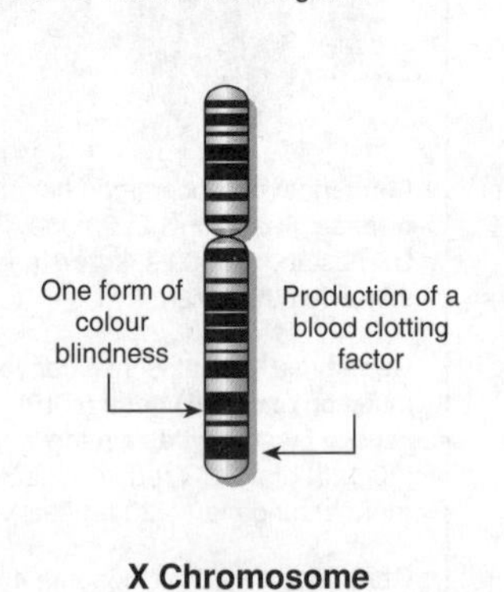

Equipment used for DNA Sequencing

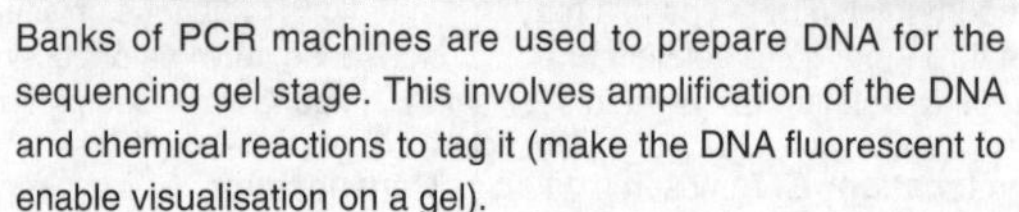
Banks of PCR machines are used to prepare DNA for the sequencing gel stage. This involves amplification of the DNA and chemical reactions to tag it (make the DNA fluorescent to enable visualisation on a gel).

Banks of DNA sequencing gels and powerful computers are used to determine the order of bases in DNA samples.

Count of Mapped Genes

The length and number of mapped genes to date for each chromosome are tabulated below. The entire human genome contains approximately 30 000 genes. By early 2003, almost half these genes had been mapped.

Chromosome	Length (Mb)	No. of Mapped Genes
1	263	1261
2	255	845
3	214	609
4	203	460
5	194	605
6	183	799
7	171	801
8	155	437
9	145	491
10	144	395
11	144	950
12	143	678
13	114	192
14	109	698
15	106	371
16	98	502
17	92	817
18	85	185
19	67	750
20	72	575
21	50	295
22	56	439
X	164	770
Y	59	89
	Total:	**14 014**

As at: 23 May 2004 For updates see:
http://gdbwww.gdb.org/gdbreports/CountGeneByChromosome.html

Examples of Mapped Genes

The positions of an increasing number of genes have been mapped onto human chromosomes (see below). Sequence variations can cause or contribute to identifiable disorders. Note that chromosome 21 (the smallest human chromosome) has a relatively low gene density, while others are gene rich. This is possibly why trisomy 21 (Down syndrome) is one of the few viable human autosomal trisomies.

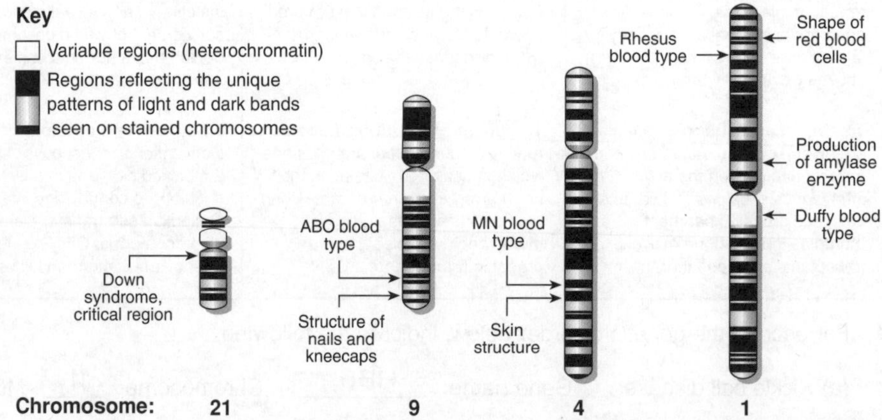

Qualities of DNA Sequence Data

The aim of the HGP is to produce a continuous block of sequence information for each chromosome. Sequence information is first obtained to draft quality. Golden standard sequence is higher quality and is obtained following more work (see below). By early 2003, four chromosomes had been sequenced to golden standard quality. A coverage of x5 means that most regions have been sequenced five times.

Quality	Error rate	Features	Coverage	Chromosomes completed
Draft	1 in 1000 bases	Significant gaps in the sequence exist. The location of some DNA sequences on chromosomes is only approximately known.	x 5	All (2000)
Golden Standard	1 in 10 000 bases	90% of the DNA is sequenced to high quality with gaps closed.	x 8-9	22, 21, 20, 14 (in order of completion date)

Benefits and ethical issues arising from the Human Genome Project

Medical benefits

- Improved **diagnosis** of disease and predisposition to disease by genetic testing.
- Better identification of disease carriers, through genetic testing.
- Better **drugs** can be designed using knowledge of protein structure (from gene sequence information) rather than by trial and error.
- Greater possibility of successfully using **gene therapy** to correct genetic disorders.

Couples can already have a limited range of genetic tests to determine the risk of having offspring with some disease-causing mutations.

Non-medical benefits

- Greater knowledge of **family relationships** through genetic testing, e.g. paternity testing in family courts.
- Advances **forensic science** through analysis of DNA at crime scenes.
- Improved knowledge of the evolutionary relationships between humans and other organisms, which will help to develop better, more accurate classification systems.

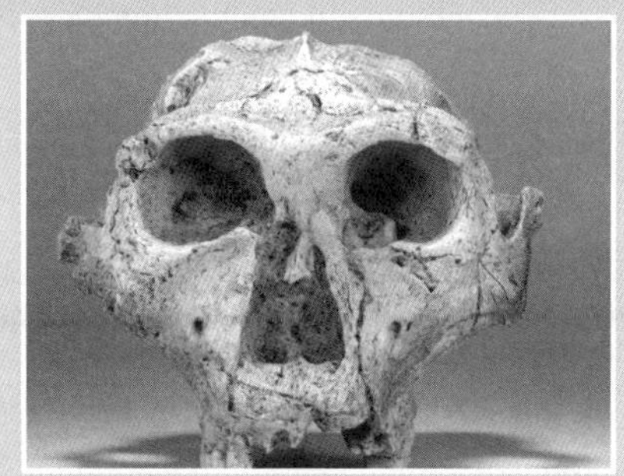

When DNA sequences are available for humans and their ancestors, comparative analysis may provide clues about human evolution.

Possible ethical issues

- It is unclear whether third parties, e.g. health insurers, have rights to genetic test results.
- If treatment is unavailable for a disease, genetic knowledge about it may have no use.
- Genetic tests are costly, and there is no easy answer as to who should pay for them.
- Genetic information is hereditary so knowledge of an individual's own genome has implications for members of their family.

Legislation is needed to ensure that there is no discrimination on the basis of genetic information, e.g. at work or for health insurance.

1. Briefly describe the objectives of the Human Genome Project (HGP): ______________________

____

____

2. Discuss the medical and non-medical benefits of the Human Genome Project (HGP): ______________________

____

____

____

____

____

3. Discuss some of the potential ethical issues associated with the Human Genome Project (HGP):

____

____

____

____

____

____

4. Explain what is meant by **proteomics** and explain the significance of this new field of research:

____

____

____

____

____

Inheritance

AQA-A	AQA-B	CIE	Edexcel	OCR
Complete:	Complete:	Complete:	Complete:	Complete:
1-21, 25-27	*1-16, 19-21, 24, 27*	*1-21*	*1-16, 19, 22-25*	*1-21*
	Extension: 17-18, 25-26	*Extension: 25-27*	*Extension: 17-18, 26-27*	*Extension: 25-27*

Learning Objectives

- ☐ 1. Compile your own glossary from the **KEY WORDS** displayed in **bold type** in the learning objectives below.

Meiosis and variation *(pages 96-97, 102-106)*

- ☐ 2. Describe the role of meiosis and fertilisation in contributing to genetic variation in the offspring. Appreciate the role of **crossing over** and **independent assortment** in meiosis, and explain when these occur. Coverage of this material, including the learning objectives is provided in "*Sources of Variation*".
- ☐ 3. Appreciate that the **phenotype** is the result of the expression of the **genotype** (the genetic constitution of an organism) and the influence of the environment. Coverage of this material, including the learning objectives is provided in "*Sources of Variation*".

The study of inheritance
(Required knowledge: see pages 120, 122, 130)

- ☐ 4. Recall Mendel's principles of inheritance, stating their importance to the understanding of heredity: the *Theory of particulate inheritance*, the *Law of segregation*, and the *Law of independent assortment*.
- ☐ 5. Appreciate that Mendelian principles were important in providing a mechanism through which natural selection could act upon the genetic variation in populations.
- ☐ 6. Define the terms: **allele** and **locus**. Distinguish between **dominant**, **recessive**, and **codominant** alleles. Appreciate how new alleles are formed.
- ☐ 7. Define the term **trait** with respect to the study of genetics.
- ☐ 8. Describe how genes, and their dominant and recessive alleles, are represented symbolically by letters of the alphabet. Explain clearly what is meant by the terms: **heterozygous** and **homozygous**.
- ☐ 9. Distinguish between **genotype** and **phenotype** and use these terms appropriately when discussing the outcomes of genetic crosses.
- ☐ 10. Demonstrate an ability to use a **Punnett square** in the solution of different inheritance problems. Define the terms: **monohybrid**, and **dihybrid cross**.
- ☐ 11. Define and demonstrate an understanding of the terms commonly used in inheritance studies: **cross**, **carrier**, **selfing**, **pure-breeding**, **test-cross**, **back-cross**, **offspring** (progeny), **F_1 generation**, **F_2 generation**.

Inheritance patterns *(pages 123-131, 142-144)*

For each of the cases below, use a Punnett square through to the F_2 generation and determine the probability of the occurrence of a particular genetic outcome (genotype and phenotype ratios).

- ☐ 12. Describe **monohybrid inheritance** involving a single trait or gene locus. Solve problems involving the inheritance of phenotypic traits that follow a simple **dominant-recessive** pattern. Define the term: **carrier**.
- ☐ 13. Distinguish between **codominance** and **incomplete dominance**, explaining how you would recognise these dominance patterns in genetic crosses.
- ☐ 14. Recognise that **multiple alleles** (alternative alleles) may exist for a single gene, e.g. ABO blood groups. Solve problems involving the inheritance of phenotypic traits involving **codominance** of multiple alleles.
- ☐ 15. EXTENSION ONLY: Describe examples of recessive and/or dominant **lethal alleles**. Recognise how lethal alleles modify the ratio of progeny resulting from a cross. If required, explain the mechanisms by which the lethal allele operates. An activity covering the inheritance of lethal alleles is available for printing from **www.biozone.co.uk/uk-extras.html**
- ☐ 16. Describe **dihybrid inheritance** involving two unlinked genes for two independent traits, where the genes described are carried on different chromosomes (**unlinked, autosomal genes**). Solve problems involving dihybrid inheritance of unlinked genes.
- ☐ 17. Demonstrate an ability to use the **chi-squared test** to test the significance of differences between the observed and expected results of genetic crosses.
- ☐ 18. Explain why experimental results may be expected only to approximate Mendelian ratios.

Sex determination in humans *(page 121)*

- ☐ 19. Understand the basis of **sex determination** in humans. Recognise humans as being of the **XX** / **XY** type. Distinguish **sex chromosomes** from **autosomes**.

Sex linkage *(pages 132-134)*

- ☐ 20. Define the term: **sex linked**. Describe examples and solve problems involving different patterns of sex linked inheritance involving sex linked genes (e.g. inheritance of red-green colour-blindness, haemophilia, and the sex linked form of rickets).
- ☐ 21. Contrast the pattern of inheritance of **sex linked recessive traits** (e.g. haemophilia) and **sex linked dominant traits** (e.g. rickets).

Gene interactions *(pages 135-144)*

- ☐ 22. Recognise and describe examples of a simple interaction between two genes: e.g. **collaboration** in the determination of **comb shape** of domestic hens. Show how genes that show collaboration influence the same trait, but produce a phenotype that could not result from the action of either gene independently. State the possible phenotypes for this type of interaction. Solve problems involving the inheritance of phenotypic traits involving collaboration.

- ☐ 23. Explain what is meant by **pleiotropy** and describe an example of a gene with pleiotropic effects, e.g. sickle cell gene mutation. Describe how the phenotype is affected by the presence of the sickle cell mutation.
- ☐ 24. Explain what is meant by **epistasis**, **epistatic gene**, and **hypostatic gene**. Solve problems involving the inheritance of phenotypic traits involving epistasis.

  Epistasis case study 1: coat colour in mice. Several genes determine coat colour in mice. Mice that are homozygous recessive for the albino gene are white regardless of their genotype for coat colour.

  Epistasis case study 2: where one gene can only be expressed in the presence of another. Such **complementary genes** control some flower colours (e.g. in sweet peas): one gene controls the production of a colourless intermediate (for a pigment) and another gene controls the transformation of that intermediate into the pigment. Both genes need to have a dominant allele present for the pigmented phenotype to be expressed.
- ☐ 25. Describe the distribution pattern of phenotypic variation produced by **polygenic inheritance**. Understand why **polygenes** are also called multiple genes. Provide examples of traits that follow this type of **quantitative inheritance** pattern.
- ☐ 26. Demonstrate genetic crosses for the inheritance of a polygenic trait (e.g. skin colour) using a Punnett square through to the F_2 generation and determine the probability of the occurrence of a particular genetic outcome (i.e. genotype and phenotype ratios).
- ☐ 27. Distinguish between **continuous** and **discontinuous variation** in phenotypes and recognise the genetic basis for each pattern.

See the 'Textbook Reference Grid' on pages 8-9 for textbook page references relating to material in this topic.

Supplementary Texts

See pages 4-6 for additional details of these texts:

- Adds, J., *et al.*, 2001. **Genetics, Evolution and Biodiversity**, (NelsonThornes), pp. 72-80, 83-86.
- Clegg, CJ., 1999. **Genetics and Evolution**, (John Murray), pp. 4-39 (in part).
- Jones, N., *et al.*, 2001. **Essentials of Genetics**, pp. 29-56, 61-74 (review), 77-110.

See page 6 for details of publishers of periodicals:

STUDENT'S REFERENCE

- **What is a Gene?** Biol. Sci. Rev., 15(2) Nov. 2002, pp. 9-11. *A good introduction to genes and their role in heredity. This article provides a useful overview and an historical perspective.*
- **What is Variation?** Biol. Sci. Rev., 13(1) Sept. 2000, pp. 30-31. *The nature of continuous and discontinuous variation in particular characters. The distribution pattern of traits that show continuous variation as a result of polygeny is discussed.*
- **It Takes Two** New Scientist, 5 October 2002, pp. 34-37. *The relative influence of genes and environment. Studies of twins may help to clarify the nature versus nurture debate.*

Gender determination

- **Determinedly Male** Biol. Sci. Rev., 7(3) Jan. 1995, pp. 28-31. *The basis of gender determination and sexual differentiation in animals.*

Inheritance patterns

- **Pay Attention Rover** New Scientist, 10 May 1997, pp. 30-33. *Genes account for about 25% of the differences between dogs in their ability to learn.*
- **Queen Victoria's Gene** Biol. Sci. Rev., 8(2) November 1995, pp. 18-22. *Inheritance of haemophilia using pedigree analysis of Queen Victoria's as an example.*
- **Secrets of The Gene** National Geographic, 196(4) Oct. 1999, pp. 42-75. *Thorough coverage of the nature of genes and the inheritance of particular genetic traits through certain populations.*
- **Stuttering Gene Goes GAA GAA** New Scientist, 16 March 1996, p. 19. *Genetic stutters underlying conditions like Huntington's and fragile X syndrome show recessive inheritance patterns.*
- **Beetles Show Why it Pays to Have Sex** New Scientist, 28 October 1995, p. 20. *New experiments with flour beetles show how sex benefits a population within a few generations.*
- **Upwardly Mobile** New Scientist, 7 Nov. 1998, p. 6. *Recent studies indicate that 70% of the variation in adult height is due to one gene.*
- **Blood Group Antigens** Biol. Sci. Rev., 9(5) May 1997, pp. 10-13. *Blood group antigens illustrate human genetic diversity at the molecular level.*

TEACHER'S REFERENCE

Gender determination

- **Why the Y is so Weird** Scientific American, Feb. 2001, pp. 42-47. *A comparison of the features of the X and Y chromosomes: why they look so different and how they behave during division.*
- **The Double Life of Women** New Scientist, 10 May 2003, pp. 42-45. *Two X chromosomes make you a girl, but why are women not overloaded by having two. A good account of the role of X inactivation in embryonic development.*
- **Decoding the Ys and Wherefores of Males** New Scientist, 21 June 2003, p. 15. *The publication of the genetic sequence of the Y chromosome has unveiled surprising information; the Y chromosome contains about 78 genes, and has mechanisms by which it can repair itself, generation to generation.*

Inheritance patterns

- **You Are What Your Mother Ate** New Scientist, 9 August 2003, pp. 14-15. *A few genes vital for development might be controlled by what your mother ate. This finding has intrigued scientists looking for the mechanisms behind epigenetic inheritance (the inheritance of phenotypic changes that are not the result of genomic changes).*
- **Fair Enough** New Scientist, 12 October 2002, pp. 34-37. *The inheritance of skin colour in humans. This article examines why humans have such varied skin pigmentation and looks at the argument for there being a selective benefit to being dark or pale in different environments.*
- **Skin Deep** Scientific American, October 2002, pp. 50-57. *This article examines the evolution of skin colour in humans and presents powerful evidence for skin colour ("race") being the end result of opposing selection forces (the need for protection of folate from UV vs the need to absorb vitamin D). Clearly written and of high interest, this is a must for student discussion and a perfect vehicle for examining natural selection.*
- **Hidden Inheritance** New Scientist, 28 Nov. 1998, pp. 27-30. *Epigenetic inheritance: the study of imprinted genes may help to explain why inheritance patterns are not always as predicted.*
- **Why Genes Have a Gender** New Scientist, 22 May 1993, pp. 34-38. *Without knowing it, parents stamp their authority on the DNA of their children.*
- **Where Did You Get Your Brains?** New Scientist, 3 May 1997, pp. 34-39. *What features do we inherit from our mothers and which from our fathers? The study of imprinted genes can make maternal/paternal inheritance clearer.*
- **Spot Marks the X** Scientific American, April 1993, pp. 12-13. *The black and yellow patches in the calico cat's fur are the outward manifestations of a more subtle genetic quirk.*

See pages 10-11 for details of how to access **Bio Links** from our web site: **www.biozone.co.uk**. From Bio Links, access sites under the topics:

GENERAL BIOLOGY ONLINE RESOURCES > **Online Textbooks and Lecture Notes**: • An on-line biology book • Kimball's biology pages • Learn.co.uk • Mark Rothery's biology web site • *... and others* > **General Online Biology resources**: • Ken's Bioweb resources *... and others* > **Glossaries**: • DNA glossary • Genetics glossary... *and others*

GENETICS: • MIT biology hypertextbook • Genetest • Gene Almanac *... and others* > **Molecular Genetics**: • DNA from the beginning *... and others* > **Inheritance**: • BioLogica • Coat color and pattern genetics of the domestic cat • Cat color genetics • Drag and drop genetics • Online Mendelian inheritance in man • Patterns of inheritance • The Biology project: Mendelian genetics • The role of genes *... and others*

Software and video resources are provided on the Teacher Resource Handbook on CD-ROM

Alleles

Sexually reproducing organisms in nearly all cases have paired sets of chromosomes, one set coming from each parent. The equivalent chromosomes that form a pair are termed **homologues**. They contain equivalent sets of genes on them. But there is the potential for different versions of a gene to exist in a population and these are termed **alleles**.

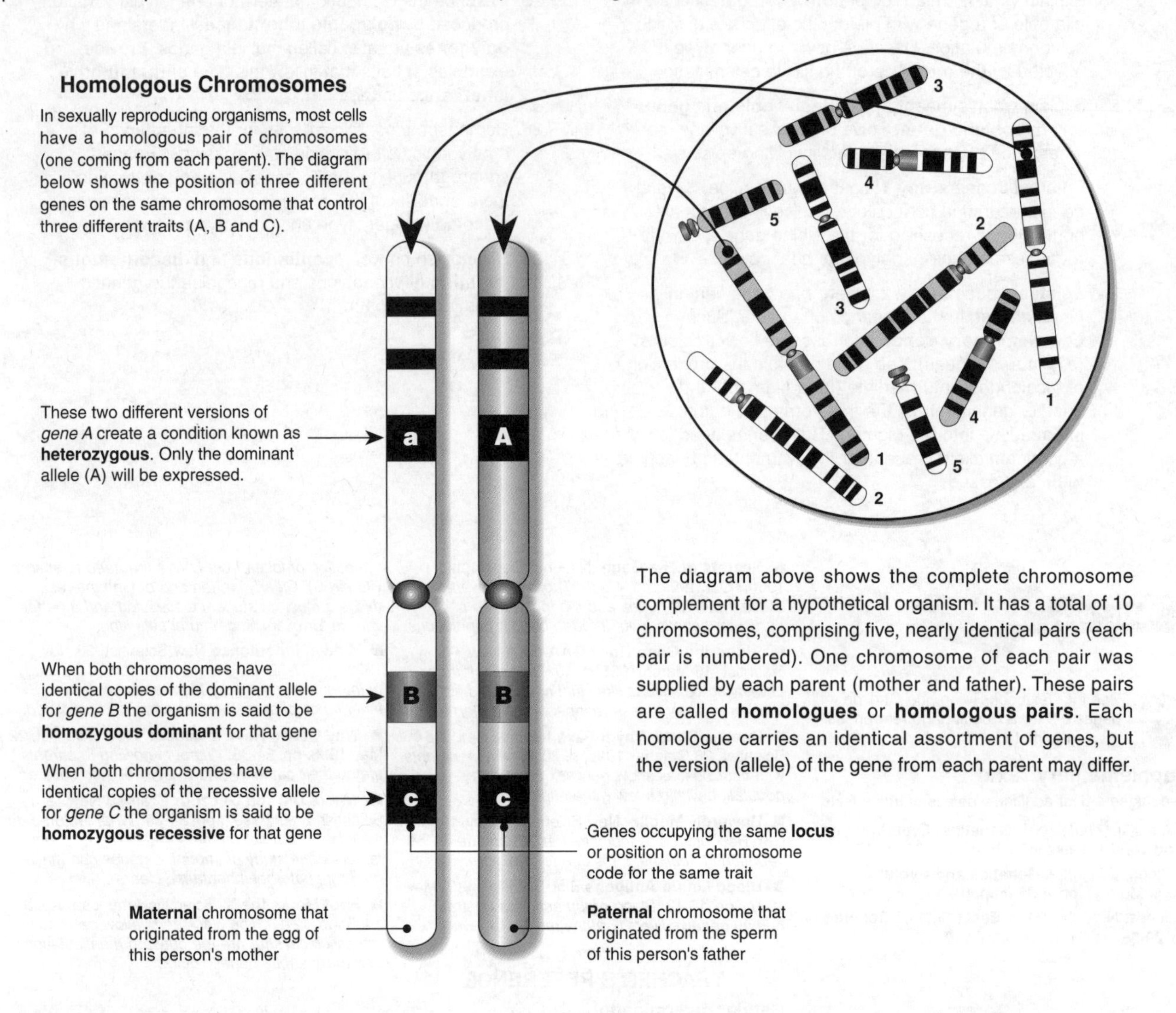

The diagram above shows the complete chromosome complement for a hypothetical organism. It has a total of 10 chromosomes, comprising five, nearly identical pairs (each pair is numbered). One chromosome of each pair was supplied by each parent (mother and father). These pairs are called **homologues** or **homologous pairs**. Each homologue carries an identical assortment of genes, but the version (allele) of the gene from each parent may differ.

1. Define the following terms used to describe the allele combinations in the genotype for a given gene:

   (a) Heterozygous: ____________________

   (b) Homozygous dominant: ____________________

   (c) Homozygous recessive: ____________________

2. For a gene given the symbol 'A', name the alleles present in an organism that is identified as:

   (a) Heterozygous: ________ (b) Homozygous dominant: ________ (c) Homozygous recessive: ________

3. Explain what a homologous pair of chromosomes is: ____________________

4. Discuss the significance of genes existing as **alleles**: ____________________

Sex Determination

The determination of the sex (gender) of an organism is controlled in most cases by the sex chromosomes provided by each parent. These have evolved to regulate the ratios of males and females produced and preserve the genetic differences between the sexes. In humans, males are referred to as the **heterogametic sex** because each somatic cell has one X chromosome and one Y chromosome. The determination of sex is based on the presence or absence of the Y chromosome. Without the Y chromosome, an individual will develop into a **homogametic** female (each somatic cell with two X chromosomes). In mammals, the male is always the hetero-gametic sex, but this is not necessarily the case in other taxa. In birds and butterflies, the female is the heterogametic sex, and in some insects the male is simply X whereas the female is XX.

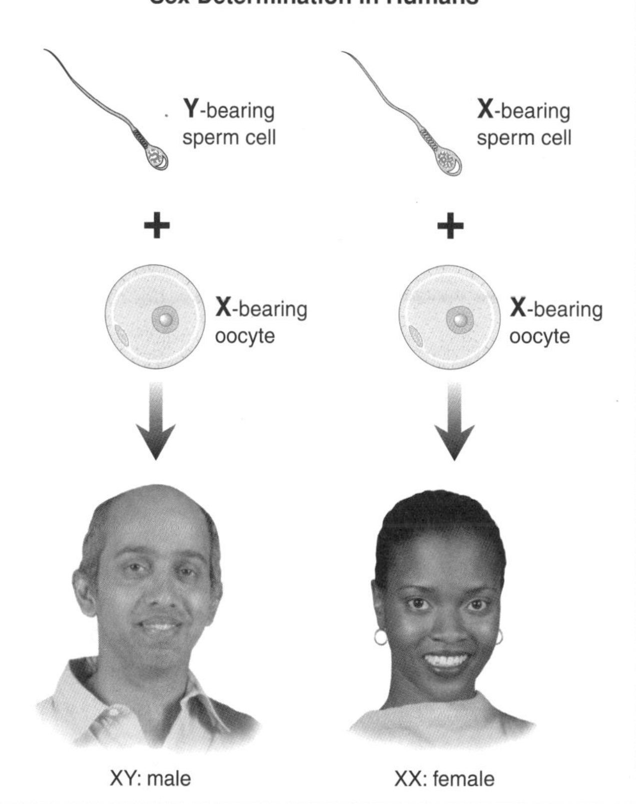

The Sex Determining Region of the Y Chromosome

Scientists have known since 1959 that the Y chromosome is associated with being male. However, it was not until 1990 that a group of researchers, working for the Medical Research Council in London, discovered the gene on the Y chromosome that determines maleness. It was named **SRY**, for **Sex Determining Region of the Y**. The SRY gene produces a type of protein called a **transcription factor**. This transcription factor switches on the genes that direct the development of male structures in the embryo.

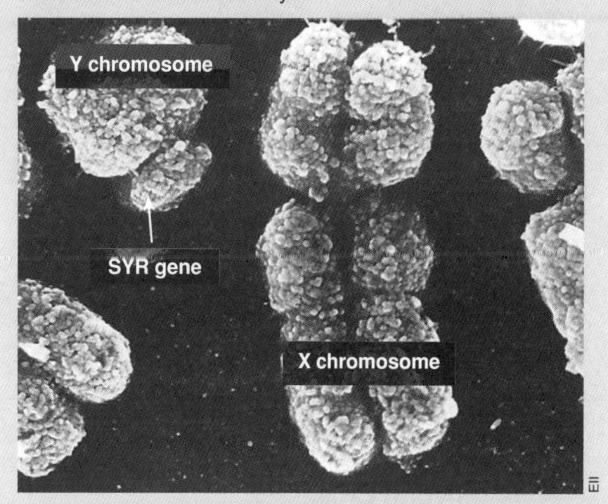

Gynandromorphism occurs when an animal is a genetic mosaic and possesses both male and female characteristics (i.e. some of its cells are genetically male and others are female). This phenomenon is found particularly in insects, but also appears in birds and mammals. Gynandromorphism occurs due to the loss of an X chromosome in a stem cell of a female (XX), so that all the tissues derived from that cell are phenotypically male.

In the pill woodlouse, *Armadillium vulgare*, sex determination is characterised by female heterogamety (ZW) and male homogamety (ZZ). However, in several wild populations this system is overridden by an infectious bacterium. This bacterium causes genetically male woodlice to change into females. The bacteria are transmitted through the egg cytoplasm of the woodlouse. Therefore the conversion of males to females increases the propagation of the bacterium.

1. Explain what determines the sex of the offspring at the moment of conception in humans: ______________________

____

2. Explain why human males are called the heterogametic sex: ______________________

____

Basic Genetic Crosses

For revision purposes, examine the diagrams below on monohybrid crosses and complete the exercise for dihybrid (two gene) inheritance. A **test cross** is also provided to show how the genotype of a dominant phenotype can be determined. A test cross will yield one of two different results, depending on the genotype of the dominant individual. A **back cross** (not shown) refers to any cross between an offspring and one of its parents (or an individual genetically identical to one of its parents).

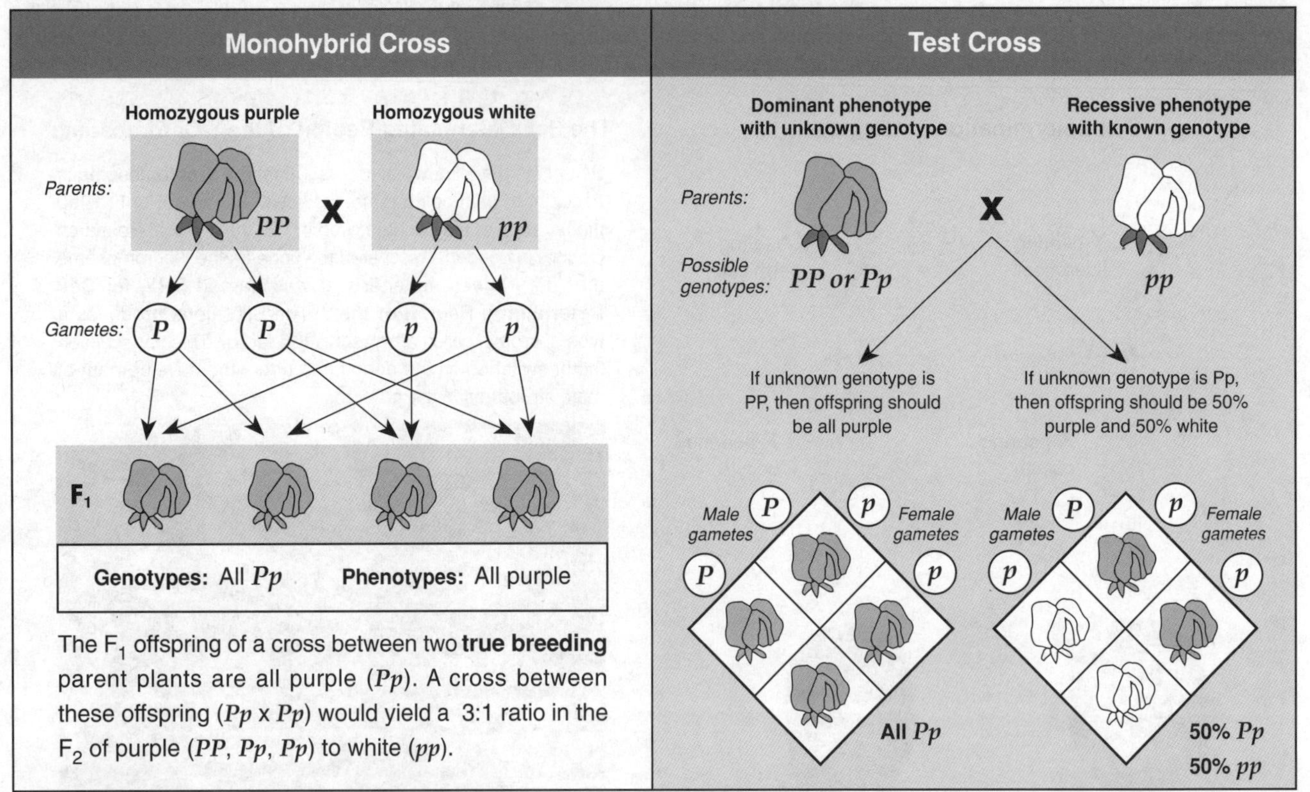

Dihybrid Cross

In pea seeds, yellow colour (Y) is dominant to green (y) and round shape (R) is dominant to wrinkled (r). Each **true breeding** parental plant has matching alleles for each of these characters ($YYRR$ or $yyrr$). F_1 offspring will all have the same genotype and phenotype (yellow-round: $YyRr$).

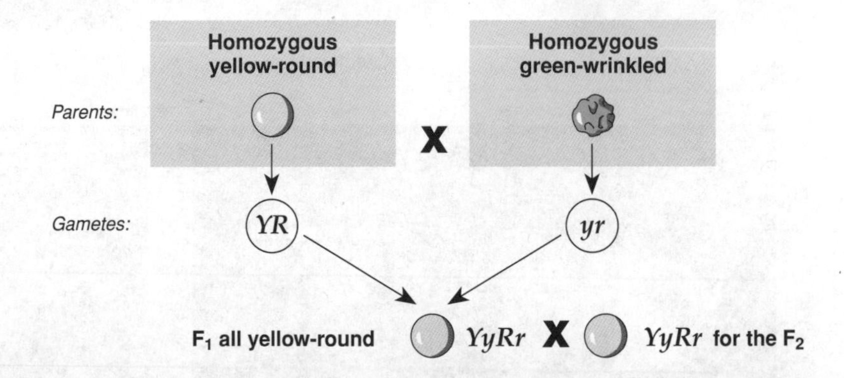

1. Fill in the Punnett square (below right) to show the genotypes of the F_2 generation.
2. In the boxes below, use fractions to indicate the numbers of each phenotype produced from this cross.

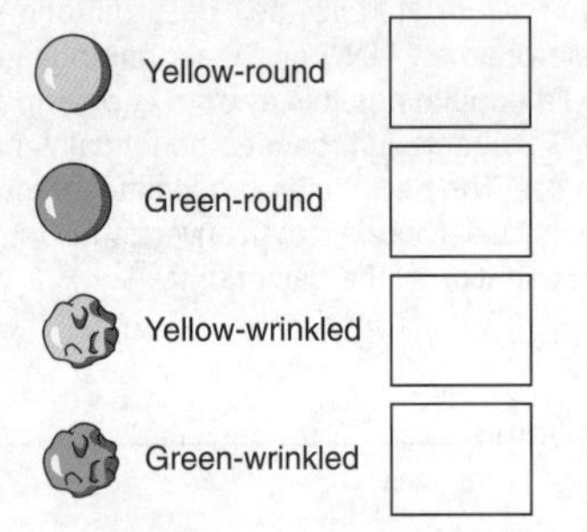

3. Express these numbers as a ratio:

____

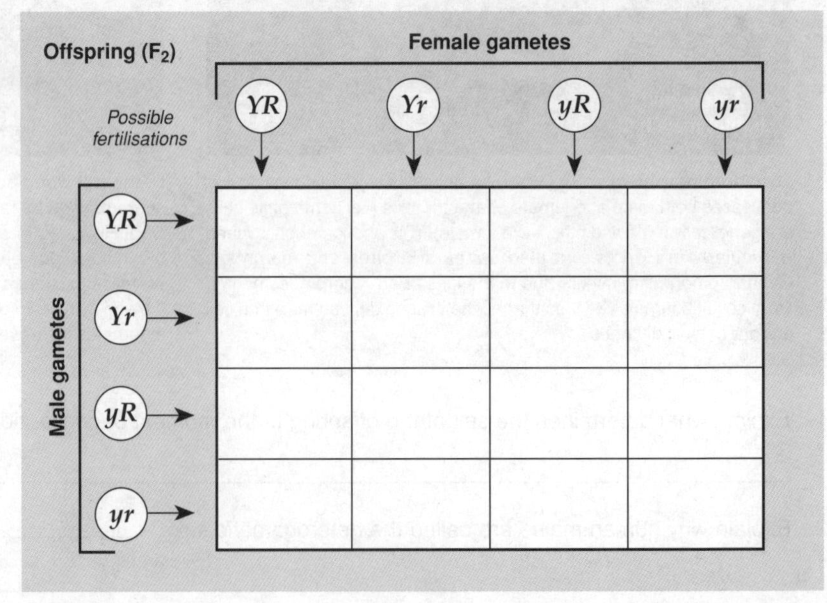

Inheritance

Monohybrid Cross

The study of **single-gene inheritance** is achieved by performing **monohybrid crosses**. The six basic types of matings possible among the three genotypes can be observed by studying a pair of alleles that govern coat colour in the guinea pig. A dominant allele: given the symbol **B** produces **black** hair, and its recessive allele: **b**, produces white. Each of the parents can produce two types of gamete by the process of **meiosis** (in reality there are four, but you get identical pairs). Determine the **genotype** and **phenotype frequencies** for the crosses below (enter the frequencies in the spaces provided). For crosses 3 to 6, you must also determine gametes produced by each parent (write these in the circles), and offspring (F_1) genotypes and phenotypes (write in the genotype inside the offspring and state if black or white).

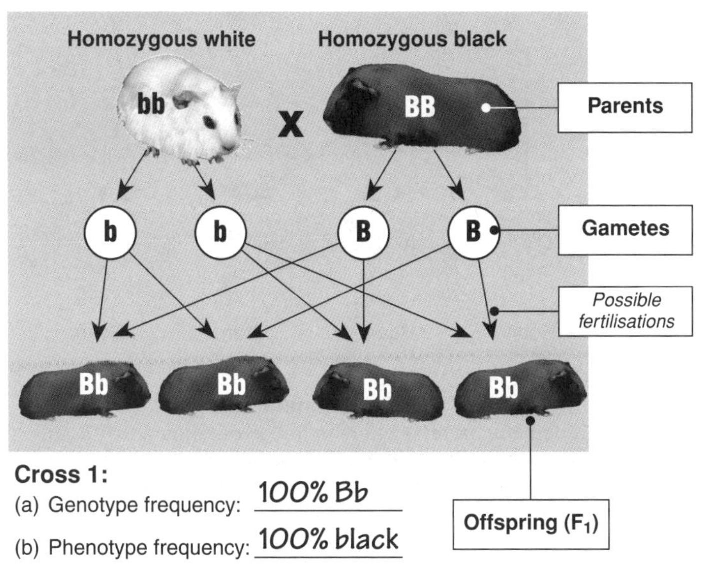

Cross 1:
(a) Genotype frequency: 100% Bb
(b) Phenotype frequency: 100% black

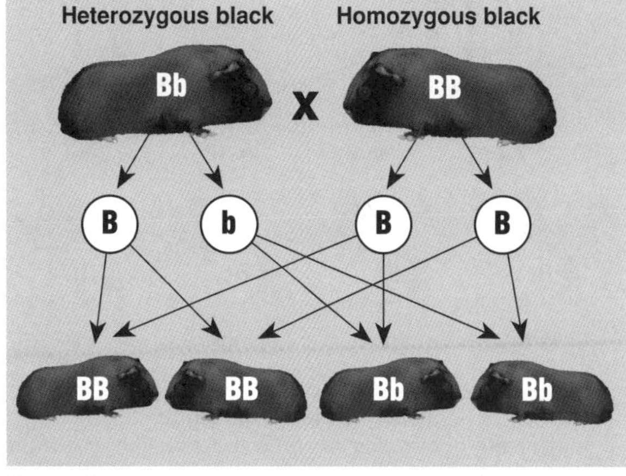

Cross 2:
(a) Genotype frequency: ______________
(b) Phenotype frequency: ______________

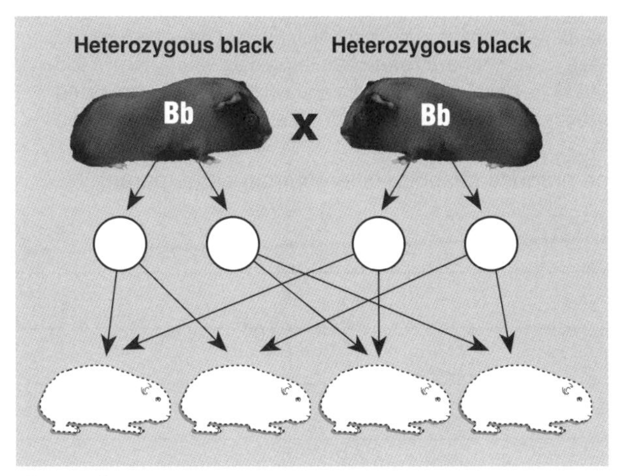

Cross 3:
(a) Genotype frequency: ______________
(b) Phenotype frequency: ______________

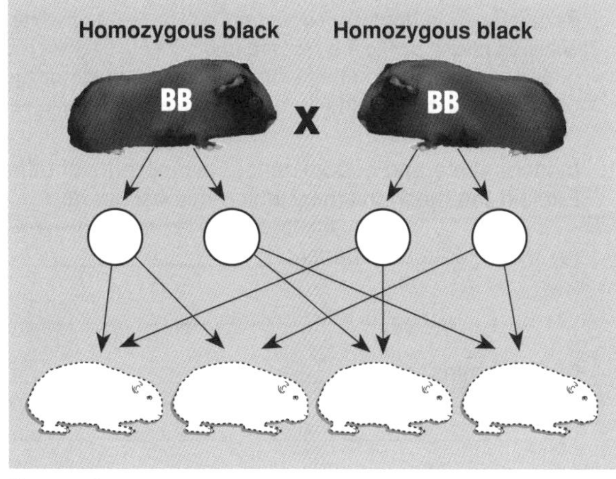

Cross 4:
(a) Genotype frequency: ______________
(b) Phenotype frequency: ______________

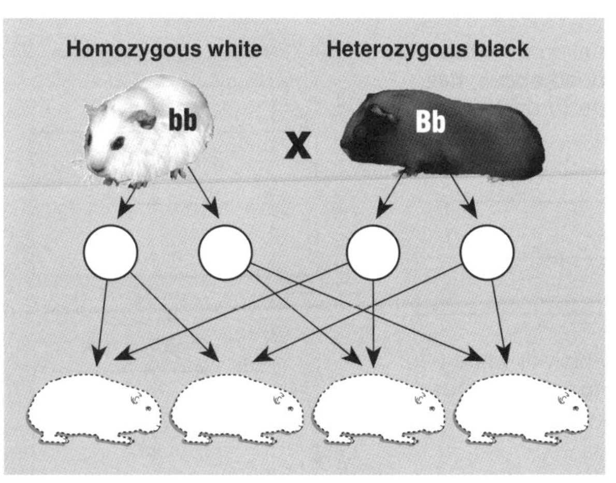

Cross 5:
(a) Genotype frequency: ______________
(b) Phenotype frequency: ______________

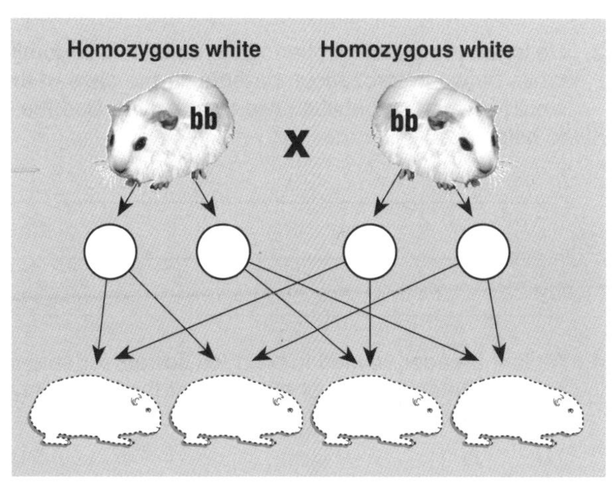

Cross 6:
(a) Genotype frequency: ______________
(b) Phenotype frequency: ______________

Dominance of Alleles

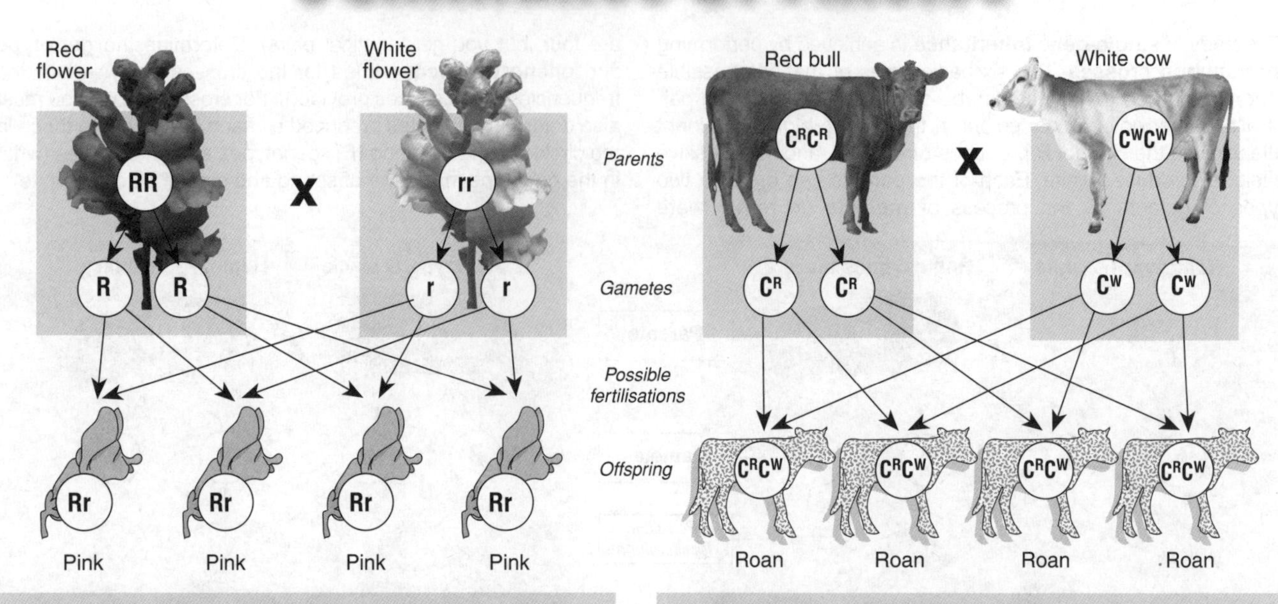

Incomplete Dominance

Incomplete dominance refers to the situation where the action of one allele does not completely mask the action of the other and neither allele has dominant control over the trait. The heterozygous offspring is **intermediate** in phenotype between the contrasting homozygous parental phenotypes. In crosses involving incomplete dominance the phenotype and genotype ratios are identical. Examples include snapdragons (*Antirrhinum*), where red and white-flowered parent plants are crossed to produce pink-flowered offspring. In this type of inheritance the phenotype of the offspring results from the partial influence of both alleles.

Codominance

Codominance refers to inheritance patterns when both alleles in a heterozygous organism contribute to the phenotype. Both alleles are **independently** and **equally expressed**. One example includes the human blood group AB which is the result of two alleles: A and B, both being equally expressed. Other examples include certain coat colours in horses and cattle. Reddish coat colour is not completely dominant to white. Animals that have both alleles have coats that are **roan**-coloured (coats with a mix of red and white hairs). The red hairs and white hairs are expressed equally and independently (not blended to produce pink).

1. In incomplete and codominance, two parents of differing phenotype produce offspring different from either parent. Explain the mechanism by which this occurs in:

   (a) Incomplete dominance: ______________________________

   ____

   (b) Codominance: ______________________________

   ____

2. For each situation below, explain how the heterozygous individuals differ in their phenotype from homozygous ones:

   (a) Incomplete dominance: ______________________________

   (b) Codominance: ______________________________

3. Describe the classical phenotypic ratio for a codominant gene resulting from the cross of two heterozygous parents (in the case of the cattle described above, this would be a cross between two roan cattle). Use the Punnett square (provided right) to help you:

   ____

   ____

   ____

Gametes from male / Gametes from female

4. A plant breeder wanted to produce flowers for sale that were only pink or white (i.e. no red). Determine the phenotypes of the two parents necessary to produce these desired offspring. Use the Punnett square (provided right) to help you:

   ____

   ____

Gametes from male / Gametes from female

In the shorthorn cattle breed coat colour is inherited. White shorthorn parents always produce calves with white coats. Red parents always produce red calves. But when a red parent mates with a white one the calves have a coat colour that is different from either parent, called roan (a mixture of red hairs and white hairs). Look at the example on the previous page for guidance and determine the offspring for the following two crosses. In the cross on the left, you are given the phenotype of the parents. From this information, their genotypes can be determined, and therefore the gametes and genotypes and phenotypes of the calves. In the cross on the right, only one parent's phenotype is known. Work out the genotype of the cow and calves first, then trace back to the unknown bull via the gametes, to determine its genotype.

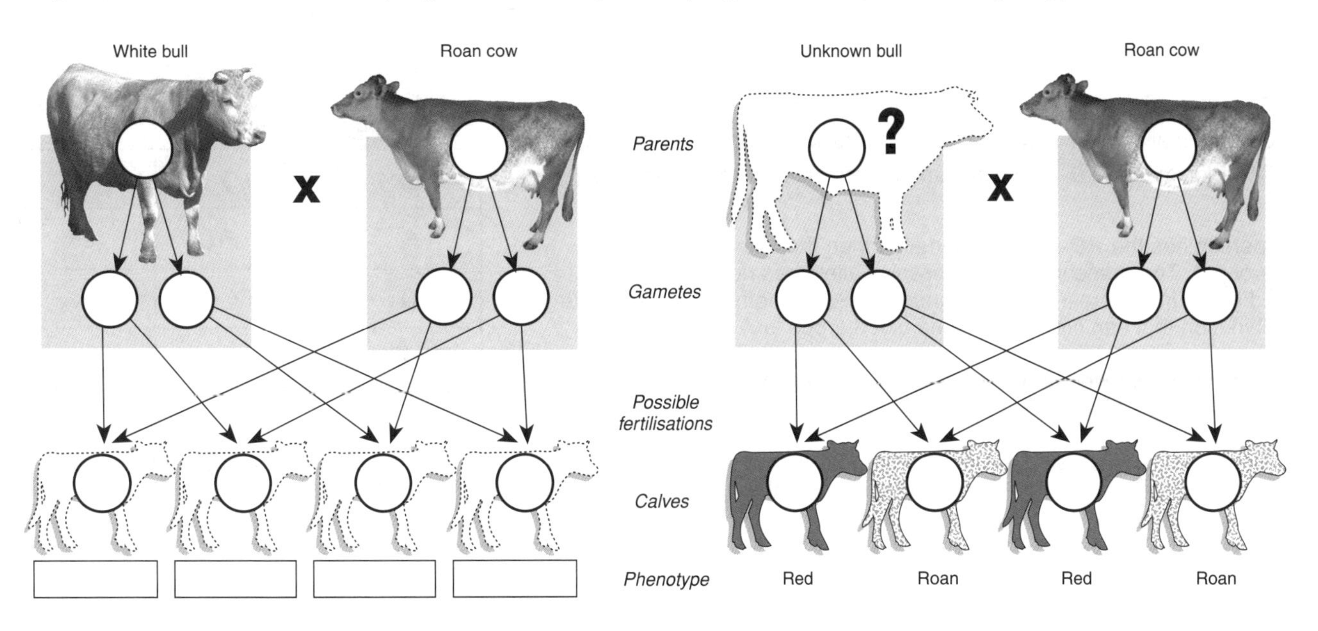

5. A white bull is mated with a roan cow (above, left).
   (a) Fill in the spaces on the diagram (above, left) to show the genotype and phenotype for parents and calves.

   (b) State the phenotype ratio for this cross: ______________________________

   (c) Suggest how the farmer who owns these cattle could control the breeding so that the herd ultimately consisted of red coloured cattle only:

   ____

   ____

   ____

6. A unknown bull is mated with a roan cow (above, right). A farmer has only roan shorthorn cows on his farm. He suspects that one of the bulls from his next door neighbours may have jumped the fence to mate with his cows earlier in the year. All the calves born were either red or roan. One neighbour has a red bull, the other has a roan bull.
   (a) Fill in the spaces on the diagram (above, right) to show the genotype and phenotype for parents and calves.

   (b) State which of the neighbour's bulls must have mated with the cows: **red** or **white** (*delete one*)

7. A plant breeder crossed two plants of the plant variety known as Japanese four o'clock. This plant is known to have its flower colour controlled by a gene which possesses incomplete dominant alleles. Pollen from a pink flowered plant was placed on the stigma of a red flowered plant.

   (a) Fill in the spaces on the diagram on the right to show the genotype and phenotype for parents and offspring.

   (b) State the phenotype ratio:

   ____

   ____

   ____

   ____

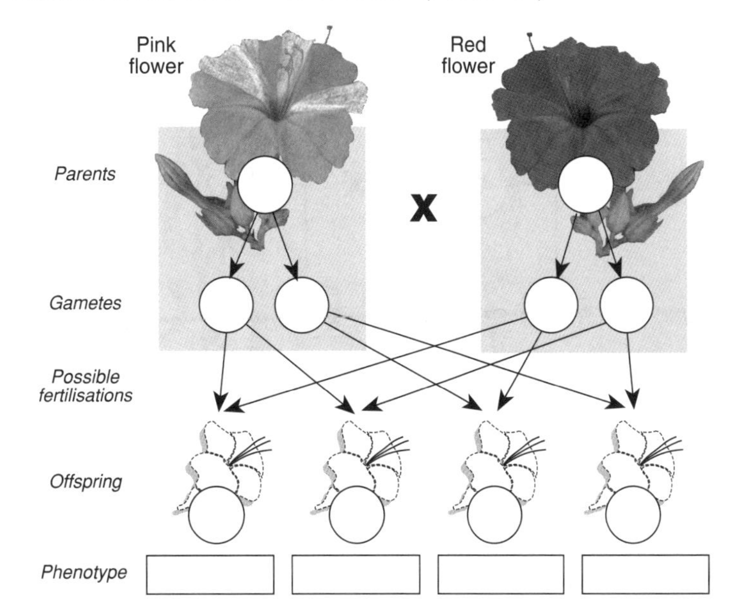

Multiple Alleles in Blood Groups

The four common blood groups of the human 'ABO blood group system' are determined by three alleles: ***A***, ***B***, and ***O*** (also represented in some textbooks as: I^A, I^B, and i^O or just **i**). This is an example of a **multiple allele** system for a gene. The ABO antigens consist of sugars attached to the surface of red blood cells. The alleles code for enzymes (proteins) that join together these sugars. The allele ***O*** produces a non-functioning enzyme that is unable to make any changes to the basic antigen (sugar) molecule. The other two alleles ***(A, B)*** are **codominant** and are expressed equally. They each produce a different functional enzyme that adds a different, specific sugar to the basic sugar molecule. The blood group A and B antigens are able to react with antibodies present in the blood from other people and must be matched for transfusion.

Recessive allele: ***O*** produces a non-functioning protein
Dominant allele: ***A*** produces an enzyme which forms **A antigen**
Dominant allele: ***B*** produces an enzyme which forms **B antigen**

Blood group (phenotype)	Possible genotypes	Frequency in the UK
O	*OO*	45%
A	*AA AO*	43%
B		9%
AB		3%

If a person has the ***AO*** allele combination then their blood group will be group **A**. The presence of the recessive allele has no effect on the blood group in the presence of a dominant allele. Another possible allele combination that can create the same blood group is ***AA***.

1. Use the information above to complete the table for the possible genotypes for blood group B and group AB.

2. Below are six crosses possible between couples of various blood group types. The first example has been completed for you. Complete the genotype and phenotype for the other five crosses below:

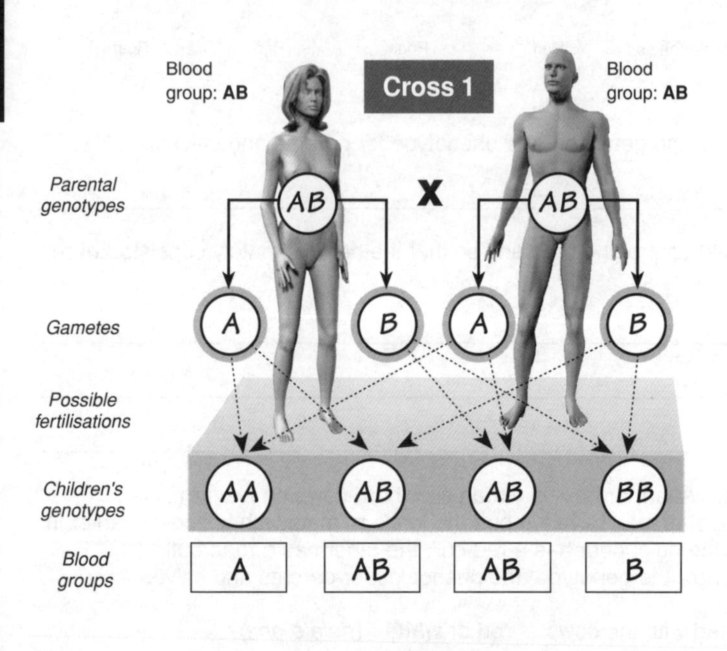

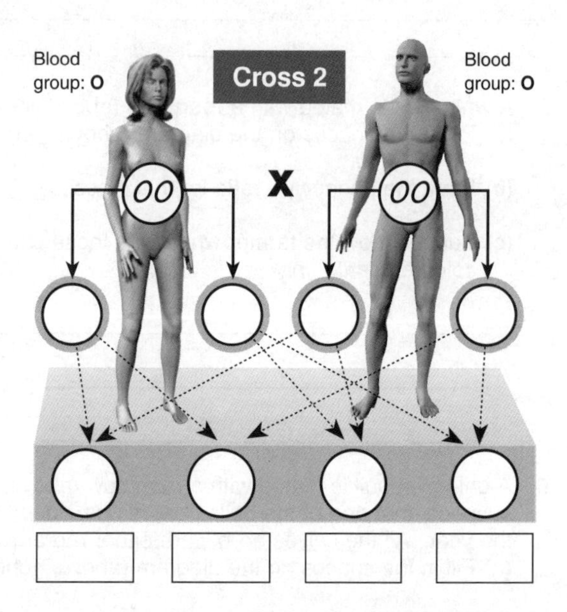

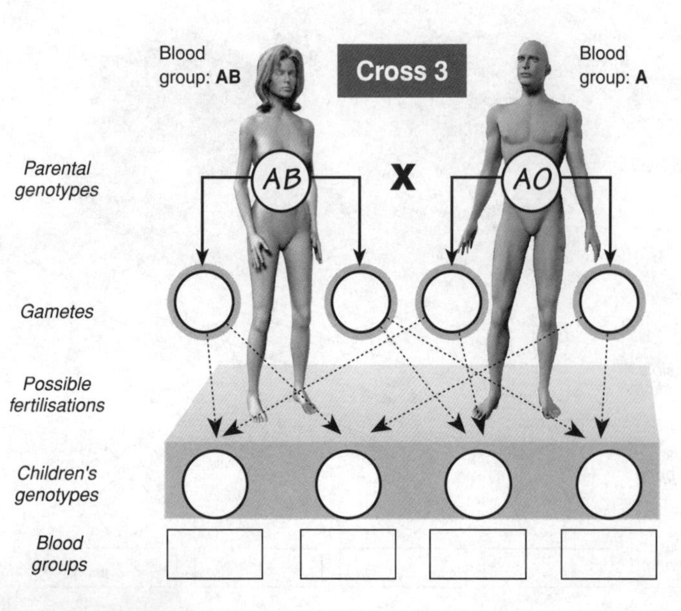

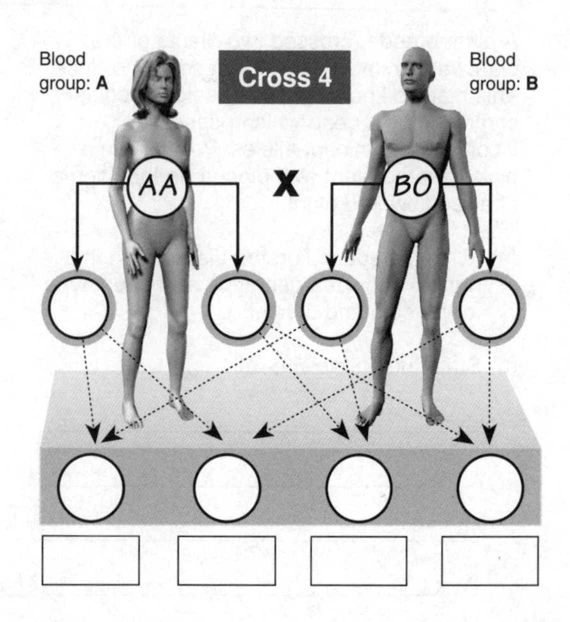

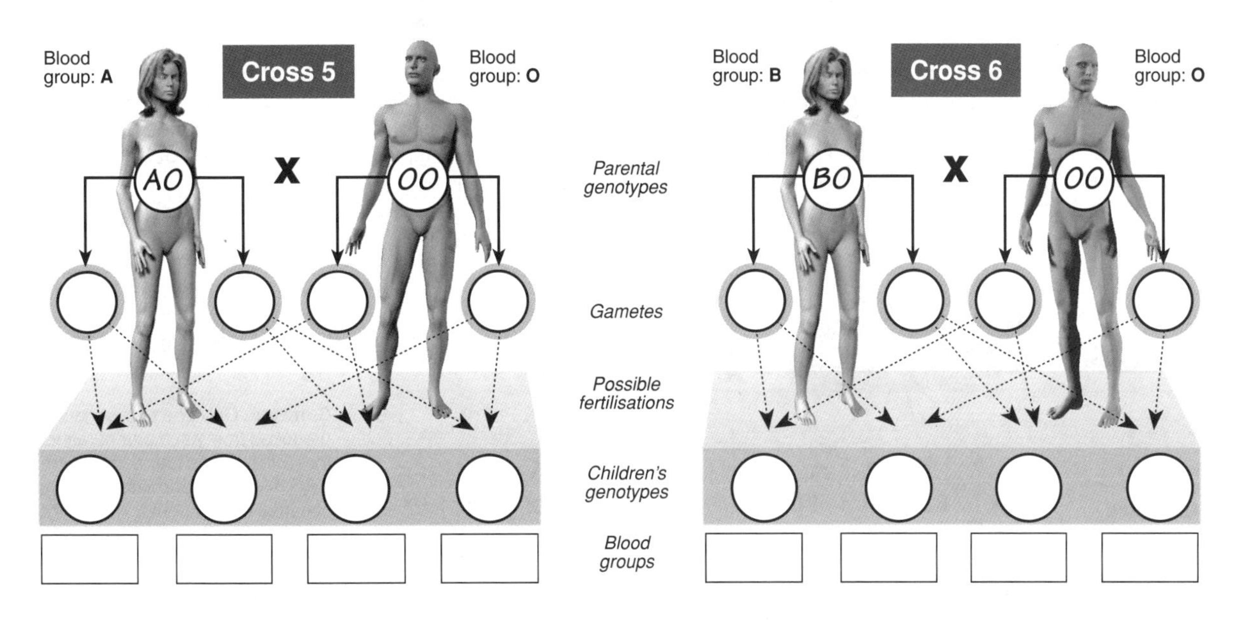

3. A wife is heterozygous for blood group **A** and the husband has blood group **O**.

   (a) Give the genotypes of each parent (fill in spaces on the diagram on the right).

   Determine the probability of:

   (b) One child having blood group **O**:

   ____

   (c) One child having blood group **A**:

   ____

   (d) One child having blood group **AB**:

   ____

Blood group **A**
Blood group **O**
X
Parental genotypes
Gametes
Possible fertilisations
Children's genotypes
Blood groups

4. In a court case involving a paternity dispute (i.e. who is the father of a child) a man claims that a male child (blood group **B**) born to a woman is his son and wants custody. The woman claims that he is not the father.

   (a) If the man has a blood group **O** and the woman has a blood group **A**, could the child be his son? Use the diagram on the right to illustrate the genotypes of the three people involved.

   (b) State with reasons whether the man can be correct in his claim:

   ____

   ____

   ____

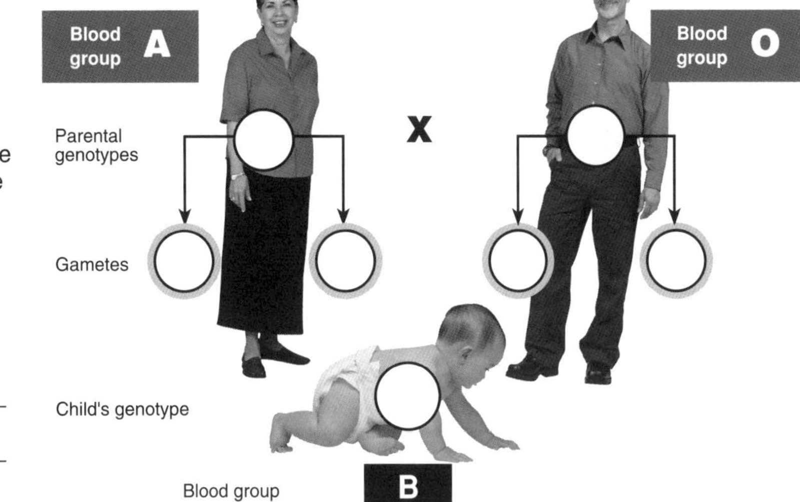
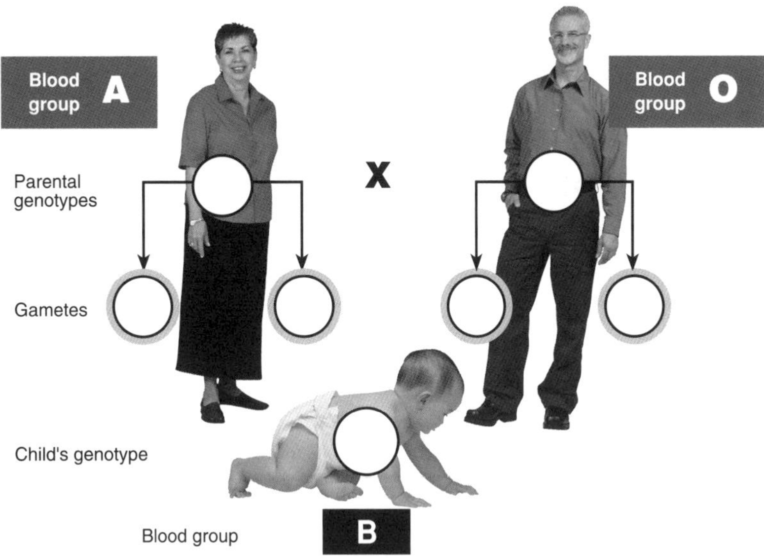

5. Give the blood groups which are possible for children of the following parents (remember that in some cases you don't know if the parent is homozygous or heterozygous).

   (a) Mother is group **AB** and father is group **O**: ______________________

   (b) Father is group **B** and mother is group **A**: ______________________

Dihybrid Cross

A cross (or mating) between two organisms where the inheritance patterns of **two genes** are studied is called a **dihybrid cross** (compared with the study of one gene in a monohybrid cross). There are a greater number of gamete types produced when two genes are considered (four types). Remember that the genes described are being carried by separate chromosomes and are sorted independently of each other during meiosis (that is why you get four kinds of gamete). The two genes below control two unrelated characteristics **hair colour** and **coat length**. Black and short are dominant.

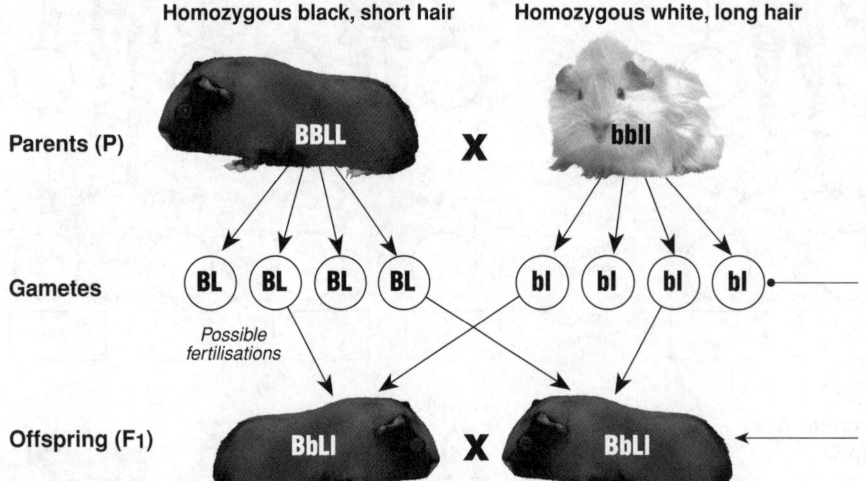

Parents: The notation **P**, is only used for a cross between **true breeding** (homozygous) parents.

Gametes: Only one type of gamete is produced from each parent (although they will produce four gametes from each oocyte or spermatocyte). This is because each parent is homozygous for both traits.

F1 offspring: There is only one **kind** of gamete from each parent, therefore only one kind of offspring produced in the first generation. The notation **F1** is only used to denote the heterozygous offspring of a cross between two true breeding parents.

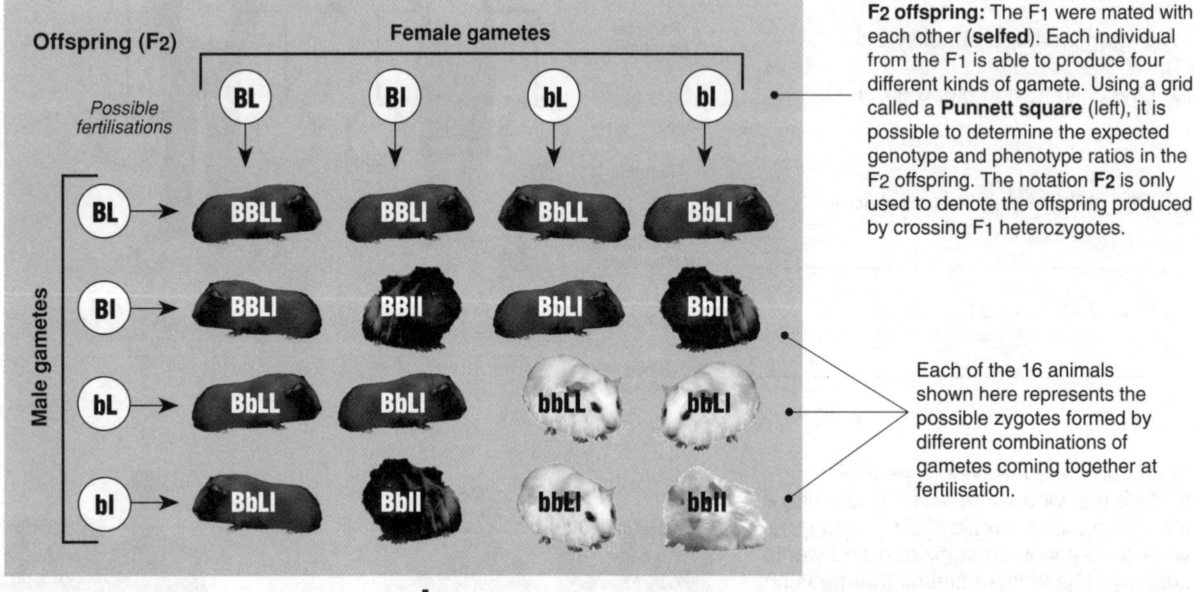

F2 offspring: The F1 were mated with each other (**selfed**). Each individual from the F1 is able to produce four different kinds of gamete. Using a grid called a **Punnett square** (left), it is possible to determine the expected genotype and phenotype ratios in the F2 offspring. The notation **F2** is only used to denote the offspring produced by crossing F1 heterozygotes.

Each of the 16 animals shown here represents the possible zygotes formed by different combinations of gametes coming together at fertilisation.

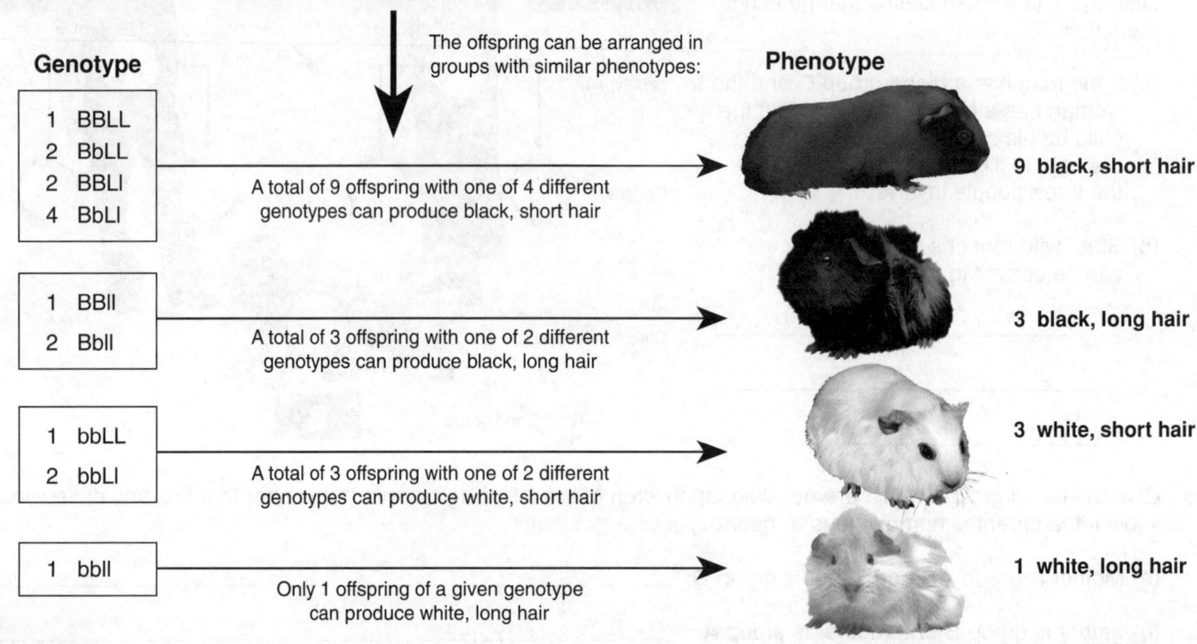

Cross No. 1

The dihybrid cross on the right has been partly worked out for you. You must determine:

1. The genotype and phenotype for each animal (write your answers in its dotted outline).
2. Genotype **ratio** of the offspring:

3. Phenotype **ratio** of the offspring:

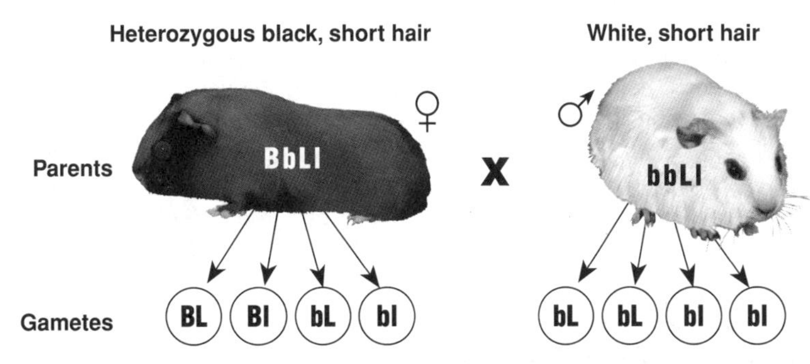

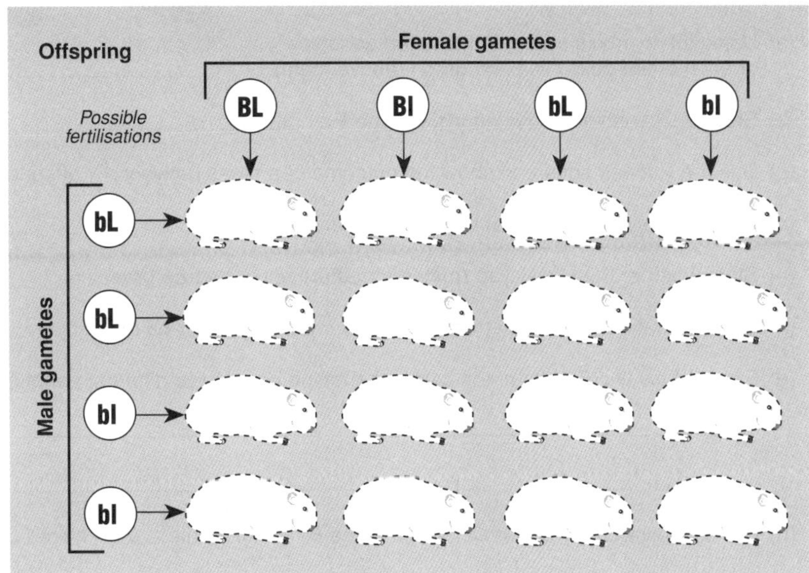

Cross No. 2

For the dihybrid cross on the right, determine:

1. Gametes produced by each parent (write these in the circles).
2. The genotype and phenotype for each animal (write your answers in its dotted outline).
3. Genotype **ratio** of the offspring:

4. Phenotype **ratio** of the offspring:

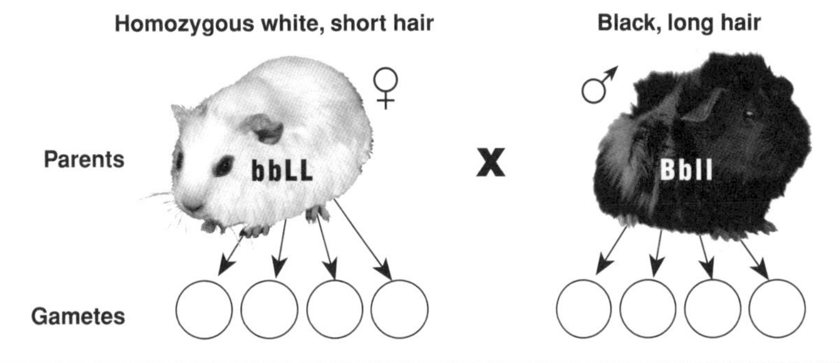

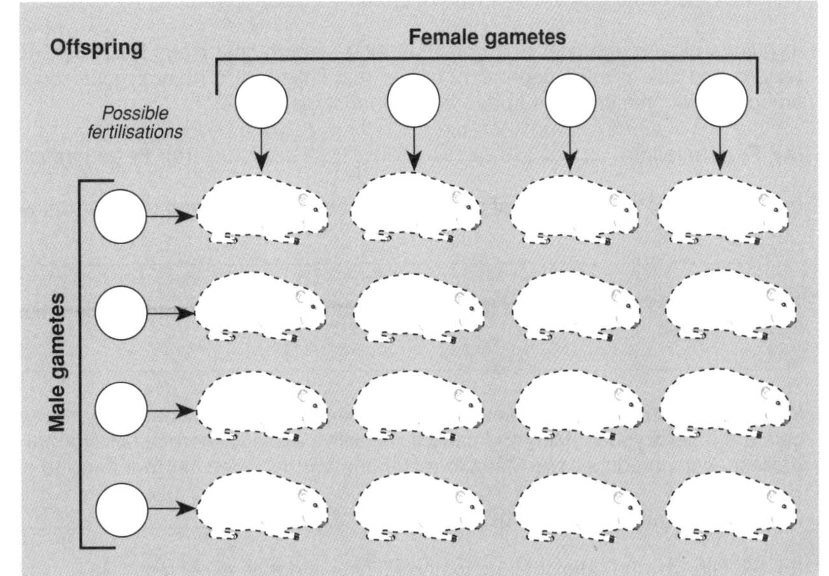

Problems in Mendelian Genetics

The following problems examine the outcome of matings in domestic breeds. They involve following the breeding through to the F2 generation. The alleles involved are associated with hair colour or length in a number of named mammals. See the earlier activity "*Basic Genetic Crosses*" if you need to review test crosses and back crosses.

1. The Himalayan colour-pointed, long-haired cat is a breed developed by crossing a pedigree (true-breeding), uniform-coloured, long-haired Persian with a pedigree colour-pointed (darker face, ears, paws, and tail) short-haired Siamese.
   The genes controlling hair colouring and length are on separate chromosomes: uniform colour **U**, colour pointed **u**, short hair **S**, long hair **s**.

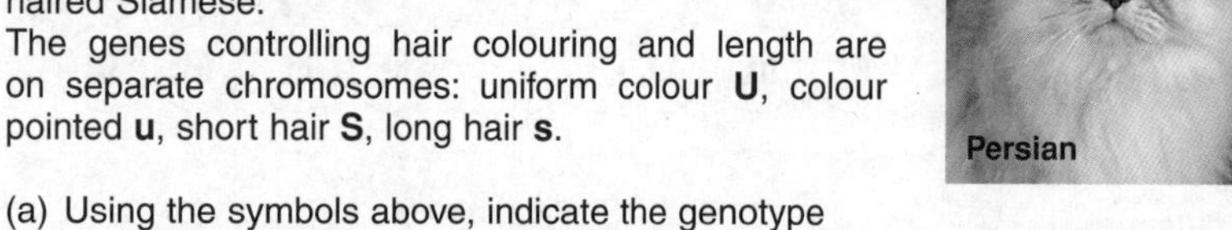

   (a) Using the symbols above, indicate the genotype of each breed below its photograph (above, right). ______ ______ ______

   (b) State the genotype and phenotype of the F_1: ______

   (c) Use the Punnett square to show the outcome of a cross between F_1 offspring (the F_2):

   (d) State the ratio of the F_2 that would be Himalayan: ______

   (e) State whether the Himalayan (mated together) would be true breeding: ______

   (f) State the ratio of the F_2 that would be colour-pointed, short-haired cats: ______

   (g) Explain how two F_2 cats of the same phenotype could have different genotypes:

   ______

   ______

   (h) Explain how would you discover which of the F_2 colour-pointed, short-haired cats were true breeding for these characters:

   ______

   ______

   ______

For the problems below, use a separate sheet to determine the outcomes of the crosses using Punnett squares. Staple the sheet into your manual to provide a permanent record of your workings:

2. In rabbits, spotted coat **S** is dominant to solid colour **s,** while for coat colour: black **B** is dominant to brown **b**. A brown spotted rabbit is mated with a solid black one and all the offspring are black spotted (the genes are not linked).

   (a) State the genotypes: Male parent: ______ Female parent: ______ Offspring: ______

   (b) Using ratios, state the phenotypes of the F_2 generation if two of these F_1 black spotted rabbits were mated:

   ______

   (c) State the name given to this type of cross: ______

3. In guinea pigs, rough coat **R** is dominant over smooth coat **r** and black coat **B** is dominant over white **b**. The genes for coat texture and colour are independent genes (not linked). If a homozygous rough, black animal is crossed with a homozygous smooth white one, state the appearance (phenotype) of the:

   (a) **F_1** generation: ______ (b) **F_2** generation: ______

   (c) Offspring of a back-cross of the **F_1** to the rough, black parent (include ratios of each phenotype if applicable):

   ______

   (d) Offspring of a test-cross of the **F_1** to the smooth, white parent (include ratios of each phenotype if applicable):

   ______

4. In Manx cats, the allele for taillessness (**M**) is incompletely dominant over the recessive allele for normal tail (**m**). Tailless Manx cats are heterozygous (**Mm**) and carry a recessive allele for normal tail. Normal tailed cats are **mm**. A cross between two Manx (tailless) cats, produces two Manx to every one normal tailed cat (not three to one as would be expected).

   (a) State the genotypes arising from this type of cross: ______

   (b) Explain why the ratio of Manx to normal cats is not as expected: ______

Using Chi-Squared in Genetics

The following problems examine the use of the chi-squared (χ^2) test in genetics. A worked example illustrating the use of the chi-squared test for a genetic cross is provided in "*Advanced Biology AS*".

1. In a tomato plant experiment, two heterozygous individuals were crossed (the details of the cross are not relevant here). The predicted Mendelian ratios for the offspring of this cross were **9:3:3:1** for each of the **four following phenotypes**: purple stem-jagged leaf edge, purple stem-smooth leaf edge, green stem-jagged leaf edge, green stem-smooth leaf edge.

   The observed results of the cross were not exactly as predicted. The numbers of offspring with each phenotype are provided below:

Observed results of the tomato plant cross			
Purple stem-jagged leaf edge	**12**	Green stem-jagged leaf edge	**8**
Purple stem-smooth leaf edge	**9**	Green stem-smooth leaf edge	**0**

   (a) State your null hypothesis for this investigation (Ho): ______________________

   (b) State the alternative hypothesis (HA): ______________________

2. Use the chi-squared (χ^2) test to determine if the differences observed between the phenotypes are significant. The table of critical values of χ^2 at different *P* values is available in "*Advanced Biology AS*", on the Teacher resource handbook on CD-ROM, or in any standard biostatistics book.

   (a) Enter the observed values (number of individuals) and complete the table to calculate the χ^2 value:

Category	O	E	O — E	$(O - E)^2$	$\frac{(O-E)^2}{E}$
Purple stem, jagged leaf					
Purple stem, smooth leaf					
Green stem, jagged leaf					
Green stem, smooth leaf					
	Σ				Σ

   (b) Calculate χ^2 value using the equation:

$$\chi^2 = \sum \frac{(O-E)^2}{E}$$

   χ^2 = ____________

   (c) Calculate the degrees of freedom: ____________

   (d) Using the appendix, state the *P* value corresponding to your calculated χ^2 value: ____________

   (e) State your decision: reject Ho / do not reject Ho (circle one)

3. Students carried out a pea plant experiment, where two heterozygous individuals were crossed. The predicted Mendelian ratios for the offspring were **9:3:3:1** for each of the **four following phenotypes**: round-yellow seed, round-green seed, wrinkled-yellow seed, wrinkled-green seed.

   The observed results were as follows:

Round-yellow seed	**441**	Wrinkled-yellow seed	**143**
Round-green seed	**159**	Wrinkled-green seed	**57**

   Use a separate piece of paper to complete the following:

   (a) State the null and alternative hypotheses (Ho and HA).

   (b) Calculate the χ^2 value.

   (c) Calculate the degrees of freedom and state the *P* value corresponding to your calculated χ^2 value.

   (d) State whether or not you reject your null hypothesis: reject Ho / do not reject Ho (circle one)

4. Comment on the whether the χ^2 values obtained above are similar. Suggest a reason for any difference:

   ____

   ____

Sex Linked Genes

Sex linkage is a special case of linkage occurring when a gene is located on a sex chromosome (usually the X). The result of this is that the character encoded by the gene is usually seen only in one sex (the heterogametic sex) and occurs rarely in the homogametic sex. In humans, recessive sex linked genes are responsible for a number of heritable disorders in males, e.g. haemophilia. Women who have the recessive alleles on their chromosomes are said to be **carriers**. In cats, one of the genes controlling coat colour in cats is sex-linked; the two alleles, black and orange, are found only on the X-chromosome.

Allele types	Genotypes	Phenotypes
X_B = Black pigment	X_BX_B, X_BY	= Black coated female, male
X_O = Orange pigment	X_OX_O, X_OY	= Orange coated female, male
	X_BX_O	= Tortoiseshell (intermingled black and orange in fur) in female cats only

1. An owner of a cat is thinking of mating her black female cat with an orange male cat. Before she does this, she would like to know what possible coat colours could result from such a cross. Use the symbols above to fill in the diagram on the right. Summarise the possible genotypes and phenotypes of the kittens in the tables below.

	Genotypes	Phenotypes
Male kittens		

	Genotypes	Phenotypes
Female kittens		

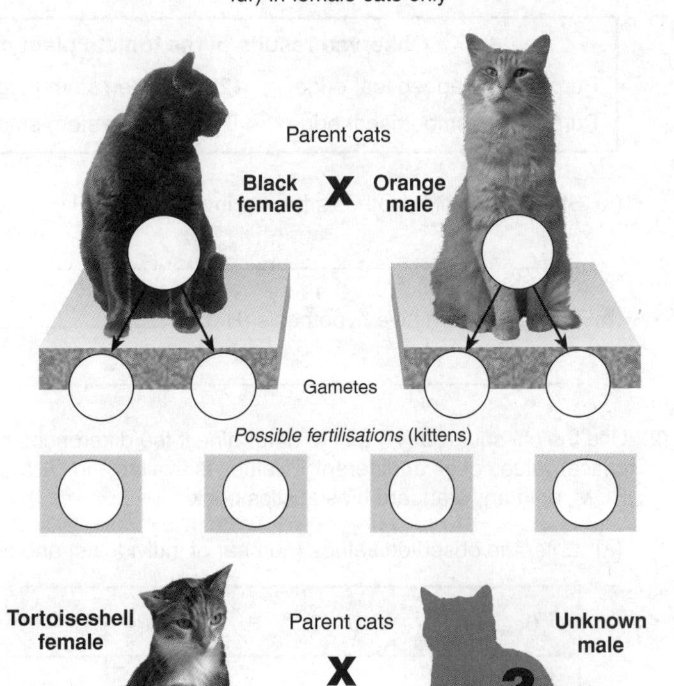

2. A female tortoiseshell cat mated with an unknown male cat in the neighbourhood and has given birth to a litter of six kittens. The owner of this female cat wants to know what the appearance and the genotype of the father was of these kittens. Use the symbols above to fill in the diagram on the right. Also show the possible fertilisations by placing appropriate arrows.

Describe the father cat's:

(a) Genotype: ______________________

(b) Phenotype: ______________________

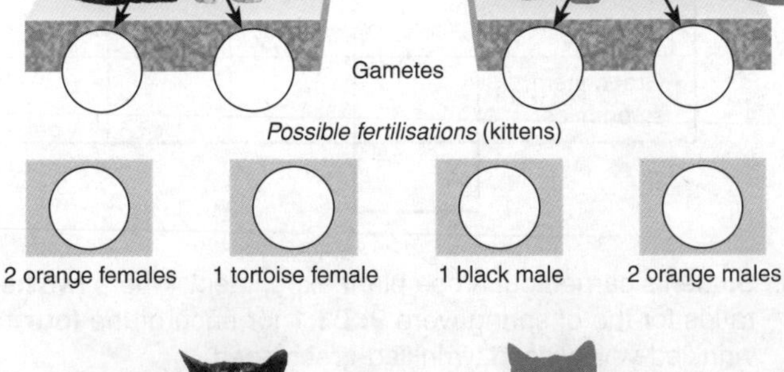

3. The owner of another cat, a black female, also wants to know which cat fathered her two tortoiseshell female and two black male kittens. Use the symbols above to fill in the diagram on the right. Show the possible fertilisations by placing appropriate arrows.

Describe the father cat's:

(a) Genotype: ______________________

(b) Phenotype: ______________________

(c) Was it the same male cat that fathered both this litter and the one above?
YES / **NO** *(delete one)*

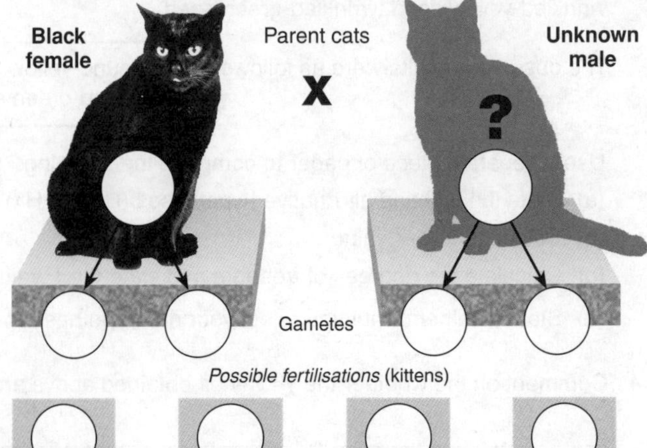

Dominant allele in humans

A rare form of rickets in humans is determined by a **dominant** allele of a gene on the **X chromosome** (it is not found on the Y chromosome). This condition is not successfully treated with vitamin D therapy. The allele types, genotypes, and phenotypes are as follows:

Allele Types		Genotypes		Phenotypes
X_R = affected by rickets		X_RX_R, X_RX	=	Affected female
X = normal		X_RY	=	Affected male
		XX, XY	=	Normal female, male

As a genetic counsellor you are presented with a married couple where one of them has a family history of this disease. The husband is affected by this disease and the wife is normal. The couple, who are thinking of starting a family, would like to know what their chances are of having a child born with this condition. They would also like to know what the probabilities are of having an affected boy or affected girl. Use the symbols above to complete the diagram right and determine the probabilities stated below (expressed as a proportion or percentage).

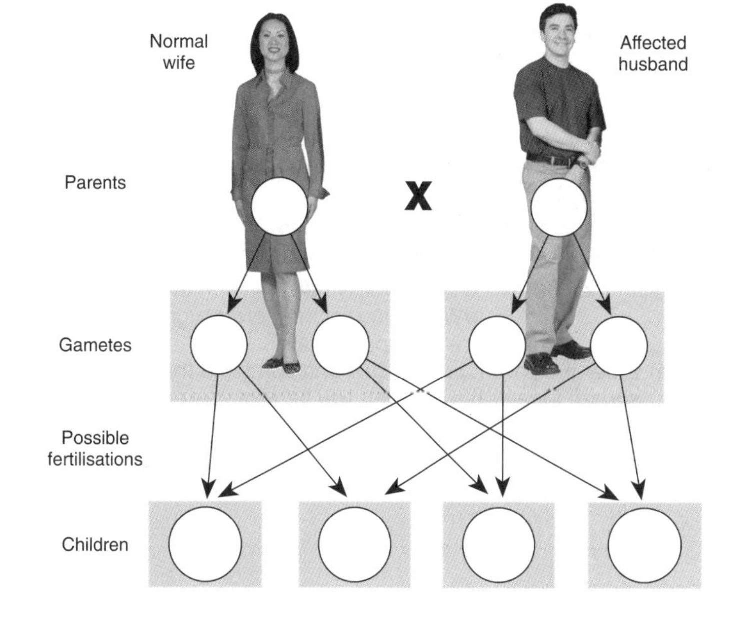

4. Determine the probability of having:

   (a) Affected children: ________________

   (b) An affected girl: ________________

   (c) An affected boy: ________________

Another couple with a family history of the same disease also come in to see you to obtain genetic counselling. In this case the husband is normal and the wife is affected. The wife's father was not affected by this disease. Determine what their chances are of having a child born with this condition. They would also like to know what the probabilities are of having an affected boy or affected girl. Use the symbols above to complete the diagram right and determine the probabilities stated below (expressed as a proportion or percentage).

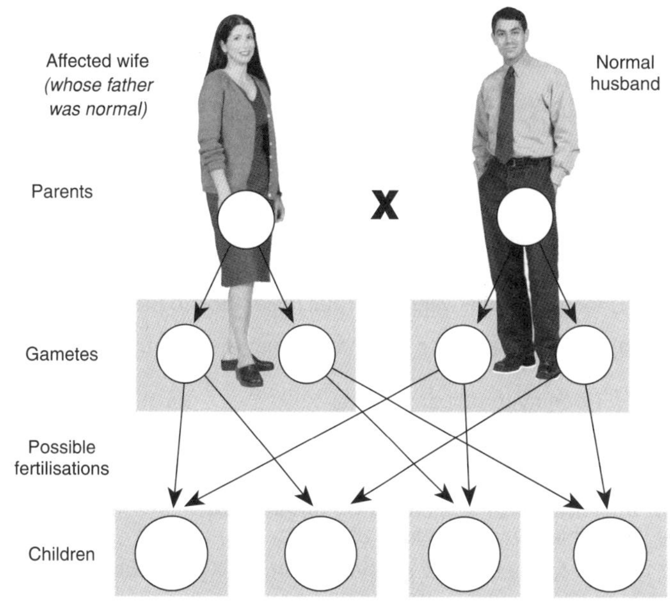

5. Determine the probability of having:

   (a) Affected children: ________________

   (b) An affected girl: ________________

   (c) An affected boy: ________________

6. Describing examples other than those above, discuss the role of **sex linkage** in the inheritance of genetic disorders:

Inheritance Patterns

Complete the following monohybrid crosses for different types of inheritance patterns in humans: autosomal recessive, autosomal dominant, sex linked recessive, and sex linked dominant inheritance.

1. **Inheritance of autosomal recessive traits**
   Example: *Albinism*

   Albinism (lack of pigment in hair, eyes and skin) is inherited as an autosomal recessive allele (not sex-linked).

   Using the codes: **PP** (normal)
   Pp (carrier)
   pp (albino)

   (a) Enter the parent phenotypes and complete the Punnett square for a cross between two carrier genotypes.

   (b) Give the ratios for the phenotypes from this cross.

   Phenotype ratios: ______________________________

   ____

2. **Inheritance of autosomal dominant traits**
   Example: *Woolly hair*

   Woolly hair is inherited as an autosomal dominant allele. Each affected individual will have at least one affected parent.

   Using the codes: **WW** (woolly hair)
   wW (woolly hair, heterozygous)
   ww (normal hair)

   (a) Enter the parent phenotypes and complete the Punnett square for a cross between two heterozygous individuals.

   (b) Give the ratios for the phenotypes from this cross.

   Phenotype ratios: ______________________________

   ____

3. **Inheritance of sex linked recessive traits**
   Example: *Haemophilia*

   Inheritance of haemophilia is sex linked. Males with the recessive (haemophilia) allele, are affected. Females can be carriers.

   Using the codes: X_oX_o (normal female)
   X_oX_h (carrier female)
   X_hX_h (haemophiliac female)
   X_oY (normal male)
   X_hY (haemophiliac male)

   (a) Enter the parent phenotypes and complete the Punnett square for a cross between a normal male and a carrier female.

   (b) Give the ratios for the phenotypes from this cross.

   Phenotype ratios: ______________________________

   ____

4. **Inheritance of sex linked dominant traits**
   Example: *Sex linked form of rickets*

   A rare form of rickets is inherited on the X chromosome.

   Using the codes: X_oX_o (normal female); X_oY (normal male)
   X_RX_o (affected heterozygote female)
   X_RX_R (affected female)
   X_RY (affected male)

   (a) Enter the parent phenotypes and complete the Punnett square for a cross between an affected male and heterozygous female.

   (b) Give the ratios for the phenotypes from this cross.

   Phenotype ratios: ______________________________

   ____

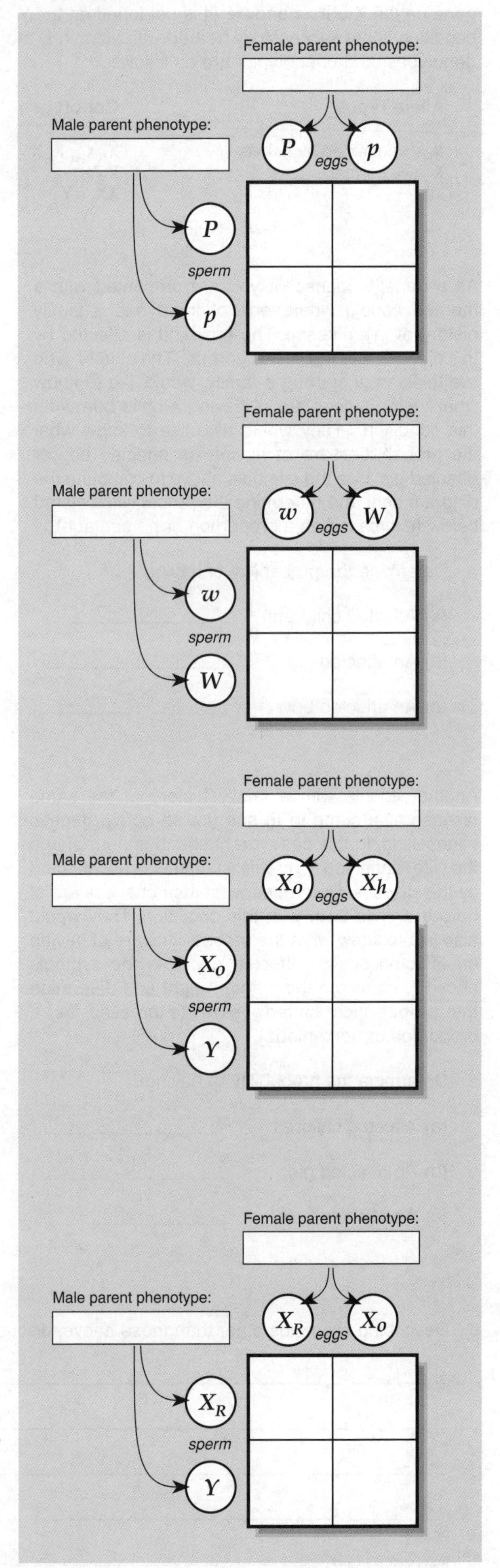

Interactions Between Genes

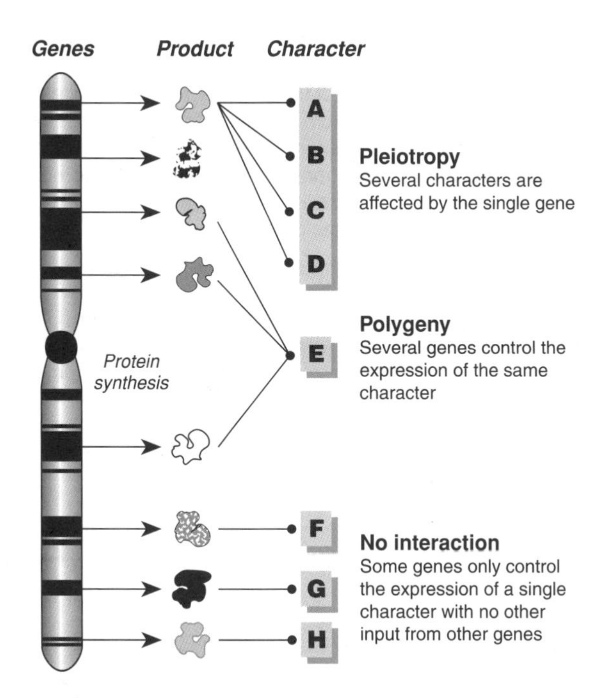

Sickle Cell Disease as an Example of Pleiotropy

A person homozygous recessive (**Hb^{S}Hb^S**) for the sickle cell allele produces mutant haemoglobin. This affects a number of organs.

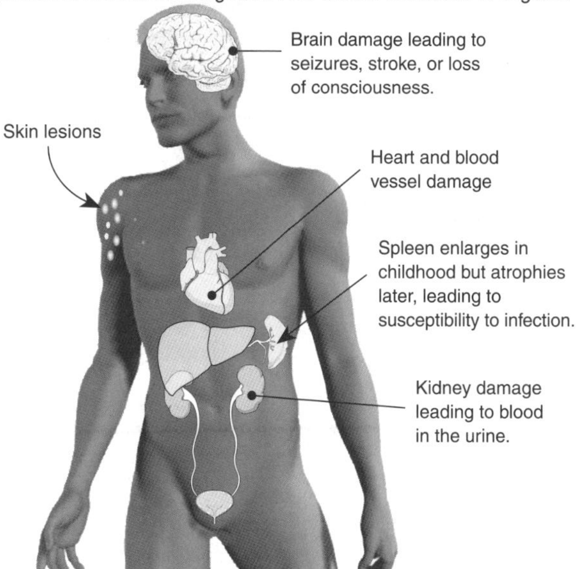

Types of Gene Interaction

Some genes may not just control a single characteristic or trait in the phenotype of an organism. It is thought that most genes probably have an effect on one or more phenotypic traits, a phenomenon known as **pleiotropy**. In some cases a single characteristic may be controlled by more than one gene; a situation known as **polygeny**. A further possible type of gene interaction, called **epistasis**, involves two non-allelic genes (at different loci), where the action of one gene masks or otherwise alters the expression of other genes. An example is albinism, which appears in rodents that are homozygous recessive for colour even if they have the alleles for agouti or black fur (the gene for colour is epistatic and the hypostatic gene determines the nature of the colour).

Pleiotropy

A single gene may produce a product that can influence a number of traits in the phenotype of an organism. Such a gene is said to be **pleiotropic**. The gene Hb codes for production of haemoglobin, an important oxygen-carrying molecule in the blood. A point mutation to this gene produces sickle cell disease. The phenotype has poor oxygen-carrying capability and deformed red blood cells leading to haemolytic anaemia. A range of other organ abnormalities (above) also occur as a result of the mutant haemoglobin. In the diagram above, **Hb^S** stands for the single gene that codes for the mutated haemoglobin molecule. A person with sickle cell disease is homozygous recessive: **Hb^{S}Hb^S**. This condition is eventually fatal. The normal genotype is HbHb.

1. Discuss the basic differences between **polygeny**, **pleiotropy**, and **epistasis**, giving examples to illustrate your answer:

2. (a) Describe the cause of sickle cell disease:

   (b) State the genotype of an affected individual:

   (c) Describe the phenotype of an individual who is homozygous recessive for the sickle cell mutation:

   (d) Explain why the sickle cell gene is regarded as pleiotropic:

Epistasis

In its narrowest definition, **epistatic genes** are those that mask the effect of other genes. In this definition, they are also known as supplementary genes. Typically there are **three possible phenotypes** for a dihybrid cross involving this type of gene interaction. One example of this type of epistasis occurs between the genes controlling coat colour in rodents and other mammals. Skin and hair colour is the result of melanin, a pigment which may be either black/brown (eumelanin) or reddish/yellow (phaeomelanin). Melanin itself is synthesised via several biochemical steps from the amino acid tyrosine. The control of coat colour and patterning in mammals is complex and involves at least five major interacting genes. One of these genes (gene C), controls the production of the pigment melanin, while another gene (gene B), is responsible for whether the colour is black or brown (this interaction is illustrated for mice, below). Epistasis literally means "standing upon". In albinism, the homozygous recessive condition, cc, "stands upon" the other coat colour genes, blocking their expression.

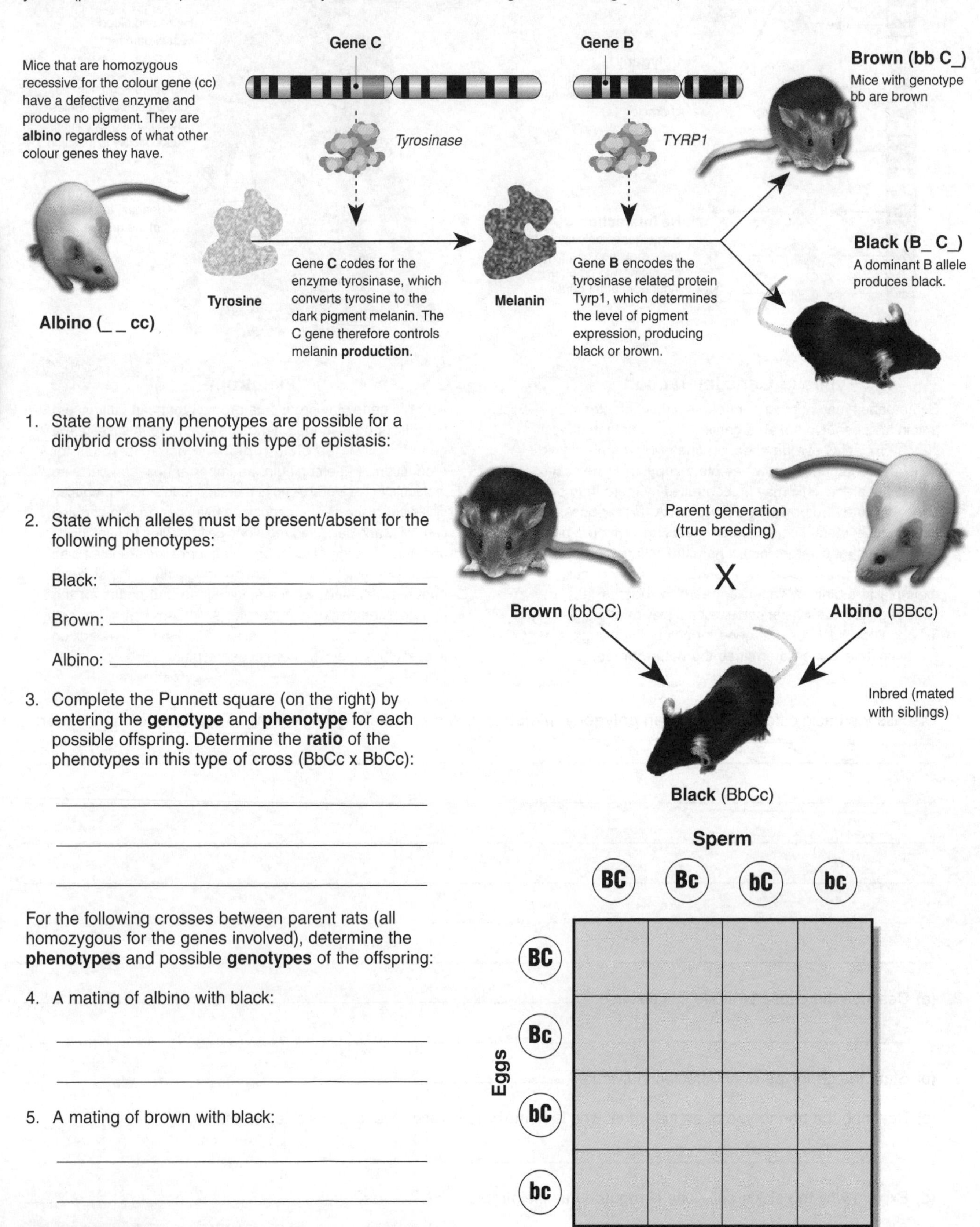

1. State how many phenotypes are possible for a dihybrid cross involving this type of epistasis:

2. State which alleles must be present/absent for the following phenotypes:

   Black:

   Brown:

   Albino:

3. Complete the Punnett square (on the right) by entering the **genotype** and **phenotype** for each possible offspring. Determine the **ratio** of the phenotypes in this type of cross (BbCc x BbCc):

For the following crosses between parent rats (all homozygous for the genes involved), determine the **phenotypes** and possible **genotypes** of the offspring:

4. A mating of albino with black:

5. A mating of brown with black:

Complementary Genes

In some cases of epistasis, some genes can only be expressed in the presence of other genes: the genes are **complementary**. Both genes must have a dominant allele present for the final end product of the phenotype to be expressed. Typically, there are two possible phenotypes for this condition. An example of such an interaction would be if one gene controls the production of a pigment (intermediate) and another gene controls the transformation of that intermediate into the pigment (by producing a controlling enzyme). Such genes have been found to control some flower colours. The diagram below right illustrates how one kind of flower colour in sweet peas is controlled by two complementary genes. The purple pigment is produced only in the presence of the dominant allele for each of the two genes. If a dominant is absent for either gene, then the flower is white.

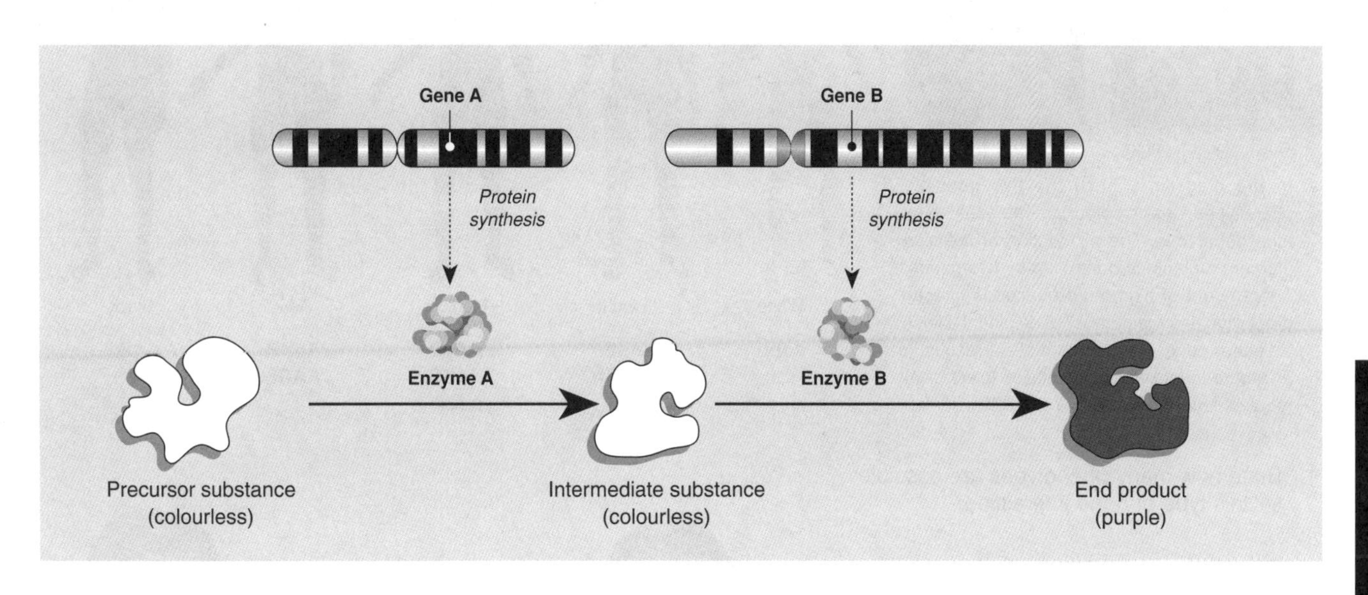

1. State how many phenotypes are possible for this type of gene interaction:

____

2. State which alleles must be present/absent for the following phenotypes:

   Purple flower: _______________

   White flower: _______________

3. Complete the Punnett square (on the right) by entering the **genotype** and **phenotype** for each possible **offspring**. Determine the ratio of the phenotypes in this type of cross, between heterozygous parents (AaBb x AaBb):

____

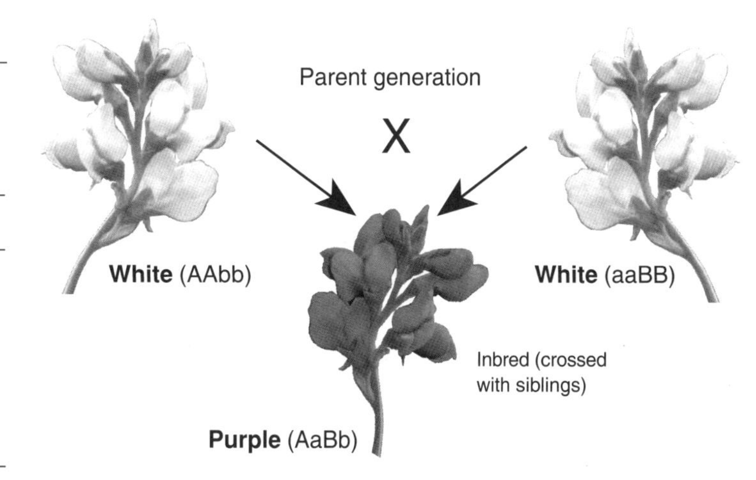

Sperm

Eggs	AB	Ab	aB	ab
AB				
Ab				
aB				
ab				

For the following three crosses of sweet peas, determine the **genotypes** of the **parents**:

4. A white-flowered plant, crossed with a purple, produces offspring of which three-eighths are purple and five-eighths are white.

____

5. A purple-flowered plant, crossed with a purple, produces offspring of which one-half are purple and one-half white.

____

6. A white-flowered plant, crossed with another white, produces offspring of which three-fourths are white and one-fourth purple.

____

Polygenes

A single phenotype may be influenced, or determined, by more than one gene. Such phenotypes exhibit continuous variation in a population. Examples are skin colour and height, although the latter has been found to be influenced mostly by a single gene.

A light-skinned person A dark-skinned person

In the diagram (right), the five possible phenotypes for skin colour are represented by nine genotypes. The production of the skin pigment melanin is controlled by two genes. The amount of pigment produced is directly proportional to the number of dominant alleles for either gene. No dominant allele results in an **albino** (aabb). Full pigmentation (black skin) requires four dominant alleles (AABB).

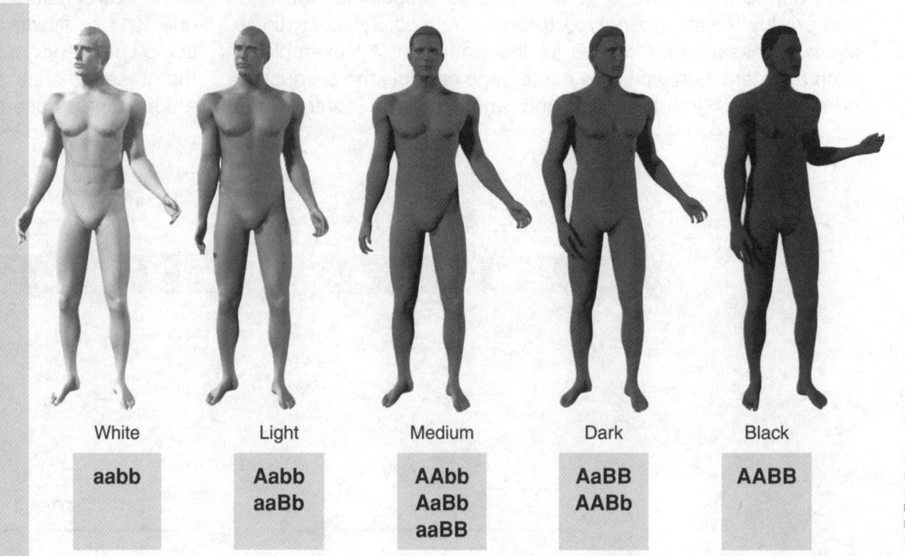

1. State how many phenotypes are possible for this type of gene interaction:

   ____

2. State which alleles must be present/absent for the following phenotypes:

   Black: ______________________________

   Medium: ______________________________

   White: ______________________________

3. Complete the Punnett square (on the right) by entering the **genotype** and **phenotype** for each possible offspring. Determine the ratio of the phenotypes in this type of cross, between heterozygous parents (AaBb x AaBb):

   ____

   ____

   ____

   ____

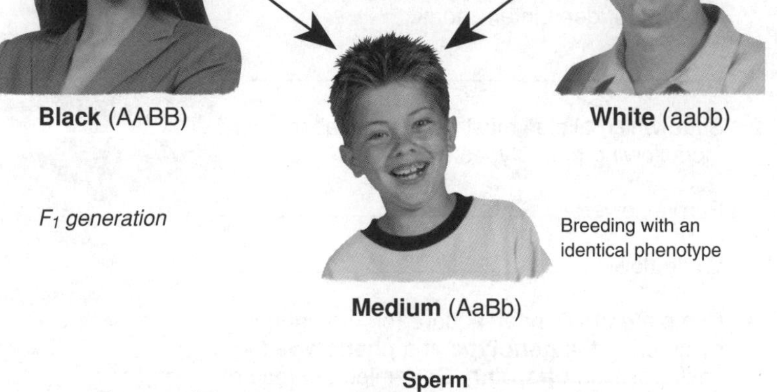

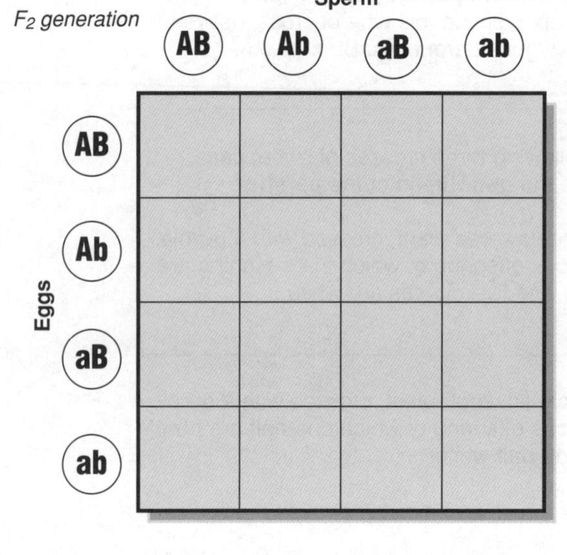

For the following two crosses between humans, determine the phenotypes of the offspring:

4. A mating of white with black:

   ____

5. A mating between two individuals of medium skin colour:

   ____

6. In a polygenic inheritance illustrated above, two genes (A and B) are able to produce 5 phenotypes. Determine how many possible phenotypes could be produced if three genes were involved (i.e. genes A, B and C produce genotypes aabbcc, Aabbcc, etc.):

   ____

   ____

   ____

7. Discuss the differences between **continuous** and **discontinuous** variation, giving examples to illustrate your answer:

8. From a sample of no less than 30 adults, collect data (by request or measurement) for one continuous variable (e.g. height, weight, shoe size, or hand span). Record and tabulate your results in the space below, and then plot a frequency histogram of the data on the grid below:

Raw data

Tally Chart (frequency table)

Variable: ....

Frequency

(a) Calculate the mean, median, and mode of your data (see "*Advanced Biology AS*" if you need help):

Mean: ______ **Mode**: ______ **Median**: ______

(b) Describe the pattern of distribution shown by the graph, giving a reason for your answer:

(c) Explain the genetic basis of this distribution:

(d) Explain the importance of a large sample size when gathering data relating to a continuous variable:

Collaboration

There are genes that may influence the same trait, but produce a phenotype that could not result from the action of either gene independently. These are termed collaborative genes (they show **collaboration**). There are typically four possible phenotypes for this condition. An example of this type of interaction can be found in the comb shape of domestic hens.

Single comb	Pea comb	Rose comb	Walnut comb
Genotypes:	Genotypes:	Genotypes:	Genotypes:
rrpp	**rrP_**	**R_pp**	**R_P_**
rrpp	rrPp, rrPP	Rrpp, RRpp	RRPP, RrPP, RrPp, RRPp

The dash (missing allele) in the bold genotypes on the right indicates that the allele may be dominant or recessive, it will not affect the phenotype (i.e. the resulting character displayed).

1. State **how many phenotypes** are possible for this type of gene interaction:

   ____

2. State which alleles must be present or absent for the following phenotypes:

   Pea comb: ______________________________

   Rose comb: ______________________________

   Single comb: ______________________________

   Walnut comb: ______________________________

3. Complete the Punnett square (on the right) by entering the **genotype** and **phenotype** for each possible **offspring**. Determine the ratio of the phenotypes in this type of cross, between heterozygous parents (RrPp x RrPp):

   ____

4. Determine the **genotype** of the **parents** for each of the following crosses:

   (a) A rose crossed with a walnut produces offspring, 3/8 of which are walnut, 3/8 rose, 1/8 pea, and 1/8 single.

   ____

   (b) A walnut crossed with a single, produces in the F_1 generation, 1/4 walnut, 1/4 pea, 1/4 rose, and 1/4 single.

   ____

   (c) A rose crossed with a pea produces six walnut and five rose offspring.

   ____

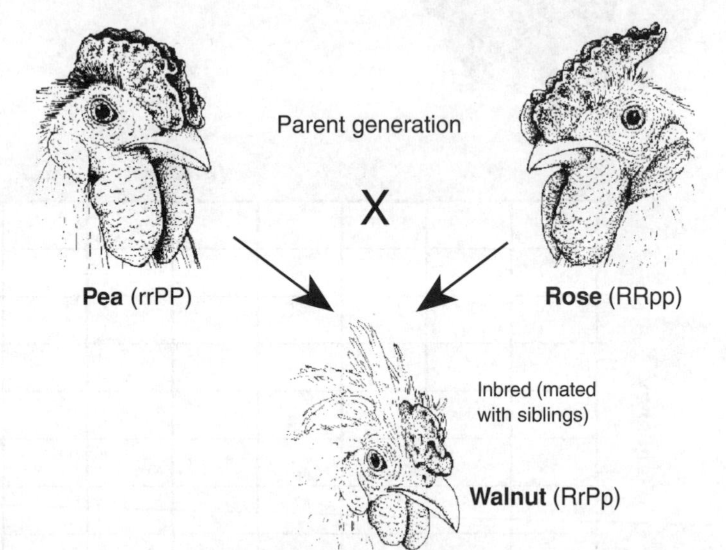

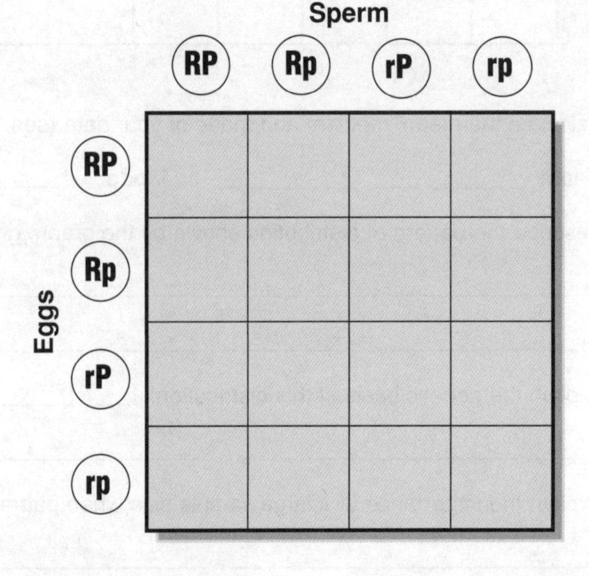

Variation in Coat Colour in Domestic Cats

Non-agouti

A completely jet black cat has no markings on it whatsoever. It would have the genotype: ***aaB–D–*** since no dominant agouti allele must be present, and the black pigment is not diluted.

Siamese

The colour pointing of Siamese cats is caused by warm temperature deactivation of a gene that produces melanin pigment. Cooler parts of the body are not affected and appear dark.

Tortoiseshell

Because this is a sex linked trait, it is normally found only in female cats (***XO***, ***Xo***). The coat is a mixture of orange and black fur irregularly blended together.

Agouti hair

Enlarged view of agouti hair. Note that the number of darkly pigmented stripes can vary on the same animal.

Stripes of dark pigment

Lighter colour

Sex linked orange

The orange (***XO***, ***XO***) cat has an orange coat with little or no patterns such as tabby showing.

Blotched tabby

Lacks stripes but has broad, irregular bands arranged in whorls (***tb***).

Wild type

Mackerel (striped) tabby (***A–B–T–***) with evenly spaced, well-defined, vertical stripes. The background colour is **agouti** with the stripes being areas of completely black hairs.

Calico

Similar to a tortoiseshell, but with substantial amounts of white fur present as well. Black, orange and white fur.

Marmalade

The orange colour (***XO***, ***XO***) is expressed, along with the alleles for the tabby pattern. The orange allele shows epistatic dominance and overrides the expression of the normal agouti colour so that the tabby pattern appears dark orange.

Other Inherited Features in Domestic Cats

Manx tail (Mm)

The Manx breed of cat has little or no tail. This dominant allele is lethal if it occurs in the homozygous condition.

Polydactylism (Pd–)

This is a dominant mutation. The number of digits on the front paw should be five, with four digits on the rear paw.

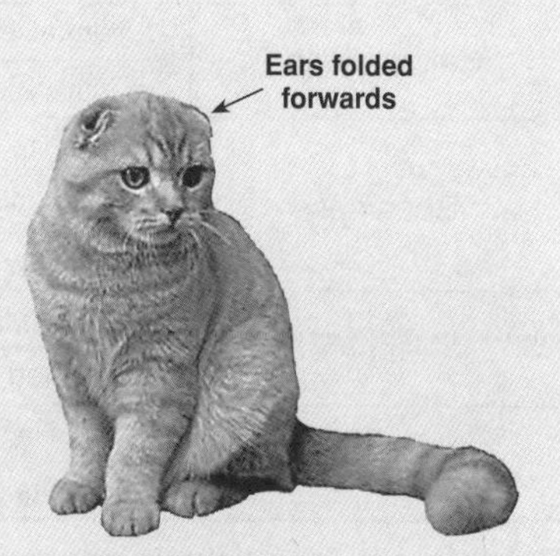

Ear fold (Fd–)

Most cats have normal pointed ears. A dominant mutation exists where the ear is permanently folded forwards.

What Genotype Has That Cat?

Consult the table of genes listed on the previous pages and enter the allele symbols associated with each of the phenotypes in the column headed 'Alleles'. For this exercise, study the appearance of real cats around your home or look at colour photographs of different cats. For each cat, complete the checklist of traits listed below by simply placing a tick in the appropriate spaces. These traits are listed in the same order as the genes for **wild forms** and **mutant forms** on page 142. On a piece of paper, write each of the cat's genotypes. Use a dash (-) for the second allele for characteristics that could be either heterozygous or homozygous dominant (see the sample at the bottom of the page).

NOTES:
1. Agouti fur colouring is used to describe black **hairs** with a light band of pigment close to its tip.
2. Patches of silver fur (called chinchilla) produces the silver tabby phenotype in agouti cats. Can also produce "smoke" phenotype in Persian long-haired cats, causing reduced intensity of the black.
3. Describes the dark extremities (face, tail and paws) with lighter body (e.g. Siamese).
4. The recessive allele makes black cats blue-grey in colour and yellow cats cream.
5. Spottiness involving less than half the surface area is likely to be heterozygous.

Phenotype Record Sheet for Domestic Cats

Gene	Phenotype	Allele	Sample	Cat 1	Cat 2	Cat 3	Cat 4
Agouti colour	Agouti[1]						
	Non-agouti		✔				
Pigment colour	Black		✔				
	Brown						
Colour present	Unicoloured		✔				
	Silver patches[2]						
	Pointed[3]						
	Albino with blue eyes						
	Albino with pink eyes						
Pigment density	Dense pigment						
	Dilute pigment[4]		✔				
Ear shape	Pointed ears		✔				
	Folded ears						
Hairiness	Normal, full coat		✔				
	Hairlessness						
Hair length	Short hair		✔				
	Long hair						
Tail length	Normal tail (long)		✔				
	Stubby tail or no tail at all						
Orange colour	Normal colours (non-orange)		✔				
	Orange						
Number of digits	Normal number of toes		✔				
	Polydactylism (extra toes)						
Hair curliness	Normal, smooth hair		✔				
	Curly hair (rex)						
Spottiness	No white spots						
	White spots (less than half)[5]		✔				
	White spots (more than half)						
Stripes	Mackerel striped (tabby)						
	Blotched stripes						
White coat	Not all white		✔				
	All white coat colour						

Sample cat (see ticks in chart above)
To give you an idea of how to read the chart you have created, here is an example genotype of the author's cat with the following features: *A smoky grey uniform-coloured cat, with short smooth hair, normal tail and ears, with 5 digits on the front paws and 4 on the rear paws, small patches of white on the feet and chest.* (Note that the stripe genotype is completely unknown since there is no agouti allele present).

GENOTYPE: **aa B– C– dd fdfd Hr– ii L– mm oo pdpd R– Ss ww**

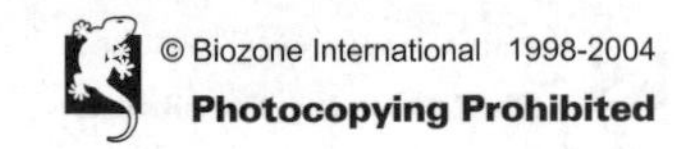

Population Genetics and Evolution

AQA-A	AQA-B	CIE	Edexcel	OCR
Complete:	Complete:	Complete:	Complete:	Complete:
1-2, 7-13, 15, 25-31	*1-2, 4-5, 11-12, 15, 24-31, 34*	*1-9, 11-14, 16-25, 27-29*	*1-2, 7-8, 11-13, 16-18, 24, 27-31*	*1-9, 11-14, 16-25, 27-29*
Extension as required	*Extension as required*	*Extension as required*	*Extension as required*	*Extension as required*

Learning Objectives

☐ 1. Compile your own glossary from the **KEY WORDS** displayed in **bold type** in the learning objectives below.

A modern synthesis of evolution *(pages 148-151)*

☐ 2. Give a precise definition of the term **evolution**, explaining how evolution is a feature of **populations** and not of individuals.

☐ 3. Name some of the main contributors to the modern theory of evolution. List some of the important observations that **Darwin** made on his voyage on the Beagle that led him to formulate his theory.

☐ 4. Outline the fundamental ideas in Darwin's *"Theory of evolution by natural selection"* and describe in which way the theory was inadequate. Recognise that Alfred Russel Wallace was following a similar line of thought.

☐ 5. Appreciate how Darwin's original theory has been modified in the **new synthesis** to take into account our understanding of genetics and inheritance. Recognise examples of recently documented evolutionary change.

☐ 6. Understand the term: **fitness** and explain how evolution through **adaptation**, equips species for survival. Recognise structural and physiological adaptations of organisms to their environment.

The concept of the gene pool *(pages 155-163)*

☐ 7. Understand the concept of the **gene pool** and explain the term **deme**. Recognise that populations may be of various sizes and geographical extent.

☐ 8. Explain the term **allele frequency** and describe how allele frequencies are expressed for a population.

☐ 9. Explain the concept of **genetic equilibrium**. Describe the criteria that must be satisfied in order to achieve genetic equilibrium in a population and explain why they are important:

(a) Large population size with random mating.
(b) No mutations (or mutations are balanced).
(c) No selective advantage for any genotype.
(d) No net migration (gene flow) between populations.
(e) Normal meiosis; chance operates in gametogenesis.

Identify the consequences of the fact that these criteria are rarely, perhaps never, met in reality.

☐ 10. Explain how the **Hardy-Weinberg equation** provides a simple mathematical model of **genetic equilibrium** in a population. Demonstrate an ability to use the Hardy-Weinberg equation to calculate the allele, genotype, and phenotype frequencies from appropriate data.

Processes in gene pools *(pages 155-156)*

☐ 11. Recognise that changes within a gene pool occur when any or all of the above criteria for genetic equilibrium are not met. Identify **microevolution** as changes in the allele frequencies of gene pools. Recognise the four forces in microevolution that may alter allele frequencies: **natural selection**, **genetic drift**, **gene flow**, and **mutation**.

Identify which of these microevolutionary processes increase genetic diversity in the gene pool and which decrease it. Recognise the role of natural selection and genetic drift in sorting the genetic variability and establishing adaptive genotypes.

Natural selection *(pages 148, 150, 152-54, 164-67)*

☐ 12. Understand that individuals within a species show **variation** and that heritable variation is the raw material for natural selection. Explain how **natural selection** is responsible for most evolutionary change by selectively reducing or changing genetic variation through differential survival and reproduction. Interpret data and/or use unfamiliar information to explain how natural selection produces change within a population.

☐ 13. Describe three types of natural selection: **stabilising**, **directional**, and **disruptive selection**. Describe the outcome of each type in a population exhibiting a normal curve in phenotypic variation.

☐ 14. With respect to natural selection, explain how environmental factors (e.g. climate) can act as a stabilising force or a force of change (evolution).

☐ 15. Describe examples of evolution by **natural selection**. Examples could include:

(a) **Transient polymorphism** as shown by **industrial melanism** in peppered moths (*Biston betularia*).
(b) The sickle cell trait as the basis for **balanced polymorphism** in malarial regions.
(c) The evolution of **antibiotic resistance** in bacteria.
(d) Changes to the size and shape of the beaks of Galapagos finches.
(e) The evolution of **pesticide resistance** in insects or **heavy metal tolerance** in plants.

Genetic drift *(pages 155-156, 175)*

☐ 16. Explain what is meant by **genetic drift** and describe the conditions under which it is important. Distinguish between genetic drift and natural selection. Explain, using diagrams or a gene pool model, how genetic drift may lead to loss or **fixation of alleles** (where a gene is represented in the population by only one allele).

Mutation *(pages 96, 148, 155-156)*

- ☐ 17. Recognise **mutations** as the ultimate source of all new alleles. Explain, using diagrams or a gene pool model, how mutations alter the genetic equilibrium of a population. Recall that recombination during meiosis reshuffles alleles into different combinations and therefore increases variation, but it does not create new alleles.

Gene flow *(pages 155-156)*

- ☐ 18. Explain, using diagrams or a gene pool model, how **migration** leads to **gene flow** between natural populations, and may affect allele frequencies.

Artificial selection *(pages 168-172)*

- ☐ 19. Define the term: **artificial selection** (**selective breeding**) and explain its genetic basis. Describe examples of artificial selection in plants and animals, e.g. crop plants, livestock, and companion animals.
- ☐ 20. Recognise **genetic improvement** in livestock breeding as the gain towards a desired **phenotype**. Explain the contribution of modern technology to the rate of genetic improvement achieved in breeding programmes.

Special events in gene pools *(pages 173-175)*

- ☐ 21. Recognise how the **founder effect** and **population** (**genetic**) **bottlenecks** may accelerate the pace of evolutionary change. Explain the importance of **genetic drift** in populations that undergo these events.
- ☐ 22. Describe the basis of the founder effect and explain its genetic and evolutionary consequences. Provide examples where the founder effect has been important in the establishment of island populations.
- ☐ 23. Describe the bottleneck effect on a population following a sudden reduction in numbers. Discuss situations where population bottlenecks are likely.

Speciation *(pages 155, 176-183)*

- ☐ 24. Recall your definition of **evolution** and distinguish between **microevolution** and **macroevolution** (the formation of completely new species, genera etc.).
- ☐ 25. Provide a clear definition of the term **species**. Describe how the nature of some species can create problems for our definition. Give examples of organisms for which our definition of a species is problematic.
- ☐ 26. Providing examples, explain the concept of a **ring species** and closely related species where the **reproductive isolation** is no longer maintained.
- ☐ 27. Explain what is meant by **speciation**. Recognise the role of **natural selection** and (reproductive) **isolation** in the formation of new species.
- ☐ 28. Explain what is meant by **reproductive isolation**. Describe how populations may become reproductively isolated through:
  (a) Altered behaviour or physiology
  (b) Geographical isolation
  (c) Polyploidy
  (d) Niche differentiation
- ☐ 29. Distinguish between and describe **prezygotic** and **postzygotic** reproductive isolating mechanisms.
  - ***Prezygotic isolating mechanisms**:* *Geographical, ecological, behavioural, structural (morphological), gamete mortality, temporal.*
  - ***Postzygotic isolating mechanisms**:* *Hybrid sterility, hybrid breakdown, hybrid inviability.*
- ☐ 30. Explain the events occurring in **allopatric speciation**, identifying situations in which it is most likely to occur.
- ☐ 31. Explain the events occurring in **sympatric speciation** and describe the situations in which it is most likely to occur. Explain why reproductive isolating mechanisms tend to be much more pronounced between sympatric (as opposed to allopatric) species. Recognise the role of **polyploidy** in instant speciation events.

Patterns of Evolution *(pages 184-192)*

- ☐ 32. Recognise the major stages in a **species life cycle**, extending from the species **origin** to **extinction**.
- ☐ 33. Distinguish patterns of species formation:
  (a) **Sequential** (**phyletic**) **speciation**
  (b) **Divergent evolution**.
  (c) **Adaptive radiation** (dichotomous) speciation.
  Describe at least one example to illustrate each pattern. Recognise adaptive radiation as a form of **divergent evolution**.
- ☐ 34. Describe how evolutionary change over time has resulted in a great diversity of forms among living organisms.
- ☐ 35. Distinguish between the two models for the pace of evolutionary change:
  (a) **Punctuated equilibrium**.
  (b) **Gradualism**.
  Discuss the evidence for each of these models and discuss the evidence for each in different taxa.
- ☐ 36. Explain **convergent evolution** and provide examples. Discuss how **analogous structures** (analogies) may arise as a result of convergence. Distinguish clearly between **analogous structures** and **homologous structures** and explain the role of homology in identifying evolutionary relationships.
- ☐ 37. Understand that some biologists also recognise **parallel evolution** (as distinct from convergence) to indicate evolution along similar lines in closely related groups (as opposed to more distantly related groups).
- ☐ 38. In a general way, describe how **comparative anatomy**, **embryology**, and physiology have contributed to an understanding of evolutionary relationships.
- ☐ 39. Using examples, distinguish between **homologous** structures and **analogous structures** arising as a result of convergent evolution (cross reference with #34). Explain the evidence for evolution provided by homologous anatomical structures, including the vertebrate pentadactyl limb.
- ☐ 40. Discuss the significance of **vestigial organs** as indicators of evolutionary trends in some groups.

See the 'Textbook Reference Grid' on pages 8-9 for textbook page references relating to material in this topic.

Supplementary Texts

See pages 4-6 for additional details of these texts:

■ Adds, J., *et al.,* 2001. **Genetics, Evolution and Biodiversity**, (NelsonThornes), pp. 101-108.

■ Clegg, C.J., 1999. **Genetics and Evolution**, (John Murray), pp. 60-78.

■ Jones, N., *et al.,* 2001. **Essentials of Genetics**, pp. 190-232.

See page 6 for details of publishers of periodicals:

STUDENT'S REFERENCE

Species, microevolution, and speciation

■ **The Species Enigma** New Scientist, 13 June 1998 (Inside Science). *An excellent article on the nature of species, ring species, and the status of hybrids. Includes species life cycles and extinction.*

■ **The Hardy-Weinberg Principle** Biol. Sci. Rev., 15(4), April 2003, pp. 7-9. *A succinct explanation of the basis of the Hardy-Weinberg principle, and its uses in estimating genotype frequencies and predicting change in populations.*

■ **Speciation** Biol. Sci. Rev., 16(2) Nov. 2003, pp. 24-28. *An excellent account of speciation. It covers the nature of species, reproductive isolation, how separated populations diverge, and sympatric speciation. Case examples include the cichlids of Lake Victoria and the founder effect in mynahs.*

■ **Polymorphism** Biol. Sci. Rev., 14(1), Sept. 2001, pp. 19-21. *An account of polymorphism in populations, with several case studies (including Biston moths) provided as illustrative examples.*

■ **Plants on the Move** New Scientist, 20 March 1999 (Inside Science). *Glaciation and warming have evolutionary consequences for flora.*

■ **Butterflies or Bitterflies** Biol. Sci. Rev., 11(3) January 1999, pp. 18-21. *Mimicry as the basis for natural selection in polymorphic populations.*

■ **Tails** *(sic)* **of Love and War** Biol. Sci. Rev., 11(1) Sept. 1998, pp. 30-33. *A discussion of the mechanisms of evolutionary change with examples of natural selection in animal populations.*

■ **Zoos - The Modern Noah's Ark** Biol. Sci. Rev., 9(3) Jan. 1997, pp. 21-23. *The consequences of genetic drift in populations of captive bred species.*

■ **The Cheetah: Losing the Race?** Biol. Sci. Rev., 14(2) Nov. 2001, pp. 7-10. *The evolutionary bottleneck experienced by cheetahs and its implications for the genetics of the species.*

■ **Unnatural Evolution** New Scientist, 2 March 1996, pp. 34-37. *The influence of chemicals in the environment on normal evolutionary rates.*

Artificial selection

■ **A Kinder, Gentler Killer** New Scientist, 1 July 2000, pp. 34-36. *South American killer bees may be used to breed a gentle, high producing honey bee.*

■ **Plant Breeding** Biol. Sci. Rev., 7(3) January 1995, pp. 32-34. *The principles of plant breeding and the role of selective breeding in crop evolution.*

Macroevolution

■ **The Rise of Life on Earth: From Fins to Feet** National Geographic, 195(5) May 1999, pp. 114-127. *The evolution of diversity in the Devonian. Includes material on the evolution of tetrapods.*

■ **The Rise of Mammals** National Geographic, 203(4), pp. April 2003, p. 2-37. *An account of the adaptive radiation of mammals and the significance of the placenta in mammalian evolution.*

TEACHER'S REFERENCE

Species, microevolution, and speciation

■ **Live and Let Live** New Scientist, 3 July 1999, pp. 32-36. *Recent research suggests that hybrids are intact entities subject to the same evolutionary pressures as pure species.*

■ **Figs and Fig Wasps** Biologist, 48(3) June 2001. *The fig and the fig wasp have coevolved; the fig wasp is the fig's only pollinator. This account details their rather tenuous relationship.*

■ **How are Species Formed?** New Scientist, 14 June 2003, pp. 36-37. *Part of an in-depth examination of evolution ("Evolution: The five big questions"). This article discusses how ideas about speciation have moved away from chance and small populations. New species may be the result of parallel evolution (ecological selection), sexual selection, or hybridisaton.*

■ **The Challenge of Antibiotic Resistance** Scientific American, March 1998, pp. 32-39. *An excellent article on the basis of antibiotic resistance in bacteria and its selective advantage.*

■ **Replaying Life** New Scientist, 13 February 1999, pp. 29-33. *Rapid evolution in bacteria driven by habitat diversity and niche differentiation.*

■ **Fair Enough** New Scientist, 12 October 2002, pp. 34-37. *The inheritance of skin colour in humans. This article examines why humans have such varied skin pigmentation and looks at the argument for there being a selective benefit to being dark or pale in different environments.*

■ **Skin Deep** Scientific American, October 2002, pp. 50-57. *This article examines the evolution of skin colour in humans and presents powerful evidence for skin colour ("race") being the end result of opposing selection forces (the need for protection of folate from UV vs the need to absorb vitamin D). Clearly written and of high interest, this is a must for student discussion and a perfect vehicle for examining natural selection.*

■ **A New Ant on the Block** New Scientist, 4 Nov. 1995, pp. 28-31. *The spread of South American fire ants offers a chance to study founder populations.*

■ **Together We're Stronger** New Scientist, 15 March 2003. (Inside Science). *The mechanisms behind the evolution of social behaviour in animals. The evolution of eusociality in hymenopteran insects is the case study provided.*

■ **Listen, We're Different** New Scientist, 17 July 1999, pp. 32-35. *An excellent account of speciation in periodic cicadas as a result of behavioural and temporal isolating mechanisms.*

■ **Cichlids of the Rift Lakes** Scientific American, February 1999, pp. 44-49. *An excellent, thorough account of the recent speciation events documented in the cichlid fishes of Lake Victoria.*

■ **How the Species Became** New Scientist, 11 Oct. 2003, pp. 32-35. *Stability in species and new ideas on speciation. Species are stable if changes in form or behaviour are damped out, but unstable if the changes escalate as new generations shuffle parental genes and natural selection discards the allele combinations that do not work well.*

■ **Adaptation - A Question of Definitions** SSR 83(304), March 2002, pp. 97-101. *The term adaptation has various meanings in biology and is an area in teaching where conceptual difficulties often arise. This article reviews the use of the term in the scientific literature and examines some of its ambiguous uses.*

Macroevolution

■ **Food for Thought: Dietary Change was a Driving Force in Human Evolution** Scientific American, Dec. 2002, pp. 74-83. *In human evolution, natural selection for improved quality of the diet may have been very important. The acquisation of higher energy food goes hand in hand with increasing use of fire, tool making, and cooperative hunting.*

■ **Which Came First?** Scientific American, Feb. 1997, pp. 12-14. *Shared features among fossils; the result from convergence or common ancestry?*

■ **We Were Meant To Be** New Scientist, 16 Nov. 2002, pp. 26-29. *Organisms faced with the same challenges repeatedly arrive at the same solutions. Increasing intelligence is one such solution.*

■ **A Waste of Space** New Scientist, 25 April 1998, pp. 38-39. *Vestigial organs: how they arise in an evolutionary sense and what role they may play.*

<u>Case study: the evolution of flight in birds</u>

■ **Birds Do It ..... Did Dinosaurs?** New Scientist 1 Feb. 1997, pp. 26-31. *Evolution of flight in the light of recent fossil finds. An excellent article.*

■ **Dinosaurs take Wing** National Geographic, 194(1) July 1998, pp. 74-99. *The evolution of birds from small theropod dinosaurs. This excellent account explores the homology between the typical dinosaur limb and the wing of a modern bird.*

■ **The Origin of Birds and their Flight** Scientific American, Feb. 1998, pp. 28-37. *More on avian evolution: a thorough and well illustrated account.*

■ **Which Came First: the Feather or the Bird?** Scientific American, March 2003, pp. 60-69. *An excellent account of the evolution of feathers: how and why did they evolve and how is their structure related to their function.*

■ **Winging It** New Scientist, 28 Aug. 1999, pp. 28-32. *Update on the evidence for the origin of bird flight.*

See pages 10-11 for details of how to access **Bio Links** from our web site: **www.biozone.co.uk**. From Bio Links, access sites under the topics:

GENERAL BIOLOGY ONLINE RESOURCES > **Online Textbooks and Lecture Notes**: • An on-line biology book • Kimball's biology pages • Learn.co.uk • Mark Rothery's biology web site • *... and others* > **General Online Biology resources**: • Ken's Bioweb resources *... and others* > **Glossaries**: • Evolutionary biology and genetics glossary*... and others*

EVOLUTION: • A history of evolutionary thought • BIO 414 evolution • Enter evolution: theory and history • Evolution • Evolution on the web for biology students • Harvard University biology links: evolution • The Talk.Origins archive *... and others* > **Charles Darwin**: • Darwin and evolution overview • Darwin's "On the Origin of Species" Overview • What is Darwinism? *... and others*

Also see the sites listed under "*Evolution Theory and Evidence*" and "*The Fossil Record*".

GENETICS > **Population Genetics**: • Introduction to evolutionary biology • Microevolution and population genetics • Random genetic drift • Population genetics • Population genetics: lecture notes *... and others*

Software and video resources are provided on the Teacher Resource Handbook on CD-ROM

Genes and Evolution

Each individual in a population is the carrier of its own particular combination of genetic material. Different combinations of genes come about because of the shuffling of the chromosomes during gamete formation. New combinations of alleles arise as a result of mate selection and the chance meeting of a vast range of different gametes from each of the two parents. Some combinations are well suited to particular environments, while others are not. Those organisms with an inferior collection of genes will have reduced reproductive success. This means that the genes (alleles) they carry will decrease in frequency and fewer will be passed on to the next generation's gene pool. Those individuals with more successful allele combinations will have higher reproductive success. The frequency of their alleles in the gene pool will increase.

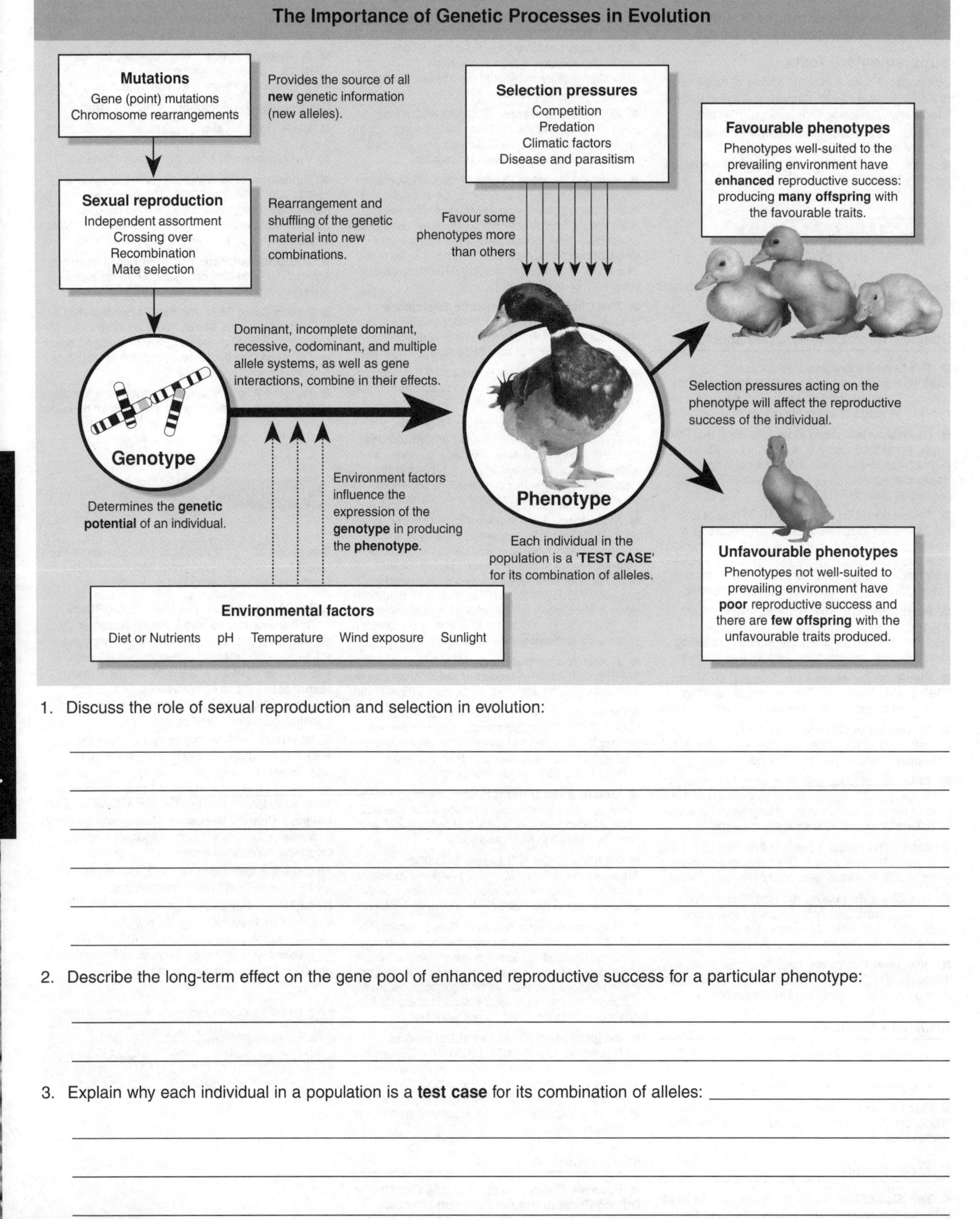

1. Discuss the role of sexual reproduction and selection in evolution:

2. Describe the long-term effect on the gene pool of enhanced reproductive success for a particular phenotype:

3. Explain why each individual in a population is a **test case** for its combination of alleles:

The Modern Theory of Evolution

Although **Charles Darwin** is credited with the development of the theory of evolution by natural selection, there were many people that contributed ideas upon which he built his own. Since Darwin first proposed his theory, aspects that were problematic (such as the mechanism of inheritance) have now been explained. The theory has undergone refinement and has been expanded to incorporate the modern developments in biology. The development of the modern theory of evolution has a history going back at least two centuries. The diagram below illustrates the way in which some of the major contributors helped to form the currently accepted model, often referred to as the **new synthesis** (or the Neo-Darwinian theory). Some of the early contributors did not have the concept of evolution in their minds when they put forward their ideas, but their work contributed toward the development of a unifying theory explaining how species can change over time.

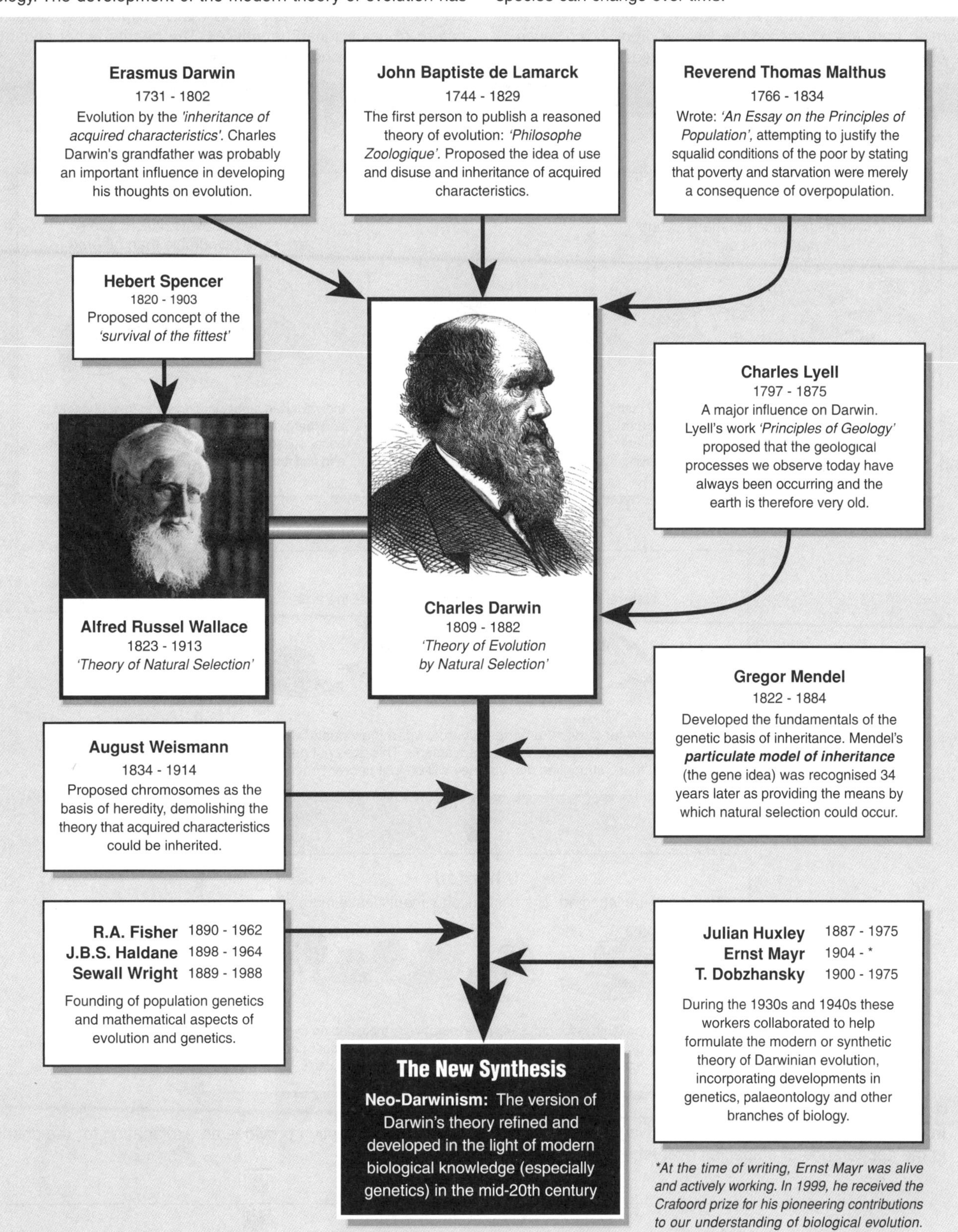

1. From the diagram above, choose one of the contributors to the development of evolutionary theory (excluding Charles Darwin himself), and write a few paragraphs discussing their role in contributing to Darwin's ideas. You may need to consult an encyclopaedia or other reference to assist you.

Darwin's Theory

In 1859, Darwin and Wallace jointly proposed that new species could develop by a process of natural selection. Natural selection is the term given to the mechanism by which better adapted organisms survive to produce a greater number of viable offspring. This has the effect of increasing their proportion in the population so that they become more common. It is Darwin who is best remembered for the theory of evolution by natural selection through his famous book: '**On the origin of species by means of natural selection**', written 23 years after returning from his voyage on the Beagle, from which much of the evidence for his theory was accumulated. Although Darwin could not explain the origin of variation nor the mechanism of its transmission (this was provided later by Mendel's work), his basic theory of evolution by natural selection (outlined below) is widely accepted today. The study of population genetics has greatly improved our understanding of evolutionary processes, which are now seen largely as a (frequently gradual) change in allele frequencies within a population. Students should be aware that scientific debate on the subject of evolution centres around the relative merits of various alternative hypotheses about the nature of evolutionary processes. The debate is not about the existence of the phenomenon of evolution itself.

Darwin's Theory of Evolution by Natural Selection

Overproduction

Populations produce too many young: many must die

Populations tend to produce more offspring than are needed to replace the parents. Natural populations normally maintain constant numbers. There must therefore be a certain number dying.

Variation

Individuals show variation: some are more favourable than others

Individuals in a population vary in their phenotype and therefore, their genotype. Some variants are better suited to the current conditions than others and find it easier to survive and reproduce.

Natural Selection

Natural selection favours the best suited at the time

The struggle for survival amongst overcrowded individuals will favour those variations which have the best advantage. This does not necessarily mean that those struggling die, but they will be in a poorer condition.

Inherited

Variations are Inherited. The best suited variants leave more offspring.

The variations (both favourable and unfavourable) are passed on to offspring. Each new generation will contain proportionally more descendents from individuals with favourable characters than those with unfavourable.

1. In your own words, describe how Darwin's theory of evolution by natural selection provides an explanation for the change in the appearance of a species over time:

Adaptations and Fitness

An **adaptation**, is any heritable trait that suits an organism to its natural function in the environment (its niche). These traits may be structural, physiological, or behavioural. The idea is important for evolutionary theory because adaptive features promote fitness. **Fitness** is a measure of how well suited an organism is to survive in its habitat and its ability to maximise the numbers of offspring surviving to reproductive age. Adaptations are distinct from *properties* which, although they may be striking, cannot be described as adaptive unless they are shown to be functional in the organism's natural habitat. Genetic adaptation must not be confused with *physiological adjustment* (acclimatisation), which refers to an organism's ability to *adapt* during its lifetime to changing environmental conditions. The physiological changes that occur when a person spends time at altitude provide a good example of acclimatisation. Examples of adaptive features arising through evolution are illustrated below.

Ear Length in Rabbits and Hares

The external ears of many mammals are used as important organs to assist in thermoregulation (controlling loss and gain of body heat). The ears of rabbits and hares native to hot, dry climates, such as the jack rabbit of south-western USA and northern Mexico, are relatively very large. The Arctic hare lives in the tundra zone of Alaska, northern Canada and Greenland, and has ears that are relatively short. This reduction in the size of the extremities (ears, limbs, and noses) is typical of cold adapted species.

Arctic hare: *Lepus arcticus*

Black-tail jackrabbit: *Lepus californicus*

Body Size in Relation to Climate

Regulation of body temperature requires a large amount of energy and mammals exhibit a variety of structural and physiological adaptations to increase the effectiveness of this process. Heat production in any endotherm depends on body volume (heat generating metabolism), whereas the rate of heat loss depends on surface area. Increasing body size minimises heat loss to the environment by reducing the surface area to volume ratio. Animals in colder regions therefore tend to be larger overall than those living in hot climates. This relationship is know as **Bergman's rule** and it is well documented in many mammalian species. Cold adapted species also tend to have more compact bodies and shorter extremities than related species in hot climates.

Fennec fox

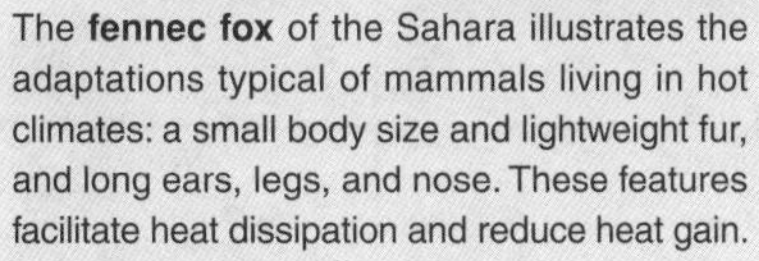

The **fennec fox** of the Sahara illustrates the adaptations typical of mammals living in hot climates: a small body size and lightweight fur, and long ears, legs, and nose. These features facilitate heat dissipation and reduce heat gain.

Arctic fox

The **Arctic fox** shows the physical characteristics typical of cold-adapted mammals: a stocky, compact body shape with small ears, short legs and nose, and dense fur. These features reduce heat loss to the environment.

Number of Horns in Rhinoceroses

Not all differences between species can be convincingly interpreted as adaptations to particular environments. Rhinoceroses charge rival males and predators, and the horn(s), when combined with the head-down posture, add effectiveness to this behaviour. Horns are obviously adaptive, but it is not clear that the possession of one (Indian rhino) or two (black rhino) horns is necessarily related directly to the environment in which those animals live.

Great Indian rhino

African black rhino

1. Distinguish between adaptive features (genetic) and acclimatisation:

2. Explain the nature of the relationship between the length of extremities (such as limbs and ears) and climate:

3. Explain the adaptive value of a larger body size at high latitude:

Natural Selection

Natural selection operates on the phenotypes of individuals, produced by their particular combinations of alleles. In natural populations, the allele combinations of some individuals are perpetuated at the expense of other genotypes. This differential survival of some genotypes over others is called **natural selection**. The effect of natural selection can vary; it can act to maintain the genotype of a species or to change it. **Stabilising selection** maintains the established favourable characteristics and is associated with stable environments. In contrast, **directional selection** favours phenotypes at one extreme of the phenotypic range and is associated with gradually changing environments. **Disruptive selection** is a much rarer form of selection favouring two phenotypic extremes, and is a feature of fluctuating environments.

Stabilising Selection

Extreme variations are culled from the population (there is selection against them). Those with the established (middle range) adaptive phenotype are retained in greater numbers. This reduces the variation for the phenotypic character. In the example right, light and dark snails are eliminated, leaving medium coloured snails. Stabilising selection can be seen in the selection pressures on human birth weights.

Directional Selection

Directional selection is associated with gradually changing conditions, where the adaptive phenotype is shifted in one direction and one aspect of a trait becomes emphasised (e.g. coloration). In the example right, light coloured snails are eliminated and the population becomes darker. Directional selection was observed in peppered moths in England during the Industrial Revolution. They responded to the air pollution of industrialisation by increasing the frequency of darker, melanic forms.

Disruptive or Diversifying Selection

Disruptive selection favours two extremes of a trait at the expense of intermediate forms. It is associated with a fluctuating environment and gives rise to **balanced polymorphism** in the population. In the example right, there is selection against medium coloured snails, which are eliminated. There is considerable evidence that predators, such as insectivorous birds, are more likely to find and eat common morphs and ignore rare morphs. This enables the rarer forms to persist in the population.

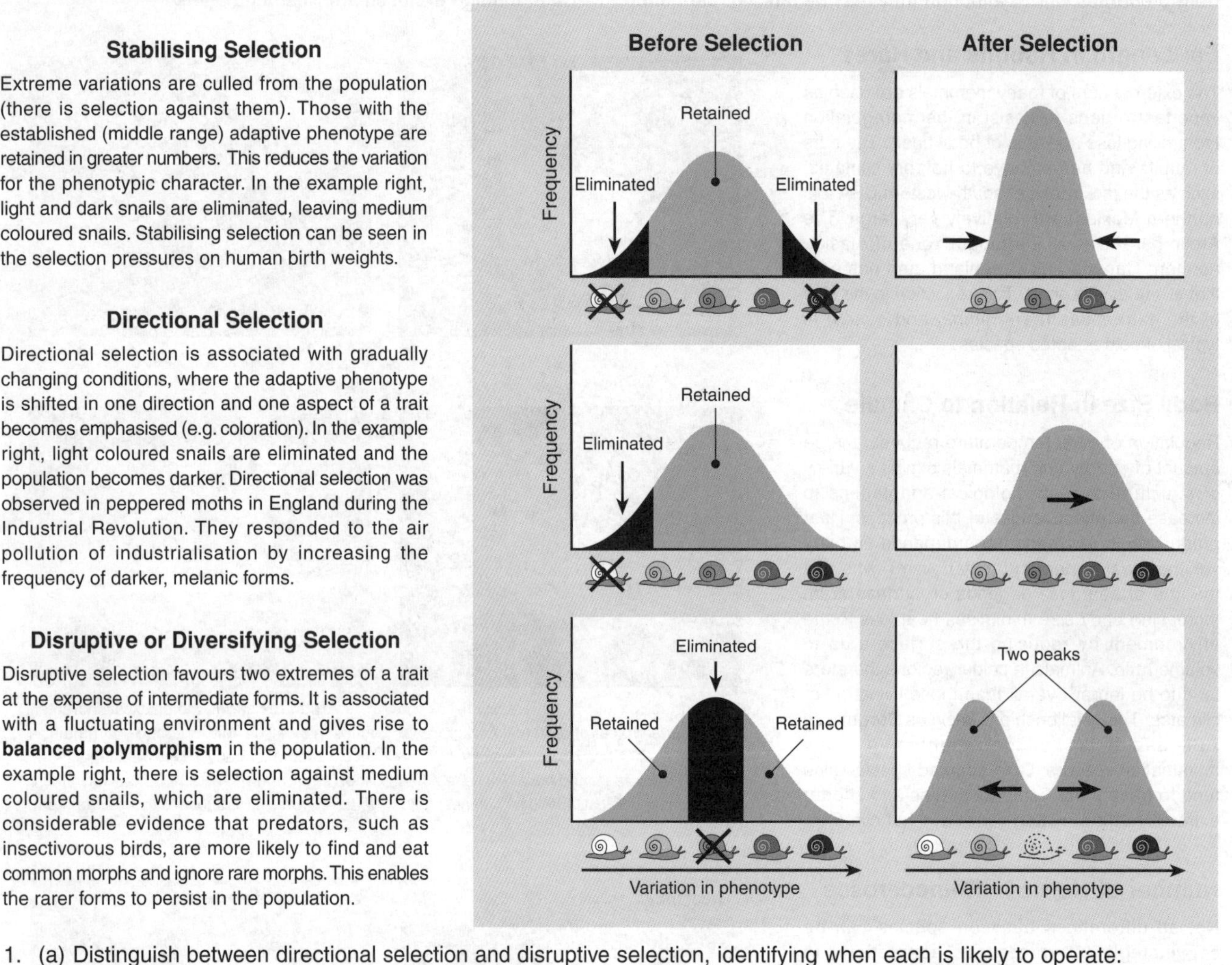

1. (a) Distinguish between directional selection and disruptive selection, identifying when each is likely to operate:

   (b) State which of the three types of selection described above will lead to evolution, and explain why:

2. Explain how a change in environment may result in selection becoming directional rather than stabilising:

3. Explain how, in a population of snails, through natural selection, shell colour could change from light to dark over time:

Darwin's Finches

The Galapagos Islands, 920 km off the west coast of Ecuador, played a major role in shaping Darwin's thoughts about natural selection and evolution. While exploring the islands in 1835, he was struck by the unique and peculiar species he found there. In particular, he was intrigued by the island's finches. The Galapagos group is home to 13 species of finch in four genera. This variety has arisen as a result of evolution from one common ancestral species. Initially, a number of small finches, probably grassquits, made their way from South America across the Pacific to the Galapagos Islands. In this new environment, which was relatively free of competitors, the colonisers underwent an adaptive radiation, producing a range of species each with its own feeding niche. Although similar in their plumage, nest building techniques, and calls, the different species of finches can easily be distinguished by the size and shape of their beaks. The beak shape of each species is adapted for a different purpose, such as crushing seeds, pecking wood, or probing flowers for nectar. Between them, the 13 species of this endemic group fill the roles of seven different families of South American mainland birds. Modern methods of DNA (genetic) analysis have confirmed Darwin's insight and have shown that all 13 species evolved from a flock of about 30 birds arriving a million years ago.

The Evolution of Darwin's Finches

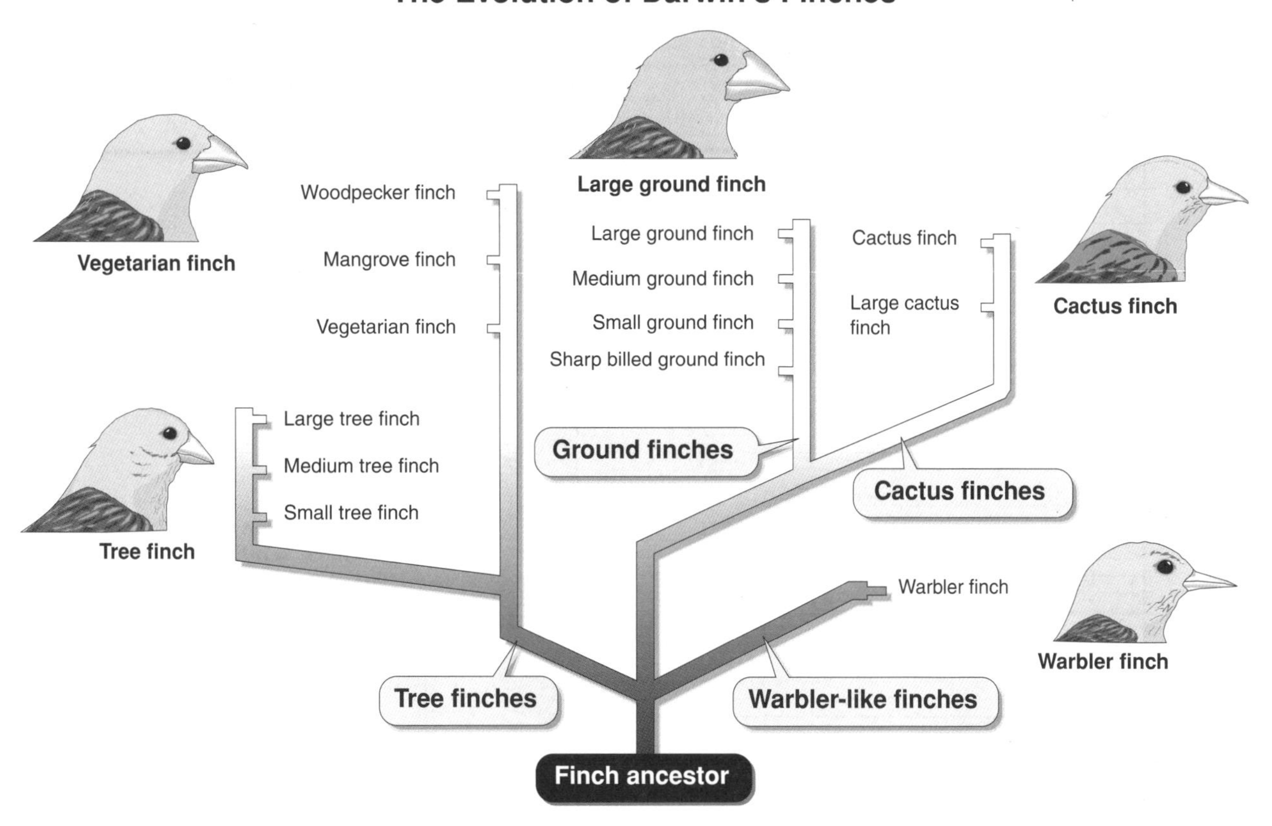

As the name implies, tree finches are largely arboreal and feed mainly on insects. The bill is sharper than in ground finches and better suited to grasp insects. Paler than ground or cactus finches, they also have streaked breasts.

Probably descended from ground finches. Beak is probing. Males are mostly black, females are streaked, like ground finches. Found in arid areas on prickly pear cactus where they eat insects on the cactus, or the cactus itself.

Four species with crushing-type bills used for seed eating. On Wolf Island, they are called vampire finches because they peck the skin of animals to draw blood, which they then drink. Such behaviour has evolved from eating parasitic insects off animals.

Named for their resemblance to the unrelated warblers, the beak of the warbler finch is the thinnest of the Galapagos finches. It is the most widespread species, found throughout the archipelago. Warbler finches prey on flying and ground dwelling insects.

1. Outline the main factors that have contributed to the adaptive radiation of Darwin's finches:

Selection for Human Birth Weight

This activity explores the selection pressures acting on the birth weight of human babies. Carry out the steps below:

Step 1: Collect the birth weights from 100 birth notices from your local newspaper (or 50 if you are having difficulty getting enough; this should involve looking back through the last 2-3 weeks of birth notices). If you cannot obtain birth weights in your local newspaper, a set of 100 sample birth weights is provided in the Model Answers booklet.

Step 2: Group the weights into each of the 12 weight classes (of 0.5 kg increments). Determine what percentage (of the total sample) fall into each weight class (e.g. 17 babies weigh 2.5-3.0 kg out of the 100 sampled = 17%)

Step 3: Graph these in the form of a **histogram** for the 12 weight classes (use the graphing grid provided below). Be sure to use the scale provided on the **left** vertical (y) axis.

Step 4: Create a second graph by plotting percentage mortality of newborn babies in relation to their birth weight. Use the scale on the **right** y axis and data provided (below right).

Step 5: Draw a **line** of 'best fit' through these points.

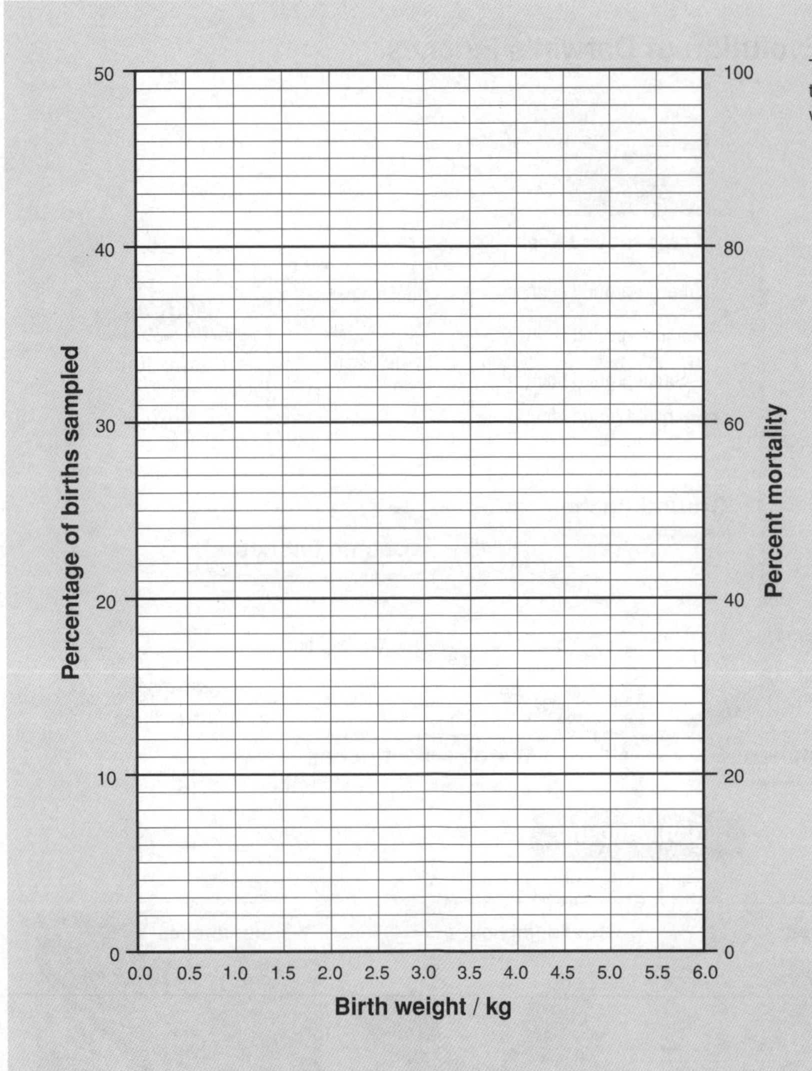

The size of the baby and the diameter and shape of the birth canal are the two crucial factors in determining whether a normal delivery is possible.

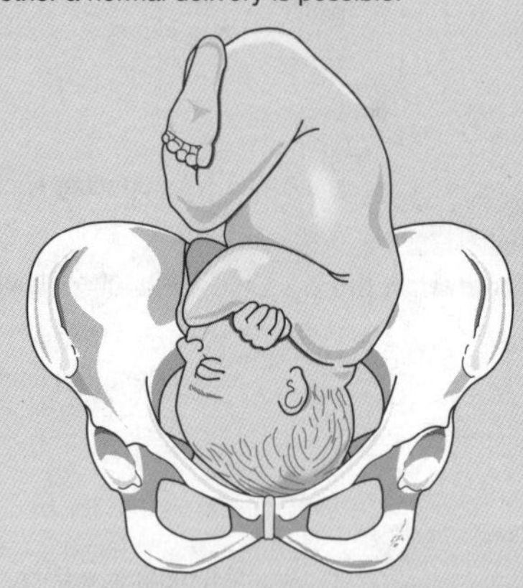

Mortality of newborn babies related to birth weight

Weight / kg	Mortality / %
1.0	80
1.5	30
2.0	12
2.5	4
3.0	3
3.5	2
4.0	3
4.5	7
5.0	15

Source: Biology: The Unity & Diversity of Life (4th ed), by Starr and Taggart

1. Describe the shape of the histogram for birth weights: ____________________

2. State the optimum birth weight in terms of the lowest newborn mortality: ____________________

3. Describe the relationship between newborn mortality and birth weight: ____________________

4. Describe the selection pressures that are operating to control the range of birth weight: ____________________

5. Describe how medical intervention methods during pregnancy and childbirth may have altered these selection pressures: ____________________

Population Genetics & Evolution

Code: PDA 2

Gene Pools and Evolution

The diagram below illustrates the dynamic nature of **gene pools**. It portrays two imaginary populations of one beetle species. Each beetle is a 'carrier' of genetic information, represented here by the alleles (A and a) for a single **codominant gene** that controls the beetle's colour. Normally, there are three versions of the phenotype: black, dark, and pale. Mutations may create other versions of the phenotype. Some of the **microevolutionary processes** that can affect the genetic composition (**allele frequencies**) of the gene pool are illustrated below.

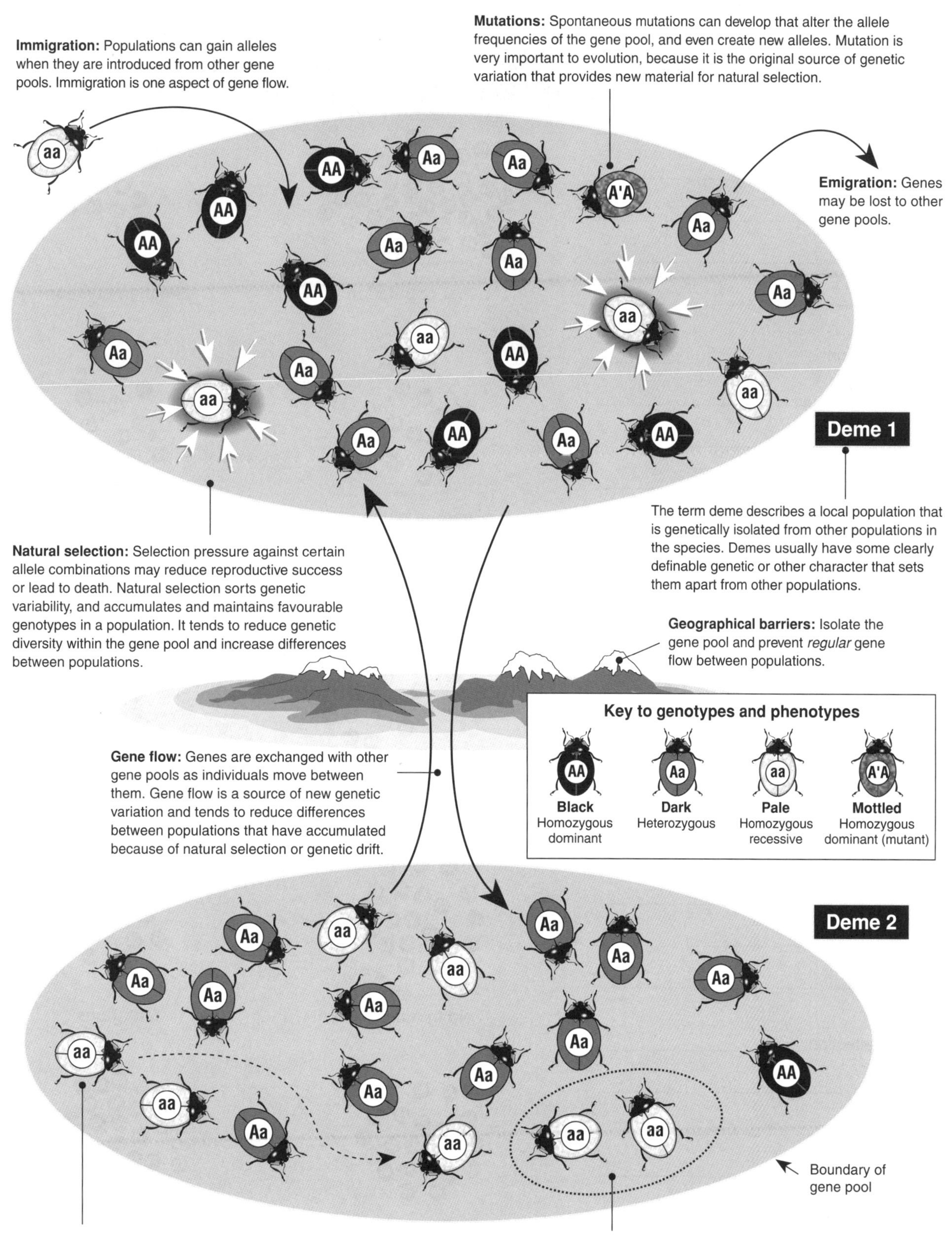

Mate selection (non-random mating): Individuals may not select their mate randomly and may seek out particular phenotypes, increasing the frequency of these "favoured" alleles in the population.

Genetic drift: Chance events can cause the allele frequencies of small populations to "drift" (change) randomly from generation to generation. Genetic drift can play a significant role in the microevolution of very small populations. The two situations most often leading to populations small enough for genetic drift to be significant are the ***bottleneck effect*** (where the population size is dramatically reduced by a catastrophic event) and the ***founder effect*** (where a small number of individuals colonise a new area).

Factors Affecting Gene Pools

One of the fundamental concepts for population genetics is stated as follows:

*For a very large, randomly mating population, the proportion of dominant to recessive alleles remains constant from one generation to the next (the population is in **genetic equilibrium**).*

In practical terms this means that, if a gene pool is to remain unchanged, it must satisfy all of the criteria listed on the left side of the diagram below (factors that favour gene pool stability). The fact that few populations can be identified as meeting all (or any) of these criteria means that they must be undergoing continual change in their genetic makeup.

For each of the five factors (numbers 1-5) below, state briefly **how** and **why** each would affect the allele frequency in a gene pool:

1. Population size: ______________________

2. Mate selection: ______________________

3. Gene flow between populations: ______________________

4. Mutations: ______________________

5. Natural selection: ______________________

Factors That Favour Gene Pool Stability	Factors That Favour Gene Pool Change
Large population	Small population
Random mating	Assortative mating
No gene flow	Gene flow
No mutation	Mutations
No natural selection	Natural selection

6. (a) List the factors that tend to increase genetic variation in populations: ______________________

(b) List the factors that tend to decrease genetic variation in populations: ______________________

Gene Pool Exercise

Cut out each of the beetles on this page and use them to reenact different events within a gene pool as described in this topic (*Gene Pools and Evolution, Changes in a Gene Pool, Founder Effect, Population Bottlenecks, Genetic Drift*).

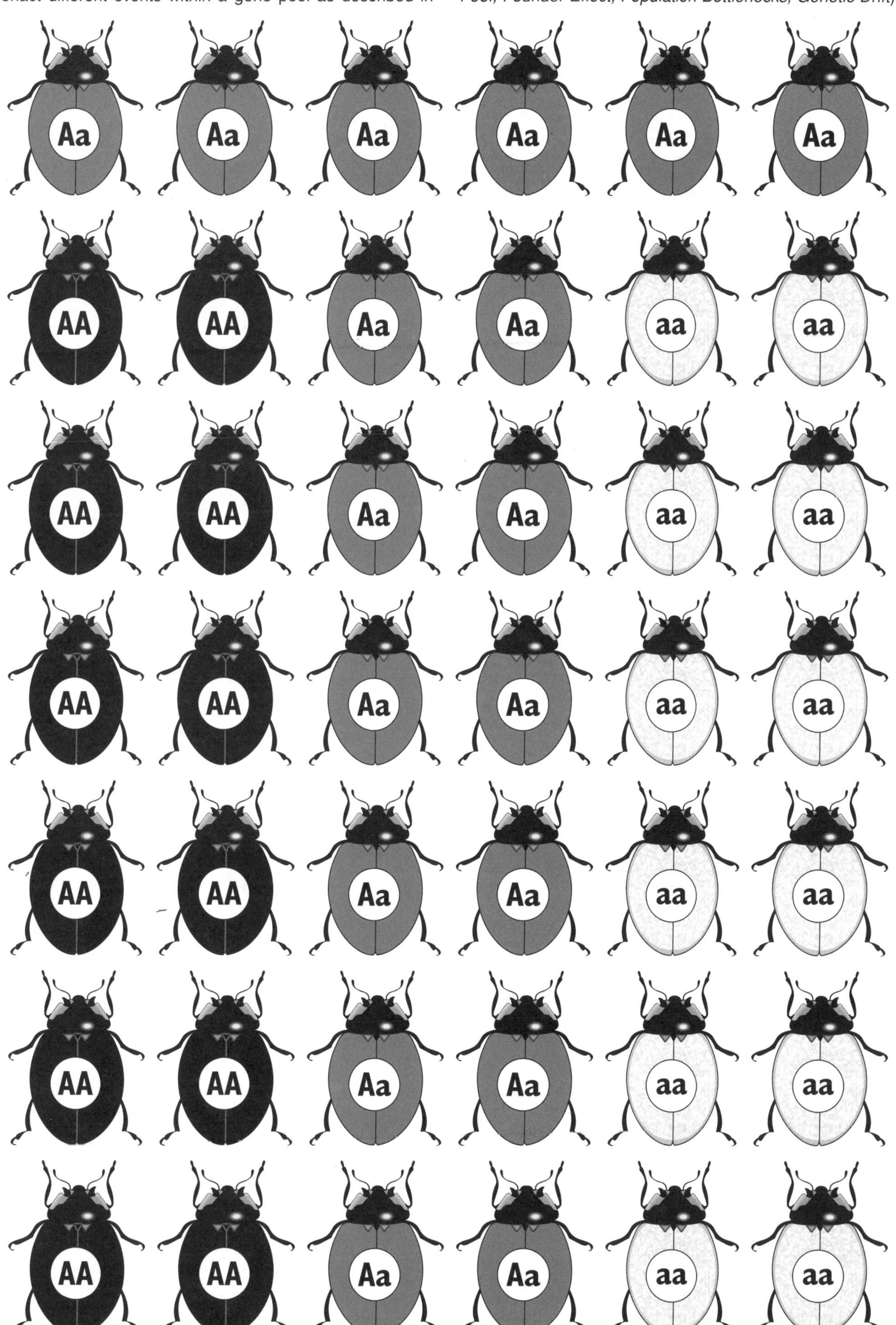

This page has deliberately been left blank

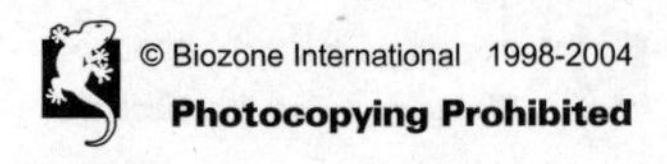

Changes in a Gene Pool

The diagram below shows an imaginary population of beetles undergoing changes as it is subjected to two 'events'. The three phases represent a progression in time (i.e. the same gene pool, undergoing change). The beetles have three phenotypes determined by the amount of pigment deposited in the cuticle. Three versions of this trait exist: black, dark, and pale. The gene controlling this character is represented by two alleles **A** and **a**. Your task is to analyse the gene pool as it undergoes changes.

Phase 1: Initial gene pool

Calculate the frequencies of the *allele types* and *allele combinations* by counting the actual numbers, then working them out as percentages.

	A	a	AA	Aa	aa
No.	27		7		
%	54		28		

Allele types (A, a) — *Allele combinations* (AA, Aa, aa)

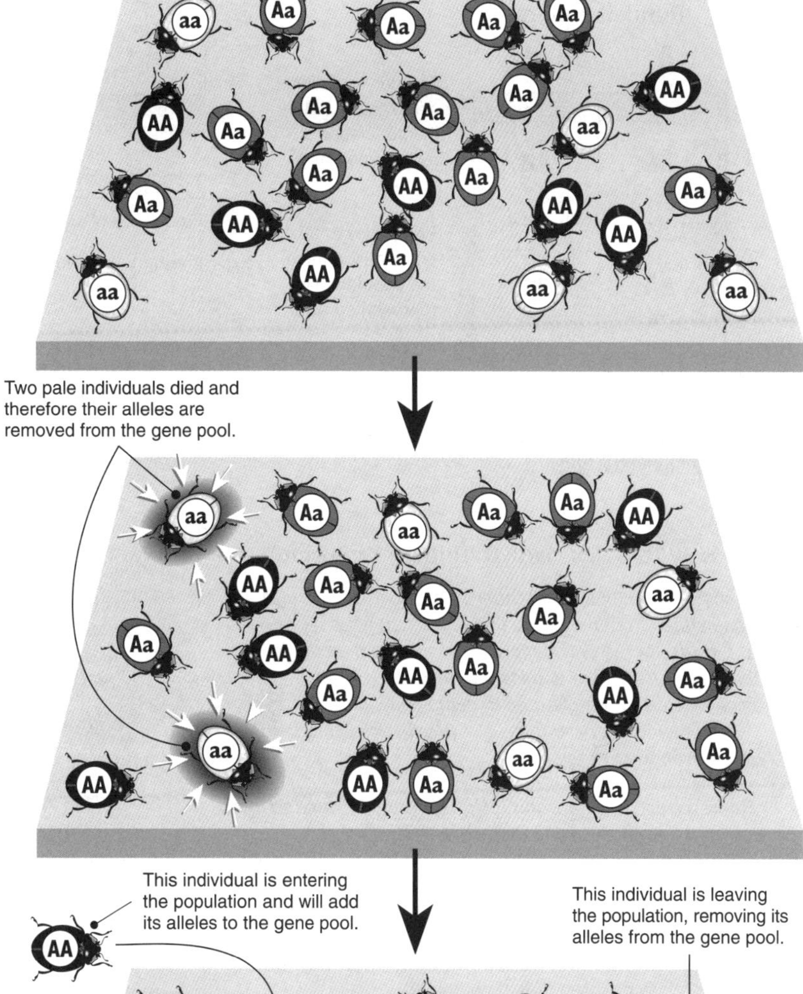

Phase 2: Natural selection

In the same gene pool at a later time there was a change in the allele frequencies. This was due to the loss of certain allele combinations due to natural selection. Some of those with a genotype of aa were eliminated (poor fitness).

Calculate as for above. Do not include the individuals surrounded by small white arrows in your calculations; they are dead!

	A	a	AA	Aa	aa
No.					
%					

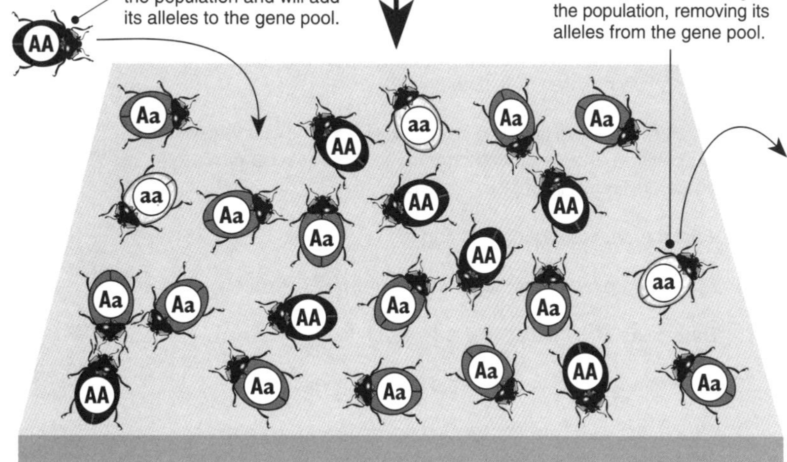

Phase 3: Immigration and emigration

This particular kind of beetle exhibits wandering behaviour. The allele frequencies change again due to the introduction and departure of individual beetles, each carrying certain allele combinations.

Calculate as above. In your calculations, include the individual coming into the gene pool (AA), but remove the one leaving (aa).

	A	a	AA	Aa	aa
No.					
%					

1. Explain how the number of dominant alleles (A) in the genotype of a beetle affects its phenotype:

2. For each phase in the gene pool above (place your answers in the tables provided; some have been done for you):
   (a) Determine the relative frequencies of the two alleles: A and a. Simply total the **A** alleles and **a** alleles separately.
   (b) Determine the frequency of how the alleles come together as allele pair combinations in the gene pool (AA, Aa, and aa). Count the number of each type of combination
   (c) For each of the above, work out the frequencies as percentages:

   Allele frequency = Number of counted alleles ÷ Total number of alleles x 100

Population Genetics Calculations

The **Hardy-Weinberg equation** provides a simple mathematical model of genetic equilibrium in a gene pool, but its main application in population genetics is in calculating allele and genotype frequencies in populations, particularly as a means of studying changes and measuring their rate. The use of the Hardy-Weinberg equation is described below.

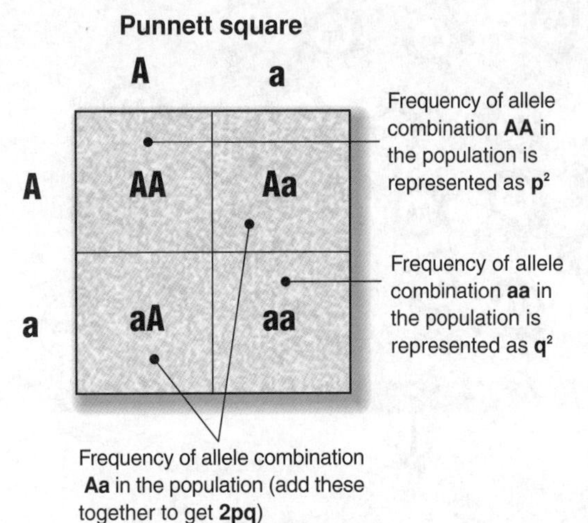

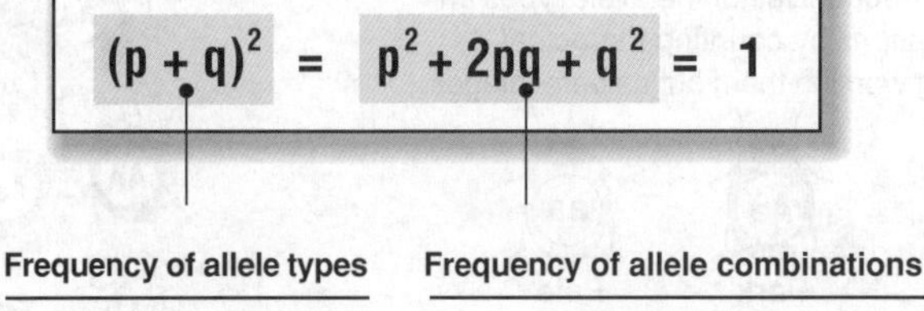

Frequency of allele types	Frequency of allele combinations
p = Frequency of allele A	p^2 = Frequency of AA (homozygous dominant)
q = Frequency of allele a	2pq = Frequency of Aa (heterozygous)
	q^2 = Frequency of aa (homozygous recessive)

The Hardy-Weinberg equation is applied to populations with a simple genetic situation: dominant and recessive alleles controlling a single trait. The frequency of all of the dominant (A) and recessive alleles (a) equals the total genetic complement, and adds up to 1 or 100% of the alleles present.

How To Solve Hardy-Weinberg Problems

In most populations, the frequency of two alleles of interest is calculated from the proportion of homozygous recessives (q^2), as this is the only genotype identifiable directly from its phenotype. If only the dominant phenotype is known, q^2 may be calculated (1 – the frequency of the dominant phenotype). The following steps outline the procedure for solving a Hardy-Weinberg problem:

Remember that all calculations must be carried out using proportions, NOT PERCENTAGES!

1. Examine the question to determine what piece of information you have been given about the population. In most cases, this is the percentage or frequency of the homozygous recessive phenotype q^2, or the dominant phenotype $p^2 + 2pq$ (see note above).
2. The first objective is to find out the value of p or q, If this is achieved, then every other value in the equation can be determined by simple calculation.
3. Take the square root of q^2 to find q.
4. Determine p by subtracting q from 1 (i.e. $p = 1 - q$).
5. Determine p^2 by multiplying p by itself (i.e. $p^2 = p \times p$).
6. Determine 2pq by multiplying p times q times 2.
7. Check that your calculations are correct by adding up the values for $p^2 + q^2 + 2pq$ (the sum should equal 1 or 100%).

Worked example

In the American white population approximately 70% of people can taste the chemical phenylthiocarbamide (PTC) (the dominant phenotype), while 30% are non-tasters (the recessive phenotype).

Determine the frequency of:	***Answers***
(a) Homozygous recessive phenotype(**q^2**).	30% - provided
(b) The dominant allele (**p**).	45.2%
(c) Homozygous tasters (**p^2**).	20.5%
(d) Heterozygous tasters (**2pq**).	49.5%

Data: The frequency of the dominant phenotype (70% tasters) and recessive phenotype (30% non-tasters) are provided.

Working:

Recessive phenotype: **q^2** = 30%
use 0.30 for calculation

therefore: **q** = 0.5477
square root of 0.30

therefore: **p** = 0.4522
$1 - q = p$
$1 - 0.5477 = 0.4523$

Use p and q in the equation (top) to solve any unknown:

Homozygous dominant **p^2** = 0.2046
($p \times p = 0.4523 \times 0.4523$)

Heterozygous: **2pq** = 0.4953

1. A population of hamsters has a gene consisting of 90% M alleles (black) and 10% m alleles (grey). Mating is random.
   Data: Frequency of recessive allele (10% m) and dominant allele (90% M).

   Determine the proportion of offspring that will be black and the proportion that will be grey (show your working).

Recessive allele:	q	=
Dominant allele:	p	=
Recessive phenotype:	q^2	=
Homozygous dominant:	p^2	=
Heterozygous:	2pq	=

2. You are working with pea plants and found 36 plants out of 400 were dwarf.
Data: Frequency of recessive phenotype (36 out of 400 = 9%)

(a) Calculate the frequency of the tall gene: ______________________

(b) Determine the number of heterozygous pea plants:

____

____

Recessive allele:	q	=	
Dominant allele:	p	=	
Recessive phenotype:	q^2	=	
Homozygous dominant:	p^2	=	
Heterozygous:	$2pq$	=	

3. In humans, the ability to taste the chemical phenylthiocarbaminde (PTC) is inherited as a simple dominant characteristic. Suppose you found out that 360 out of 1000 college students could not taste the chemical.
Data: Frequency of recessive phenotype (360 out of 1000).

(a) State the frequency of the gene for tasting PTC:

____

(b) Determine the number of heterozygous students in this population:

____

Recessive allele:	q	=	
Dominant allele:	p	=	
Recessive phenotype:	q^2	=	
Homozygous dominant:	p^2	=	
Heterozygous:	$2pq$	=	

4. A type of deformity appears in 4% of a large herd of cattle. Assume the deformity was caused by a recessive gene.
Data: Frequency of recessive phenotype (4% deformity).

(a) Calculate the percentage of the herd that are carriers of the gene:

____

(b) Determine the frequency of the dominant gene in this case:

____

Recessive allele:	q	=	
Dominant allele:	p	=	
Recessive phenotype:	q^2	=	
Homozygous dominant:	p^2	=	
Heterozygous:	$2pq$	=	

5. Assume you placed 50 pure bred black guinea pigs (dominant allele) with 50 albino guinea pigs (recessive allele) and allowed the population to attain genetic equilibrium (several generations have passed).
Data: Frequency of recessive allele (50%) and dominant allele (50%).

Determine the proportion (%) of the population that becomes white:

____

Recessive allele:	q	=	
Dominant allele:	p	=	
Recessive phenotype:	q^2	=	
Homozygous dominant:	p^2	=	
Heterozygous:	$2pq$	=	

6. It is known that 64% of a large population exhibit the recessive trait of a characteristic controlled by two alleles (one is dominant over the other).
Data: Frequency of recessive phenotype (64%). Determine:

(a) The frequency of the recessive allele: ______________________

(b) The percentage that are heterozygous for this trait: ______________________

(c) The percentage that exhibit the dominant trait: ______________________

(d) The percentage that are homozygous for the dominant trait: ______________________

(e) The percentage that has one or more recessive alleles: ______________________

7. Albinism is recessive to normal pigmentation in humans. The frequency of the albino allele was 10% in a population.
Data: Frequency of recessive allele (10% albino allele).

Determine the proportion of people that you would expect to be albino:

____

____

____

Recessive allele:	q	=	
Dominant allele:	p	=	
Recessive phenotype:	q^2	=	
Homozygous dominant:	p^2	=	
Heterozygous:	$2pq$	=	

Analysis of a Squirrel Gene Pool

In Olney, Illinois, in the United States, there is a unique population of albino (white) and grey squirrels. Between 1977 and 1990, students at Olney Central College carried out a study of this population. They recorded the frequency of grey and albino squirrels. The albinos displayed a mutant allele expressed as an albino phenotype only in the homozygous recessive condition. The data they collected are provided in the table below. Using the **Hardy-Weinberg equation** for calculating genotype frequencies, it was possible to estimate the frequency of the normal 'wild' allele (G) providing grey fur colouring, and the frequency of the mutant albino allele (g) producing white squirrels. This study provided real, first hand data that students could use to see how genotype frequencies can change in a real population.

Thanks to **Dr. John Stencel**, Olney Central College, Olney, Illinois, US, for providing the data for this exercise.

Grey squirrel, usual colour form

Albino form of grey squirrel

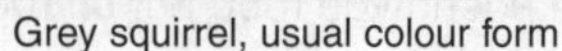

Population of grey and white squirrels in Olney, Illinois (1977-1990)

Year	Grey	White	Total	GG	Gg	gg	Freq. of g	Freq. of G
1977	602	182	784	26.85	49.93	23.21	48.18	51.82
1978	511	172	683	24.82	50.00	25.18	50.18	49.82
1979	482	134	616	28.47	49.77	21.75	46.64	53.36
1980	489	133	622	28.90	49.72	21.38	46.24	53.76
1981	536	163	699	26.74	49.94	23.32	48.29	51.71
1982	618	151	769	31.01	49.35	19.64	44.31	55.69
1983	419	141	560	24.82	50.00	25.18	50.18	49.82
1984	378	106	484	28.30	49.79	21.90	46.80	53.20
1985	448	125	573	28.40	49.78	21.82	46.71	53.29
1986	536	155	691	27.71	49.86	22.43	47.36	52.64
1987	*No data collected this year*							
1988	652	122	774	36.36	47.88	15.76	39.70	60.30
1989	552	146	698	29.45	49.64	20.92	45.74	54.26
1990	603	111	714	36.69	47.76	15.55	39.43	60.57

1. **Graph population changes**: Use the data in the first three columns of the table above to plot a line graph. This will show changes in the phenotypes: numbers of grey and white (albino) squirrels, as well as changes in the total population. Plot: **grey**, **white**, and **total** for each year:

   (a) By how much have total population numbers fluctuated over the sampling period (as a %):

   ____

   (b) Describe the overall trend in total population numbers and any pattern that may exist:

   ____

   ____

   ____

   ____

   ____

   ____

   ____

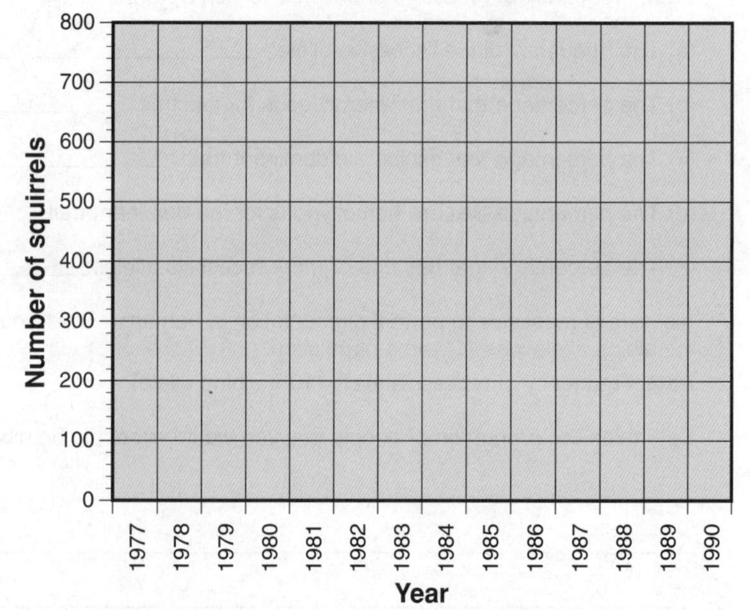

2. **Graph genotype changes**: Use the data in the genotype columns of the table on the previous page to plot a line graph. This will show changes in the allele combinations (**GG**, **Gg**, **gg**). Plot: **GG**, **Gg**, and **gg** for each year:

Describe the overall trend in the frequency of:

(a) Homozygous dominant (**GG**) genotype:

(b) Heterozygous (**Gg**) genotype:

(c) Homozygous recessive (**gg**) genotype:

Frequency of genotype: 0%, 10%, 20%, 30%, 40%, 50%, 60%

Year: 1977, 1978, 1979, 1980, 1981, 1982, 1983, 1984, 1985, 1986, 1987, 1988, 1989, 1990

3. **Graph allele changes**: Use the data in the last two columns of the table on the previous page to plot a line graph. This will show changes in the *allele frequencies* for each of the dominant (**G**) and recessive (**g**) alleles.
Plot: the frequency of **G** and the frequency of **g**:

(a) Describe the overall trend in the frequency of the dominant allele (**G**):

(b) Describe the overall trend in the frequency of the recessive allele (**g**):

Frequency of allele: 0%, 10%, 20%, 30%, 40%, 50%, 60%, 70%

Year: 1977, 1978, 1979, 1980, 1981, 1982, 1983, 1984, 1985, 1986, 1987, 1988, 1989, 1990

4. (a) State which of the three graphs best indicates that a significant change may be taking place in the gene pool of this population of squirrels:

(b) Give a reason for your answer:

5. Describe a possible cause of the changes in allele frequencies over the sampling period:

Industrial Melanism

Natural selection may act on the frequencies of phenotypes (and hence genotypes) in populations in one of three different ways (through stabilising, directional, or disruptive selection). Over time, natural selection may lead to a permanent change in the genetic makeup of a population. The increased prevalence of melanic forms of the peppered moth, *Biston betularia*, during the Industrial Revolution, is one of the best known examples of directional selection following a change in environmental conditions. Although the protocols used in the central experiments on *Biston*, and the conclusions drawn from them, have been queried, it remains one of the clearest documented examples of phenotypic change in a polymorphic population.

Industrial melanism in peppered moths, *Biston betularia*

The **peppered moth**, *Biston betularia*, occurs in two forms (morphs): the grey mottled form, and a dark melanic form. Changes in the relative abundance of these two forms was hypothesised to be the result of selective predation by birds, with pale forms suffering higher mortality in industrial areas because they are more visible. The results of experiments by H.D. Kettlewell supported this hypothesis but did not confirm it, since selective predation by birds was observed but not quantified. Other research indicates that predation by birds is not the only factor determining the relative abundance of the different colour morphs.

Grey or mottled morph: vulnerable to predation in industrial areas where the trees are dark.

Melanic or carbonaria morph: dark colour makes it less vulnerable to predation in industrial areas.

Museum collections of the peppered moth made over the last 150 years show a marked change in the frequency of the melanic form. Moths collected in 1850 (above left), prior to the major onset of the industrial revolution in England. Fifty years later (above right) the frequency of the darker melanic forms had greatly increased. Even as late as the mid 20th century, coal-based industries predominated in some centres, and the melanic form occurred in greater frequency in these areas (see map, right).

Frequency of peppered moth forms in 1950

This map shows the relative frequencies of the two forms of peppered moth in the UK in 1950; a time when coal-based industries still predominated in some major centres.

Glasgow
Newcastle
Belfast
Middlesbrough
Leeds
Hull
Manchester
Liverpool
Sheffield
Nottingham
Leicester
Coventry
Birmingham
Cardiff
Bristol
London
Portsmouth
Plymouth

Scale 60 km / 60 miles

Key to frequency graphs

Grey or speckled form
Melanic or carbonaria form
Industrial areas
Non-industrial areas

A grey (mottled) form of *Biston*, camouflaged against a lichen covered bark surface. In the absence of soot pollution, mottled forms appear to have the selective advantage.

A melanic form of *Biston*, resting on a dark branch, so that it appears as part of the branch. Note that the background has been faded out so that the moth can be seen.

Changes in frequency of melanic peppered moths

In the 1940s and 1950s, coal burning was still at intense levels around the industrial centres of Manchester and Liverpool. During this time, the melanic form of the moth was still very dominant. In the rural areas further south and west of these industrial centres, the grey or speckled forms increased dramatically. With the decline of coal burning factories and the Clean Air Acts in cities, the air quality improved between 1960 and 1980. Sulphur dioxide and smoke levels dropped to a fraction of their previous levels. This coincided with a sharp fall in the relative numbers of melanic moths.

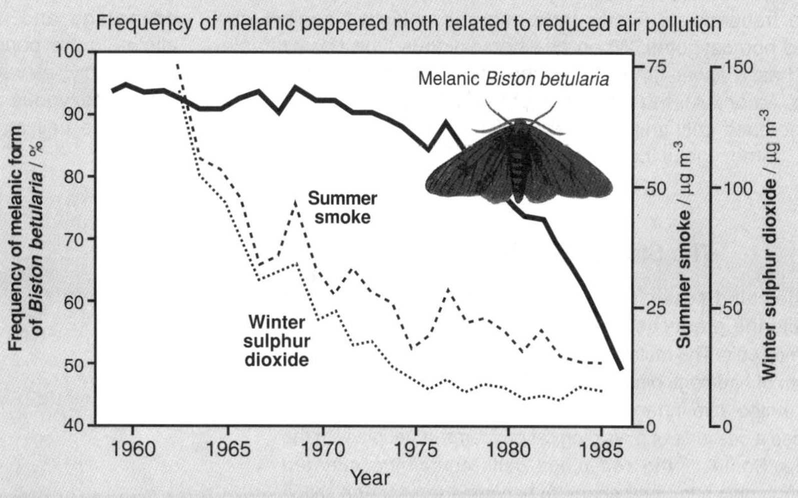

1. The populations of peppered moth in England have undergone changes in the frequency of an obvious phenotypic character over the last 150 years. Describe the phenotypic character that changed in its frequency:

2. (a) Identify the (proposed) selective agent for phenotypic change in *Biston*:

   (b) Describe how the selection pressure on the light coloured morph has changed with changing environmental conditions over the last 150 years:

3. The industrial centres for England in 1950 were located around London, Birmingham, Liverpool, Manchester, and Leeds. Glasgow in Scotland also had a large industrial base. Comment on how the relative frequencies of the two forms of peppered moth were affected by the geographic location of industrial regions:

4. The level of pollution dropped around Manchester and Liverpool between 1960 and 1985.

   (a) State how much the pollution dropped by:

   (b) Describe how the frequency of the darker melanic form changed during the period of reduced pollution:

5. In the example of the peppered moths, state whether the selection pressure is disruptive, stabilising, or directional:

6. Outline the key difference between natural and artificial selection:

7. Discuss the statement "the environment directs natural selection":

Heterozygous Advantage

There are two mechanisms by which natural selection can affect allele frequencies. Firstly, there may be selection against one of the homozygotes. When one homozygous type (for example, aa), has a lower fitness than the other two genotypes (in this case, Aa or AA), the frequency of the deleterious allele will tend to decrease until it is completely eliminated. In some situations, both homozygous conditions (aa and AA) have lower fitness than the heterozygote; a situation that leads to **heterozygous advantage** and may result in the stable coexistence of both alleles in the population (**balanced polymorphism**). There are remarkably few well-documented examples in which the evidence for heterozygous advantage is conclusive. The maintenance of the sickle cell mutation in malaria-prone regions is one such example.

The Sickle Cell Allele (Hb^S)

Sickle cell disease is caused by a mutation to a gene that directs the production of the human blood protein called haemoglobin. The mutant allele is known as **Hb^S** and produces a form of haemoglobin that differs from the normal form by just one amino acid in the β-chain. This minute change however causes a cascade of physiological problems in people with the allele. Some of the red blood cells containing mutated haemoglobin alter their shape to become irregular and spiky; the so-called **sickle cells**.

Sickle cells have a tendency to clump together and work less efficiently. In people with just one sickle cell allele plus a normal allele (the heterozygote condition **Hb^SHb**), there is a mixture of both red blood cell types and they are said to have the sickle cell trait. They are generally unaffected by the disease except in low oxygen environments (e.g. climbing at altitude). People with two Hb^S genes (**Hb^SHb^S**) suffer severe illness and even death. For this reason Hb^S is considered **a lethal gene**.

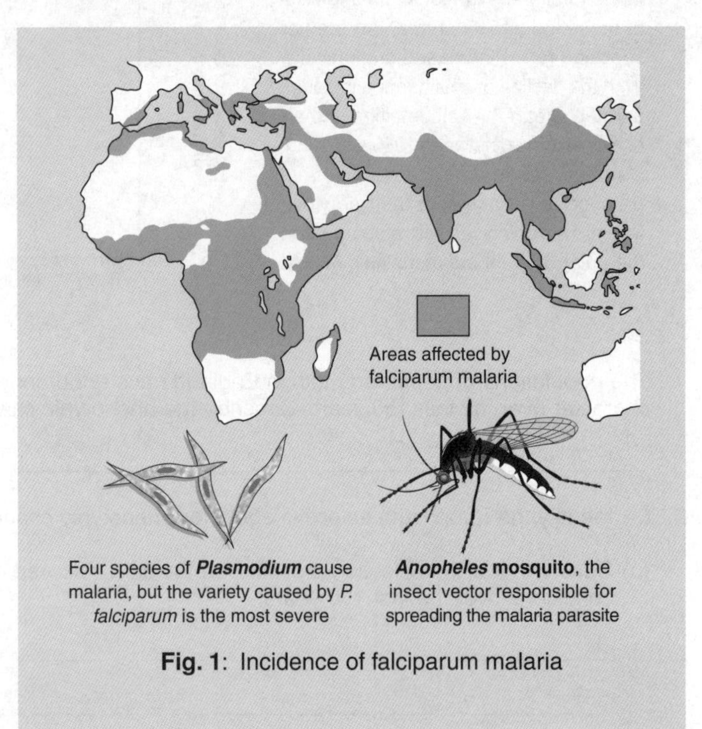

Four species of ***Plasmodium*** cause malaria, but the variety caused by *P. falciparum* is the most severe

***Anopheles* mosquito**, the insect vector responsible for spreading the malaria parasite

Fig. 1: Incidence of falciparum malaria

Heterozygous Advantage in Malarial Regions

Falciparum malaria is widely distributed throughout central Africa, the Mediterranean, Middle East, and tropical and semi-tropical Asia (Fig. 1). It is transmitted by the *Anopheles* mosquito, which spreads the protozoan *Plasmodium falciparum* from person to person as it feeds on blood.

SYMPTOMS: These appear 1-2 weeks after being bitten, and include headache, shaking, chills, and fever. Falciparum malaria is more severe than other forms of malaria, with high fever, convulsions, and coma. It can be fatal within days of the first symptoms appearing.

THE PARADOX: The Hb^S allele offers considerable protection against malaria. Sickle cells have low potassium levels, which causes plasmodium parasites inside these cells to die. Those with a normal phenotype are very susceptible to malaria, but heterozygotes (**Hb^SHb**) are much less so. This situation, called **heterozygous advantage**, has resulted in the Hb^S allele being present in moderately high frequencies in parts of Africa and Asia despite its harmful effects (Fig. 2). This is a special case of balanced polymorphism, called a **balanced lethal system** because neither of the homozygotes produces a phenotype that survives, but the heterozygote is viable.

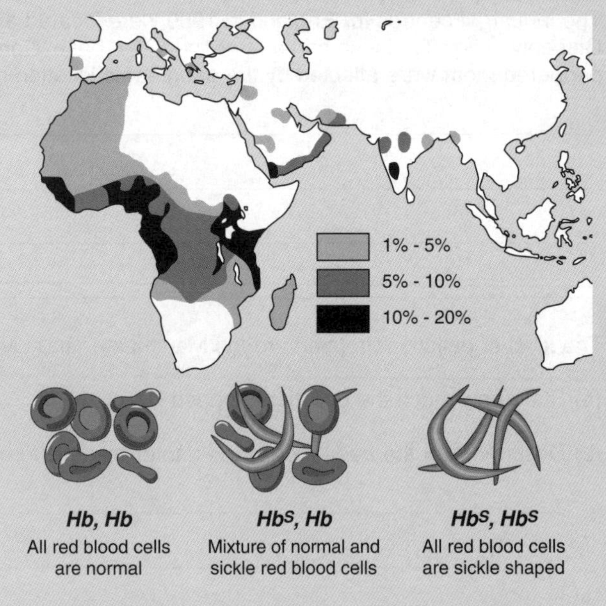

Hb*, *Hb
All red blood cells are normal

Hb^S*, *Hb
Mixture of normal and sickle red blood cells

Hb^S*, *Hb^S
All red blood cells are sickle shaped

Fig. 2: Frequency of the sickle cell allele

1. With respect to the sickle cell allele, explain how **heterozygous advantage** can lead to **balanced polymorphism**:

Evolution in Bacteria

As a result of their short generation times, bacterial populations can show significant evolutionary change within relatively short periods of time. One such evolutionary change is the acquisition of **antibiotic resistance**. Antibiotics are drugs that fight bacterial infections. After being discovered in the 1940s, they rapidly transformed medical care and dramatically reduced illness and death from bacterial disease. With the increased and often indiscriminate use of antibiotics, many bacteria quickly developed drug resistance. The increasing number of multi-drug resistant bacterial strains is particularly worrying; resistant infections inhibit the treatment of patients and increase patient mortality. Moreover, antibiotic resistance adds considerably to the costs of treating disease and, as resistance spreads between bacterial strains, new drugs have an increasingly limited life span during which they are effective.

The Evolution of Drug Resistance in Bacteria

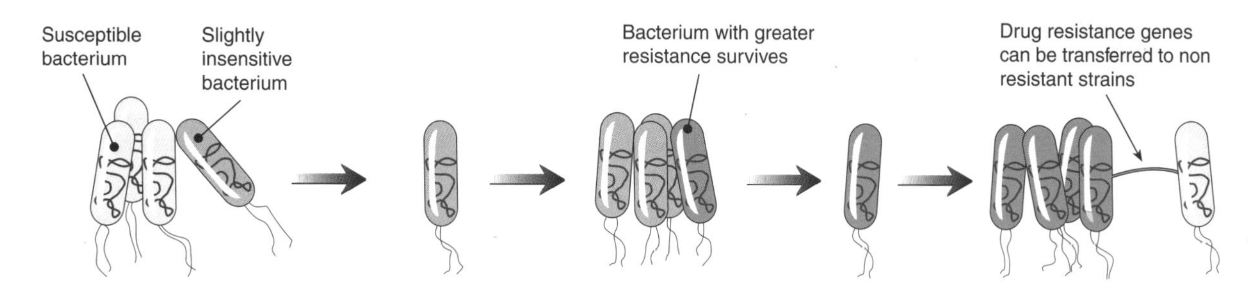

Within any population, there is genetic variation. In this case, the susceptibility of the bacterial strain is normally distributed, with some cells being more susceptible than others.

If the amount of antibiotic delivered is too low, or the full course of antibiotics is not completed, only the most susceptible bacteria will die.

Now a population of insensitive bacteria has developed. Within this population there will also be variation in the susceptibility to antibiotics. As treatment continues, some of the bacteria may acquire greater resistance.

A highly resistant population has evolved. The resistant cells can exchange genetic material with other bacteria, passing on the resistance genes. The antibiotic that was initially used against this bacterial strain will now be ineffective against it.

Observing Adaptive Radiation

Recently, scientists have demonstrated rapid evolution in bacteria. *Pseudomonas fluorescens* was used in the experiment and propagated in a simple heterogeneous environment consisting of a 25 cm^3 glass container containing 6 cm^3 of broth medium. Over a short period of time, the bacteria underwent morphological diversification, with a number of new morphs appearing. These morphs were shown to be genetically distinct. A striking feature of the evolved species is their niche specificity, with each new morph occupying a distinct habitat (below, left). In a follow up experiment (below, right), the researchers grew the same original bacterial strain in the same broth under identical incubation conditions, but in a homogenous environment (achieved by shaking the broth). Without the different habitats offered by an undisturbed environment, no morphs emerged. The experiment illustrated the capacity of bacteria to evolve to utilise available niches.

Heterogeneous environment

WS bacteria (wrinkly morphology) evolved to colonise the air-broth interface.

The FS species (fuzzy morphology) colonised the bottom of the container.

Homogenous environment

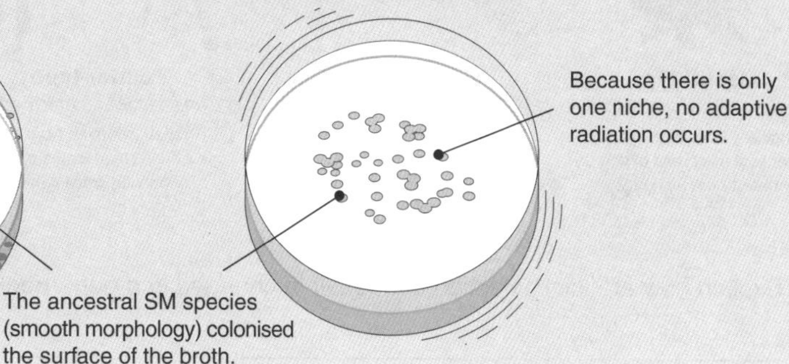

1. Using an illustrative example, explain why evolution of new properties in bacteria can be very rapid: ______________

____

____

2. (a) In the example above, suggest why the bacteria evolved when grown in a heterogeneous environment:

____

____

(b) Predict what would happen if the FS morph was cultured in the homogeneous environment:

____

____

Artificial Selection

The ability of people to control the breeding of domesticated animals and crop plants has resulted in an astounding range of phenotypic variation over relatively short time periods. Most agricultural plants and animals, as well as pets, have undergone **artificial selection** (selective breeding). The dog is a striking example of this, as there are now over 400 different breeds. Artificial selection involves breeding from individuals with the most desirable phenotypes. The aim of this is to alter the average phenotype within the species. As well as selecting for physical characteristics, desirable behavioural characteristics (e.g. the ability to 'read' the body language of humans) has also been selected for in dogs. All breeds of dog are members of the same species, ***Canis familiaris***. This species descended from a single wild species, the grey wolf ***Canis lupus***, over 15 000 years ago. Five ancient dog breeds are recognised, from which all other breeds are thought to have descended by artificial selection.

Grey wolf *Canis lupus pallipes*

The grey wolf is distributed throughout Europe, North America, and Asia. Amongst members of this species, there is a lot of variation in coat coloration. This accounts for the large variation in coat colours of dogs today.

The Ancestor of Domestic Dogs

Until recently, it was unclear whether the ancestor to the modern domestic dogs was the desert wolf of the Middle East, the woolly wolf of central Asia, or the grey wolf of Northern Hemisphere. Recent genetic studies (mitochondrial DNA comparisons) now provide strong evidence that the ancestor of domestic dogs throughout the world is the grey wolf. It seems likely that this evolutionary change took place in a single region, most probably China.

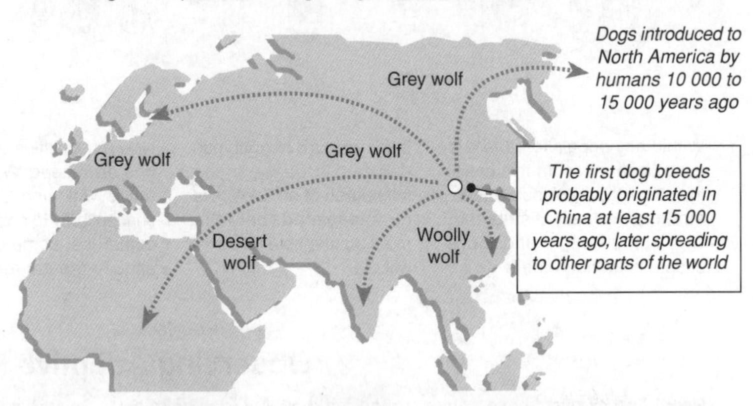

Mastiff-type
Canis familiaris inostranzevi
Originally from Tibet, the first records of this breed of dog go back to the Stoneage.

Greyhound
Canis familiaris leineri
Drawings of this breed on pottery dated from 8000 years ago in the Middle East make it one of the oldest.

Pointer-type
Canis familiaris intermedius
Probably derived from the greyhound breed for the purpose of hunting small game.

Sheepdog
Canis familiaris metris optimae
Originating in Europe, this breed has been used to guard flocks from predators for thousands of years.

Wolf-like
Canis familiaris palustris
Found in snow covered habitats in northern Europe, Asia (Siberia), and North America (Alaska).

1. Explain how artificial selection can result in changes in a gene pool over time: ______

2. Describe the behavioural tendency of wolves that predisposed them to becoming a domesticated animal: ______

3. List the physical and behavioural traits that would be desirable (selected for) in the following uses of a dog:

   (a) Hunting large game (e.g. boar and deer): ______

   (b) Game fowl dog: ______

   (c) Stock control (sheep/cattle dog): ______

   (d) Family pet (house dog): ______

   (e) Guard dog: ______

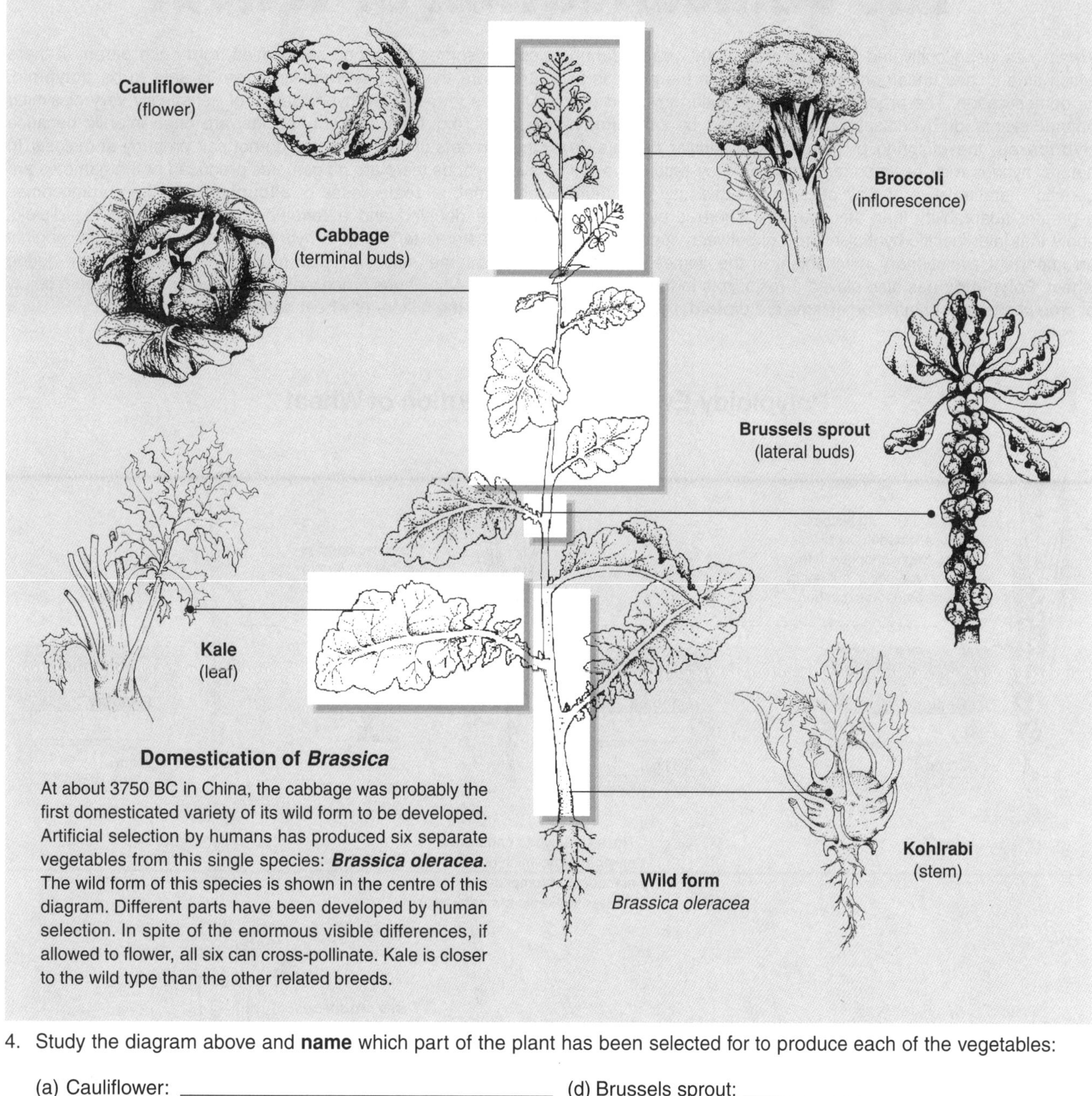

Domestication of *Brassica*

At about 3750 BC in China, the cabbage was probably the first domesticated variety of its wild form to be developed. Artificial selection by humans has produced six separate vegetables from this single species: ***Brassica oleracea***. The wild form of this species is shown in the centre of this diagram. Different parts have been developed by human selection. In spite of the enormous visible differences, if allowed to flower, all six can cross-pollinate. Kale is closer to the wild type than the other related breeds.

4. Study the diagram above and **name** which part of the plant has been selected for to produce each of the vegetables:

   (a) Cauliflower: ______________________ (d) Brussels sprout: ______________________

   (b) Kale: ______________________ (e) Cabbage: ______________________

   (c) Broccoli: ______________________ (f) Kohlrabi: ______________________

5. Describe the feature of these vegetables that suggests they are members of the same species: ______________________

6. Human artificial selection pressures can also influence the development of characteristics in 'unwanted' species. Suggest how human weed control measures may inadvertently select for weed plants that have a resistance to the measures:

7. Explain how a farmer thousands of years ago was able to improve the phenotypic character of a cereal crop:

The Domestication of Wheat

Wheat has been cultivated for more than 9000 years, during which time it has undergone many changes in the process of its domestication. The process of wheat evolution involved two natural events of hybridisation, accompanied by **polyploidy**. **Hybrids** are the offspring of genetically dissimilar parents. In nature, hybrids may be important because they recombine the genetic characteristics of their parents and this may give them a greater adaptability than their parents. There is evidence to show that interspecific hybridisation (i.e. between species) was an important evolutionary mechanism in the domestication of wheat. **Polyploidy** has also played a major role in the evolution of crop plants. Most higher organisms are **diploid**, i.e. two set of chromosomes (2N), one set derived from each parent. If there are more than two sets the organism is said to be **polyploid**. Diploids formed from hybridisation of genetically very dissimilar parents, e.g. from different species, are often infertile because the two sets of chromosomes do not pair properly at meiosis. In such hybrids there are no gametes produced or the gametes are abnormal. In some cases of **allopolyploidy** the chromosomes can be doubled and a tetraploid is formed from the diploid. This restores fertility to a hybrid, because each of the original chromosome sets can pair properly with each other during meiosis. All of these processes are outlined in the diagram below showing the history of wheat domestication.

Polyploidy Events in the Evolution of Wheat

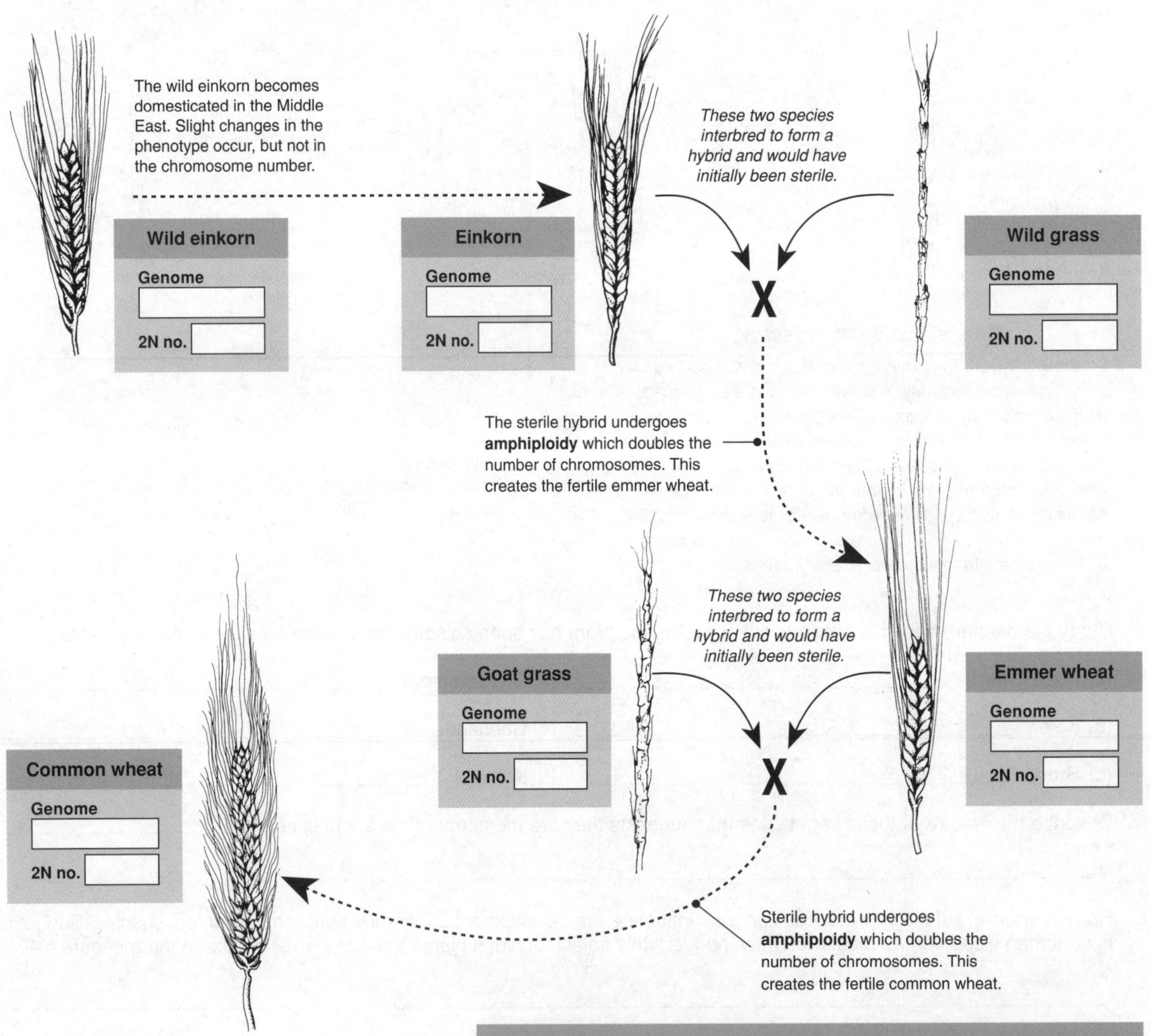

The table on the right and the diagram above show the evolution of the common wheat. Common wheat is thought to have resulted from two sets of crossings between different species to produce hybrids. Wild einkorn (7 chromosomes, genome AA) evolved into einkorn, which crossed with a wild grass (7 chromosomes, genome BB) and gave rise to emmer wheat (14 chromosomes, genome AABB). Common wheat arose when emmer wheat was crossed with another type of grass (goat grass).

Common name	Taxonomic name	Genome	Chromosomes (1N)	Chromosomes (2N)
Wild einkorn	*Triticum aegilopiodes*	**AA**	7	14
Einkorn	*Triticum monococcum*	**AA**	7	14
Wild grass	*Aegilops speltoides*	**BB**	7	14
Emmer wheat	*Triticum dicoccum*	**AABB**	14	28
Goat grass	*Aegilops squarrosa*	**DD**	7	14
Common wheat	*Triticum aestivum*	**AABBDD**	21	42

Ancient cereal grasses had heads which shattered readily so that seeds would be scattered widely.

Modern wheat has been selected for its non shattering heads, high yield, and high gluten content.

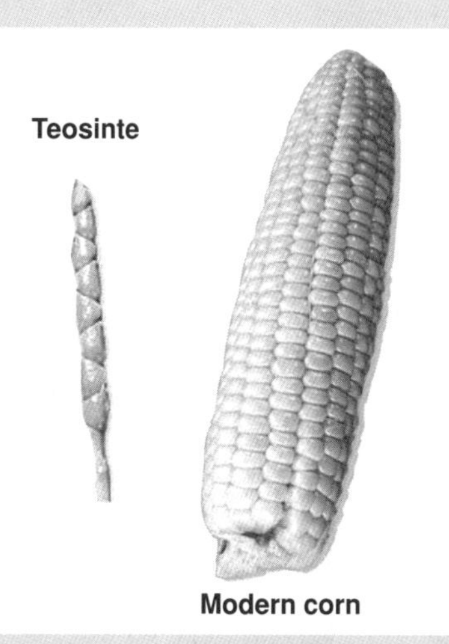

Corn has also evolved during its domestication. Teosinte is thought to be the ancestor to both corn and maize.

1. Using the table on the left, label each of the wheats and grasses in the diagram with the correct **genome** and **2N** chromosome number for each plant.

2. Explain what is meant by **F_1 hybrid vigour** (heterosis):

3. Discuss the role of **polyploidy** and **interspecific hybridisation** in the evolution of wheat:

4. Cultivated wheat arose from wild, weedy ancestors through the selection of certain characters.

   (a) Identify the phenotypic traits that are desirable in modern wheat cultivars:

   (b) Suggest how ancient farmers would have carried out a selective breeding programme:

5. Cultivated American cotton plants have a total of 52 chromosomes (2N = 52). In each cell there are 26 large chromosomes and 26 small chromosomes. Old World cotton plants have 26 chromosomes (2N = 26), all large. Wild American cotton plants have 26 chromosomes, all small. Briefly explain how cultivated American cotton may have originated from Old World cotton and wild American cotton:

6. Discuss the need to maintain the biodiversity of wild plants and ancient farm breeds:

Livestock Improvement

The domestication of livestock has a long history dating back at least 8 000 years. Modern sheep and cattle breeds have been developed over centuries of breeding for particular qualities, e.g. milk or meat production, tolerance to climate or terrain, or as draught animals. Furthermore, different countries have different criteria for selection, based on their local environments and consumer preferences. Cattle breeding in the UK today centres around selection for milk or meat production, with particular breeds being suited to different uses. Sheep are bred for meat, wool, and/or milk production. Traditional **selective breeding** involves choosing breeding stock with desirable qualities. This method produces a steady but slow gain in the desired qualities of the line (the **genetic gain**). This gain has accelerated in recent times with the advent of more reliable ways in which to assess genetic value and assist reproduction. New technologies refine the selection process and increase the rate at which stock improvements are made. Rates are predicted to accelerate further, as technologies improve and become less costly.

Artificial Selection and Genetic Gain in Cattle Breeds

Cattle are selected on the basis of particular desirable traits. Some breeds are selected primarily for milk production and others primarily for beef. Most of the genetic improvement in dairy cattle has relied on selection of high quality progeny from proven stock and extensive use of superior sires through artificial insemination (AI). In beef cattle, AI is useful for introducing new breeds. Consumer demand has led to the shift towards continental breeds such as the Charolais because they are larger, with a higher proportion of lean muscle. Many mixed breeds, e.g. Hereford-Friesian crosses, combine the favourable qualities of two breeds and are suitable for mixed dairying/beef production.

Beef breeds: Aberdeen-Angus, Hereford (above), Simmental, Galloway, Charolais. **Desirable traits**: high muscle to bone ratio, rapid growth and weight gain, hardy, easy calving, docile temperament.

Dairy breeds: Jersey, Friesian (above), Holstein, Aryshire. **Desirable traits**: high yield of milk with high butterfat, milking speed, docile temperament, and udder characteristics such as teat placement.

Special breeds: Some cattle are bred for their suitability for climate or terrain. Scottish highland cattle (above) are a hardy, long coated breed and produce well where other breeds cannot thrive.

A breed is defined as a group of animals that, through selection and breeding, have come to resemble one another and pass on their traits reliably to their offspring. Improved breeding techniques accelerate the genetic progress (the gain toward a desirable phenotype).

The graph (right) illustrates the **predicted** gains based on artificial insemination and standard selection techniques (based on criteria such as production or temperament). These are compared with the predicted gains using breeding values and reproductive technologies e.g. embryo multiplication and transfer (EMT) of standard and transgenic stock, marker (gene) assisted selection, and sib-selection (selecting bulls on the basis of their sisters' performance).

A **breeding value** is a score assigned to a stud animal, derived from the sum of individual scores for different characteristics e.g. milk yield or temperament. Accurate assessment of breeding values makes the selection process less subjective and more value based.

Sources: Breeds of Livestock, Oklahoma State University (web site) and Genetics Australia (web site). Access both via Biolinks from Biozone's web site

Predicted genetic gains based on different breeding methods

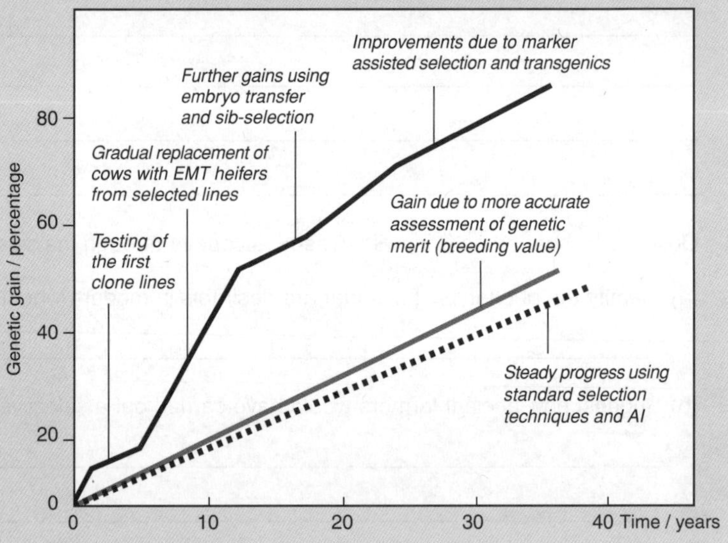

1. Identify the two methods by which most of the genetic progress in dairy cattle has been achieved:

(a) ______________________ (b) ______________________

2. Explain what is meant by the term **genetic gain** as it applies to livestock breeding: ______________________

____

3. Describe the contribution that new reproductive technologies are making to the selective breeding in cattle:

____

4. Describe two features that would be desirable in:

(a) A dairy breed: ______________________

(b) A beef breed: ______________________

The Founder Effect

Occasionally, a small number of individuals from a large population may migrate away, or become isolated from, their original population. If this colonising or 'founder' population is made up of only a few individuals, it will probably have a *non-representative sample* of alleles from the parent population's gene pool. As a consequence of this **founder effect**, the colonising population may evolve differently from that of the parent population, particularly since the environmental conditions for the isolated population may be different. In some cases, it may be possible for certain alleles to be missing altogether from the individuals in the isolated population. Future generations of this population will not have this allele.

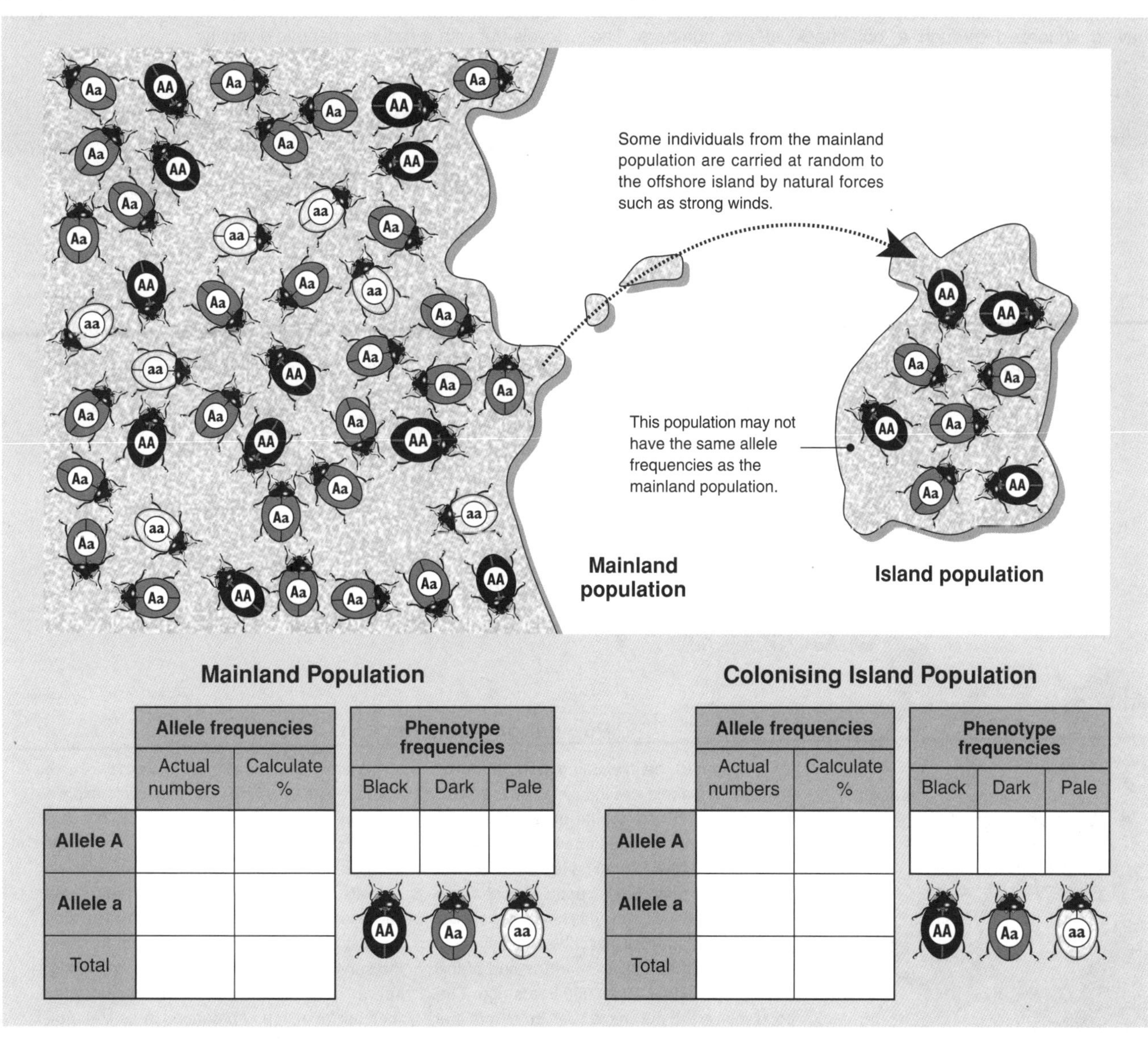

Mainland Population

	Allele frequencies		Phenotype frequencies		
	Actual numbers	Calculate %	Black	Dark	Pale
Allele A					
Allele a					
Total					

Colonising Island Population

	Allele frequencies		Phenotype frequencies		
	Actual numbers	Calculate %	Black	Dark	Pale
Allele A					
Allele a					
Total					

1. Compare the mainland population to the population which ended up on the island (use the spaces in the tables above):
   (a) Count the **phenotype** numbers for the two populations (ie. the number of black, dark and pale beetles).
   (b) Count the **allele** numbers for the two populations: the number of dominant alleles (A) and recessive alleles (a). Calculate these as a percentage of the total number of alleles for each population.

2. Describe how the allele frequencies of the two populations are different: ______

3. Describe some possible ways in which various types of organism can be carried to an offshore island:

   (a) Plants: ______

   (b) Land animals: ______

   (c) Non-marine birds: ______

4. Since founder populations are often very small, describe another process that may further alter the allele frequencies:

______

Population Bottlenecks

Populations may sometimes be reduced to low numbers by predation, disease, or periods of climatic change. A population crash may not be 'selective': it may affect all phenotypes equally. Large scale catastrophic events, such as fire or volcanic eruption, are examples of such non-selective events. Humans may severely (and selectively) reduce the numbers of some species through hunting and/or habitat destruction. These populations may recover, having squeezed through a 'bottleneck' of low numbers. The diagram below illustrates how population numbers may be reduced as a result of a catastrophic event. Following such an event, the small number of individuals contributing to the gene pool may not have a representative sample of the genes in the pre-catastrophe population, i.e. the allele frequencies in the remnant population may be severely altered. Genetic drift may cause further changes to allele frequencies. The small population may return to previous levels but with a reduced genetic diversity.

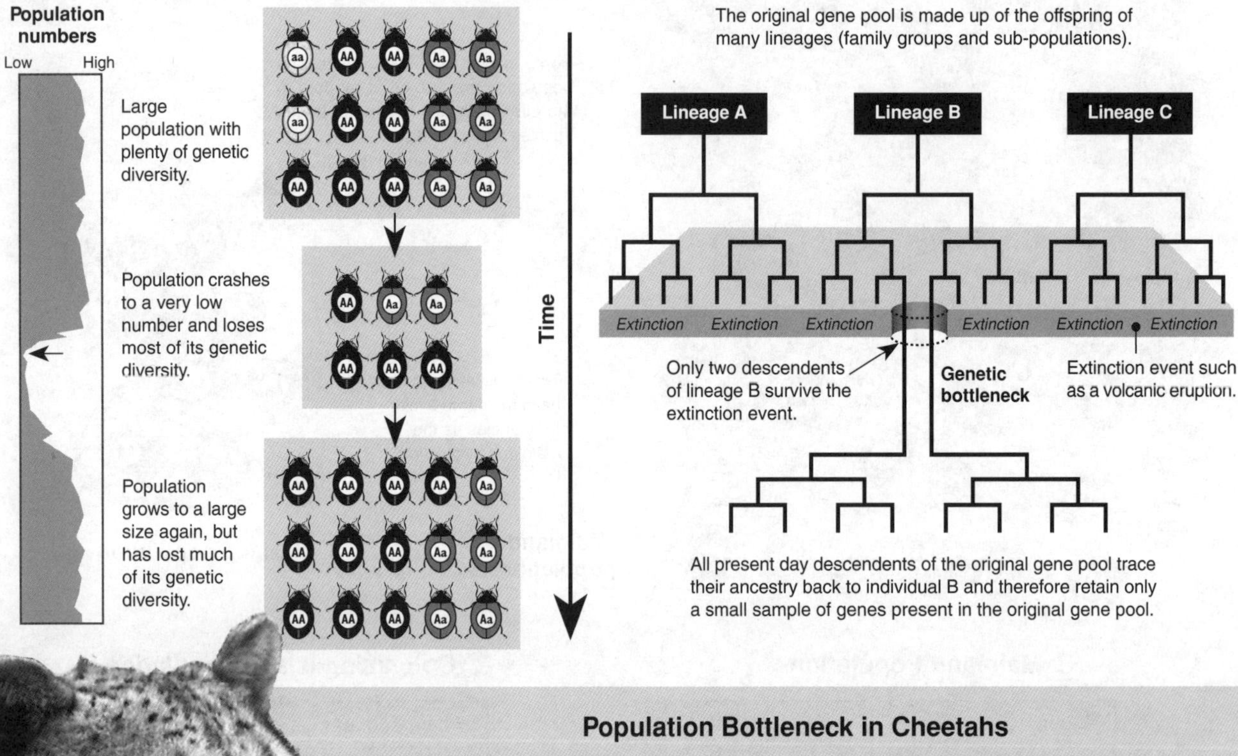

Population Bottleneck in Cheetahs

Until recently, the dwindling population of cheetahs in the wild was thought to be the result of over-hunting and habitat destruction. The world population of cheetahs has declined to fewer than 20 000. Recent genetic analysis has found that the total cheetah population has very little genetic diversity (they all have very similar genotypes). It appears that cheetahs may have narrowly escaped extinction at the end of the last ice age, about 10-20 000 years ago. The population crash may have been so severe that the total species may have been reduced to a single family group. If all modern cheetahs arose from a single surviving litter, this would explain the lack of genetic diversity. This is not a surprising finding, since 75% of all large mammals perished at this time (including well-known animals such as mammoths, cave bears and sabre-tooth tigers). The lack of genetic variation has led to a number of features that threaten the survival of the cheetah species, including: sperm abnormalities, decreased fecundity (number of offspring produced in its lifetime), high cub mortality, and sensitivity to disease.

1. Endangered species are often subjected to population bottlenecks. Explain how population bottlenecks affect the ability of a population of an endangered species to recover from its plight:

2. Explain why the lack of genetic diversity in cheetahs has increased their sensitivity to disease:

3. Describe the effect of a population bottleneck on the potential of a species to adapt to changes (i.e. its ability to evolve):

Genetic Drift

Not all individuals, for various reasons, will be able to contribute their genes to the next generation. **Genetic drift** (also known as the Sewell-Wright Effect) refers to the *random changes in allele frequency* that occur in all populations, but are much more pronounced in small populations. In a small population, the effect of a few individuals not contributing their alleles to the next generation can have a great effect on allele frequencies. Alleles may even become **lost** from the gene pool altogether (frequency becomes 0%) or **fixed** as the only allele for the gene present (frequency becomes 100%).

The genetic makeup (allele frequencies) of the population changes randomly over a period of time

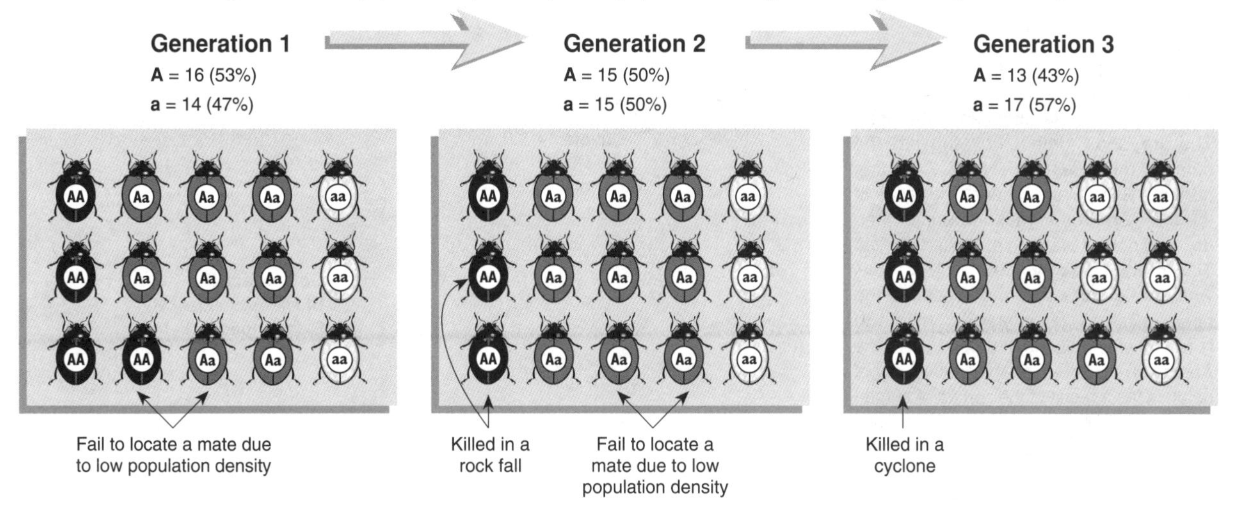

This diagram shows the gene pool of a hypothetical small population over three generations. For various reasons, not all individuals contribute alleles to the next generation. With the random loss of the alleles carried by these individuals, the allele frequency changes from one generation to the next. The change in frequency is directionless as there is no selecting force. The allele combinations for each successive generation are determined by how many alleles of each type are passed on from the preceding one.

Computer Simulation of Genetic Drift

Below are displayed the change in allele frequencies in a computer simulation showing random genetic drift. The breeding population progressively gets smaller from left to right. Each simulation was run for 140 generations.

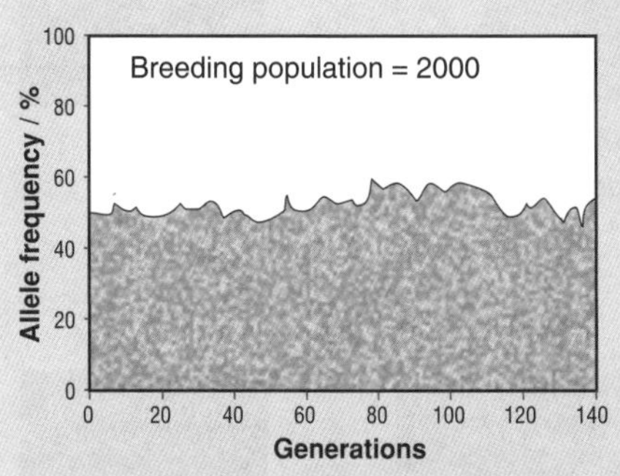

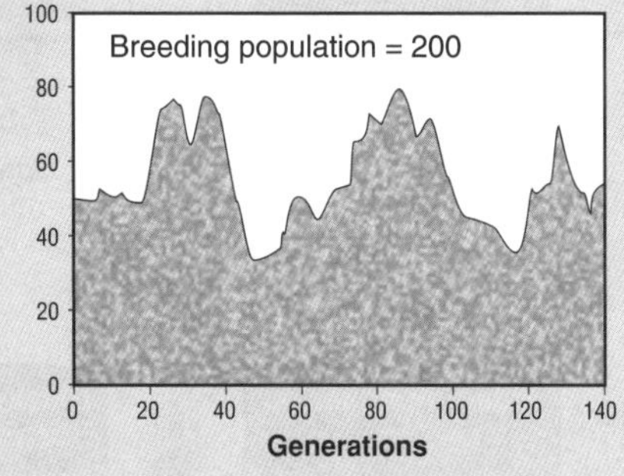

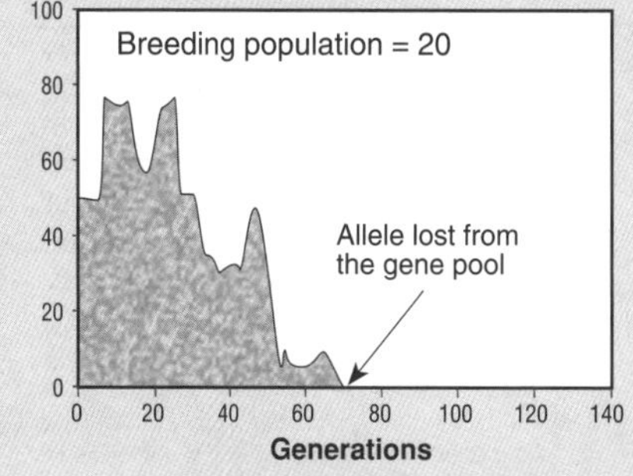

Large breeding population

Fluctuations are minimal in large breeding populations because the large numbers buffer the population against random loss of alleles. On average, losses for each allele type will be similar in frequency and little change occurs.

Small breeding population

Fluctuations are more severe in smaller breeding populations because random changes in a few alleles cause a greater percentage change in allele frequencies.

Very small breeding population

Fluctuations in very small breeding populations are so extreme that the allele can become fixed (frequency of 100%) or lost from the gene pool altogether (frequency of 0%).

1. Explain what is meant by **genetic drift**: ______________________________

____

2. Describe how genetic drift affects the amount of genetic variation within very small populations: ______________

____

____

____

3. Name a small breeding population of animals or plants in Britain or Europe in which genetic drift could be occurring:

____

The Species Concept

The concept of a species is not as simple as it may first appear. Interbreeding between closely related species, such as the dog family below and 'ring species' on the facing page, suggest that the boundaries of a species gene pool can be somewhat unclear. One of the best recognised definitions for a species has been proposed by the respected zoologist, Ernst Mayr: "***A species is a group of actually or potentially interbreeding natural populations that is reproductively isolated from other such groups***". Each species is provided with a unique classification name to assist with future identification.

Geographical distribution of selected *Canis* species

The global distribution of most of the species belonging to the genus *Canis* (dogs and wolves) is shown on the map to the right. The **grey wolf** (timber wolf) inhabits the cold, damp forests of North America, northern Europe and Siberia. The range of the three species of **jackal** overlap in the dry, hot, open savannah of Eastern Africa. The now-rare **red wolf** is found only in Texas, while the **coyote** is found inhabiting the open grasslands of the prairies. The **dingo** is found widely distributed throughout the Australian continent inhabiting a variety of habitats. As a result of the spread of human culture, distribution of the domesticated **dog** is global. The dog has been able to interbreed with all other members of the genus listed here, to form fertile hybrids.

Interbreeding between *Canis* species

Members of the genus to which all dogs and wolves belong present problems with the species concept. The domesticated dog is able to breed with numerous other members of the same genus to produce fertile hybrids. The coyote and red wolf in North America have ranges that overlap. They are also able to produce fertile hybrids, although these are rare. By contrast, the ranges of the three distinct species of jackal overlap in the Serengeti of Eastern Africa. These animals are highly territorial, but simply ignore members of the other jackal species and no interbreeding takes place.

For an excellent discussion of species definition among dogs see the article "The Problematic Red Wolf" in Scientific American, July 1995, pp. 26-31. This discusses whether or not the red wolf is a species or a long established hybrid of the grey wolf and coyote.

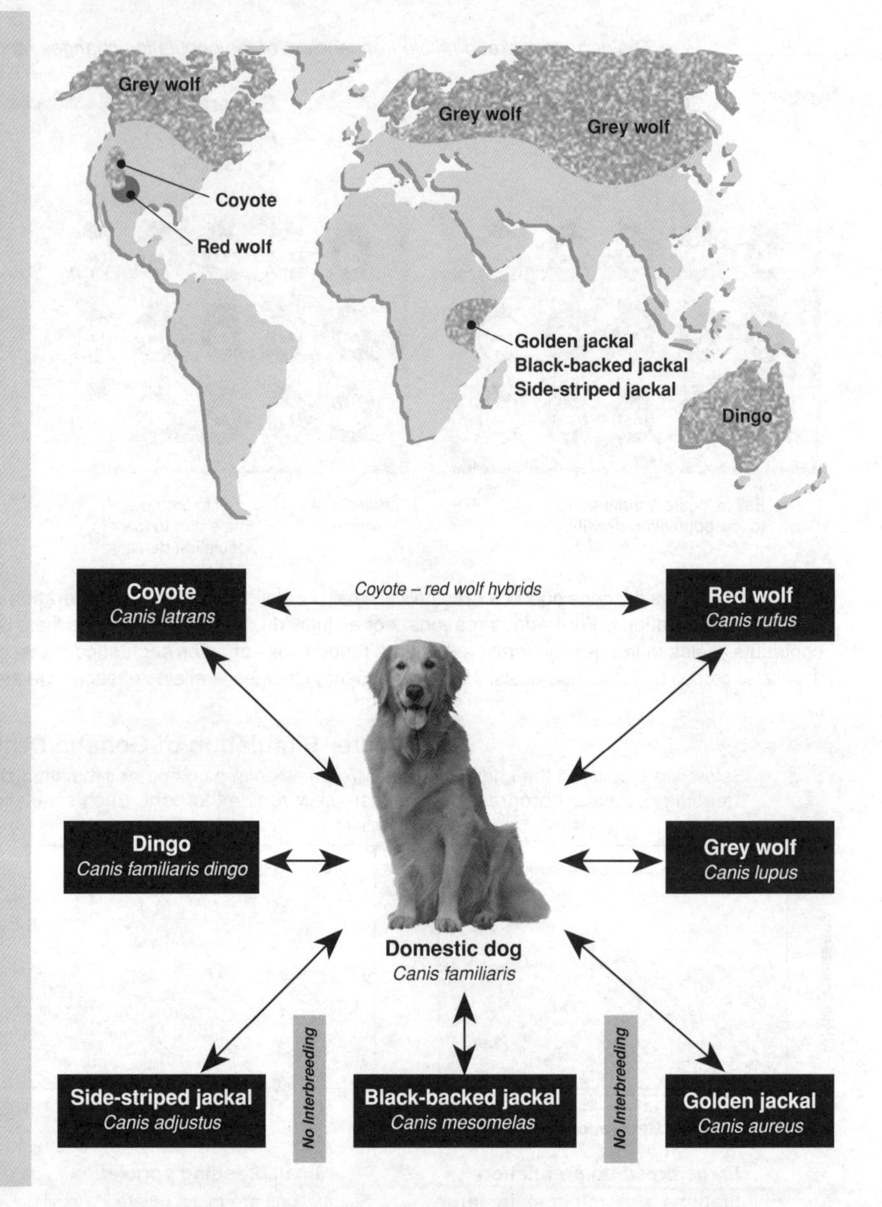

1. Describe the type of barrier that prevents the three species of jackal from interbreeding:

2. Describe the factor that has prevented the dingo from interbreeding with other *Canis* species (apart from the dog):

3. Describe a possible contributing factor to the occurrence of interbreeding between the coyote and red wolf:

4. The grey wolf is a widely distributed species. Explain why the North American population is considered to be part of the same species as the northern European and Siberian populations:

Gene Pool of the Lesser Black-backed Gull and the Herring Gull

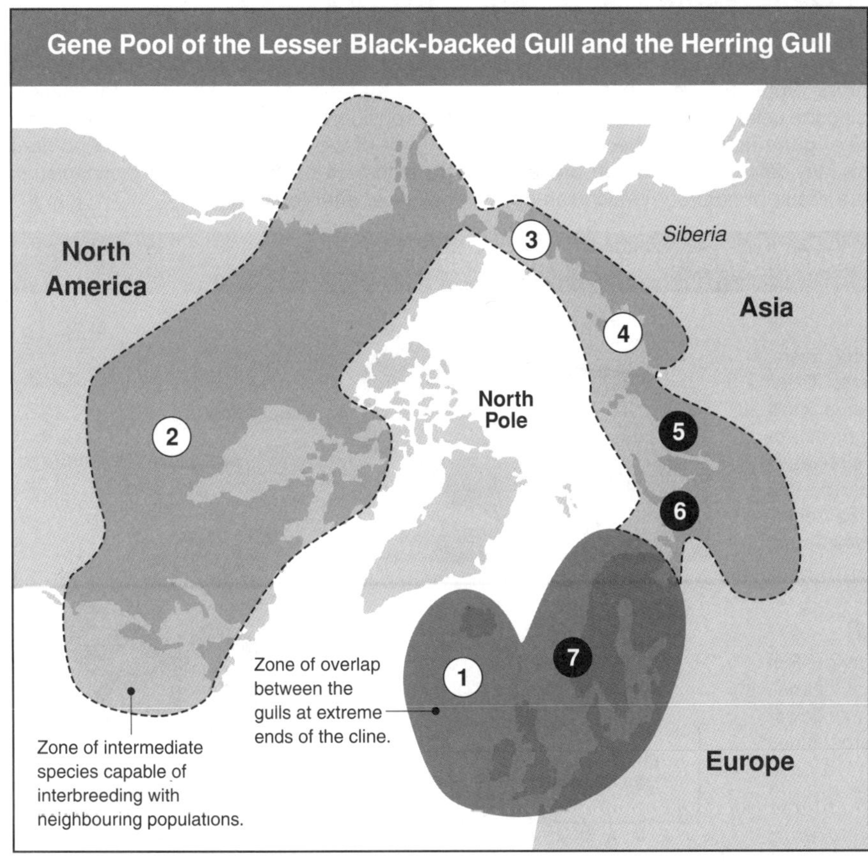

Herring gull *Larus argentatus*

(1) → (4) Gulls are recognisable as herring gulls. Species are classified as subspecies of *L. argentatus*.

Lesser black-backed gull *Larus fuscus*

Gulls are recognisable as lesser-backed gulls. Species are classified as subspecies of *L. fuscus*.

Species may show a gradual change in phenotype over a geographical area. Such a continuous gradual change is called a **cline**, and often occurs along the length of a country or continent. All the populations are of the same species as long as interbreeding populations link them together. **Ring species** are a special type of cline that has a circular or looped geographical distribution, resulting in the two ends of the cline overlapping. Adjacent populations can interbreed but not where the arms of the loop overlap. In the example above, four subspecies of the herring gull, and three of the lesser black-backed gull are currently recognised, forming a chain that circles the North Pole. The evidence suggests that all subspecies were derived from a single ancestral population that originated in Siberia. Members of this ancestral population migrated in opposite directions, and at the same time evolved so that, at various stages, new subspecies could be identified. Each subspecies can breed with those on either side of it. For instance, subspecies 2 can breed with subspecies 1 and 3, subspecies 3 can breed with subspecies 4 and 2, and so on. The two populations at the ends of the cline, which overlaps in northern Europe, rarely interbreed (i.e. subspecies 1 and 7 behave as two distinct species) even though they are connected by a series of intermediate interbreeding populations.

5. Explain what you understand by the term species, identifying examples where the definition is problematic:

____

____

____

6. The **ring species** (above) do not fit comfortably with the standard definition of a species. Describe the aspects of the population of gulls that:

   (a) Supports the idea that they are a single species: ____

   ____

   ____

   ____

   ____

   (b) Does not agree with the standard definition of a species:

   ____

   ____

   ____

   ____

Reproductive Isolation

Any factor that prevents two species from producing fertile hybrids contributes to **reproductive isolation**. Reproductive isolating mechanisms are important in preserving the uniqueness of a gene pool. They prevent the dilution effect of **gene flow** *into* the pool from other populations. Such gene flow may detract from the good combinations already developed as a result of natural selection. Single barriers may not completely stop gene flow, so most species have more than one type of barrier. Geographical barriers are sometimes considered not to be isolating mechanisms because they are not part of the species' biology. Such barriers often precede the development of other isolating mechanisms, which can operate before or after fertilisation.

Prezygotic Isolating Mechanisms (Before-Fertilisation)

Spatial (geographical)

Includes physical barriers such as: mountains, rivers, altitude, oceans, isthmuses, deserts, ice sheets. There are many examples of speciation occurring as a result of isolation by oceans or by geological changes in lake basins (e.g. the proliferation of cichlid fish species in Lake Victoria). The many species of iguana from the Galapagos Islands are now quite distinct from the Central and South American species from which they arose.

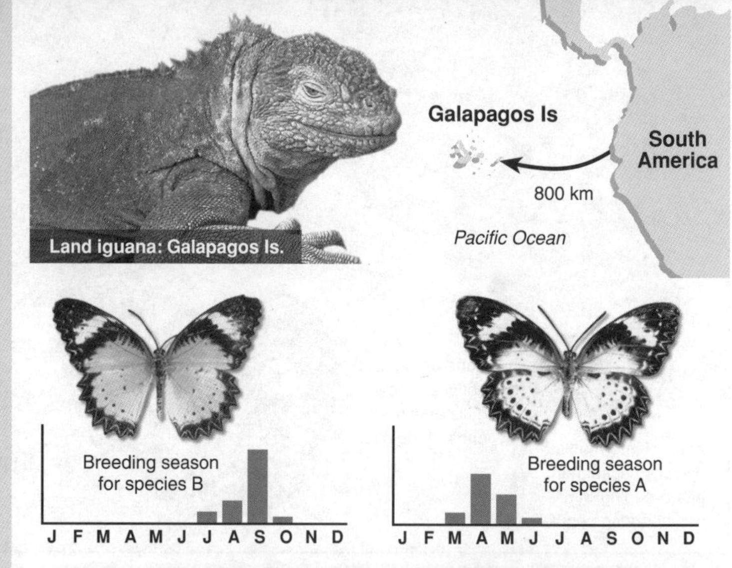

Land iguana: Galapagos Is.

Temporal (including seasonal)

Timing of mating activity for an organism may prevent contact with closely related species: nocturnal, diurnal, spring, summer, autumn, spring tide etc. Plants flower at different times of the year or even at different times of the day. Closely related animals may have quite different breeding seasons.

Ecological (habitat)

Closely related species may occupy different habitats even within the same general area. In the USA, geographically isolated species of antelope squirrels occupy different ranges either side of the Grand Canyon. The white tailed antelope squirrel inhabits the desert to the north of the canyon, while the smaller Harris's antelope squirrel has a much more limited range to the south of the canyon.

Grand Canyon

Harris's antelope squirrel

Gamete mortality

Sperm and egg fail to unite. Even if mating takes place, most gametes will fail to unite. The sperm of one species may not be able to survive in the reproductive tract of another species. Gamete recognition may be based on the presence of species specific molecules on the egg or the egg may not release the correct chemical attractants for sperm of another species.

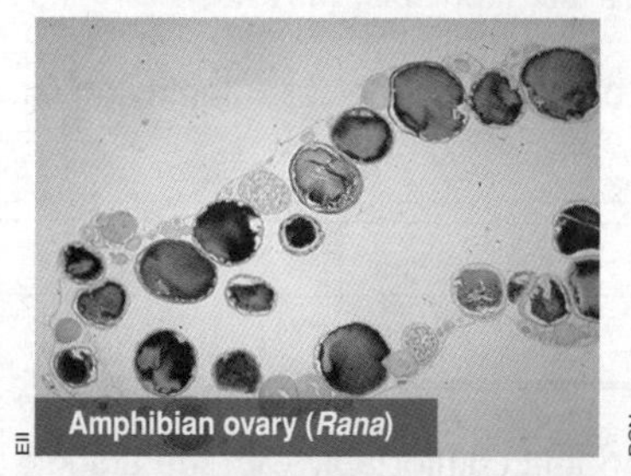

Amphibian ovary (*Rana*)

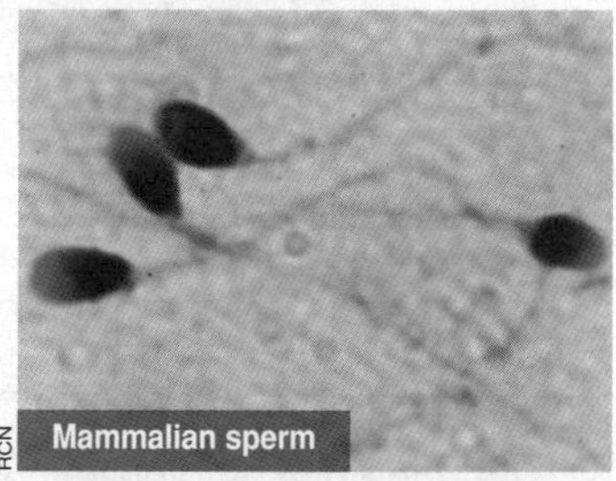

Mammalian sperm

Behavioural (ethological)

Animals attract mates with calls, rituals, dances, body language, etc. Complex displays, such as the flashes of fireflies, are quite specific. In animals, behavioural responses are a major isolating factor, preserving the integrity of mating within species. Birds exhibit a remarkable range of courtship displays that are often quite species-specific.

Peacock display of tail

Blue footed boobies courtship

Structural (morphological)

Shape of the copulatory (mating) apparatus, appearance, coloration, insect attractants. Insects have a lock-and-key arrangement for their copulatory organs. Pheromone chemical attractants, which may travel many kilometres with the aid of the wind, are quite specific, attracting only members of the same species.

Beetles mating

Damselflies mating

Postzygotic Isolating Mechanisms

Hybrid sterility

Even if two species mate and produce hybrid offspring that are vigorous, the species are still reproductively isolated if the hybrids are sterile (genes cannot flow from one species' gene pool to the other). Such cases are common among the horse family (such as the zebra and donkey shown on the right). One cause of this sterility is the failure of meiosis to produce normal gametes in the hybrid. This can occur if the chromosomes of the two parents are different in number or structure (see the **"zebronkey"** karyotype on the right). The **mule**, a cross between a donkey stallion and a horse mare, is also an example of **hybrid vigour** (they are robust) as well as **hybrid sterility**. Female mules sometimes produce viable eggs but males are infertile.

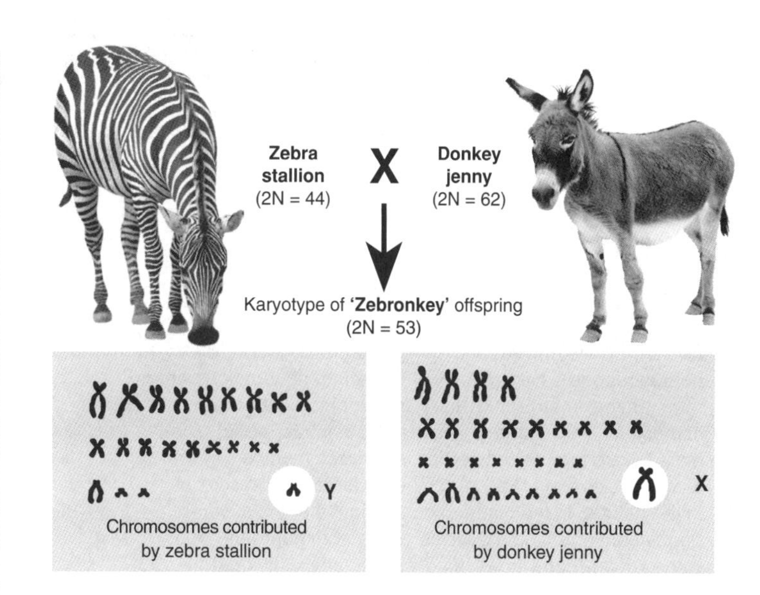

Hybrid inviability

Mating between individuals of two different species may sometimes produce a zygote. In such cases, the genetic incompatibility between the two species may stop development of the fertilised egg at some embryonic stage. Fertilised eggs often fail to divide because of unmatched chromosome numbers from each gamete (a kind of aneuploidy between species). Very occasionally, the hybrid zygote will complete embryonic development but will not survive for long.

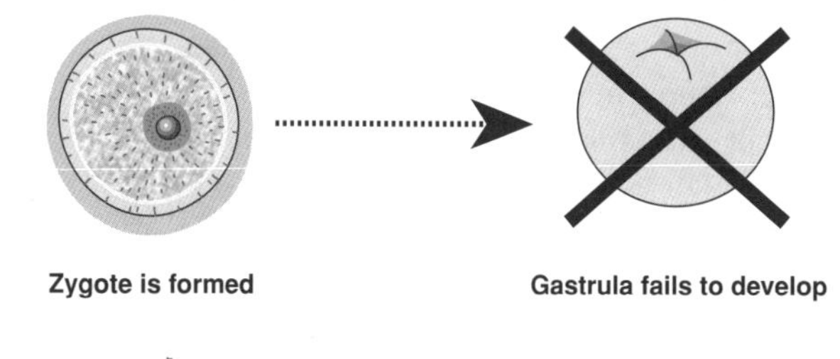

Hybrid breakdown

First generation (F_1) are fertile, but the second generation (F_2) are infertile or inviable. Conflict between the genes of two species sometimes manifests itself in the second generation.

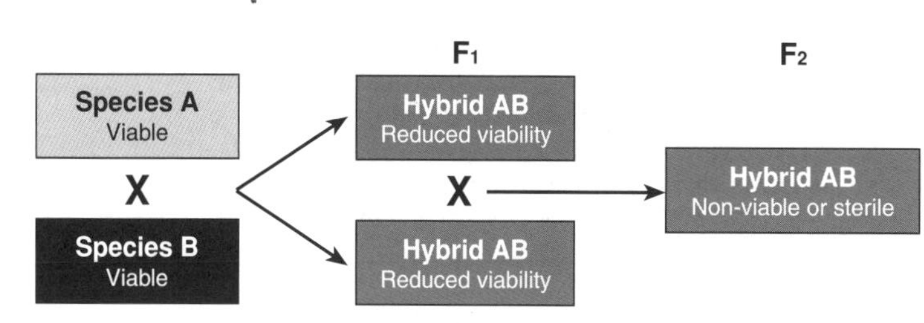

1. In general terms, explain the role of reproductive isolating mechanisms in maintaining the integrity of a species:

____

____

2. In the following examples, classify the reproductive isolating mechanism as either **prezygotic** or **postzygotic** and describe the mechanisms by which the isolation is achieved (e.g. temporal isolation, hybrid sterility etc.):

   (a) Some different cotton species can produce fertile hybrids, but breakdown of the hybrid occurs in the next generation when the offspring of the hybrid die in their seeds or grow into defective plants:

   Prezygotic / postzygotic (delete one) Mechanism of isolation: ____________________

   (b) Many plants have unique arrangements of their floral parts that stops transfer of pollen between plants:

   Prezygotic / postzygotic (delete one) Mechanism of isolation: ____________________

   (c) Three species of orchid living in the same rainforest do not hybridise because they flower on different days:

   Prezygotic / postzygotic (delete one) Mechanism of isolation: ____________________

   (d) Several species of the frog genus *Rana*, live in the same regions and habitats, where they may occasionally hybridise. The hybrids generally do not complete development, and those that do are weak and do not survive long:

   Prezygotic / postzygotic (delete one) Mechanism of isolation: ____________________

3. Postzygotic isolating mechanisms are said to reinforce prezygotic ones. Explain why this is the case:

____

____

Allopatric Speciation

Allopatric speciation is a process thought to have been responsible for a great many instances of species formation. It has certainly been important in countries which have had a number of cycles of geographical fragmentation. Such cycles can occur as the result of glacial and interglacial periods, where ice expands and then retreats over a land mass. Such events are also accompanied by sea level changes which can isolate populations within relatively small geographical regions.

Stage 1: Moving into new environments

There are times when the range of a species expands for a variety of different reasons. A single population in a relatively homogeneous environment will move into new regions of their environment when they are subjected to intense competition (whether it is interspecific or intraspecific). The most severe form of competition is between members of the same species since they are competing for identical resources in the habitat. In the diagram on the right there is a 'parent population' of a single species with a common gene pool with regular 'gene flow' (theoretically any individual has access to all members of the opposite sex for mating purposes).

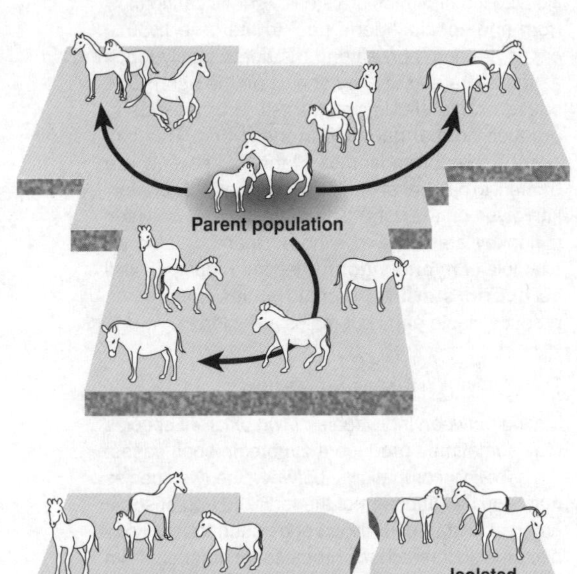

Stage 2: Geographical isolation

Isolation of parts of the population may occur due to the formation of **physical barriers**. These barriers may cut off those parts of the population that are at the extremes of the species range and gene flow is prevented or rare. The rise and fall of the sea level has been particularly important in functioning as an isolating mechanism. Climatic change can leave 'islands' of habitat separated by large inhospitable zones that the species cannot traverse.

Example: In mountainous regions, alpine species are free to range widely over extensive habitat during cool climatic periods. During warmer periods, however, they may become isolated because their habitat is reduced to 'islands' of high ground surrounded by inhospitable lowland habitat.

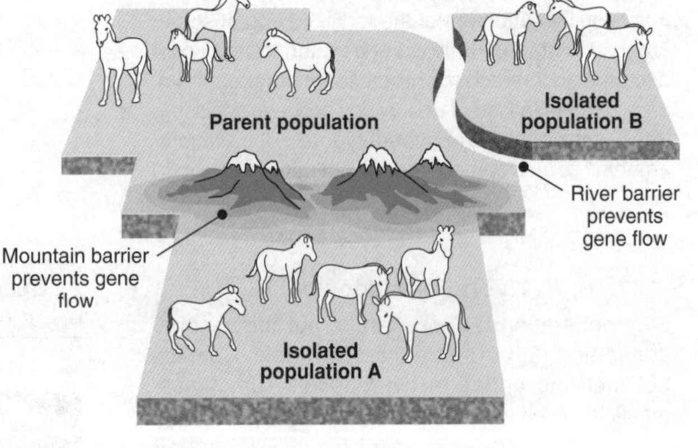

Stage 3: Different selection pressures

The isolated populations (A and B) may be subjected to quite different selection pressures. These will favour individuals with traits that suit each particular environment. For example, population A will be subjected to selection pressures that relate to drier conditions. This will favour those individuals with phenotypes (and therefore genotypes) that are better suited to dry conditions. They may for instance have a better ability to conserve water. This would result in improved health, allowing better disease resistance and greater reproductive performance (i.e. more of their offspring survive). Finally, as allele frequencies for certain genes change, the population takes on the status of a **subspecies**. Reproductive isolation is not yet established but the subspecies are significantly different genetically from other related populations.

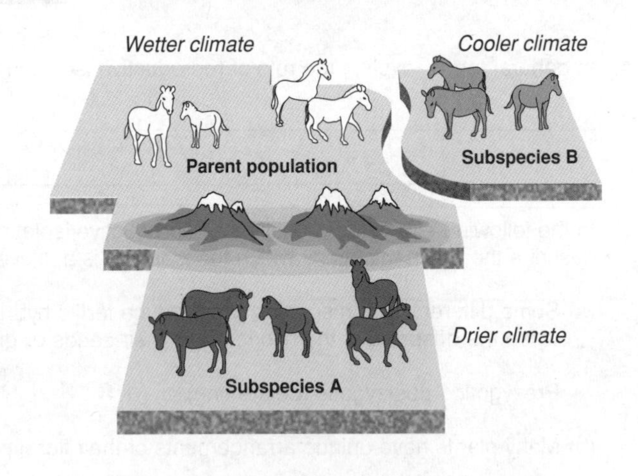

Stage 4: Reproductive isolation

The separated populations (isolated subspecies) will often undergo changes in their genetic makeup as well as their behaviour patterns. These ensure that the gene pool of each population remains isolated and 'undiluted' by genes from other populations, even if the two populations should be able to remix (due to the removal of the geographical barrier). Gene flow does not occur. The arrows (in the diagram to the right) indicate the zone of overlap between two species after the new Species B has moved back into the range inhabited by the parent population. Closely-related species whose distribution overlaps are said to be **sympatric species**. Those that remain geographically isolated are called **allopatric species**.

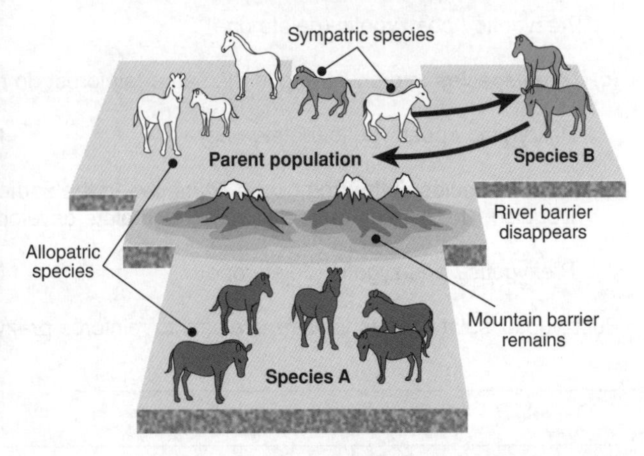

1. Describe why some animals, given the opportunity, move into new environments:

2. (a) Plants are unable to move. Explain how plants might disperse to new environments:

   (b) Describe the amount of **gene flow** within the parent population prior to and during this range expansion:

3. Identify the **process** that causes the formation of new **mountain ranges**:

4. Identify the event that can cause large changes in **sea level** (up to 200 metres):

5. Describe six **physical barriers** that could isolate different parts of the same population:

6. Describe the effect that physical barriers have on **gene flow**:

7. (a) Describe four different types of **selection pressure** that could have an effect on a gene pool:

   (b) Describe briefly how these selection pressures affect the isolated gene pool in terms of **allele frequencies**:

8. Describe two types of **prezygotic** and two types of **postzygotic** reproductive isolating mechanisms (see previous pages):

   (a) Prezygotic:

   (b) Postzygotic:

9. Distinguish between **allopatry** and **sympatry** in populations:

Sympatric Speciation

New species may be formed even where there is no separation of the gene pools by physical barriers. Called **sympatric speciation**, it is rarer than allopatric speciation, although not uncommon in plants which form **polyploids**. There are two situations where sympatric speciation is thought to occur. These are described below:

Speciation Through Niche Differentiation

Niche isolation

In a heterogeneous environment (one that is not the same everywhere), a population exists within a diverse collection of **microhabitats**. Some organisms prefer to occupy one particular type of 'microhabitat' most of the time, only rarely coming in contact with fellow organisms that prefer other microhabitats. Some organisms become so dependent on the resources offered by their particular microhabitat that they never meet up with their counterparts in different microhabitats.

Reproductive isolation

Finally, the individual groups have remained genetically isolated for so long because of their microhabitat preferences, that they have become reproductively isolated. They have become new species that have developed subtle differences in behaviour, structure, and physiology. Gene flow (via sexual reproduction) is limited to organisms that share a similar microhabitat preference (as shown in the diagram on the right).

Example: Some beetles prefer to find plants identical to the species they grew up on, when it is time for them to lay eggs. Individual beetles of the same species have different preferences.

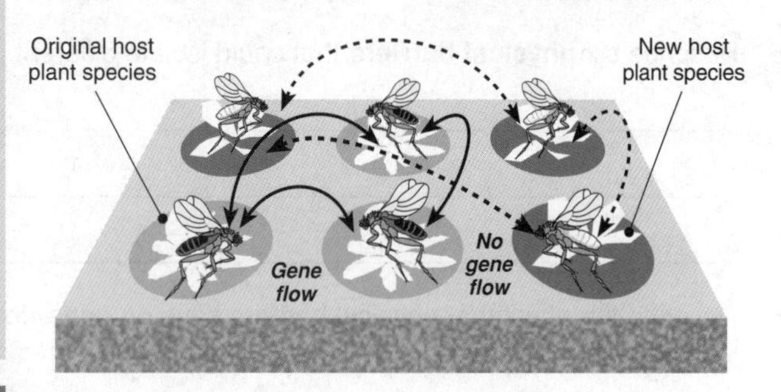

Instant Speciation by Polyploidy

Polyploidy may result in the formation of a new species without isolation from the parent species. This event, occurring during meiosis, produces sudden reproductive isolation for the new group. Because the sex-determining mechanism is disturbed, animals are rarely able to achieve new species status this way (they are effectively sterile, e.g. tetraploid XXXX). Many plants, on the other hand, are able to reproduce vegetatively, or carry out self pollination. This ability to reproduce on their own enables such polyploid plants to produce a breeding population.

Speciation by allopolyploidy

This type of polyploidy usually arises from the doubling of chromosomes in a hybrid between two different species. The doubling often makes the hybrid fertile.

Examples: Modern wheat. Swedes are polyploid species formed from a hybrid between a type of cabbage and a type of turnip.

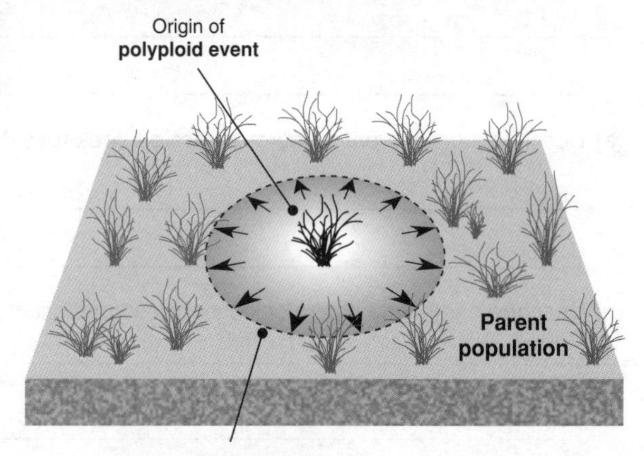

1. Explain what is meant by **sympatric speciation** (do not confuse this with sympatric species):

2. Explain briefly how polyploidy may cause the formation of a new species:

3. Identify an example of a species that has been formed by polyploidy:

4. Explain briefly how niche differentiation may cause the formation of a new species:

Population Genetics & Evolution

Stages in Species Development

The diagram below represents a possible sequence of genetic events involved in the origin of two new species from an ancestral population. As time progresses (from top to bottom of the diagram) the amount of genetic variation increases and each group becomes increasingly isolated from the other. The mechanisms that operate to keep the two gene pools isolated from one another may begin with **geographical barriers**. This may be followed by **prezygotic** mechanisms which protect the gene pool from unwanted dilution by genes from other pools. A longer period of isolation may lead to **postzygotic** mechanisms (see the page on reproductive isolating mechanisms). As the two gene pools become increasingly isolated and different from each other, they are progressively labelled: population, race, and subspecies. Finally they attain the status of separate species.

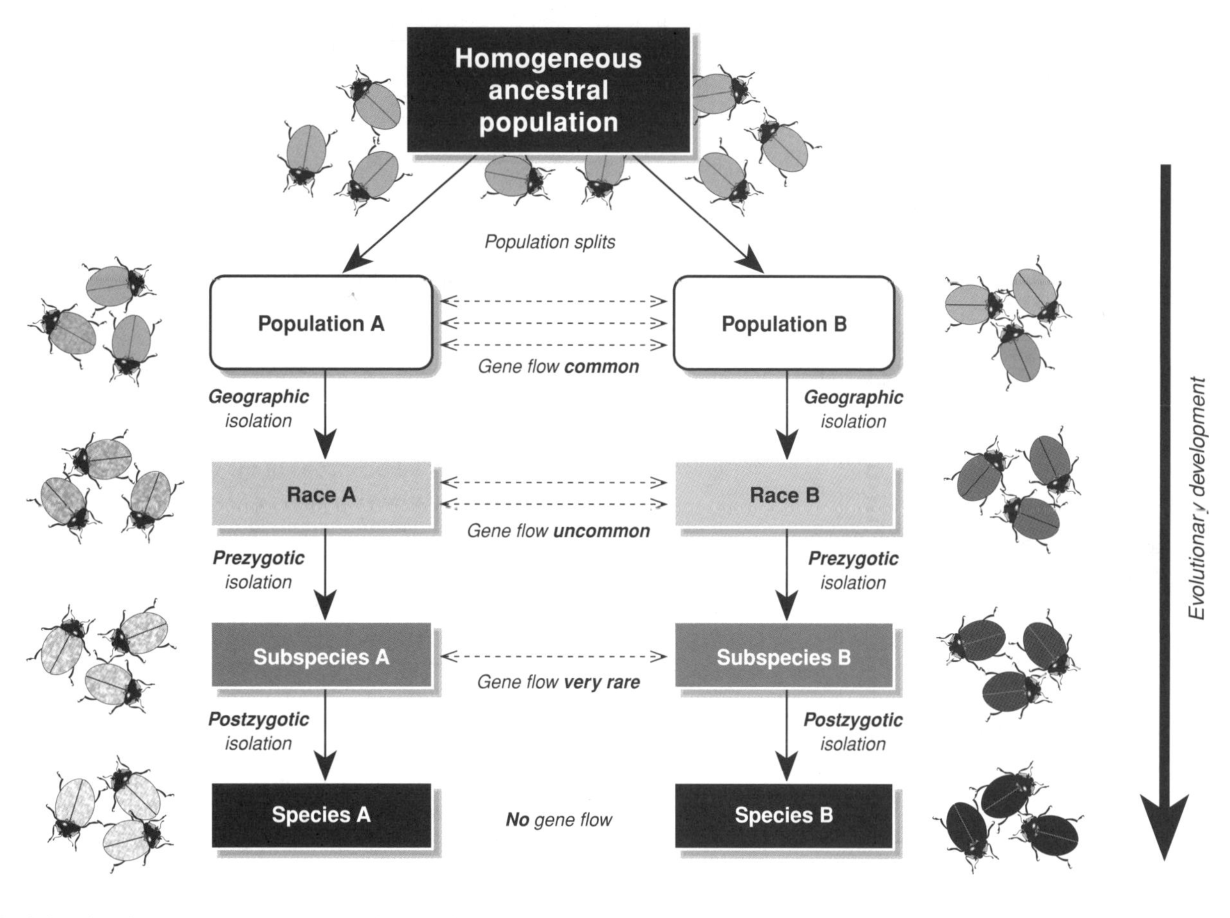

1. Explain what happens to the extent of gene flow between two diverging populations that are gradually attaining species status (as in the diagram above):

2. Early human populations about 500 000 years ago were scattered across the continents of Africa, Europe, and Asia. This was a time of many regional variants, collectively called archaic *Homo sapiens* (or *Homo heidelbergensis*). The fossil skulls from different regions showed odd mixtures of characteristics, some modern and some considered 'primitive'. These regional populations are generally given the status of subspecies. Suggest reasons why gene flow may have been rare between these populations, but still occasionally took place:

3. In the southern hemisphere, the native grey duck and the introduced mallard duck (from the Northern hemisphere) are undergoing 'species breakdown'. These two closely related species can interbreed to form hybrids.

   (a) Describe the factor preventing the two species interbreeding before the introduction of the mallards:

   (b) Describe the factor that may be deterring some of the ducks from interbreeding with the other species:

Patterns of Evolution

The diversification of an ancestral group into two or more species in different habitats is called **divergent evolution**. This process is illustrated in the diagram below, where two species have diverged from a **common ancestor**. Note that another species budded off, only to become extinct. Divergence is common in evolution. When divergent evolution involves the formation of a large number of species to occupy different niches, this is called an **adaptive radiation**. The example below (right) describes the radiation of the mammals that occurred after the extinction of the dinosaurs; an event that made niches available for exploitation. Note that the evolution of species may not necessarily involve branching: a species may accumulate genetic changes that, over time, result in the emergence of what can be recognised as a different species. This is known as **sequential evolution** (below).

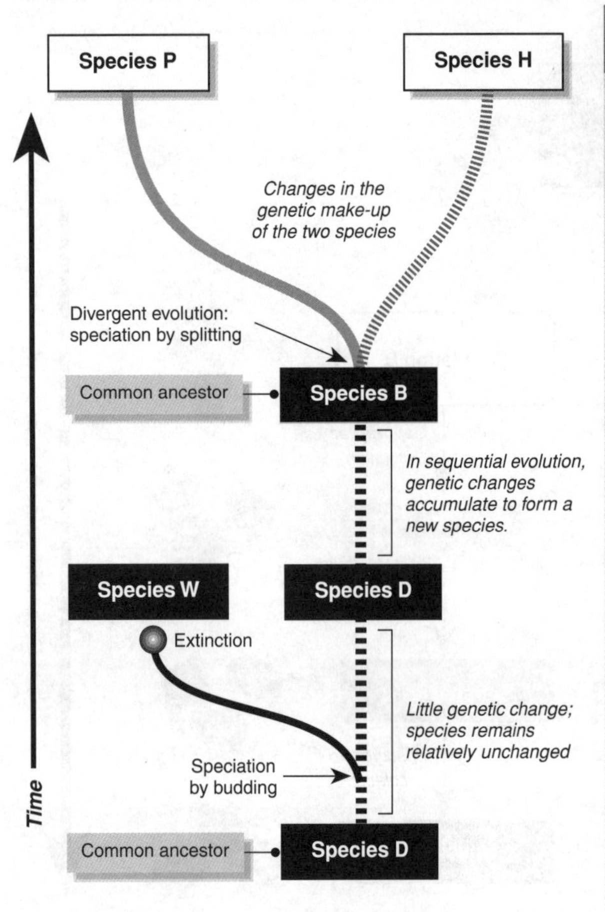

Mammalian Adaptive Radiation

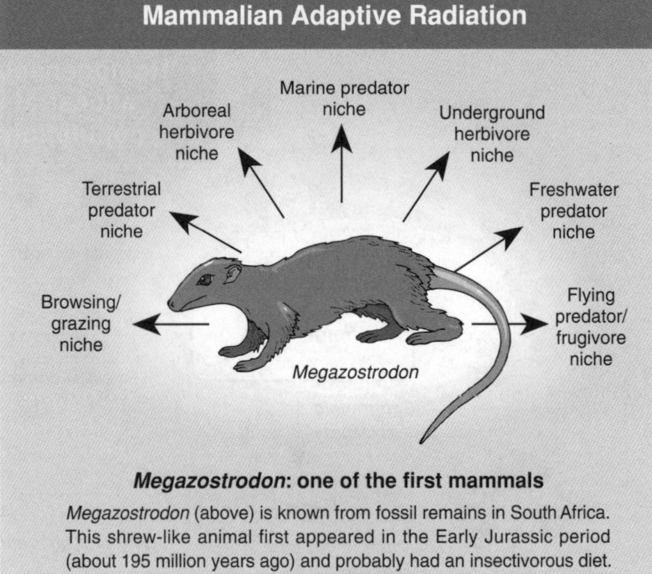

***Megazostrodon*: one of the first mammals**

Megazostrodon (above) is known from fossil remains in South Africa. This shrew-like animal first appeared in the Early Jurassic period (about 195 million years ago) and probably had an insectivorous diet.

The earliest true mammals evolved about 195 million years ago, long before they underwent their major adaptive radiation some 65-50 million years ago. These ancestors to the modern forms were very small (12 cm), many were nocturnal and fed on insects and other invertebrate prey. It was climatic change as well as the extinction of the dinosaurs (and their related forms) that suddenly left many niches vacant for exploitation by such an adaptable 'generalist'. All modern mammal orders developed very quickly and early.

1. In the hypothetical example of divergent evolution illustrated above, left:

   (a) Classify the type of evolution that produced species B from species D: ____________________

   (b) Classify the type of evolution that produced species P and H from species B: ____________________

   (c) Name all species that evolved from: **Common ancestor D**: __________ **Common ancestor B**: __________

   (d) Suggest why species B, P, and H all possess a physical trait not found in species D or W: ____________________

2. (a) Explain the distinction between **divergence** and **adaptive radiation**: ____________________

   (b) Discuss the differences between **sequential evolution** and **divergent evolution**: ____________________

The Rate of Evolutionary Change

There has been debate in recent years over the pace of evolution, with two theories being proposed: **gradualism** and **punctuated equilibrium**. Some scientists believe that both mechanisms may operate: the pace of evolution may be gradual and steady in certain instances and abrupt in others.

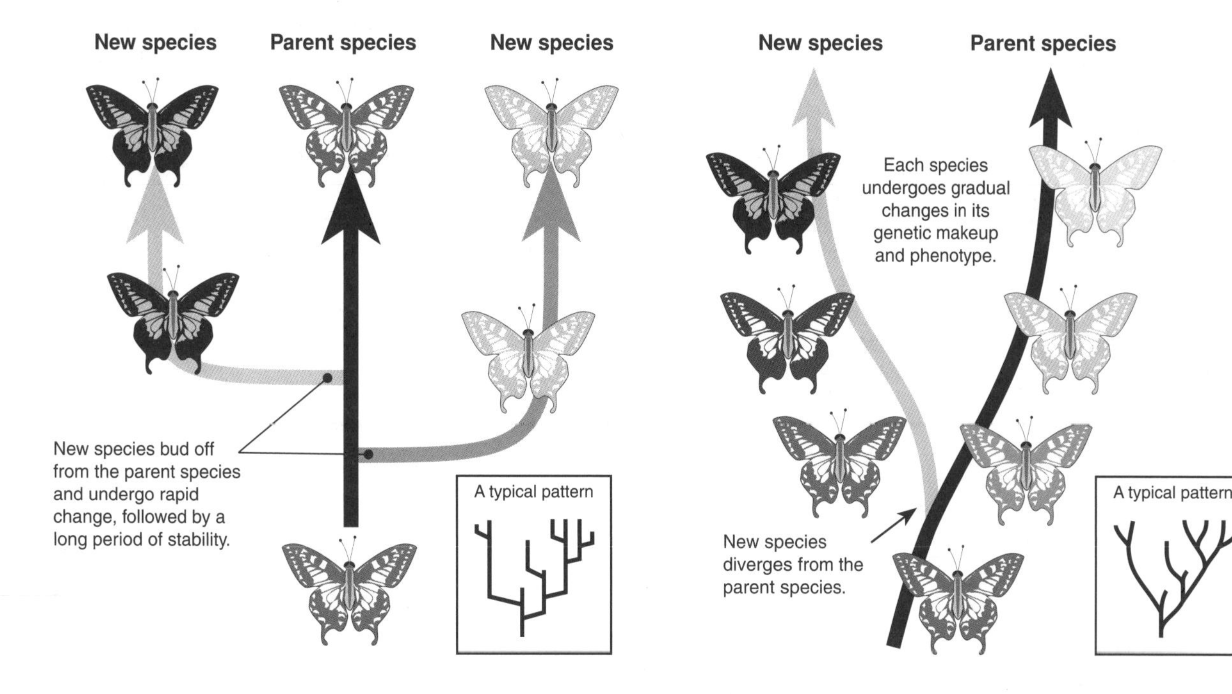

Punctuated Equilibrium

There is abundant evidence in the fossil record that, instead of gradual change, species stayed much the same for long periods of time (called stasis). These periods were punctuated by short bursts of evolution which produce new species quite rapidly. According to the punctuated equilibrium theory, most of a species' existence is spent in stasis and little time is spent in active evolutionary change. The stimulus for evolution occurs when some crucial factor in the environment changes.

Gradualism

Gradualism assumes that populations slowly diverge from one another by accumulating adaptive characteristics in response to different selective pressures. If species evolve by gradualism, there should be transitional forms seen in the fossil record, as is seen with the evolution of the horse. Trilobites, an extinct marine arthropod, are another group of animals that have exhibited gradualism. In a study in 1987 a researcher found that they changed gradually over a three million year period.

1. Discuss the nature of the environments that would support each of the following paces of evolutionary change:

   (a) Punctuated equilibrium: ______________________

   (b) Gradualism: ______________________

2. In the fossil record of early human evolution, species tend to appear suddenly, linger for often very extended periods before disappearing suddenly. There are few examples of smooth inter-gradations from one species to the next. Explain which of the above models best describes the rate of human evolution:

   ____

3. Some species apparently show little evolutionary change over long periods of time (hundreds of millions of years).

   (a) Name two examples of such species: ______________________

   (b) State the term given to this lack of evolutionary change: ______________________

   (c) Explain why such species have changed little over evolutionary time: ______________________

Population Genetics & Evolution

Code: RA 2

Homologous Structures

The evolutionary relationships between groups of organisms is determined mainly by structural similarities called **homologous structures** (homologies), which suggest that they all descended from a common ancestor with that feature. The bones of the forelimb of air-breathing vertebrates are composed of similar bones arranged in a comparable pattern. This is indicative of a common ancestry. The early land vertebrates were amphibians and possessed a limb structure called the **pentadactyl limb**: a limb with five fingers or toes (below left). All vertebrates that descended from these early amphibians, including reptiles, birds and mammals, have limbs that have evolved from this same basic pentadactyl pattern. They also illustrate the phenomenon known as **adaptive radiation**, since the basic limb plan has been adapted to meet the requirements of different niches.

Generalised Pentadactyl Limb

The forelimbs and hind limbs have the same arrangement of bones but they have different names. In many cases bones in different parts of the limb have been highly modified to give it a specialised locomotory function.

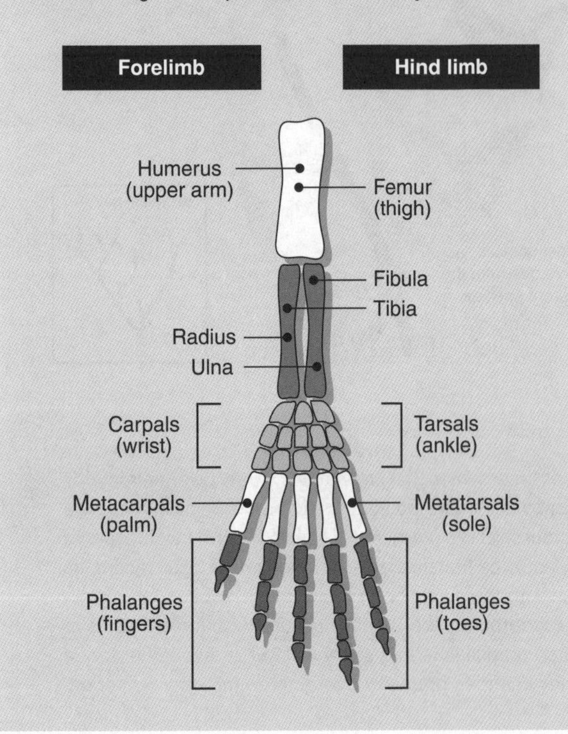

Specialisations of Pentadactyl Limbs

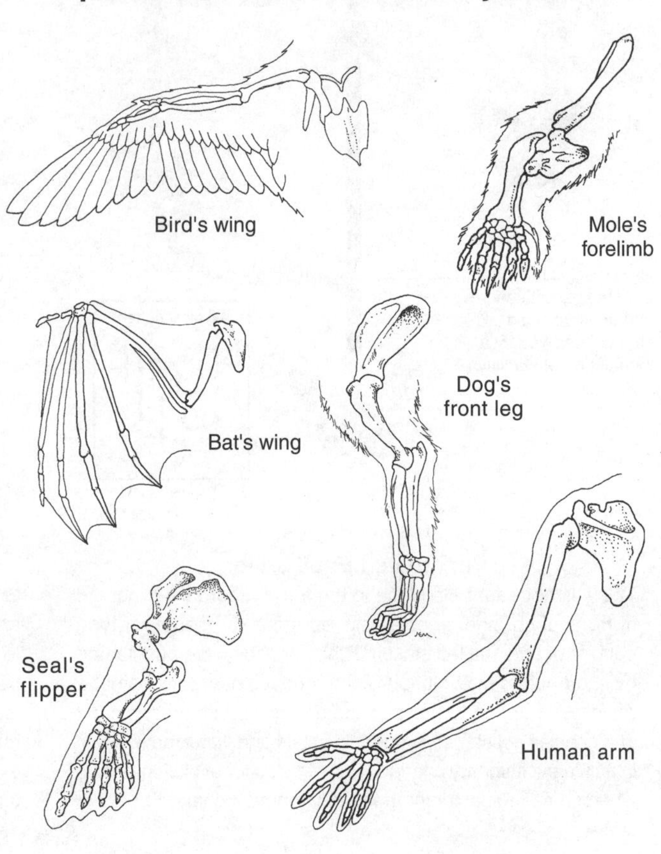

1. Explain how homology in the pentadactyl limb is evidence for adaptive radiation:

2. Homology in the behaviour of animals (for example, sharing similar courtship or nesting rituals) is sometimes used to indicate the degree of relatedness between groups. Suggest how behaviour could be used in this way:

3. Describe the natural selection pressure that would have encouraged a generalised pentadactyl limb to develop into the following structures:

(a) Flipper:

(b) Human hand:

(c) Bird's wing:

(d) Digging forelimb (mole):

Vestigial Organs

Some classes of characters are more valuable than others as reliable indicators of common ancestry. Often, the less any part of an animal is used for specialised purposes, the more important it becomes for classification. Vestigial organs are an example of this. If vestigial features have no clear function and are no longer subject to natural selection, the common ancestry between different species is not clouded by later adaptation to particular purposes. It is sometimes argued that some vestigial organs are not truly vestigial, i.e. they may perform some small function. While this may be true in some cases, the features can still be considered vestigial if their new role is a minor one, unrelated to their original function.

Ancestors of Modern Whales

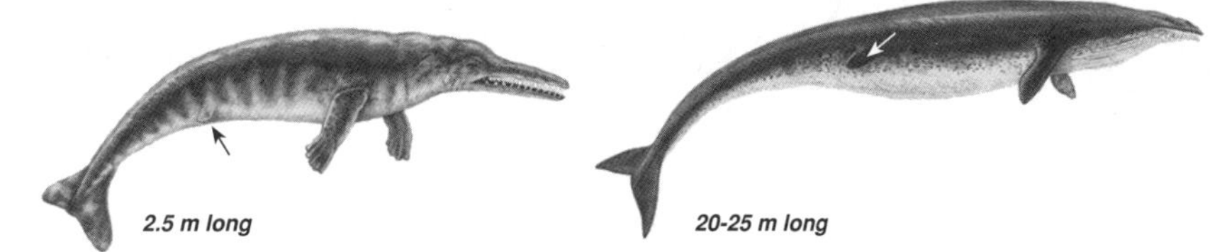

Pakicetus (early Eocene) a carnivorous, four limbed, early Eocene whale ancestor, probably rather like a large otter. It was still partly terrestrial and not fully adapted for aquatic life.

Protocetus (mid Eocene). Much more whale-like than *Pakicetus*. The hindlimbs were greatly reduced and although they still protruded from the body (arrowed), they were useless for swimming.

Basilosaurus (late Eocene). A very large ancestor of modern whales. The hind limbs contained all the leg bones, but were vestigial and located entirely within the main body, leaving a tissue flap on the surface (arrowed).

Vestigial organs are common in nature. The vestigial hindlimbs of modern whales (right) provide anatomical evidence for their evolution from a carnivorous, four footed, terrestrial ancestor. The oldest known whale, *Pakicetus*, from the early Eocene (~54 mya) still had four limbs. By the late Eocene (~40 mya), whales were fully marine and had lost almost all traces of their former terrestrial life. For fossil evidence, see *Whale Origins* at: www.neoucom.edu/Depts/Anat/whaleorigins.htm

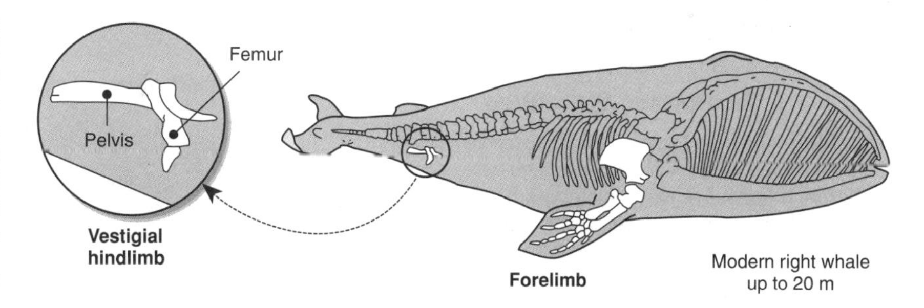

Vestigial organs in birds and reptiles

In all snakes (far left), one lobe of the lung is vestigial (there is not sufficient room in the narrow body cavity for it). In some snakes there are also vestiges of the pelvic girdle and hind limbs of their walking ancestors. Like all ratites, kiwis (left) are flightless. However, more than in other ratites, the wings of kiwis are reduced to tiny vestiges. Kiwis have evolved in the absence of predators to a totally ground dwelling existence.

1. In terms of natural selection explain how structures, that were once useful to an organism, could become vestigial:

2. Suggest why a vestigial structure, once it has been reduced to a certain size, may not disappear altogether:

3. Whale evolution shows the presence of **transitional forms** (fossils that are intermediate between modern forms and very early ancestors). Suggest how vestigial structures indicate the common ancestry of these forms:

Convergent Evolution

Not all similarities between species are a result of common ancestry. Species from different evolutionary lines may come to resemble each other if they have similar ecological roles and natural selection has shaped similar adaptations. This is called **convergent evolution (convergence)**. Analogous structures (below) may arise as result of convergence.

Convergence in Swimming Form

Although similarities in body form and function can arise because of common ancestry, it may also be a result of **convergent evolution**. Selection pressures in a particular environment may bring about similar adaptations in unrelated species. These selection pressures require the solving of problems in particular ways, leading to the similarity of body form or function. The development of succulent forms in unrelated plant groups (*Euphorbia* and the cactus family) is an example of convergence in plants. In the example (right), the selection pressures of the aquatic environment have produced a similar **streamlined** body shape in unrelated vertebrate groups. Icthyosaurs, penguins, and dolphins each evolved from terrestrial species that took up an aquatic lifestyle. Their general body form has evolved to become similar to that of the shark, which has always been aquatic. Note that flipper shape in mammals, birds, and reptiles is a result of convergence, but its origin from the pentadactyl limb is an example of **homology**.

Analogous Structures

Analogous structures are those that have the same function and often the same basic external appearance, but **quite different origins**. The example on the right illustrates how a complex eye structure has developed independently in two unrelated groups. The appearance of the **eye** is similar, but there is no genetic relatedness between the two groups (mammals and cephalopod molluscs). The **wings** of birds and insects are also an example of analogy. The wings perform the same function, but the two groups share no common ancestor. *Longisquama*, a lizard-like creature that lived about 220 million years ago, also had 'wings' that probably allowed gliding between trees. These 'wings' were not a modification of the forearm (as in birds), but highly modified long scales or feathers extending from its back.

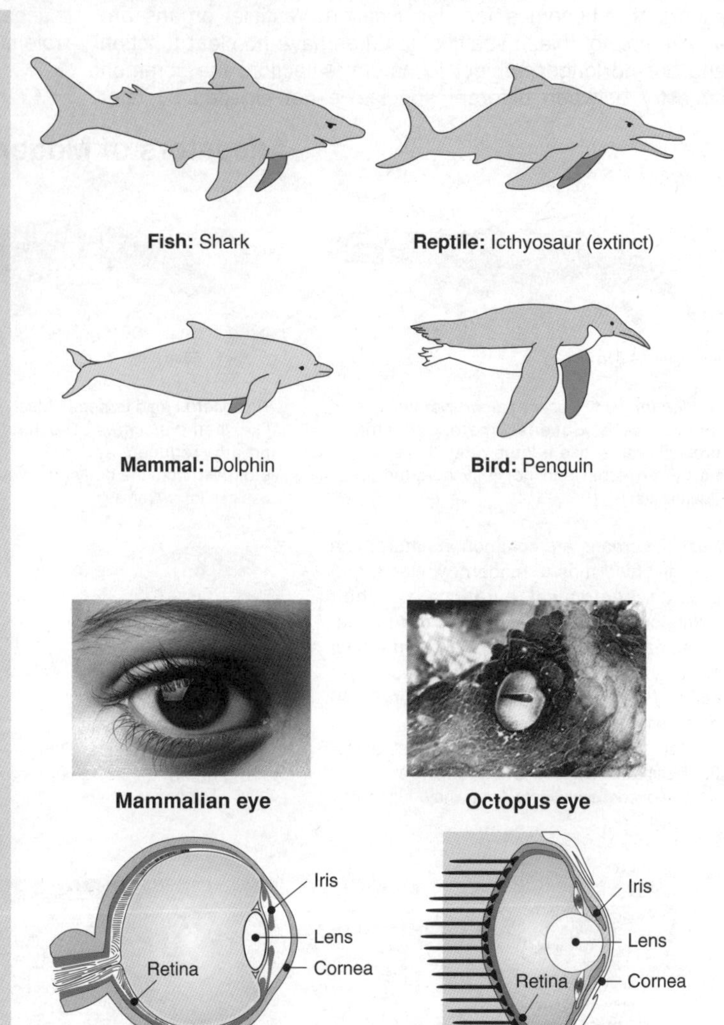

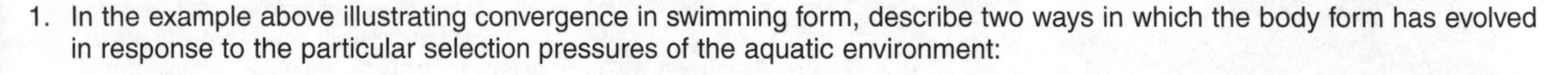

1. In the example above illustrating convergence in swimming form, describe two ways in which the body form has evolved in response to the particular selection pressures of the aquatic environment:

   (a) ______________________________

   ____

   (b) ______________________________

   ____

2. Describe two of the selection pressures that have influenced the body form of the swimming animals above:

   (a) ______________________________

   (b) ______________________________

3. When early taxonomists encountered new species in the Pacific region and the Americas, they were keen to assign them to existing taxonomic families based on their apparent similarity to European species. In recent times, many of the new species have been found to be quite unrelated to the European families they were assigned to. Explain why the traditional approach did not reveal the true evolutionary relationships of the new species:

   ____

   ____

   ____

   ____

4. For each of the paired examples (b)-(f), briefly describe the adaptations of body shape, diet, and locomotion that appear similar in both forms, and the likely selection pressures that are acting on these mammals to produce similar body forms:

Convergence Between Marsupials and Placentals

Marsupials and **placental** mammals were separated from each other very early in mammalian evolution (about 120 mya). Marsupials were initially widely distributed throughout the ancient supercontinent of Gondwana, and there are some modern species still living in the American continent. Gondwana split up about 100 million years ago. As the placentals developed, they displaced the marsupials in most habitats around the world. The island continent of Australia, because of its early isolation by the sea, escaped this competition and placentals did not reach the continent until the arrival of humans 35 000 to 50 000 years ago. The Australian marsupials evolved into a wide variety of forms (below left) that bear a remarkable resemblance to ecologically equivalent species of North American placentals (below right).

Marsupial mammals		Placental mammals
Wombat	(a) Adaptations: Both have rodent-like teeth, eat roots and above ground plants, and excavate burrows. Selection pressures: Diet requires chisel-like teeth for gnawing. The need to seek safety from predators on open grassland.	Wood chuck
Flying phalanger	(b) Adaptations: Selection pressures:	Flying squirrel
Marsupial mole	(c) Adaptations: Selection pressures:	Mole
Marsupial mouse	(d) Adaptations: Selection pressures:	Mouse
Tasmanian wolf (tiger)	(e) Adaptations: Selection pressures:	Wolf
Long-eared bandicoot	(f) Adaptations: Selection pressures:	Jack rabbit

Adaptive Radiation in Mammals

Adaptive radiation is diversification (both structural and ecological) among the descendants of a single ancestral group to occupy different niches. Immediately following the sudden extinction of the dinosaurs, the mammals underwent an adaptive radiation. Most of the modern mammalian groups became established very early. The diagram below shows the divergence of the mammals into major orders; many occupying niches left vacant by the dinosaurs. The vertical extent of each grey shape shows the time span for which that particular mammal order has existed (note that the scale for the geological time scale in the diagram is not linear). Those that reach the top of the chart have survived to the present day. The width of a grey shape indicates how many species were in existence at any given time (narrow means there were few, wide means there were many). The dotted lines indicate possible links between the various mammal orders for which there is no direct fossil evidence.

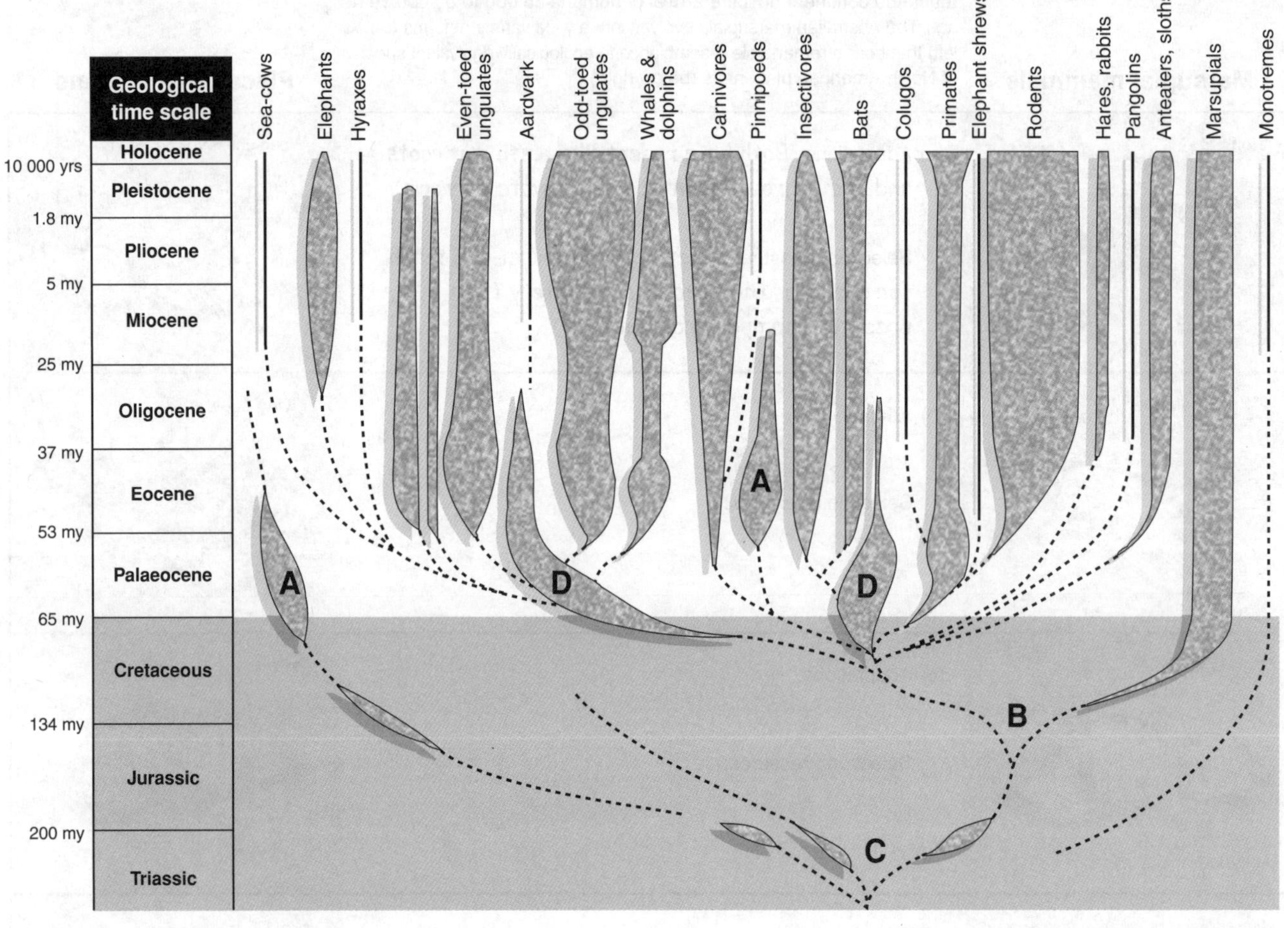

1. In general terms, discuss the **adaptive radiation** that occurred in mammals:

2. Name the term that you would use to describe the animal groups at point **C** (above):

3. Explain what occurred at point **B** (above):

4. Describe two things that the animal orders labelled **D** (above) have in common:

   (a)

   (b)

5. Identify the two orders that appear to have been most successful in terms of the number of species produced:

6. Explain what has happened to the mammal orders labelled **A** in the diagram above:

7. Identify the **epoch** during which there was the most adaptive radiation:

8. Describe two key features that distinguish mammals from other vertebrates:

(a) ______________________ (b) ______________________

9. Describe the principal reproductive features distinguishing each of the major mammalian lines (sub-classes):

(a) Monotremes: ______________________

(b) Marsupials: ______________________

(c) Placentals: ______________________

10. There are 18 orders of placental mammals (or 17 in schemes that include the pinnipeds within the Carnivora). Their names and a brief description of the type of mammal belonging to each group is provided below. Identify and label each of the diagrams with the correct name of their Order:

Order	Description
Insectivora	Insect-eating mammals
Macroscelidae	Elephant shrews (formerly classified with insectivores)
Chiroptera	Bats
Cetacea	Whales and dolphins
Pholidota	Pangolins
Rodentia	Rodents
Probiscidea	Elephants
Sirenia	Sea-cows (manatees)
Artiodactyla	Even-toed hoofed mammals
Dermoptera	Colugos
Primates	Primates
Edontata	Anteaters, sloths, and armadillos
Lagomorpha	Pikas, hares, and rabbits
Carnivora	Flesh-eating mammals (canids, raccoons, bears, cats)
Pinnipedia	Seals, sealions, walruses. (Often now included as a sub-order of Carnivora).
Tubulidentata	Aardvark
Hyracoidea	Hyraxes
Perissodactyla	Odd-toed hoofed mammals

Orders of Placental Mammals

1 ________ 2 ________ 3 ________

4 ________ 5 ________ 6 ________

7 ________ 8 ________ 9 ________ 10 ________ 11 ________ 12 ________

13 ________ 14 ________ 15 ________ 16 ________ 17 ________ 18 ________

11. For each of three named **orders** of placental mammal, describe one **adaptive feature** that allows it to exploit a different niche from other placentals, and describe a **biological advantage** conferred by the adaptation:

(a) Order: ____________ Adaptive feature: ______________________

Biological advantage: ______________________

(b) Order: ____________ Adaptive feature: ______________________

Biological advantage: ______________________

(c) Order: ____________ Adaptive feature: ______________________

Biological advantage: ______________________

Geographical Distribution

The camel family, Camelidae, consists of six modern-day species that have survived on three continents: Asia, Africa and South America. They are characterised by having only two functional toes, supported by expanded pads for walking on sand or snow. The slender snout bears a cleft upper lip. The recent distribution of the camel family is fragmented. Geophysical forces such as plate tectonics and the ice age cycles have controlled the extent of their distribution. South America, for example, was separated from North America until the end of the Pliocene, about 2 million years ago. Three general principles about the dispersal and distribution of land animals are:

- When very closely related animals (as shown by their anatomy) were present at the same time in widely separated parts of the world, it is highly probable that there was no barrier to their movement in one or both directions between the localities in the past.
- The most effective barrier to the movement of land animals (particularly mammals) was a sea between continents (as was caused by changing sea levels during the ice ages).
- The scattered distribution of modern species may be explained by the movement out of the area they originally occupied, or extinction in those regions between modern species.

Origin and Dispersal of the Camel Family

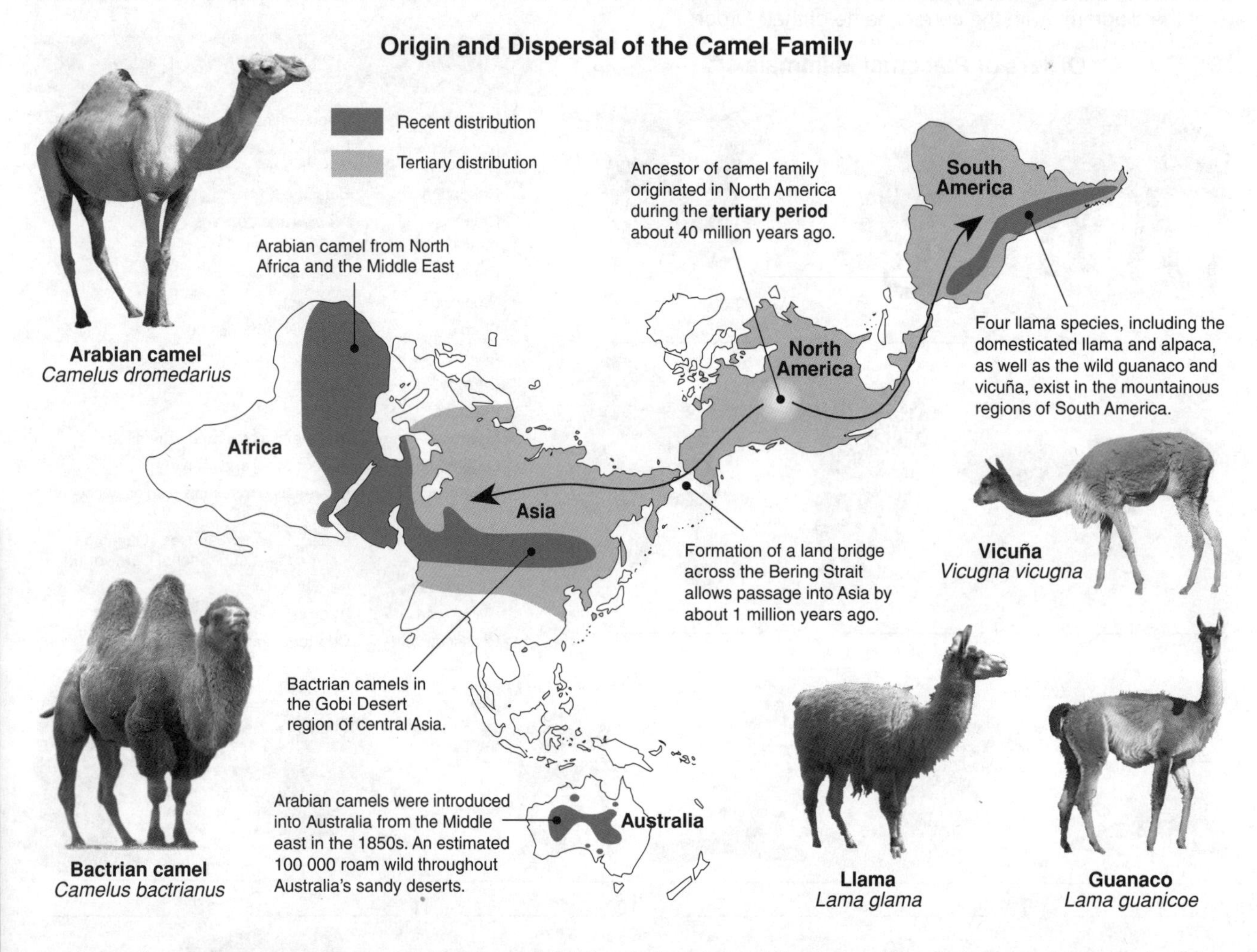

1. The early camel ancestors were able to move into the tropical regions of Central and South America. Explain why this did not happen in southern Asia and southern Africa:

2. Arabian camels are found wild in the Australian outback. Explain how they got there and why they were absent during prehistoric times:

3. The camel family originated in North America. Explain why there are no camels in North America now:

4. Explain how early camels managed to get to Asia from North America:

5. Describe the present distribution of the camel family and explain why it is scattered (discontinuous):

Biodiversity and Classification

AQA-A	AQA-B	CIE	Edexcel	OCR
Complete: *1-7, 9* *Extension: 8, 10-13*	Complete: *1-7, 9-10* *Extension: 8, 12-13*	*Does not apply to this course*	Complete: *1-9* *Extension: 10-13*	Complete: *1-10* *Extension: 12-13*

Learning Objectives

☐ 1. Compile your own glossary from the **KEY WORDS** displayed in **bold type** in the learning objectives below.

Biodiversity *(pages 176-179)*

☐ 2. Explain what is meant by **biodiversity** and explain the importance of accurate classification in recognising, appreciating, and conserving the biodiversity on Earth.

☐ 3. Understand the concept of a **species** in terms of their reproductive isolation and potential for breeding.

Classification systems

The five kingdoms *(pages 194, 196-209)*

☐ 4. Explain what is meant by classification. Recognise **taxonomy** as the study of the theory and practice of classification. Describe the principles and importance of scientific classification.

☐ 5. Explain what is meant by the **five kingdom classification system**. Describe the **distinguishing features** of each of the five kingdoms:
- **Prokaryotae** (Monera): bacteria and cyanobacteria.
- **Protoctista**: includes the algae and protozoans.
- **Fungi**: includes yeasts, moulds, and mushrooms.
- **Plantae**: includes mosses, liverworts, tracheophytes.
- **Animalia**: all invertebrate phyla and the chordates.

☐ 6. Demonstrate a working knowledge of taxonomy by classifying familiar organisms. Recognise at least seven major **taxonomic categories**: **kingdom**, **phylum**, **class**, **order**, **family**, **genus**, and **species**.

☐ 7. Appreciate that taxonomic categories should not be confused with **taxa** (sing. **taxon**), which are groups of real organisms: "genus" is a taxonomic category, whereas the genus *Drosophila* is a taxon.

☐ 8. Understand the basis for assigning organisms into different taxonomic categories. Recall what is meant by a **distinguishing feature**. Explain that species are usually classified on the basis of **shared derived characters** rather than primitive (ancestral) characters. *For example, within the subphylum Vertebrata, the presence of a backbone is a derived, therefore a distinguishing, feature. However, within the class Mammalia, the backbone is an ancestral feature and is not distinguishing, whereas mammary glands (a distinguishing feature) are derived.*

☐ 9. Explain how **binomial nomenclature** is used to classify organisms. Appreciate the problems associated with using **common names** to describe organisms.

☐ 10. Explain the relationship between classification and **phylogeny**. Appreciate that newer classification schemes (below) attempt to better reflect the true phylogeny of organisms.

New classification schemes *(pages 194-195)*

☐ 11. Recognise the recent reclassification of organisms into three **domains**: **Archaea**, **Eubacteria**, and **Eukarya**. Explain the basis and rationale for this classification.

☐ 12. Appreciate that **cladistics** provides a method of classification based on relatedness, and that it emphasises the presence of **shared derived characters**. Discuss the benefits and disadvantages associated with cladistic schemes.

Classification keys *(pages 210-211)*

☐ 13. Explain what a **classification key** is and what it is used for. Describe the essential features of a classification key. Use a simple taxonomic key to recognise and classify some common organisms.

See the 'Textbook Reference Grid' on pages 8-9 for textbook page references relating to material in this topic.

See page 6 for details of publishers of periodicals:

STUDENT'S REFERENCE

■ **How Many Mammals in Britain?** Biol. Sci. Rev., 12(4) March 2000, pp. 18-22. *This article examines the nature of species definitions and the diversity and abundance of mammalian species in Britain (both historically and present day).*

■ **The Species Enigma** New Scientist, 13 June 1998 (Inside Science). *An account of the nature of species, ring species, and the status of hybrids.*

■ **Biological Taxonomy** Biol. Sci. Rev., 8(3) Jan. 1996, pp. 34-37. *Taxonomy, including a clear, concise treatment of cladistics and phenetics.*

■ **Taxonomy: The Naming Game Revisited** Biol. Sci. Rev., 9(5) May 1997, pp. 31-35. *New tools for taxonomy and how they are used (includes the exemplar of the reclassification of the kingdoms).*

TEACHER'S REFERENCE

■ **Family Feuds** New Scientist, 24 January 1998, pp. 36-40. *Molecular and morphological analysis used for determining species inter-relatedness.*

■ **The Problematic Red Wolf** Scientific American, July 1995, pp. 26-31. *Is the red wolf a species or a long-established hybrid? Correctly naming and recognising species can affect conservation efforts.*

■ **The Loves of the Plants** Scientific American, February 1996, pp. 98-103. *The classification of plants and the development of effective keys to plant identification.*

See pages 10-11 for details of how to access **Bio Links** from our web site: **www.biozone.co.uk**. From Bio Links, access sites under the topics:

BIODIVERSITY > **Taxonomy and Classification**:
• Birds and DNA • Taxonomy: Classifying life • The phylogeny of life... *and others*

MICROBIOLOGY > **General Microbiology**:
• British Mycological Society • Major groups of prokaryotes • The microbial world ... *and others*

PLANT BIOLOGY > **Classification and Diversity**:
• Flowering plant diversity ... *and others*

Software and video resources are provided on the Teacher Resource Handbook on CD-ROM

The New Tree of Life

With the advent of more efficient genetic (DNA) sequencing technology, the genomes of many bacteria began to be sequenced. In 1996, the results of a scientific collaboration examining DNA evidence confirmed the proposal that life comprises three major evolutionary lineages (domains) and not two as was the convention. The recognised lineages were the **Eubacteria**, the **Eukarya** and the **Archaea** (formerly the Archaebacteria). The new classification reflects the fact that there are very large differences between the archaea and the eubacteria. All three domains probably had a distant common ancestor.

A Five Kingdom World (right)

The diagram on the right represents the five kingdom system of classification. It recognises two basic cell types: prokaryote and eukaryote. The domain Prokaryota includes the prokaryotes: all bacteria and cyanobacteria. Domain Eukaryota includes protoctists, fungi, plants, and animals. This is the system most commonly represented in modern biology textbooks.

A New View of the World (below)

In 1996 a large collaboration of scientists deciphered the full DNA sequence of every gene of a strange type of bacteria called *Methanococcus jannaschii.* Termed an extremophile, this methane-producing archaebacterium lives at 85°C; a temperature lethal for regular bacteria as well as multicellular plants and animals. The DNA sequence confirmed that life consists of three major evolutionary lineages, not the two that have been routinely described in textbooks. Only 44% of this archaebacterium's genes resemble those in bacteria or eukaryotes, or both.

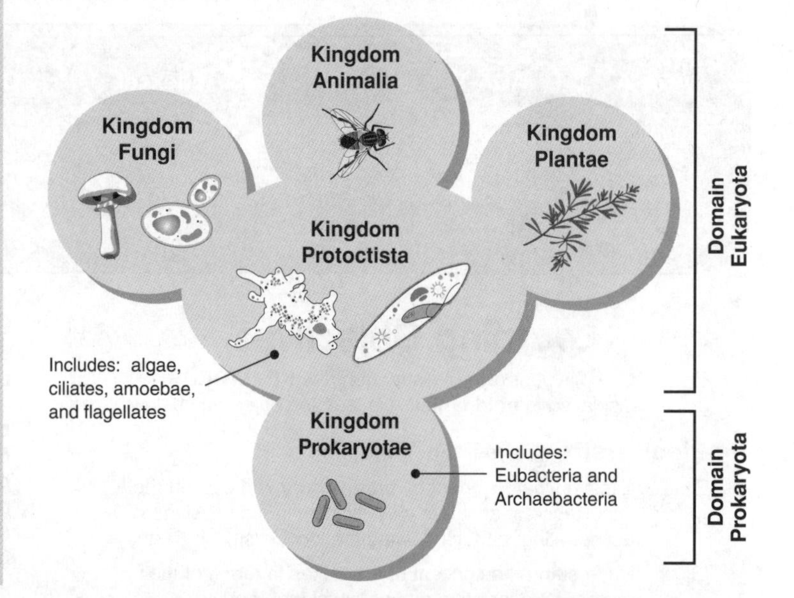

Domain Eubacteria

Lack a distinct nucleus and cell organelles. Generally prefer less extreme environments than Archaea. Includes well-known pathogens, many harmless and beneficial species, and the cyanobacteria (photosynthetic bacteria containing the pigments chlorophyll a and phycocyanin).

Domain Archaea

Closely resemble eubacteria in many ways but cell wall composition and aspects of metabolism are very different. Live in extreme environments similar to those on primeval Earth. They may utilise sulphur, methane, or halogens (chlorine, fluorine), and many tolerate extremes of temperature, salinity, or pH.

Domain Eukarya

Complex cell structure with organelles and nucleus. This group contains four of the kingdoms classified under the more traditional system. Note that Kingdom Protoctista is separated into distinct groups: e.g. amoebae, ciliates, flagellates.

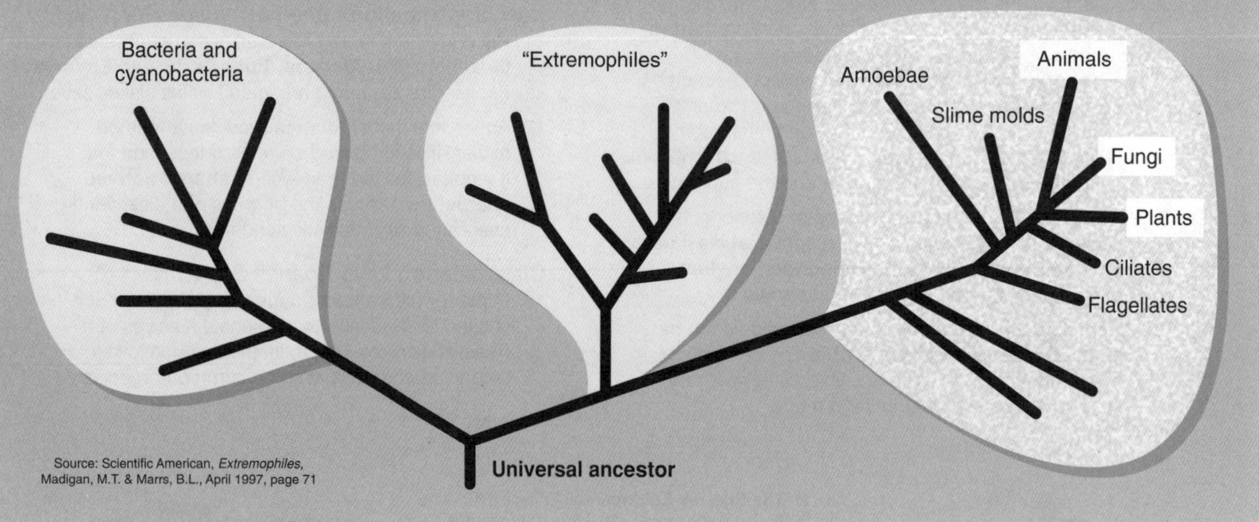

Source: Scientific American, *Extremophiles*, Madigan, M.T. & Marrs, B.L., April 1997, page 71

1. Explain why some scientists have recommended that the conventional classification of life be revised so that the Archaea, Eubacteria and Eukarya are three separate domains:

____

____

2. Describe two features of the new classification scheme that are very different from the five kingdom classification:

(a) ____

____

(b) ____

____

New Classification Schemes

Taxonomy is the study of classification. Ever since Darwin, the aim of classification has been to organise species, and to reflect their evolutionary history (**phylogeny**). Each successive group in the taxonomic hierarchy should represent finer and finer branching from a common ancestor. In order to reconstruct evolutionary history, phylogenetic trees must be based on features that are due to shared ancestry (homologies). Traditional taxonomy has relied mainly on **morphological characters** to do this. Modern technology has assisted taxonomy by providing **biochemical evidence** (from proteins and DNA) for the relatedness of species. The most familiar approach to classifying organisms is to use **classical evolutionary taxonomy**. It considers branching sequences and overall likeness. A more recent approach has been to use **cladistics**: a technique which emphasises phylogeny or relatedness, usually based on biochemical evidence (and largely ignoring morphology or appearance). Each branch on the tree marks the point where a new species has arisen by evolution. Traditional and cladistic schemes do not necessarily conflict, but there have been reclassifications of some taxa (notably the primates, but also the reptiles, dinosaurs, and birds). Traditional taxonomists criticise cladistic schemes because they do not recognise the amount of visible change in morphology that occurs in species after their divergence from a common ancestor. Popular classifications will probably continue to reflect similarities and differences in appearance, rather than a strict evolutionary history. In this respect, they are a compromise between phylogeny and the need for a convenient filing system for species diversity.

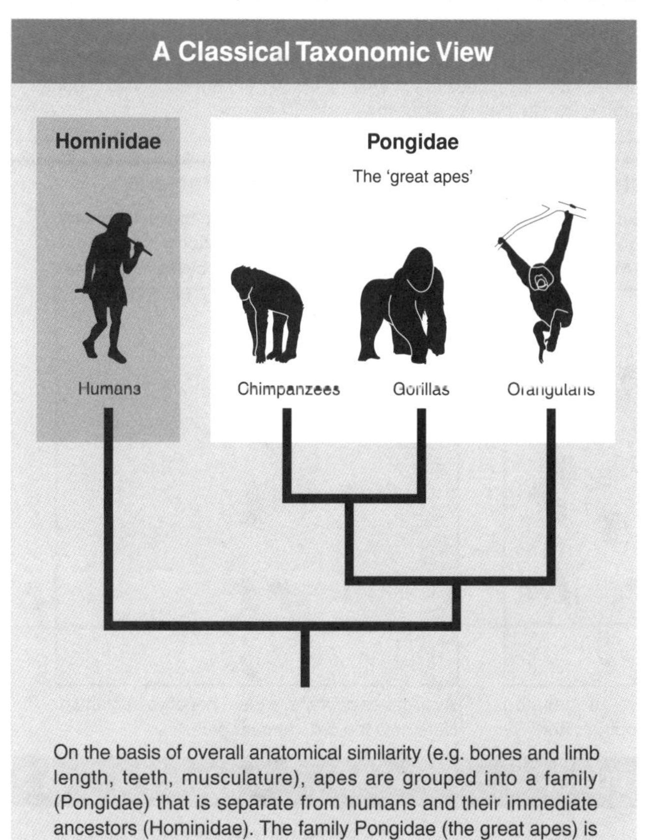

On the basis of overall anatomical similarity (e.g. bones and limb length, teeth, musculature), apes are grouped into a family (Pongidae) that is separate from humans and their immediate ancestors (Hominidae). The family Pongidae (the great apes) is not monophyletic (of one phylogeny), because it stems from an ancestor that also gave rise to a species in another family (i.e. humans). This traditional classification scheme is now at odds with schemes derived after considering genetic evidence.

A Cladistic View

Hominidae

Homininae

Ponginae

Humans

Chimpanzees

Gorillas

Orangutans

1.4%

1.8%

3.6%

A small genetic difference implies a recent common ancestor

A greater genetic difference implies that two groups are more distantly related

Based on the evidence of genetic differences (% values above), chimpanzees and gorillas are more closely related to humans than to orangutans, and chimpanzees are more closely related to humans than they are to gorillas. Under this scheme there is no true family of great apes. The family Hominidae includes two subfamilies: Ponginae and Homininae (humans, chimpanzees, and gorillas). This classification is monophyletic: the Hominidae includes all the species that arise from a common ancestor.

1. Briefly explain the benefits of classification schemes based on:

   (a) Morphological characters: ____________________

   (b) Relatedness in time (from biochemical evidence): ____________________

2. Describe the contribution of biochemical evidence to taxonomy: ____________________

3. Based on the diagram above, state the family to which the chimpanzees belong under:

   (a) A traditional scheme: __________ (b) A cladistic scheme: __________

Biodiversity & Classification

Code: A 2

Features of Taxonomic Groups

In order to distinguish organisms, it is desirable to classify and name them (a science known as **taxonomy**). An effective classification system requires features that are distinctive to a particular group of organisms. The distinguishing features of some major taxonomic groups are provided in the following pages by means of diagrams and brief summaries. Revised classification systems, recognising three domains (rather than five kingdoms) are now recognised as better representations of the true diversity of life. However, for the purposes of describing the groups with which we are most familiar, the five kingdom system (used here) is still appropriate. Note that most animals show **bilateral symmetry** (body divisible into two halves that are mirror images). **Radial symmetry** (body divisible into equal halves through various planes) is a characteristic of cnidarians and ctenophores. Definitions of specific terms relating to features of structure or function can be found in any general biology text.

Kingdom: PROKARYOTAE (Bacteria)

- Also known as monerans or prokaryotes.
- Two major bacterial lineages are recognised: the primitive **Archaebacteria** and the more advanced **Eubacteria**.
- All have a prokaryotic cell structure: they lack the nuclei and chromosomes of eukaryotic cells, and have smaller (70S) ribosomes.
- Have a tendency to spread genetic elements across species barriers by sexual conjugation, viral transduction and other processes.
- Can reproduce rapidly by binary fission in the absence of sex.
- Have evolved a wider variety of metabolism types than eukaryotes.
- Bacteria grow and divide or aggregate into filaments or colonies of various shapes.
- They are taxonomically identified by their appearance (form) and through biochemical differences.

Species diversity: 10 000 + Bacteria are rather difficult to classify to the species level because of their relatively rampant genetic exchange, and because their reproduction is usually asexual.

Eubacteria

- Also known as 'true bacteria', they probably evolved from the more ancient Archaebacteria.
- Distinguished from Archaebacteria by differences in cell wall composition, nucleotide structure, and ribosome shape.
- Very diverse group comprises most bacteria.
- The **gram stain** provides the basis for distinguishing two broad groups of bacteria. It relies on the presence of peptidoglycan (unique to bacteria) in the cell wall. The stain is easily washed from the thin peptidoglycan layer of gram negative walls but is retained by the thick peptidoglycan layer of gram positive cells, staining them a dark violet colour.

Gram-Positive Bacteria

The walls of gram positive bacteria consist of many layers of peptidoglycan forming a thick, single-layered structure that holds the gram stain.

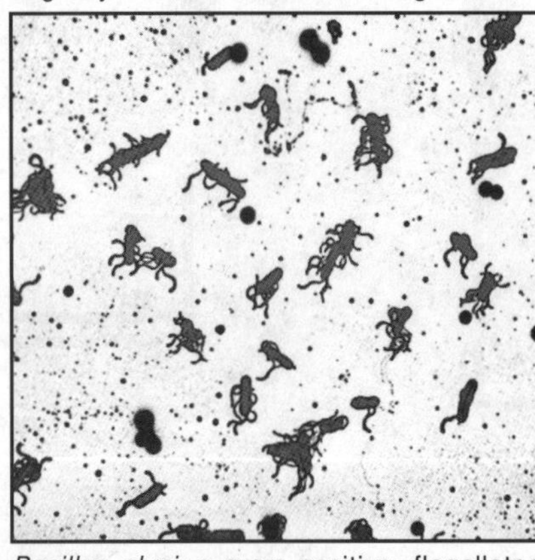

Bacillus alvei: a gram positive, flagellated bacterium. Note how the cells appear dark.

Gram-Negative Bacteria

The cell walls of gram negative bacteria contain only a small proportion of peptidoglycan, so the dark violet stain is not retained by the organisms.

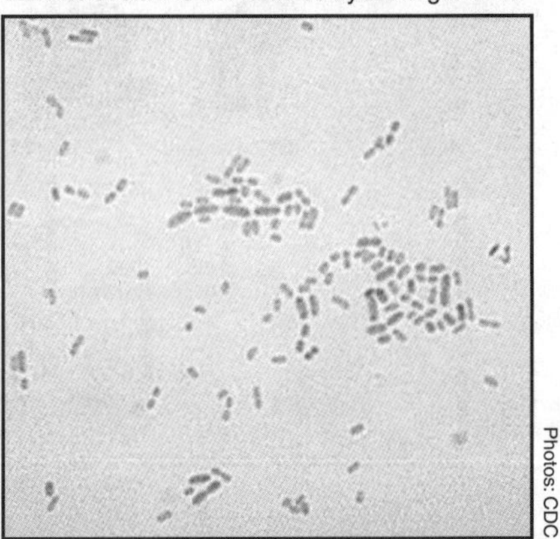

Alcaligenes odorans: a gram negative bacterium. Note how the cells appear pale.

Photos: CDC

Kingdom: FUNGI

- Heterotrophic.
- Rigid cell wall made of chitin.
- Vary from single celled to large multicellular organisms.
- Mostly saprotrophic (i.e. feeding on dead or decaying material).
- Terrestrial and immobile.

Examples:
Mushrooms/toadstools, yeasts, truffles, morels, moulds, and lichens.

Species diversity: 80 000 +

- **Lichens** are symbiotic associations of a fungus (provides protection) and an alga (provides the food).

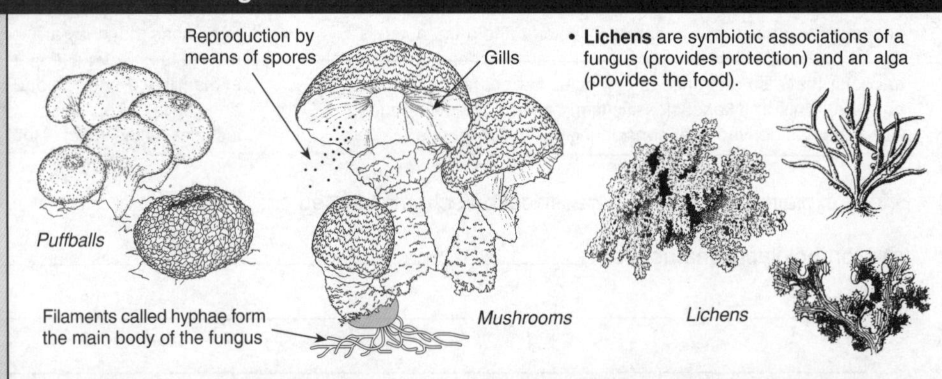

Kingdom: PROTOCTISTA

- A diverse group of organisms that do not fit easily into other taxonomic groups.
- Unicellular or simple multicellular.
- Widespread in moist or aquatic environments.

Examples of algae: green, brown, and red algae, dinoflagellates, diatoms.

Examples of protozoa: amoebas, foraminiferans, radiolarians, ciliates.

Species diversity: 55 000 +

Algae 'plant-like' protoctists

- Autotrophic (photosynthesis)
- Characterised by the type of chlorophyll present

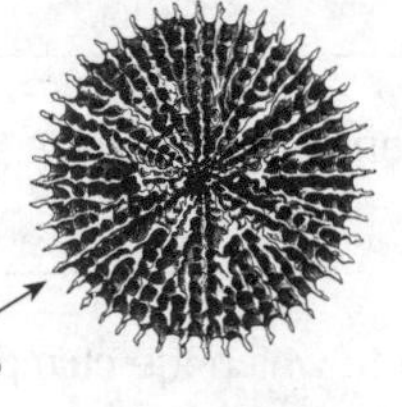

Protozoa 'animal-like' protoctists

- Heterotrophic nutrition and feed via ingestion
- Most are microscopic (5 μm-250 μm)

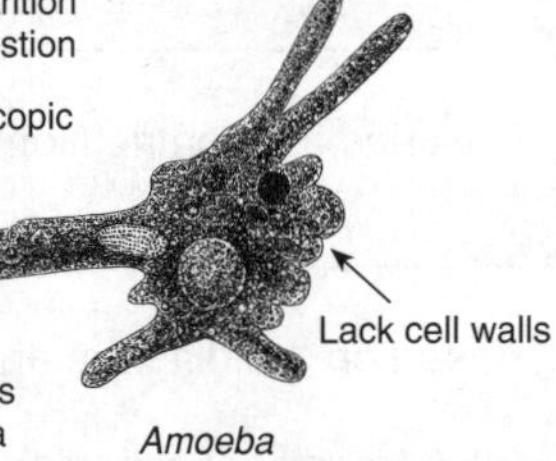

Biodiversity & Classification

Kingdom: PLANTAE

- Multicellular organisms (the majority are photosynthetic and contain chlorophyll).
- Cell walls made of cellulose; Food is stored as starch.
- Subdivided into two major divisions based on tissue structure: ***Bryophytes*** (non-vascular) and ***Tracheophytes*** (vascular) plants.

Non-Vascular Plants:

- Non vascular, lacking transport tissues (no xylem or phloem).
- They are small and restricted to moist, terrestrial environments.
- Do not possess 'true' roots, stems or leaves

Phylum Bryophyta: Mosses, liverworts, and hornworts.

Species diversity: 18 600 +

Phylum: Bryophyta

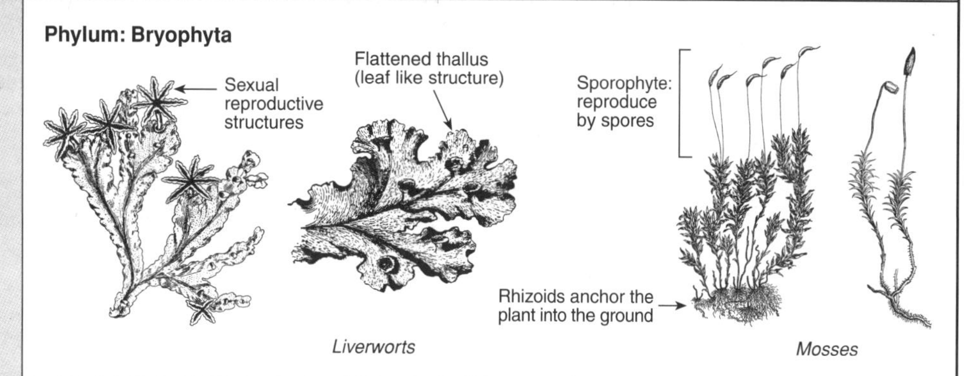

Liverworts *Mosses*

Vascular Plants:

- Vascular: possess transport tissues.
- Possess true roots, stems, and leaves, as well as stomata.
- Reproduce via spores, not seeds.
- Clearly defined *alternation of sporophyte and gametophyte generations.*

Seedless Plants:

Spore producing plants, includes:

Phylum Filicinophyta: Ferns

Phylum Sphenophyta: Horsetails

Phylum Lycophyta: Club mosses

Species diversity: 13 000 +

Phylum: Lycophyta

Leaves

Club moss

Phylum: Sphenophyta

Leaves

Horsetail

Phylum: Filicinophyta

Reproduce via spores on the underside of leaf

Large dividing leaves called fronds

Rhizome

Adventitious roots

Fern

Seed Plants:

Also called Spermatophyta. Produce seeds housing an embryo. Includes:

Gymnosperms

- Lack enclosed chambers in which seeds develop.
- Produce seeds in cones which are exposed to the environment.

Phylum Cycadophyta: Cycads

Phylum Ginkgophyta: Ginkgoes

Phylum Coniferophyta: Conifers

Species diversity: 730 +

Phylum: Cycadophyta

Palm-like leaves

Cone

Cycad

Phylum: Ginkophyta

Ginkgo

Phylum: Coniferophyta

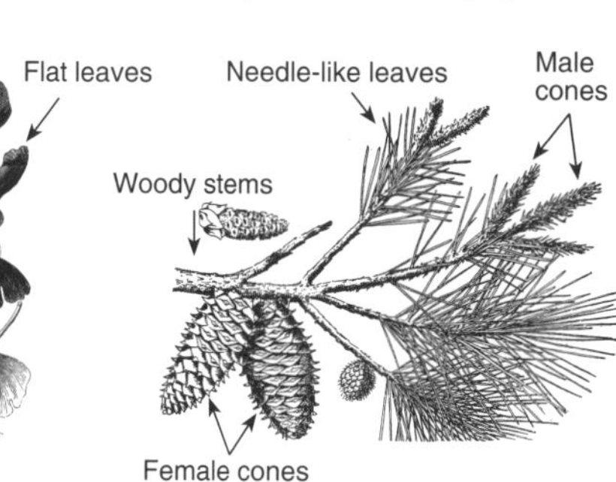

Conifer

Angiosperms

Phylum: Angiospermophyta

- Seeds in specialised reproductive structures called flowers.
- Female reproductive ovary develops into a fruit.
- Pollination usually via wind or animals.

Species diversity: 260 000 +

The phylum Angiospermophyta may be subdivided into two classes:

Class *Monocotyledoneae* (Monocots)

Class *Dicotyledoneae* (Dicots)

Angiosperms: Monocotyledons

- Only have one cotyledon (food storage organ)
- Normally herbaceous (non-woody) with no secondary growth

Examples: cereals, lilies, daffodils, palms, grasses.

Angiosperms: Dicotyledons

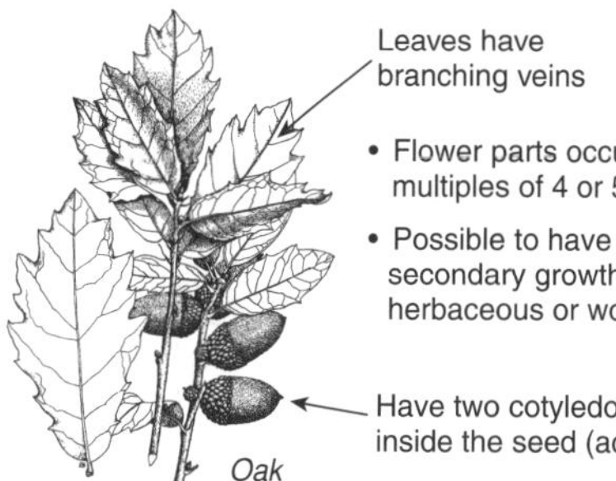

- Flower parts occur in multiples of 4 or 5
- Possible to have secondary growth (either herbaceous or woody)

Examples: many annual plants, trees and shrubs.

Biodiversity & Classification

Kingdom: ANIMALIA

- Over 800 000 species described in 33 existing phyla.
- Multicellular, heterotrophic organisms.
- Animal cells lack cell walls.
- Further subdivided into various major phyla on the basis of body symmetry, type of body cavity, and external and internal structures.

Phylum: Rotifera

- A diverse group of small organisms with sessile, colonial, and planktonic forms.
- Most freshwater, a few marine.
- Typically reproduce via cyclic parthenogenesis.
- Characterised by a wheel of cilia on the head used for feeding and locomotion, a large muscular pharynx (mastax) with jaw like trophi, and a foot with sticky toes.

Species diversity: 1500+

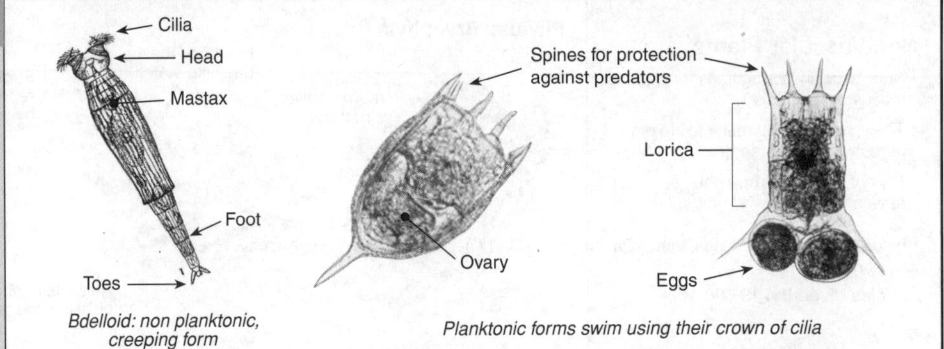

Bdelloid: non planktonic, creeping form

Planktonic forms swim using their crown of cilia

Phylum: Porifera

- Lack organs.
- All are aquatic (mostly marine).
- Asexual reproduction by budding.
- Lack a nervous system.

Examples: sponges.

Species diversity: 8000 +

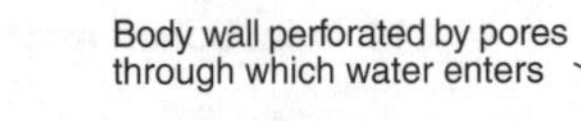

- Capable of regeneration (the replacement of lost parts)
- Possess spicules (needle-like internal structures) for support and protection

Tube sponge

Sponge

Phylum: Cnidaria

- Two basic body forms:
  Medusa: umbrella shaped and free swimming by pulsating bell.
  Polyp: cylindrical, some are sedentary, others can glide, or somersault or use tentacles as legs.
- Some species have a life cycle that alternates between a polyp stage and a medusa stage.
- All are aquatic (most are marine).

Examples: Jellyfish, sea anemones, hydras, and corals.

Species diversity: 11 000 +

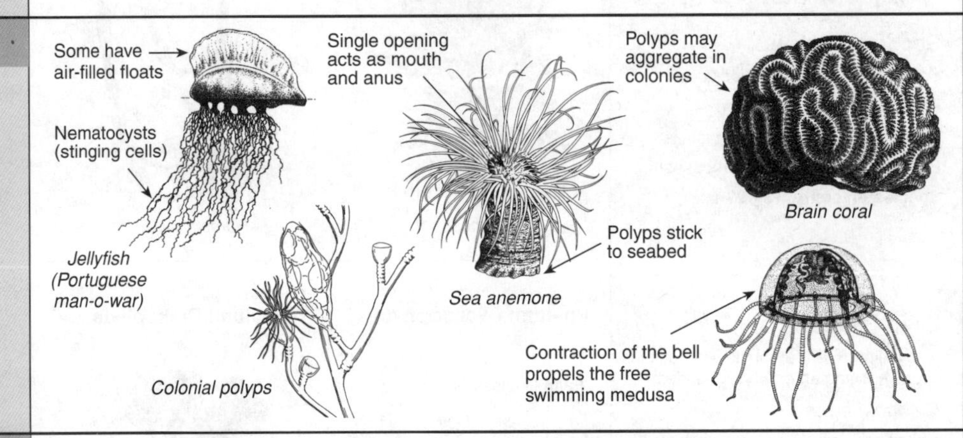

Jellyfish (Portuguese man-o-war)

Sea anemone

Brain coral

Colonial polyps

Phylum: Platyhelminthes

- Unsegmented body.
- Flattened body shape.
- Mouth, but no anus.
- Many are parasitic.

Examples: Tapeworms, planarians, flukes.

Species diversity: 20 000+

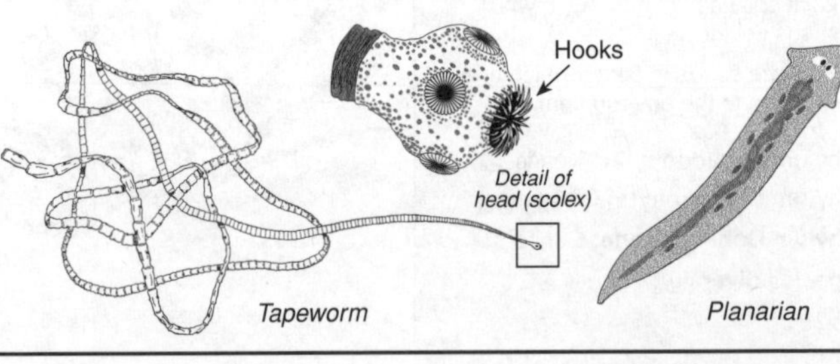

Liver fluke *Tapeworm* *Planarian*

Phylum: Nematoda

- Tiny, unsegmented round worms.
- Many are plant/animal parasites

Examples: Hookworms, stomach worms, lung worms, filarial worms

Species diversity: 80 000 - 1 million

Muscular pharynx, Ovary, Anus, Mouth, Intestine

A general nematode body plan

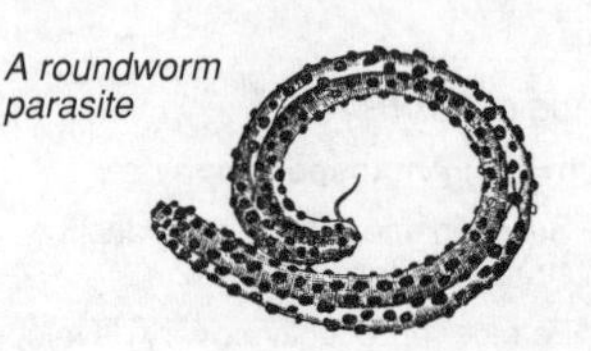

A roundworm parasite

Phylum: Annelida

- Cylindrical, segmented body with chaetae (bristles).
- Move using hydrostatic skeleton and/or parapodia (appendages).

Examples: Earthworms, leeches, polychaetes (including tubeworms).

Species diversity: 15 000 +

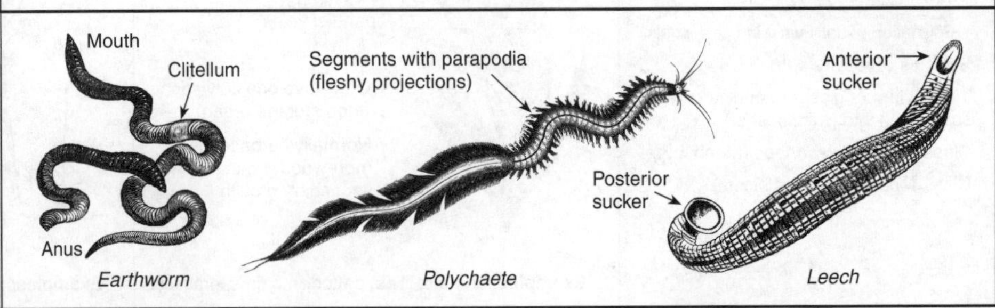

Earthworm *Polychaete* *Leech*

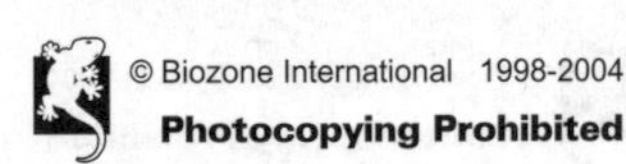

Kingdom: ANIMALIA *(continued)*

Phylum: Mollusca

- Soft bodied and unsegmented.
- Body comprises head, muscular foot, and visceral mass (organs).
- Most have radula (rasping tongue).
- Aquatic and terrestrial species.
- Aquatic species possess gills.

Examples: Snails, mussels, squid.

Species diversity: 110 000 +

Class: Bivalvia

Radula lost in bivalves

Mantle secretes shell

Two shells hinged together

Class: Gastropoda

Mantle secretes shell

Muscular foot for locomotion

Tentacles with eyes

Head

Class: Cephalopoda

Well developed eyes

Foot divided into tentacles

Phylum: Arthropoda

- Exoskeleton made of chitin.
- Grow in stages after moulting.
- Jointed appendages.
- Segmented bodies.
- Heart found on dorsal side of body.
- Open circulation system.
- Most have compound eyes.

Species diversity: 1 million +
Make up 75% of all living animals.

Arthropods are subdivided into the following classes:

Class: Crustacea (crustaceans)
- Mainly marine.
- Exoskeleton impregnated with mineral salts.
- Gills often present.
- Includes: Lobsters, crabs, barnacles, prawns, shrimps, isopods, amphipods
- **Species diversity**: 35 000 +

Class: Arachnida (chelicerates)
- Almost all are terrestrial.
- 2 body parts: cephalothorax and abdomen (except horseshoe crabs).
- Includes: spiders, scorpions, ticks, mites, horseshoe crabs.
- **Species diversity**: 57 000 +

Class: Insecta (insects)
- Mostly terrestrial.
- Most are capable of flight.
- 3 body parts: head, thorax, abdomen.
- Include: Locusts, dragonflies, cockroaches, butterflies, bees, ants, beetles, bugs, flies, and more
- **Species diversity**: 800 000 +

Class: Myriapoda (=many legs)
Diplopods (millipedes)
- Terrestrial.
- Have a rounded body.
- Eat dead or living plants.
- **Species diversity**: 2000 +

Chilopods (centipedes)
- Terrestrial.
- Have a flattened body.
- Poison claws for catching prey.
- Feed on insects, worms, and snails.
- **Species diversity**: 7000 +

Class: Crustacea

2 pairs of antennae

Cephalothorax (fusion of head and thorax)

Abdomen

3 pairs of mouthparts

Cheliped (first leg)

Shrimp

Swimmerets

Walking legs

Crab

Amphipod

Class: Arachnida

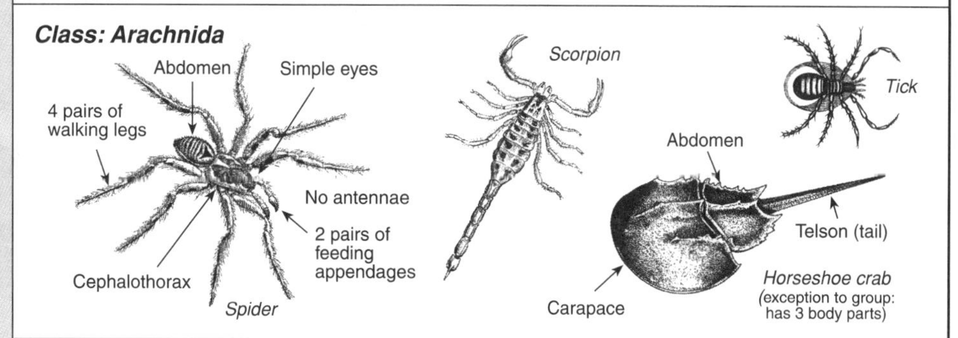

Class: Insecta

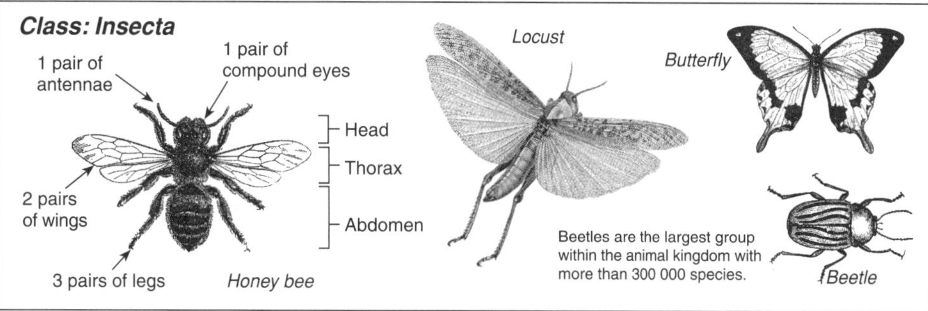

Class: Myriapoda

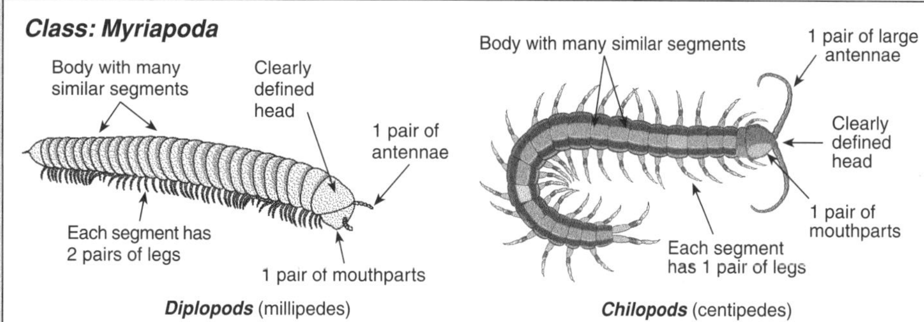

Phylum: Echinodermata

- Rigid body wall, internal skeleton made of calcareous plates.
- Many possess spines.
- Ventral mouth, dorsal anus.
- External fertilisation.
- Unsegmented, marine organisms.
- Tube feet for locomotion.
- Water vascular system.

Examples: Starfish, brittlestars, feather stars, sea urchins, sea lilies.

Species diversity: 6000 +

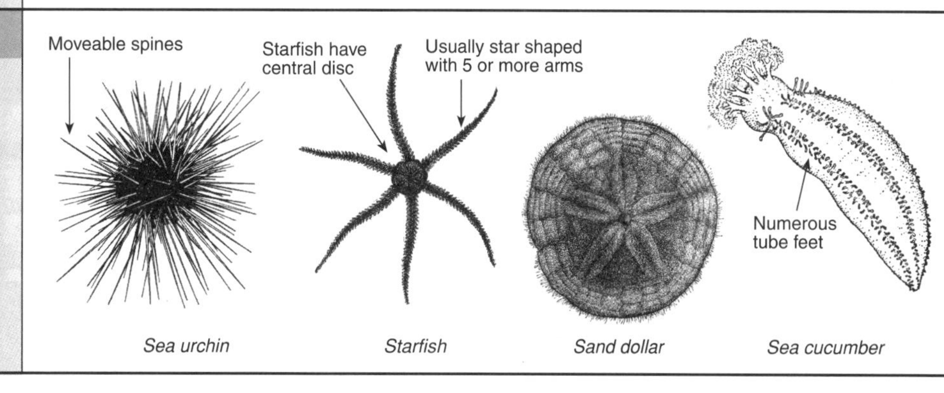

Kingdom: ANIMALIA *(continued)*

Phylum: Chordata

- Dorsal notochord (flexible, supporting rod) present at some stage in the life history.
- Post-anal tail present at some stage in their development.
- Dorsal, tubular nerve cord.
- Pharyngeal slits present.
- Circulation system closed in most.
- Heart positioned on ventral side.

Species diversity: 48 000 +

- A very diverse group with several sub-phyla:
  - Urochordata (sea squirts, salps)
  - Cephalochordata (lancelet)
  - Craniata (vertebrates)

Sub-Phylum Craniata (vertebrates)
- Internal skeleton of cartilage or bone.
- Well developed nervous system.
- Vertebral column replaces notochord.
- Two pairs of appendages (fins or limbs) attached to girdles.

Further subdivided into:

Class: Chondrichthyes (cartilaginous fish)
- Skeleton of cartilage (not bone).
- No swim bladder.
- All aquatic (mostly marine).
- Include: Sharks, rays, and skates.

Species diversity: 850 +

Class: Osteichthyes (bony fish)
- Swim bladder present.
- All aquatic (marine and fresh water).

Species diversity: 21 000 +

Class: Amphibia (amphibians)
- Lungs in adult, juveniles may have gills (retained in some adults).
- Gas exchange also through skin.
- Aquatic and terrestrial (limited to damp environments).
- Include: Frogs, toads, salamanders, and newts.

Species diversity: 3900 +

Class Reptilia (reptiles)
- Ectotherms with no larval stages.
- Teeth are all the same type.
- Eggs with soft leathery shell.
- Mostly terrestrial.
- Include: Snakes, lizards, crocodiles, turtles, and tortoises.

Species diversity: 7000 +

Class: Aves (birds)
- Terrestrial endotherms.
- Eggs with hard, calcareous shell.
- Strong, light skeleton.
- High metabolic rate.
- Gas exchange assisted by air sacs.

Species diversity: 8600 +

Class: Mammalia (mammals)
- Endotherms with hair or fur.
- Mammary glands produce milk.
- Glandular skin with hair or fur.
- External ear present.
- Teeth are of different types.
- Diaphragm between thorax/abdomen.

Species diversity: 4500 +

Subdivided into three subclasses: *Monotremes, marsupials, placentals.*

Class: Chondrichthyes (cartilaginous fish)

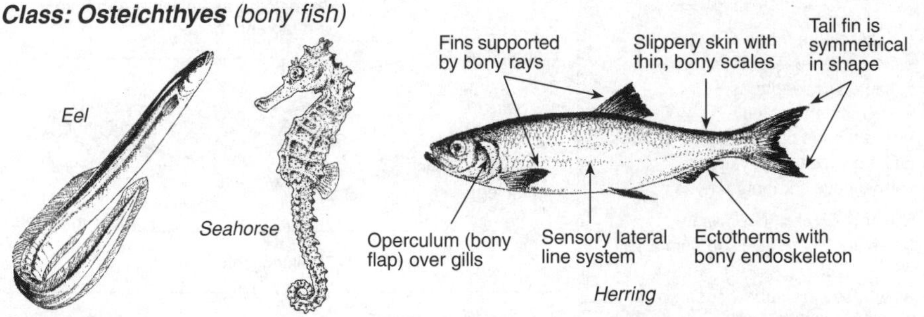

Class: Osteichthyes (bony fish)

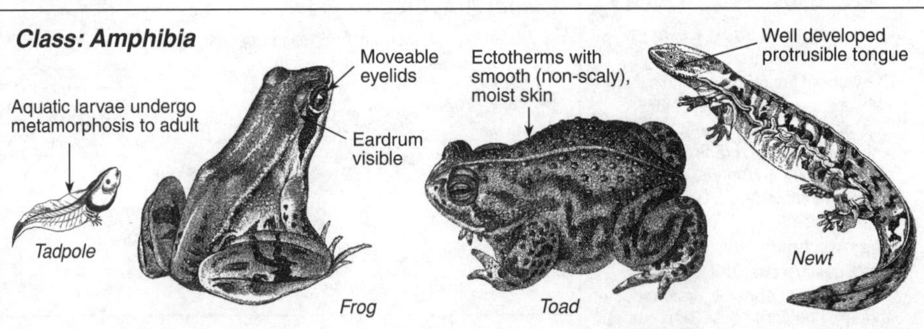

Class: Amphibia

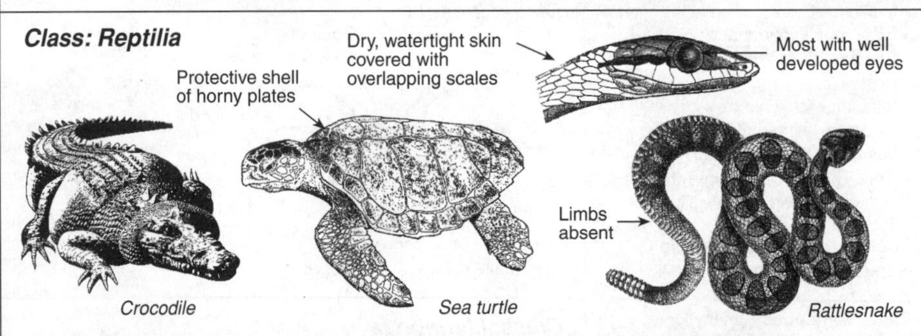

Class: Reptilia

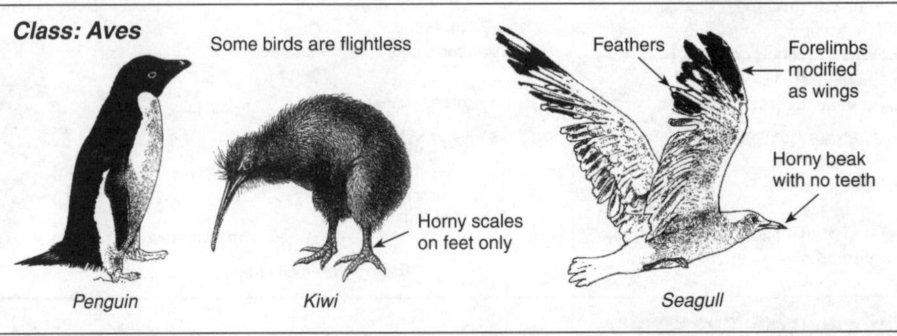

Class: Aves

Some birds are flightless

Feathers

Forelimbs modified as wings

Horny beak with no teeth

Horny scales on feet only

Penguin *Kiwi* *Seagull*

Class: Mammalia

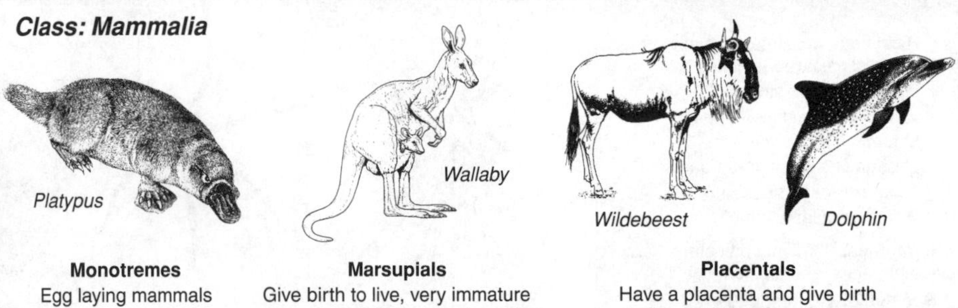

Monotremes
Egg laying mammals

Marsupials
Give birth to live, very immature young which then develop in a pouch

Placentals
Have a placenta and give birth to live, well developed young

Biodiversity & Classification

Features of the Five Kingdoms

The classification of living things into taxonomic groups is based on how biologists believe they are related in an evolutionary sense. Organisms in a taxonomic group share some common features that set the group apart from others. By identifying these features, it is possible to gain an understanding of the evolutionary development of the group. The focus of this activity is to summarise the characteristic features of each of the five kingdoms.

1. Distinguishing features of Kingdom **Prokaryotae**:

____

____

____

____

____

2. Distinguishing features of Kingdom **Protoctista**:

____

____

____

____

____

3. Distinguishing features of Kingdom **Fungi**:

____

____

____

____

____

4. Distinguishing features of Kingdom **Plantae**:

____

____

____

____

____

5. Distinguishing features of Kingdom **Animalia**:

____

____

____

____

____

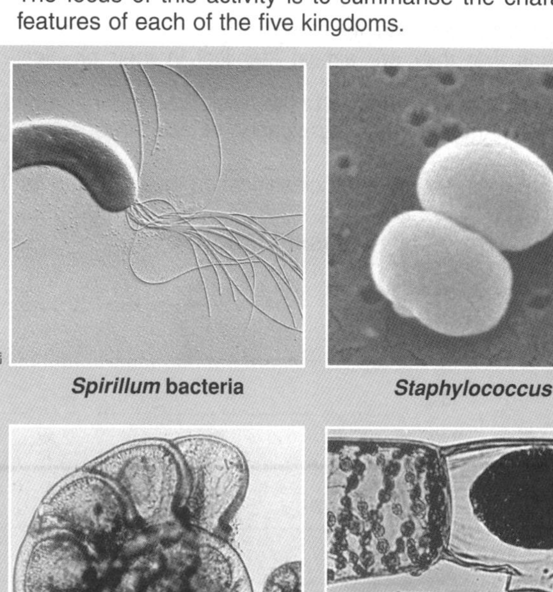

Spirillum bacteria

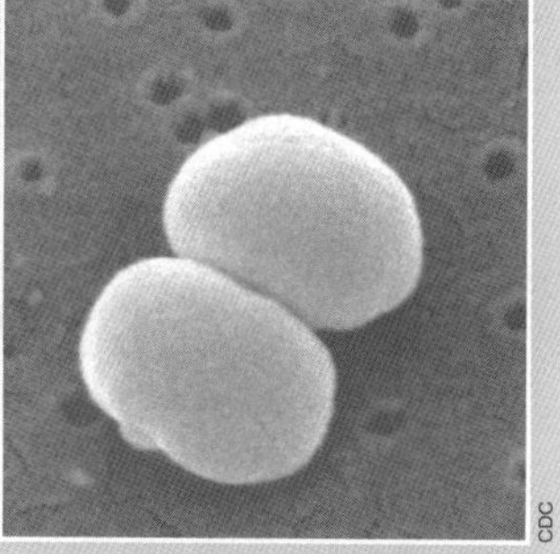

Staphylococcus

Foraminiferan

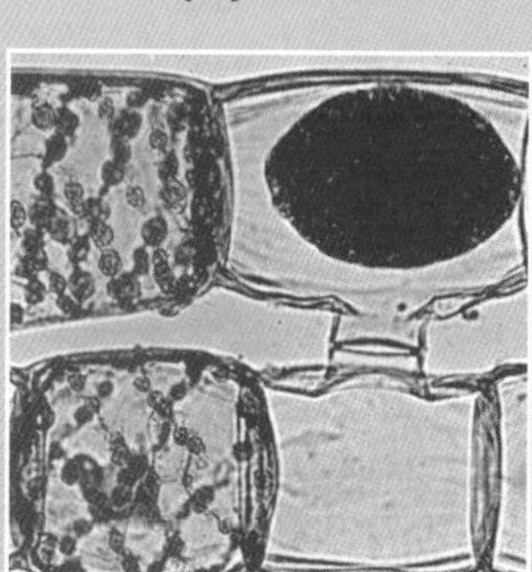

Spirogyra algae

Mushrooms

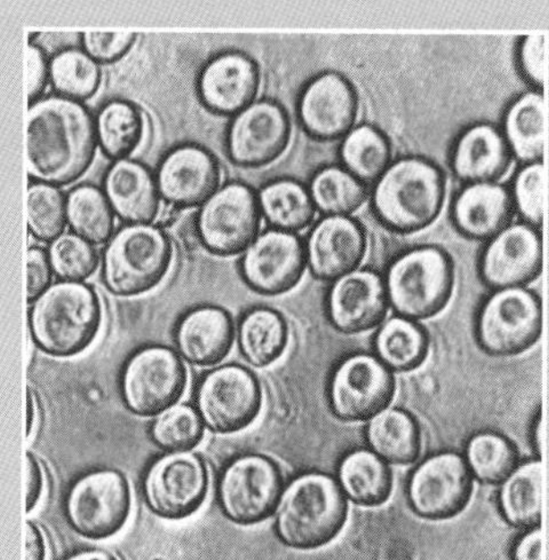

Yeast cells in solution

Moss

Pea plants

Cicada moulting

Gibbon

Classification System

The classification of organisms is designed to reflect how they are related to each other. The fundamental unit of classification of living things is the **species**. Its members are so alike genetically that they can interbreed. This genetic similarity also means that they are almost identical in their physical and other characteristics. Species are classified further into larger, more comprehensive categories (higher taxa). It must be emphasised that all such higher classifications are human inventions to suit a particular purpose.

1. The table below shows part of the classification for humans using the seven major levels of classification. For this question, use the example of the classification of the European hedgehog, on the facing page, as a guide.

   (a) Complete the list of the classification levels on the left hand side of the table below:

	Classification level	Human classification
1.		
2.		
3.		
4.		
5.	Family	Hominidae
6.		
7.		

   (b) The name of the Family that humans belong to has already been entered into the space provided. Complete the classification for humans (*Homo sapiens*) on the table above.

2. Describe the two-part scientific naming system (called the **binomial system**) which is used to name organisms:

3. Give two reasons why the classification of organisms is important:

   (a)

   (b)

4. Traditionally, the classification of organisms has been based largely on similarities in physical appearance. More recently, new methods involving biochemical comparisons have been used to provide new insights into how species are related. Describe an example of a biochemical method for comparing how species are related:

5. As an example of physical features being used to classify organisms, mammals have been divided into three major sub-classes: monotremes, marsupials, and placentals. Describe the main physical feature distinguishing each of these taxa:

   (a) Monotreme:

   (b) Marsupial:

   (c) Placental:

Classification of the European Hedgehog

Below is the classification for the **European hedgehog**. Only one of each group is subdivided in this chart showing the levels that can be used in classifying an organism. Not all possible subdivisions have been shown here. For example, it is possible to indicate such categories as **super-class** and **sub-family**. The only natural category is the **species**, often separated into geographical **races**, or **sub-species**, which generally differ in appearance.

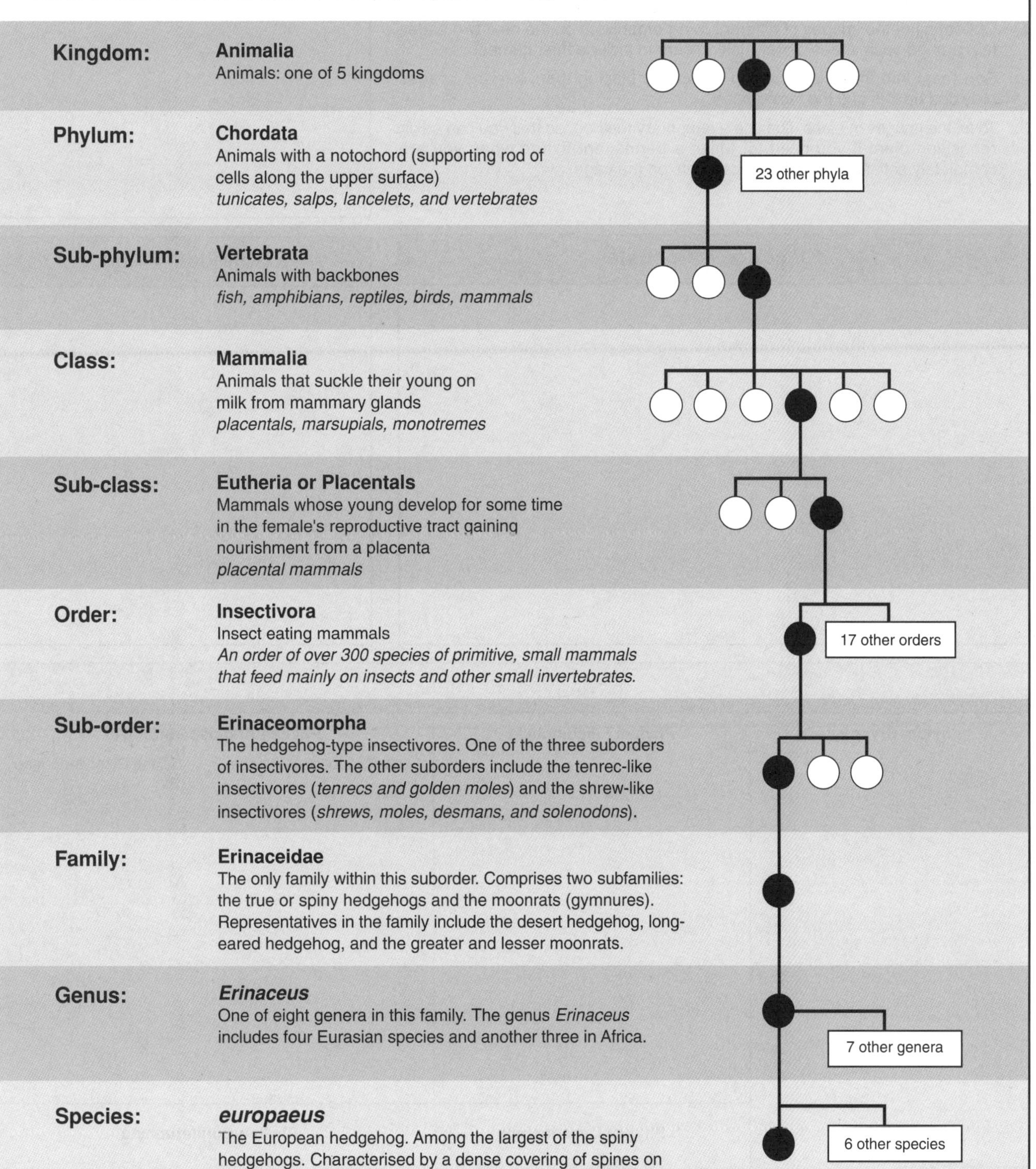

The order *Insectivora* was first introduced to group together shrews, moles, and hedgehogs. It was later extended to include tenrecs, golden moles, desmans, tree shrews, and elephant shrews, and the taxonomy of the group became very confused. Recent reclassification of the elephant shrews and tree shrews into their own separate orders has made the Insectivora a more cohesive group taxonomically.

European hedgehog
Erinaceus europaeus

The Classification of Life

For this activity, cut away the next two pages of diagrams from your book. The five kingdoms that all living things are grouped into, are listed on the two facing pages that are left remaining.

1. Cut out all of the images of different living organisms on the next two pages (cut around each shape closely, taking care to include their names).
2. Sort them into their classification groups by placing them into the spaces provided on this and the facing page.
3. To fix the images in place, first use a temporary method, so that you can easily reposition them if you need to. Make a permanent fixture when you are completely satisfied with your placements on the page.

Kingdom Prokaryotae (Monera)

Kingdom Protoctista

Kingdom Fungi

Kingdom Plantae

Phylum Bryophyta

Phylum Filicinophyta

Phylum Angiospermophyta

Class Monocotyledoneae

Class Dicotyledoneae

Phylum Cycadophyta

Phylum Coniferophyta

This page has been deliberately left blank

Cut out the images
on the other side of this page

Cut out the organisms on this and the facing page and paste them into the spaces provided in the remaining pages of this activity. Organisms are not to scale.

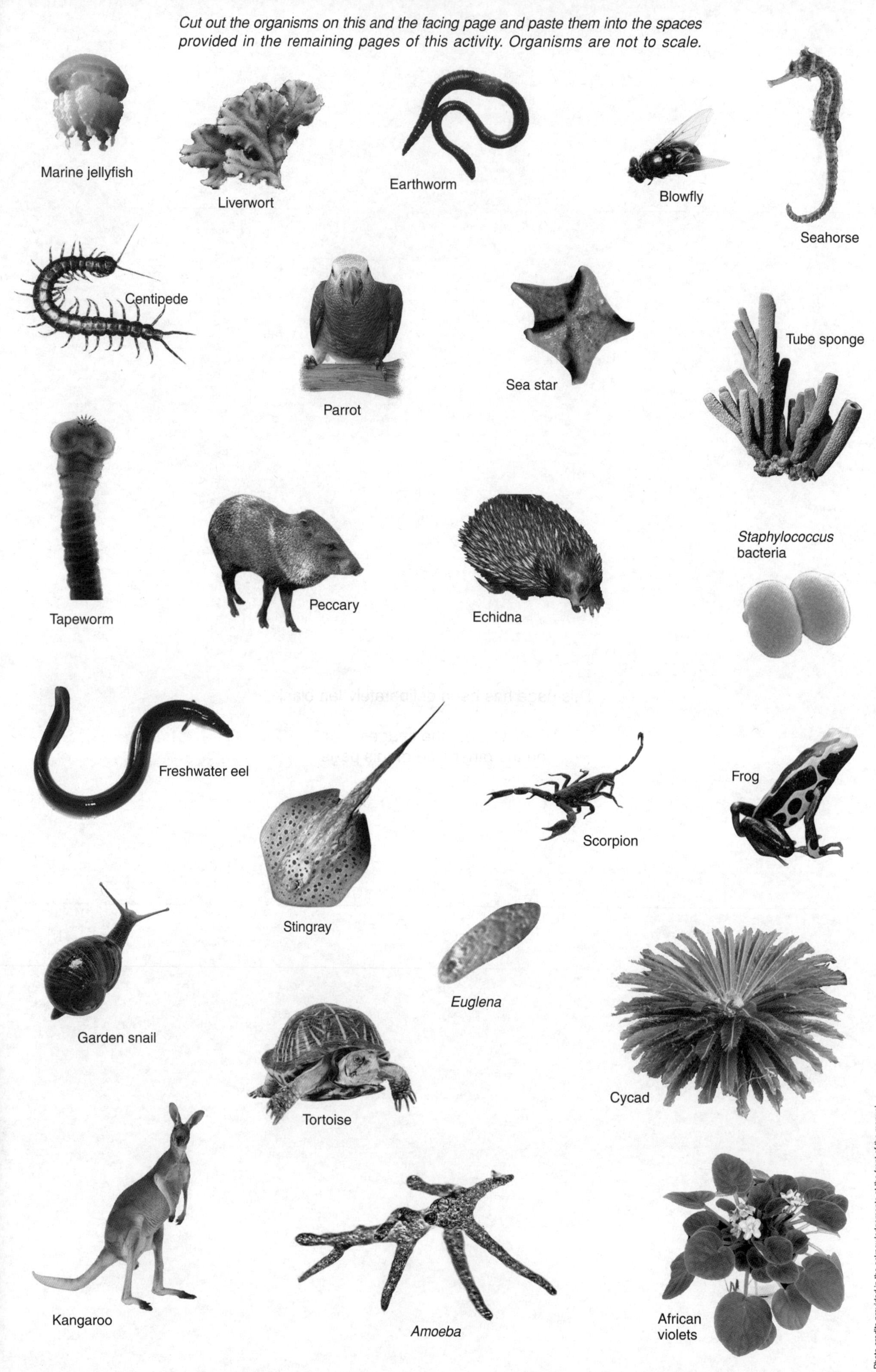

Remove this page by tearing along the perforation

Photo credits provided in the acknowledgements at the front of the manual

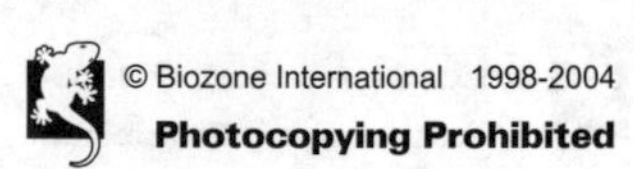

Cut out the organisms on this and the facing page and paste them into the spaces provided in the remaining pages of this activity. Organisms are not to scale.

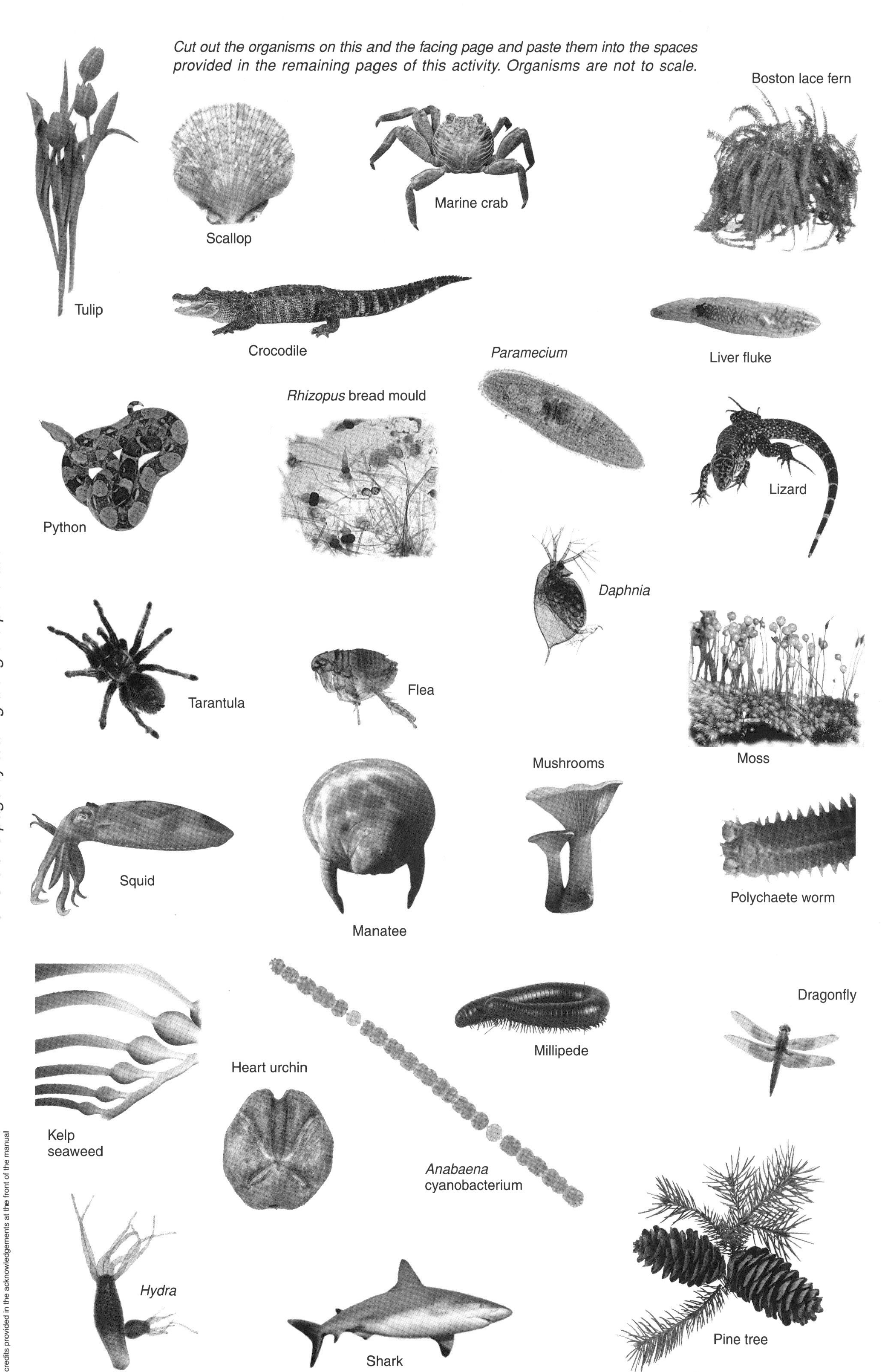

Remove this page by tearing along the perforation

Photo credits provided in the acknowledgements at the front of the manual

This page has been deliberately left blank

Cut out the images
on the other side of this page

Kingdom Animalia			
Phylum Porifera	**Phylum Cnidaria**	**Phylum Platyhelminthes**	**Phylum Annelida**

Phylum Mollusca			
Class Gastropoda	**Class Bivalvia**	**Class Cephalopoda**	**Phylum Echinodermata**

Phylum Arthropoda Superclass Crustacea	Classes Chilopoda/Diplopoda	Class Arachnida	Class Insecta

Class Chondrichthyes	**Phylum Chordata**	Class Osteichthyes	Class Amphibia
Class Reptilia Order Squamata	Order Crocodilia	Order Chelonia	Class Aves

Class Mammalia Subclass Prototheria	Subclass Metatheria	Subclass Eutheria

Classification Keys

Classification systems provide biologists with a way in which to identify species. They also indicate how closely related, in an evolutionary sense, each species is to others. An organism's classification should include a clear, unambiguous **description**, an accurate **diagram**, and its unique name, denoted by the **genus** and **species**. Classification keys are used to identify an organism and assign it to the correct species (assuming that the organism has already been formally classified and is included in the key). Typically, keys use a series of linked questions highlighting contrasting characters. The key is followed until an identification is made. If the organism cannot be identified with the established key, it may be a newly discovered species and the key may need revision. Two examples of classification keys are provided here. The first key (below) describes features for identifying the larvae of various genera within the order Trichoptera (caddisflies). From this key you should be able to assign a generic name to each of the caddisfly larvae pictured. The key opposite identifies aquatic insect orders.

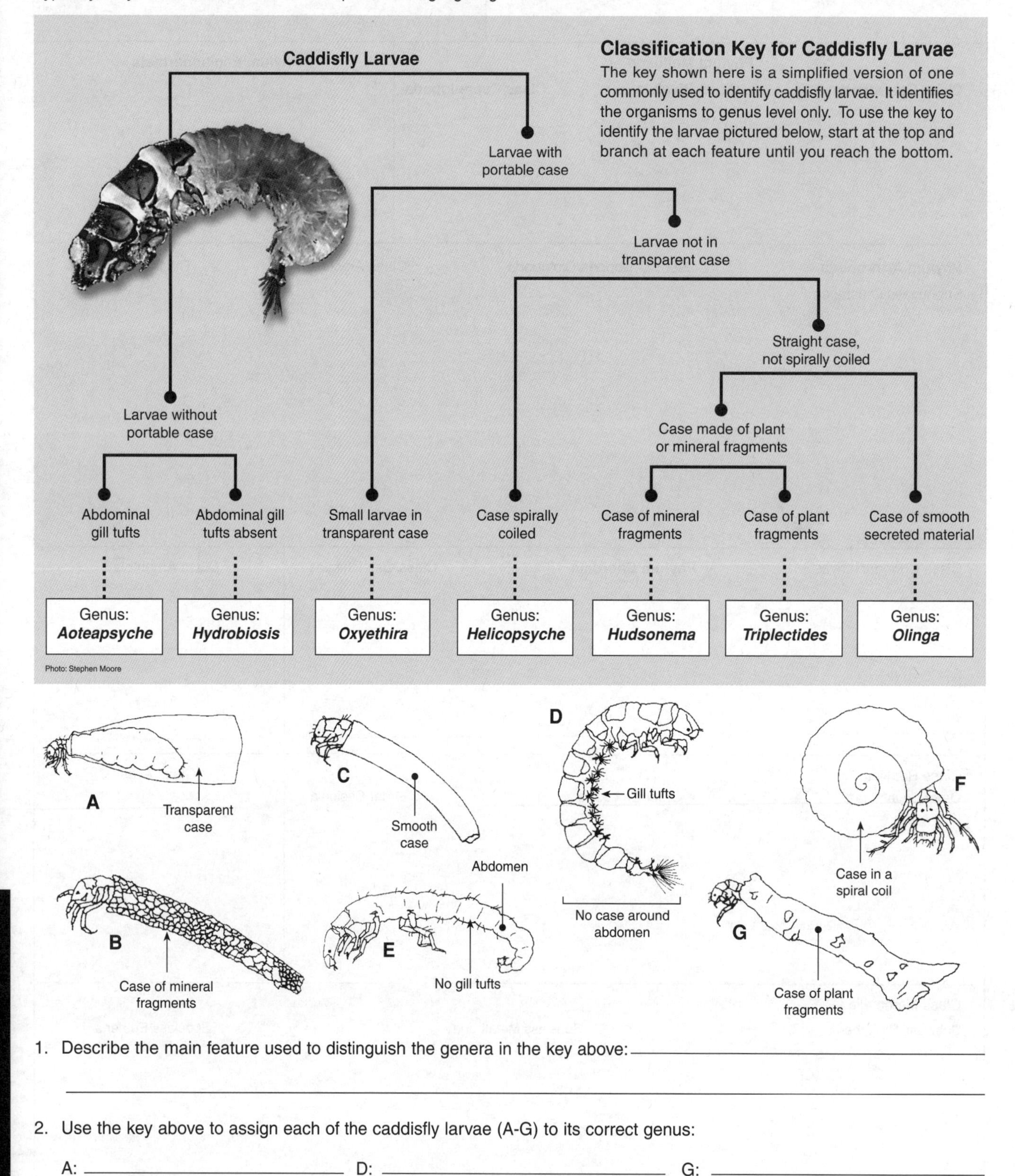

1. Describe the main feature used to distinguish the genera in the key above: ______

2. Use the key above to assign each of the caddisfly larvae (A-G) to its correct genus:

A: ______ D: ______ G: ______

B: ______ E: ______

C: ______ F: ______

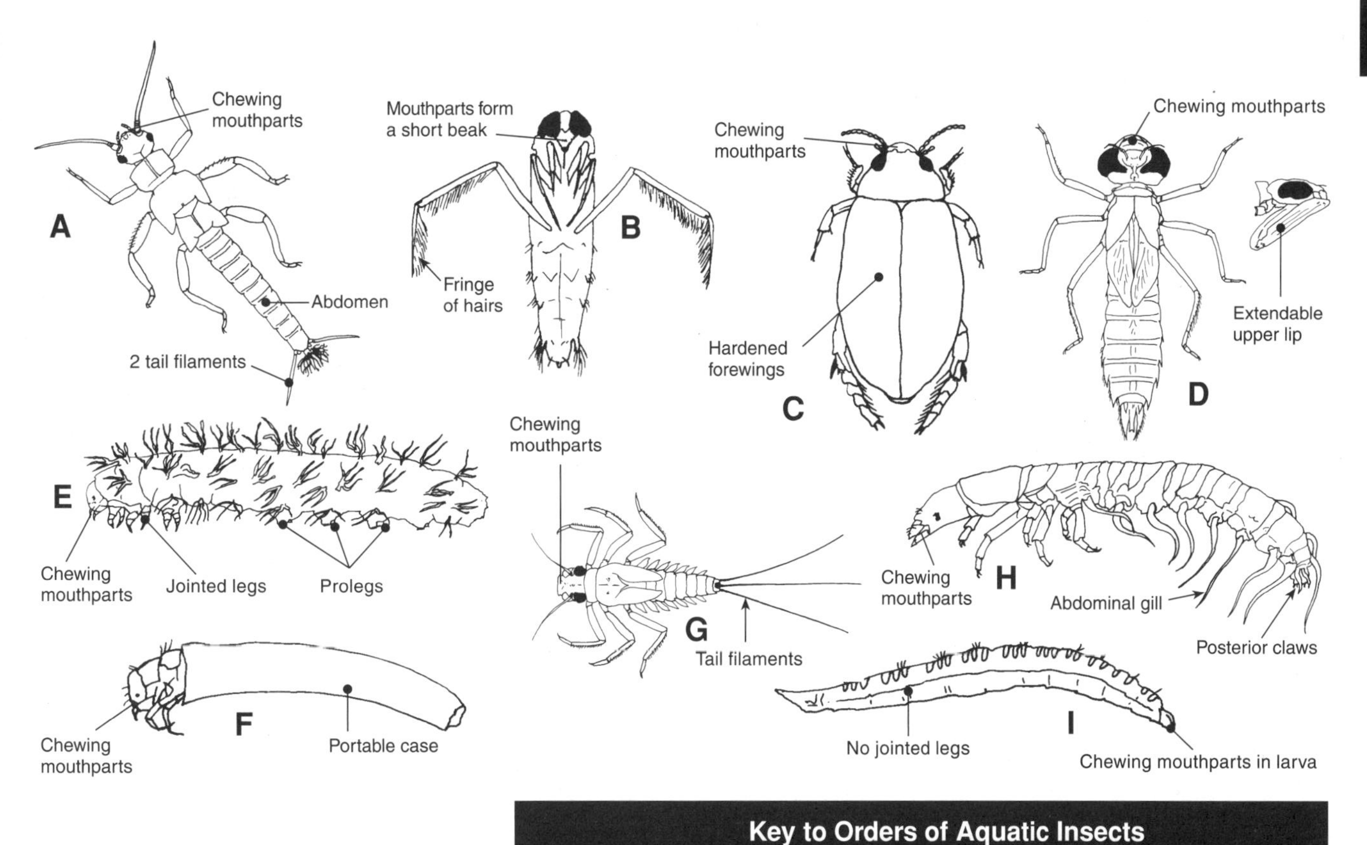

3. Use the simplified key to identify each of the orders (by order or common name) of aquatic insects (A-I) pictured above:

   (a) Order of insect A: ______________________

   (b) Order of insect B: ______________________

   (c) Order of insect C: ______________________

   (d) Order of insect D: ______________________

   (e) Order of insect E: ______________________

   (f) Order of insect F: ______________________

   (g) Order of insect G: ______________________

   (h) Order of insect H: ______________________

   (i) Order of insect I: ______________________

Key to Orders of Aquatic Insects

1	Insects with chewing mouthparts; forewings are hardened and meet along the midline of the body when at rest (they may cover the entire abdomen or be reduced in length).	**Coleoptera** (beetles)
	Mouthparts piercing or sucking and form a pointed cone With chewing mouthparts, but without hardened forewings	*Go to 2* *Go to 3*
2	Mouthparts form a short, pointed beak; legs fringed for swimming or long and spaced for suspension on water.	**Hemiptera** (bugs)
	Mouthparts do not form a beak; legs (if present) not fringed or long, or spaced apart.	*Go to 3*
3	Prominent upper lip (labium) extendable, forming a food capturing structure longer than the head.	**Odonata** (dragonflies & damselflies)
	Without a prominent, extendable labium	*Go to 4*
4	Abdomen terminating in 3 tail filaments which may be long and thin, or with fringes of hairs.	**Ephemeroptera** (mayflies)
	Without 3 tail filaments	*Go to 5*
5	Abdomen terminating in 2 tail filaments	**Plecoptera** (stoneflies)
	Without long tail filaments	*Go to 6*
6	With 3 pairs of jointed legs on thorax	*Go to 7*
	Without jointed, thoracic legs (although non-segmented prolegs or false legs may be present).	**Diptera** (true flies)
7	Abdomen with pairs of non-segmented prolegs bearing rows of fine hooks.	**Lepidoptera** (moths and butterflies)
	Without pairs of abdominal prolegs	*Go to 8*
8	With 8 pairs of finger-like abdominal gills; abdomen with 2 pairs of posterior claws.	**Megaloptera** (dobsonflies)
	Either, without paired, abdominal gills, or, if such gills are present, without posterior claws.	*Go to 9*
9	Abdomen with a pair of posterior prolegs bearing claws with subsidiary hooks; sometimes a portable case.	**Trichoptera** (caddisflies)

Homeostasis

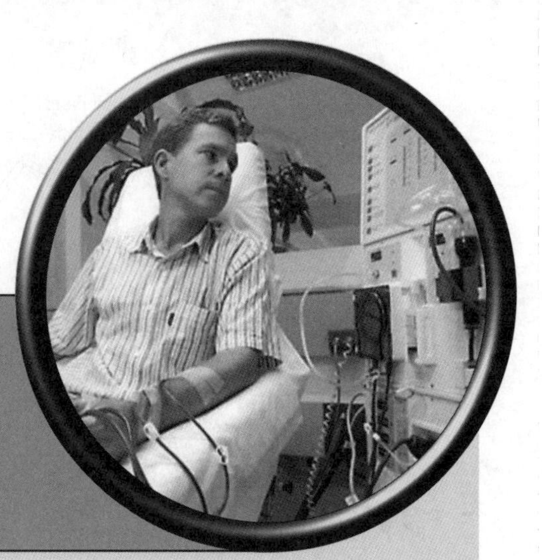

AQA-A	AQA-B	CIE	Edexcel	OCR
Complete:	Complete:	Complete:	Complete:	Complete:
1-9, 10, 12-14, 16, 18-19, 21-22, 24-25, 28-30, 32	*1-8, 12-13, 16(a)-(b), 18, 22, 24-25, 28-30, 32*	*1-8, 10-12, 14-18, 22-24, 27-30, 32*	*1-8, 10, 12, 16(a)-(b), 18, 24-25, 28-30, 32*	*1-8, 10-12, 14-18, 22-24, 27-30, 32*

Learning Objectives

- ☐ 1. Compile your own glossary from the **KEY WORDS** displayed in **bold type** in the learning objectives below.

Principles of homeostasis *(pages 214-217, 222)*

- ☐ 2. Explain what is meant by **homeostasis**. Understand the role of homeostasis in providing independence from the fluctuating external environment. Identify factors that require regulation in order for an organism to maintain a steady state.
- ☐ 3. Explain, using examples, the principle of **negative feedback**, identifying how it stabilises systems against excessive change. Explain the role of **receptors**, **effectors**, and negative feedback in homeostasis.
- ☐ 4. Appreciate the interdependence and general roles of the two regulatory systems (hormonal and nervous) with which mammals achieve homeostasis.

Nervous regulation *(page 218)*

- ☐ 5. Outline the general structure of a **nervous system**. Relate this structure to the way in which animals receive **stimuli** and generate a **response**.
- ☐ 6. Understand the principles of nerve impulse transmission. Explain the importance of negative feedback in nervous systems. Contrast the speed of nervous and endocrine responses (see #8).

Hormonal regulation *(pages 219-221)*

- ☐ 7. Explain what is meant by the terms: **endocrine gland**, **hormone**, and **target tissue**. Describe the general structure and role of the **endocrine system** in the maintenance of homeostasis. Appreciate the role of the **hypothalamus** and **pituitary** in homeostasis.
- ☐ 8. Understand how hormones exert their effects and why they are able to have wide-ranging effects on the body. Explain the role of negative feedback in the control of hormone release. Contrast the speed of hormonal and nervous responses (see #6).

Case studies in homeostasis

- ☐ 9. Demonstrate your understanding of homeostasis by discussing the control of body temperature (see the activities on the "Teacher Resource Handbook") or the control of blood glucose. Consider the following:
  - The role of negative feedback
  - The greater control gained by the possession of separate mechanisms to regulate different departures from the stable state.
  - The requirement to coordinate control mechanisms.

Control of blood glucose *(pages 222-224)*

- ☐ 10. Understand the factors that lead to variation in blood glucose levels. Understand the normal range over which blood glucose levels fluctuate.
- ☐ 11. Describe the cellular structure of an **islet of Langerhans** from the pancreas, and outline the role of the pancreas as an **endocrine organ**.
- ☐ 12. Explain, using a diagram, how the regulation **blood glucose** level is achieved in humans, including reference to the role of the following (as required):
  - (a) Negative feedback mechanisms.
  - (b) The hormones **insulin** and **glucagon**.
  - (c) The role of the liver in glucose-glycogen conversions.
  - (d) The hormone adrenaline.
- ☐ 13. Understand how the actions of insulin and glucagon are mediated, with reference to:
  - The role of membrane receptors
  - The effect of hormone activation on the enzyme controlled reactions that alter blood glucose level.
- ☐ 14. Describe the causes and symptoms of diabetes mellitus (Types I and II). Describe the control of diabetes through the use of insulin injection and regulation of carbohydrate intake.
- ☐ 15. Explain the advantages of treating diabetes with human insulin produced using **genetic engineering**. Identify the source of insulin for diabetics in the past and describe problems associated with it.

The homeostatic role of the liver *(pages 224-225)*

- ☐ 16. Appreciate the homeostatic role of the liver, including:
  - (a) Its role in **carbohydrate metabolism** and the production of glucose from amino acids.
  - (b) Its role in protein metabolism: **deamination**, **transamination**, and the formation of **urea**.
  - (c) Its role in **fat metabolism**, including the use of fats in respiration, the synthesis of triglycerides and cholesterol, and the transport of lipids.
- ☐ 17. Describe the gross structure and histology of the liver. Identify the following: liver lobule, hepatocytes, portal triad, Küpffer cells, sinusoids, and central vein.

Excretion and water balance in animals

- ☐ 18. Clearly explain what is meant by the terms **excretion** and **osmoregulation**. Appreciate why organisms must regulate water balance and dispose of the toxic waste products of metabolism.

Osmoregulation *(pages 228-229)*

- ☐ 19. Describe the different homeostatic problems associated with living in salt and fresh water. Distinguish between **osmoregulators** and **osmoconformers**, and provide examples of each.

☐ 20. Describe, with examples, some of the mechanisms by which animals in fresh and salt water regulate their water and ion balance.

Water budget in a desert mammal *(page 230)*

☐ 21. Describe the control of water budget in a desert rodent with reference to:
- Physiological mechanisms that minimise water loss.
- How water requirements are met without drinking.
- Behavioural adaptations that reduce need for water.

Excretion *(pages 226-227, 231-235)*

☐ 22. Identify the major **nitrogenous waste products** excreted by animals and identify their origin. Appreciate how the excretory product is related to the life history and the environment of the animal.

☐ 23. Identify the non-nitrogenous excretory products in mammals. Name the organs involved in their disposal.

☐ 24. The mammalian **kidney** provides a good example of a homeostatic organ. Outline two main functions of the **kidney** in the regulation of body fluids in mammals.

☐ 25. Describe what happens in deamination of amino acids and the subsequent production of urea. Identify the name of this metabolic cycle and where it takes place (see #16b). Identify the portion of the amino acid that is used as a respiratory substrate.

☐ 26. On a diagram, identify the main structures of the mammalian urinary system: kidneys, ureters, renal arteries and veins, bladder, urethra.

☐ 27. Describe the gross structure of the mammalian kidney to include the cortex, medulla, and renal pelvis. Recognise and interpret features of the histology of the kidney from sections viewed with a light microscope.

☐ 28. Using a labelled diagram, describe the structure and arrangement of a nephron and its associated blood vessels in relation to its function in producing urine. Include reference to **glomerulus**, proximal and distal **convoluted tubules**, and **collecting duct**.

☐ 29. Explain concisely how the kidney nephron produces urine. Include reference to:
(a) The process of **ultrafiltration** in the **glomerulus**
(b) The ultrastructure of the glomerulus and **renal capsule** in relation to ultrafiltration.
(c) The **selective reabsorption** of water and solutes in the **proximal convoluted tubule**.
(d) The ultrastructure of the proximal convoluted tubule in relation to reabsorption.
(e) The role of the **loop of Henle** and the **counter-current multiplier system** in creating and maintaining the ionic (salt) gradient in the kidney.
(f) The role of the ionic gradient in the kidney in producing a concentrated urine and in fluid balance.

☐ 30. Explain how the water and solute content of the blood is controlled. Include reference to the role of **osmoreceptors** in the hypothalamus, release of antidiuretic hormone (**ADH**) from the **posterior pituitary**, the action of the ADH on the kidney, and the role of **negative feedback** in regulating ADH output.

☐ 31. Recognise the role of **aldosterone** in promoting sodium reabsorption in the kidney

☐ 32. EXTENSION ONLY: In cases of renal (kidney) failure, renal dialysis may take over the role of the kidneys. Explain under what circumstances **kidney dialysis** is used and what is involved in the process.

See the 'Textbook Reference Grid' on pages 8-9 for textbook page references relating to material in this topic.

Supplementary Texts

See pages 4-6 for additional details of these texts:

■ Clegg, C.J., 1998. **Mammals: Structure and Function** (John Murray), pp. 42-57, 70-71.

■ Jones, M. and G. Jones, 2002. **Mammalian Physiology & Behaviour**, (CUP), pp. 17-28 (liver).

■ Rowett, H., 1999. **Basic Anatomy & Physiology**, (John Murray), pp. 62, 70, 91-92, 99, 117-123.

See page 6 for details of publishers of periodicals:

STUDENT'S REFERENCE

■ **Homeostasis** Biol. Sci. Rev., 12(5) May 2000, pp. 2-5. *Homeostasis: what it is, how it is achieved through feedback mechanisms, the involvement of the autonomic nervous system, and adaptations for homeostasis in different environments.*

■ **Metabolic Powerhouse** New Scientist, 11 November 2000 (Inside Science). *The myriad roles of the liver in metabolic processes, including discussion of amino acid and glucose metabolism.*

■ **Diabetes** Biol. Sci. Rev., 15(2), Nov. 2002, pp. 30-35. *The homeostatic imbalance that results in diabetes. The role of the pancreas in the hormonal regulation of blood glucose is discussed.*

■ **The Liver in Health and Disease** Biol. Sci. Rev., 14(2) Nov. 2001, pp. 14-20. *The various roles of the liver: production of bile, and metabolism of protein, lipids, carbohydrates, and drugs.*

■ **The Kidney** Biol. Sci. Rev., 16(2) Nov. 2003, pp. 2-6. *The structure of the kidneys, and their essential role in regulating extracellular fluid volume, blood pressure, acid-base balance, and metabolic waste products such as urea. The operation of the kidney nephron as a countercurrent multiplier is also discussed.*

■ **A Fair Exchange** Biol. Sci. Rev., 13(1), Sept. 2000, pp. 2-5. *Formation and reabsorption of tissue fluid (includes disorders of fluid balance).*

■ **Nitrogen Excretion in Animals** Biol. Sci. Rev., 8(5) May 1996, pp. 27-31. *Excretory products in animals, including a discussion of the urea cycle.*

■ **Uric Acid - Life Saver and Liability** Biol. Sci. Rev., 9(1) Sept. 1996, pp. 22-24. *Nitrogen excretion and the situations in which uric acid is produced.*

■ **Countercurrent Exchange Mechanisms** Biol. Sci. Rev., 9(1) Sept. 1996, pp. 2-6. *The role of countercurrent multipliers in biological systems: including operation in the kidney nephron.*

■ **Basement Membranes** Biol. Sci. Rev., 13(4) March 2001, pp. 36-39. *The structure, function, and diversity of basement membranes, with an account of their structural role in the glomerulus.*

TEACHER'S REFERENCE

■ **The Heart as an Endocrine Organ** Biologist, 49(6) December 2002. *The role of the kidney in sodium and water excretion is well known, but the heart has an important role in salt and water regulation. A peptide released from the heart muscle in response to high blood pressure signals to the kidneys to excrete more salt and water.*

■ **Chemical Toxins and Body Defences** Biologist, 49(1) February 2002. *The various and fascinating ways in which mammals are able to eliminate toxic substances from their bodies.*

See pages 10-11 for details of how to access **Bio Links** from our web site: **www.biozone.co.uk**. From Bio Links, access sites under the topics:

GENERAL BIOLOGY ONLINE RESOURCES > Online Textbooks and Lecture Notes • An on-line biology book • Biology online.org • Gondar design sciences • Kimball's biology pages • Learn.co.uk • Nova: science in the news • Welcome to the biology web ... *and others*

ANIMAL BIOLOGY: • Anatomy and physiology • Comparative vertebrate anatomy lecture notes • Human physiology lecture notes • Frog dissection for biology 110 • FROGUTS.COM • Insect mouthparts • Insects on the web • Human physiology lecture notes • Netfrog ... *and others* > **Excretion**: • Comparative physiology of vertebrate kidneys • Excretory system • The kidney • Urinary system > **Homeostasis**: • Ask the experts: What is homeostasis? • Homeostasis • Homeostasis: general principles • Physiological homeostasis

Software and video resources are provided on the Teacher Resource Handbook on CD-ROM

Principles of Homeostasis

Homeostasis is the condition where the body's internal environment remains relatively constant, within narrow limits. For the body's cells to survive and function properly, the composition and temperature of the fluids around the cells must remain much the same. An organism is said to be in homeostasis when the internal environment contains the optimal concentration of gases, nutrients, ions and water, at the optimal temperature. **Negative feedback mechanisms** are involved in the control of homeostasis. Feedback mechanisms provide information about the state of a system to its control centre. In negative feedback, movement away from an ideal state causes a return back to the ideal state (the set point). The intensity of the corrective action is reduced as the system returns to this set point. Using such control systems the body acts to counteract disturbances and restore homeostasis. The system operates through a combination of nervous and hormonal mechanisms (see below).

Negative feedback and regulatory control systems

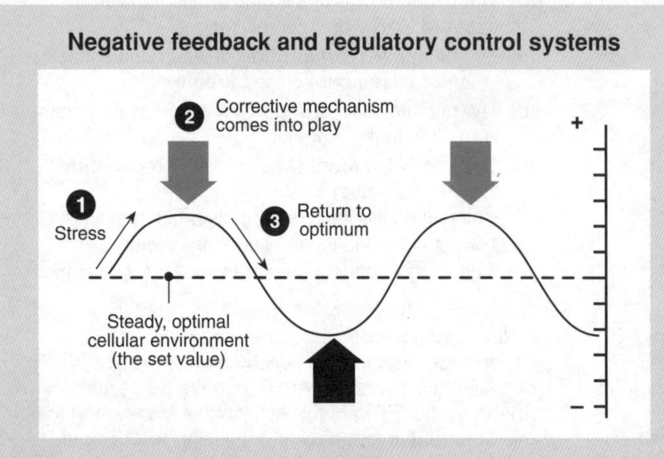

1. A stress or disturbance takes the internal environment away from optimum
2. Stress is detected by receptors and corrective mechanisms are activated
3. The corrective mechanisms act to restore conditions back to the set value

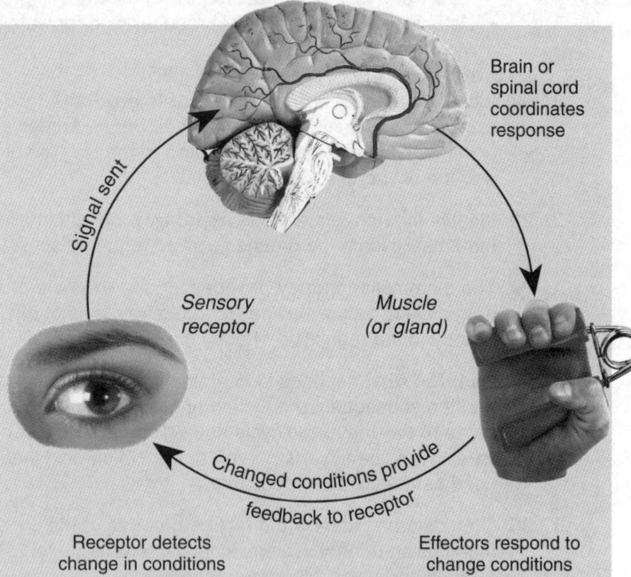

Negative feedback acts to eliminate any deviation from preferred conditions. It is part of almost all the control systems in living things. The diagram (above left) shows how a stress or disturbance is counteracted by corrective mechanisms that act to restore conditions back to an optimum value. The diagram (above right) illustrates this principle for a biological system.

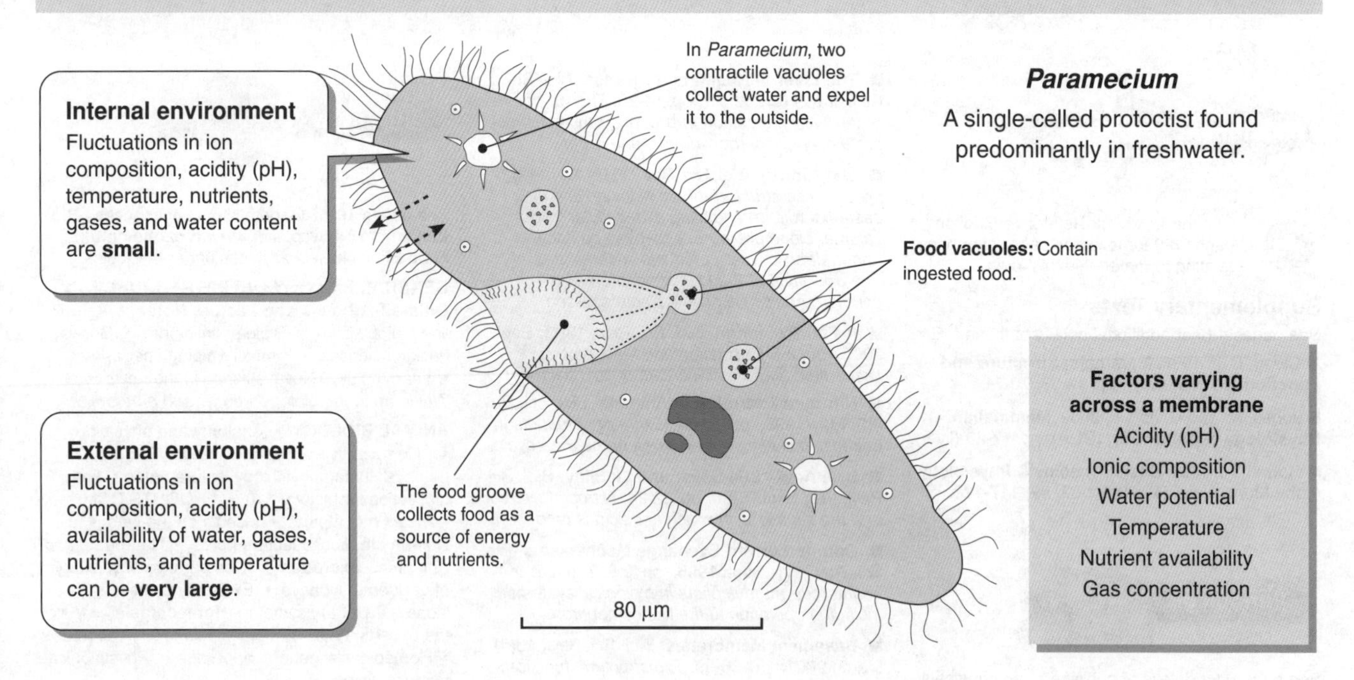

1. Identify the three main components of a regulatory control system in the human body: ______

______

2. Briefly describe the effect of negative feedback mechanisms in biological systems: ______

______

______

______

Detecting Changing States

A stimulus is any physical or chemical change in the environment capable of provoking a response in an organism. Animals respond to stimuli in order to survive. This response is adaptive; it acts to maintain the organism's state of homeostasis. Stimuli may be either external (outside the organism) or internal (within its body). Some of the stimuli to which humans and other mammals respond are described below, together with the sense organs that detect and respond to these stimuli. Note that sensory receptors respond only to specific stimuli. The sense organs an animal possesses therefore determine how it perceives the world.

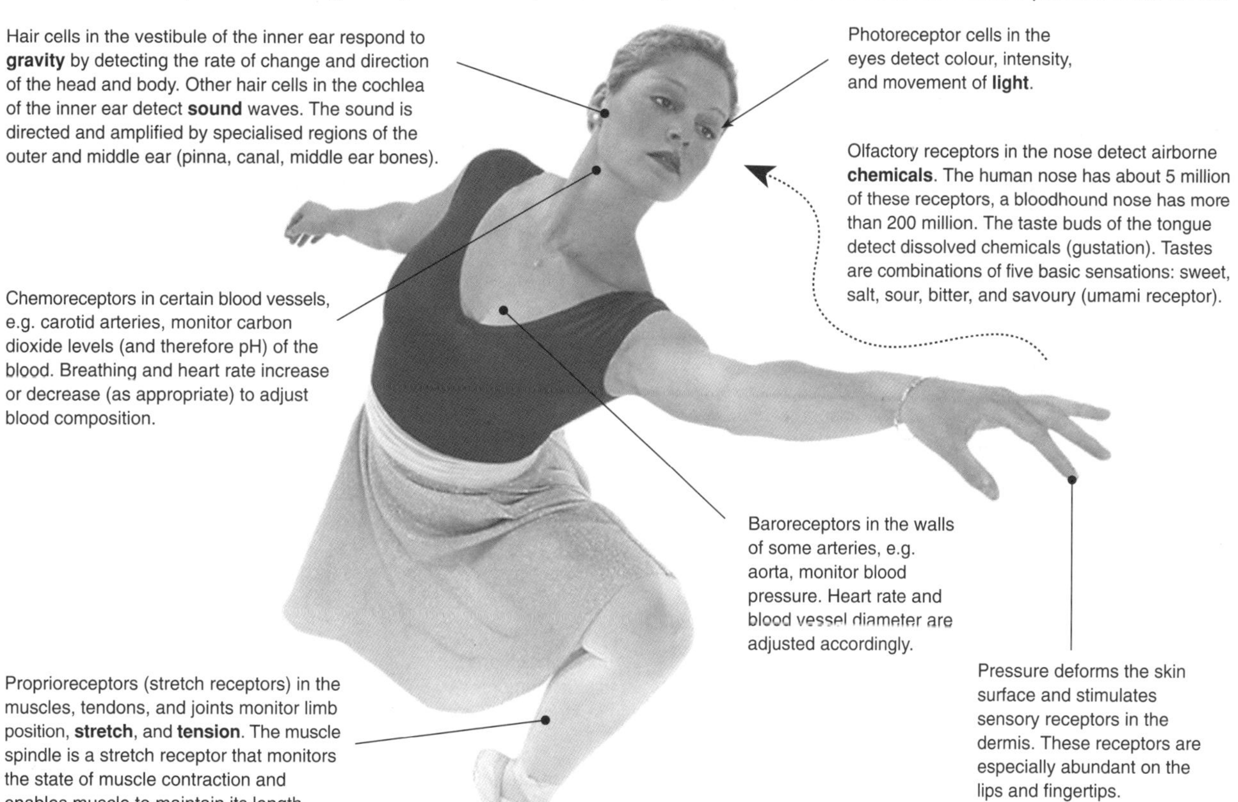

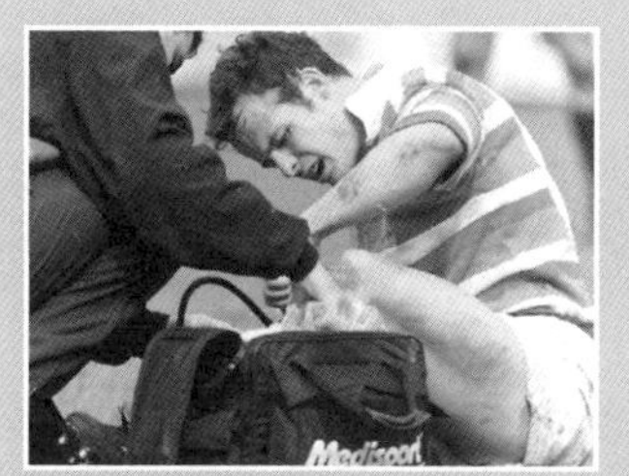

Pain and temperature are detected by simple nerve endings in the skin. Deep tissue injury is sometimes felt on the skin as referred pain.

Humans rely heavily on their hearing when learning to communicate; without it, speech and language development are more difficult.

Breathing and heart rates are regulated in response to sensory input from chemoreceptors.

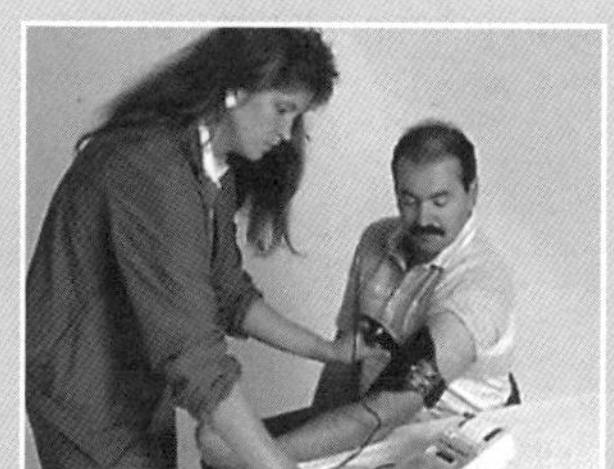

Baroreceptors and osmoreceptors act together to keep blood pressure and volume within narrow limits.

1. Provide a concise definition of a stimulus: ______________________________

____

2. Using humans as an example, discuss the need for communication systems to respond to changes in the environment:

____

____

____

3. (a) Name one internal stimulus and its sensory receptor: ______________________________

(b) Describe the role of this sensory receptor in contributing to homeostasis: ______________________________

____

____

Maintaining Homeostasis

The various organ systems of the body act to maintain homeostasis through a combination of hormonal and nervous mechanisms. In everyday life, the body must regulate respiratory gases, protect itself against agents of disease (pathogens), maintain fluid and salt balance, regulate energy and nutrient supply, and maintain a constant body temperature. All these must be coordinated and appropriate responses made to incoming stimuli. In addition, the body must be able to repair itself when injured and be capable of reproducing (leaving offspring).

Regulating Respiratory Gases

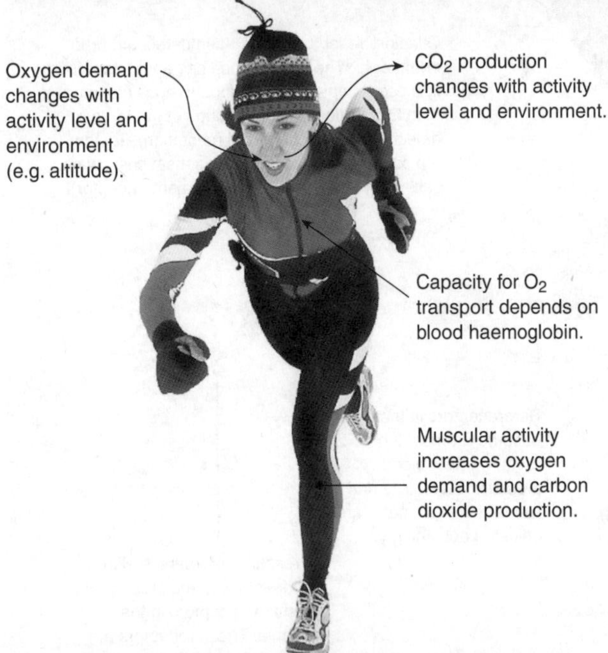

Oxygen must be delivered to all cells and carbon dioxide (a waste product of cellular respiration) must be removed. Breathing (inhalation and exhalation) brings in oxygen and expels CO_2. The rate of breathing is varied according to the oxygen requirement. Both gases are transported around the body in the blood; the oxygen mostly bound to haemoglobin.

Coping with Pathogens

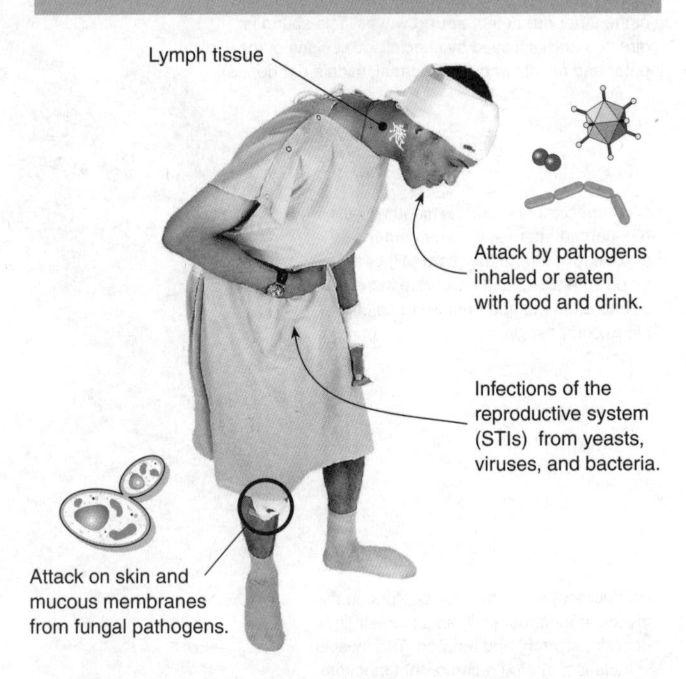

All of us are under constant attack from pathogens (disease causing organisms). The body has a number of mechanisms that help to prevent the entry of pathogens and limit the damage they cause if they do enter the body. The skin, the digestive system and the immune system are all involved in limiting damage.

Maintaining Nutrient Supply

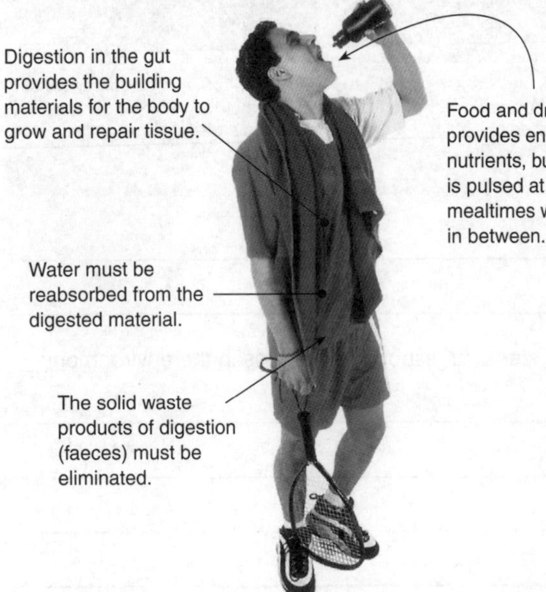

Food and drink must be taken in to maintain the body's energy supplies. Steady levels of energy (as glucose) is available to cells through hormonal regulation of blood sugar levels. Insulin, released by the endocrine cells of the pancreas, causes cells to take up glucose after a meal. Glucagon causes the release of glucose from the liver.

Repairing Injuries

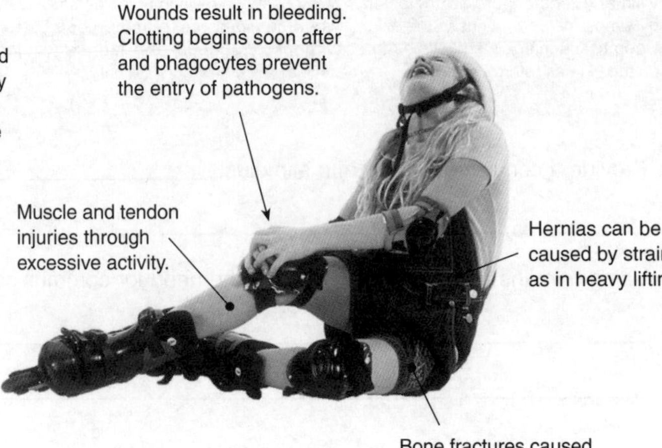

Damage to body tissues triggers the inflammatory response. There is pain, swelling, redness, and heat. Phagocytes and other white blood cells move to the injury site. The inflammatory response is started (and ended) by chemical signals (e.g. from histamine and prostaglandins) released when tissue is damaged.

Maintaining Fluid and Ion Balance

Coordinating Responses

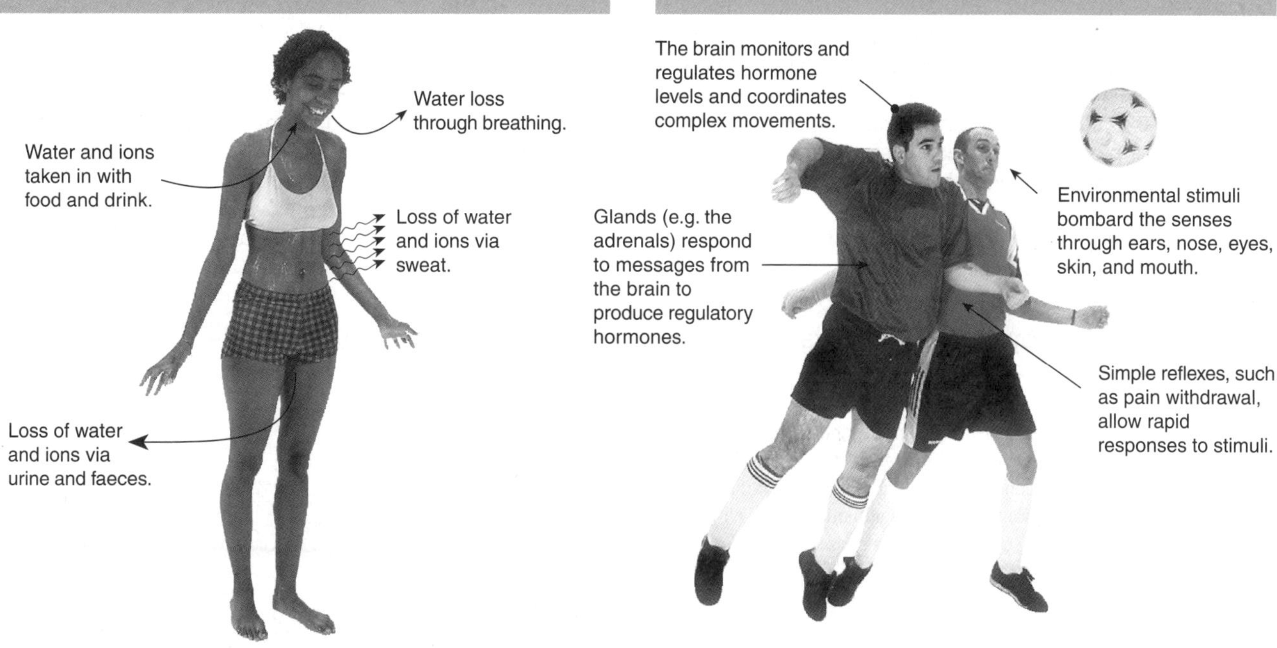

The levels of water and ions in the body are maintained mainly by the kidneys, although the skin is also important. Osmoreceptors monitor the fluid and ion levels of the blood and bring about the release of regulatory hormones; the kidneys regulate reabsorption of water and sodium from blood in response to levels of the hormones ADH and aldosterone.

The body is constantly bombarded by stimuli from the environment. The brain sorts these stimuli into those that require a response and those that do not. Responses are coordinated via nervous or hormonal controls. Simple nervous responses (reflexes) act quickly. Hormonal responses take longer to produce a response and the response is more prolonged.

1. Describe two mechanisms that operate to restore homeostasis after infection by a pathogen:

   (a) ______________________________

   (b) ______________________________

2. Describe two mechanisms by which responses to stimuli are brought about and coordinated:

   (a) ______________________________

   (b) ______________________________

3. Explain two ways in which water and ion balance are maintained. Name the organ(s) and any hormones involved:

   (a) ______________________________

   (b) ______________________________

4. Explain two ways in which the body regulates its respiratory gases during exercise:

   (a) ______________________________

   (b) ______________________________

Nervous Regulatory Systems

An essential feature of living organisms is their ability to coordinate their activities. In multicellular animals, such as mammals, detecting and responding to environmental change, and regulating the internal environment (homeostasis) is brought about by two coordinating systems: the nervous and endocrine systems. Although structurally these two systems are quite different, they frequently interact to coordinate behaviour and physiology. The nervous system contains cells called neurones (or nerve cells). Neurones are specialised to transmit information in the form of electrochemical impulses (action potentials). The nervous system is a signalling network with branches carrying information directly to and from specific target tissues. Impulses can be transmitted over considerable distances and the response is very precise and rapid. Whilst it is extraordinarily complex, comprising millions of neural connections, its basic plan (below) is quite simple. Further detail on nervous system structure and function is provided in the following topic "*Responses and Coordination*".

Coordination by the Nervous System

The vertebrate nervous system consists of the central nervous system (brain and spinal cord), and the nerves and receptors outside it (peripheral nervous system). Sensory input to receptors comes via stimuli. Information about the effect of a response is provided by feedback mechanisms so that the system can be readjusted. The basic organisation of the nervous system can be simplified into a few key components: the sensory receptors, a central nervous system processing point, and the effectors which bring about the response (below):

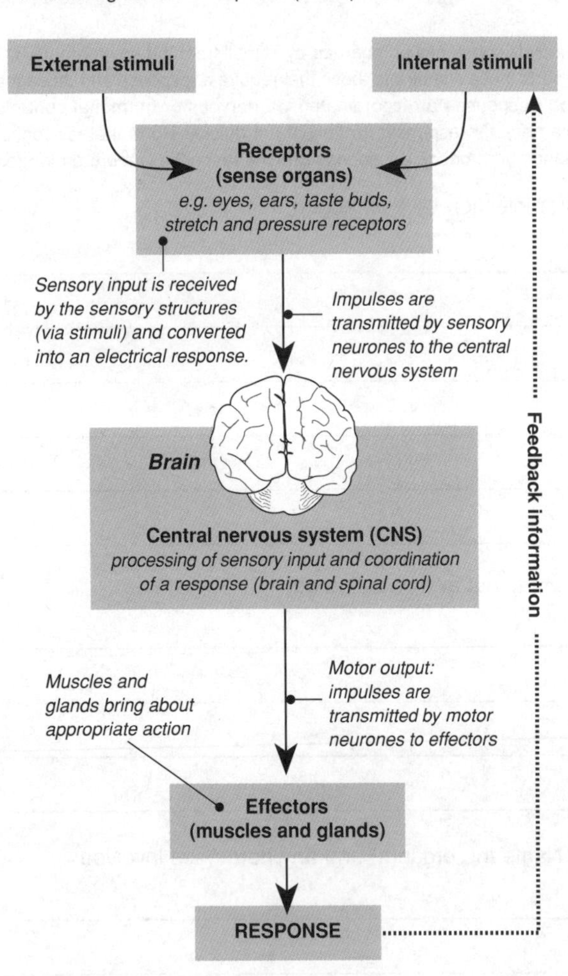

In the example above, the approach of the frisbee is perceived by the eye. The motor cortex of the brain integrates the sensory message. Coordination of hand and body orientation is brought about through motor neurones to the muscles.

Comparison of nervous and hormonal control

	Nervous control	Hormonal control
Communication	*Impulses across synapses*	*Hormones in the blood*
Speed	*Very rapid (within a few milliseconds)*	*Relatively slow (over minutes, hours, or longer)*
Duration	*Short term and reversible*	*Longer lasting effects*
Target pathway	*Specific (through nerves) to specific cells*	*Hormones broadcast to target cells everywhere*
Action	*Causes glands to secrete or muscles to contract*	*Causes changes in metabolic activity*

1. Identify the three basic components of a nervous system and explain how they function to maintain homeostasis:

2. Describe two differences between nervous control and endocrine control of body systems:

(a)

(b)

Hormonal Regulatory Systems

The endocrine system regulates the body's processes by releasing chemical messengers (hormones) into the bloodstream. Hormones are potent chemical regulators: they are produced in minute quantities yet can have a large effect on metabolism. The endocrine system comprises endocrine cells (organised into endocrine glands), and the hormones they produce. Unlike exocrine glands (e.g. sweat and salivary glands), endocrine glands are ductless glands, secreting hormones directly into the bloodstream rather than through a duct or tube. Some organs (e.g. the pancreas) have both endocrine and exocrine regions, but these are structurally and functionally distinct. The basis of hormonal control is described below. The basis of hormonal control is described below. Further examples of hormone action and regulation of hormone levels through negative feedback (e.g. in regulation of blood glucose level) are provided throughout this topic.

The Mechanism of Hormone Action

Endocrine cells produce hormones (chemical messengers) and secrete them into the bloodstream where they are distributed throughout the body. Although hormones are circulated throughout the body, they affect only specific **target cells**. These target cells have receptors on the plasma membrane that recognise and bind the hormone (see right). The binding of the hormone and its receptor triggers the response in the target cell. Cells are unresponsive to a hormone if they do not have the appropriate receptors. Hormones exert their effects at the cellular level through activation of a **second messenger** (right). In non-steroid hormones, such as insulin, cyclic AMP (CAMP) links the hormone to the cellular response. Cellular concentration of CAMP increases markedly once a hormone binds and the cascade of enzyme-driven reactions is initiated.

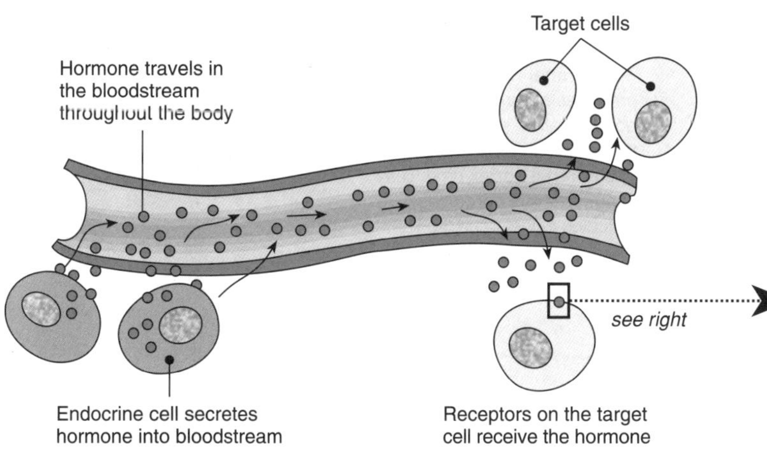

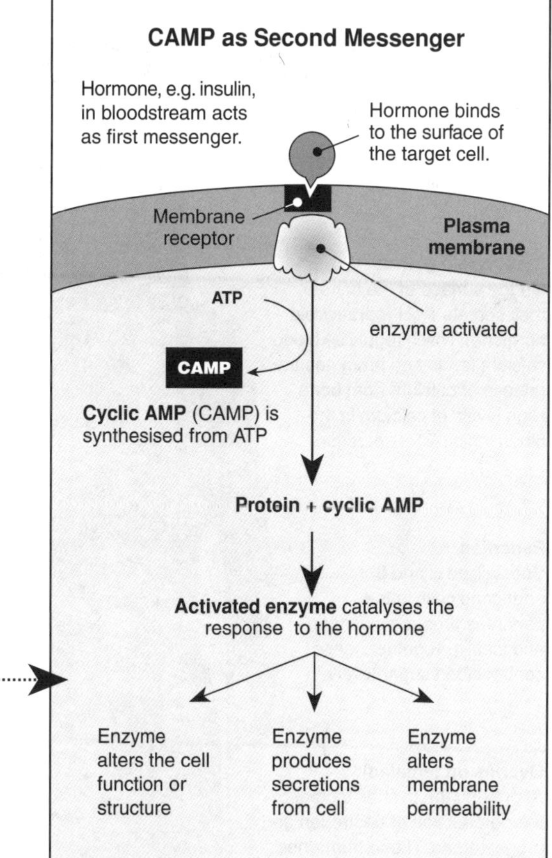

1. (a) Explain what is meant by a **hormone**:

   (b) Explain the role of feedback mechanisms in adjusting hormone levels (explain using an example if this is helpful):

2. Explain how a hormone can bring about a response in target cells even though all cells may receive the hormone:

3. Explain why hormonal control differs from nervous system control with respect to the following:

   (a) The speed of hormonal responses is slower:

   (b) Hormonal responses are generally longer lasting:

4. Summarise how a non-steroid hormone, such as insulin, brings about a cellular response:

The Endocrine System

Homeostasis is achieved through the activity of the nervous and endocrine systems, which interact in the regulation of the body's activities. The nervous system is capable of rapid responses to stimuli. Slower responses, and long term adjustments of the body (growth, reproduction, and adaptation to stress), are achieved through endocrine control. The endocrine system comprises **endocrine glands** and their **hormones**. Endocrine glands are ductless glands and are distributed throughout the body. Under appropriate stimulation, they secrete chemical messengers, called **hormones**, which are carried in the blood to **target** cells, where they have a specific metabolic effect. After exerting their effect, hormones are broken down and excreted from the body. Although a hormone circulates in the blood, only the targets will respond. Hormones may be amino acids, peptides, proteins (often modified), fatty acids, or steroids. Some basic features of the human endocrine system are explained below.

Hypothalamus
Coordinates nervous and endocrine systems. Secretes releasing hormones, which regulate the hormones of the anterior pituitary. Produces oxytocin and ADH, which are released from the posterior pituitary.

Parathyroid glands
On the surface of the thyroid, they secrete PTH (parathyroid hormone), which regulates blood calcium levels and promotes the release of calcium from bone. High levels of calcium in the blood inhibit PTH secretion.

Pancreas
Specialised α and β endocrine cells in the pancreas produce glucagon and insulin. Together, these control blood sugar levels.

Ovaries (in females)
At puberty the ovaries increase their production of oestrogen and progesterone. These hormones control and maintain female characteristics (breast development and pelvic widening), stimulate the menstrual cycle, maintain pregnancy, and prepare the mammary glands for lactation.

Pituitary gland
The pituitary is located below the hypothalamus. It secretes at least nine hormones that regulate the activities of other endocrine glands.

Thyroid gland
Secretes thyroxine, an iodine containing hormone needed for normal growth and development. Thyroxine stimulates metabolism and growth via protein synthesis.

Adrenal glands
The **adrenal medulla** produces adrenaline and noradrenaline; responsible for the fight or flight response. The **adrenal cortex** produces various steroid hormones, including aldosterone (sodium regulation) and cortisol (response to stress).

Testes (in males)
At puberty (the onset of sexual maturity) the testes of males produce testosterone in greater amounts. Testosterone controls and maintains "maleness" (muscular development and deeper voice), and promotes sperm production.

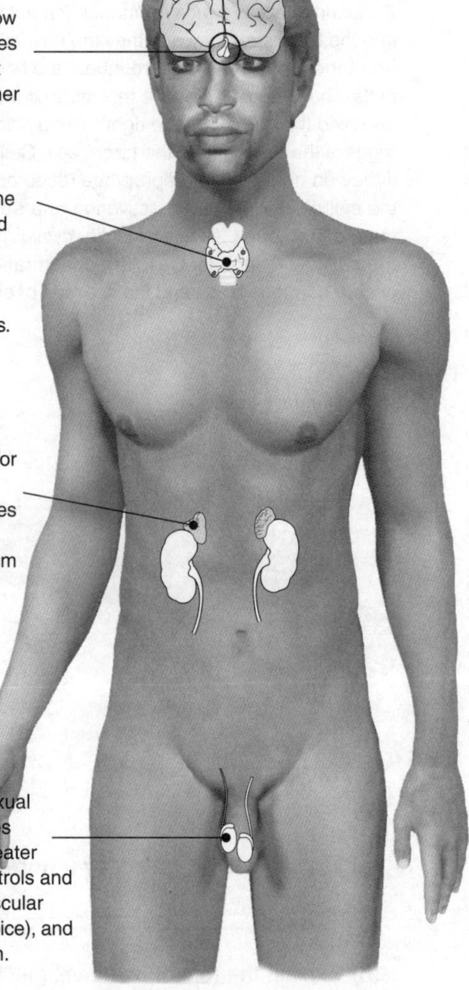

1. Explain why it is an advantage for hormones to be carried in the blood: ______________________________

____

____

2. (a) Name an endocrine gland and identify a hormone that it produces: ______________________________

(b) Name the target tissue for this hormone: ______________________________

(c) Outline the homeostatic function of this hormone and explain how it controls the activity of the target tissue:

____

____

(d) Briefly explain how the release of this hormone is regulated: ______________________________

____

____

____

The Hypothalamus and Pituitary

The **hypothalamus** is located at the base of the brain, just above the pituitary gland. Information comes to the hypothalamus through sensory pathways from the sense organs. On the basis of this information, the hypothalamus controls and integrates many basic physiological activities (e.g. temperature regulation, food and fluid intake, and sleep), including the reflex activity of the **autonomic nervous system**. The pituitary gland comprises two regions: the **posterior pituitary**, which is neural in origin and is essentially an extension of the hypothalamus, and the **anterior pituitary**, which is connected to the hypothalamus by blood vessels. The hypothalamus regulates pituitary activity and is the principal centre for coordinating the activity of the body's nervous and endocrine systems. The hypothalamus contains several distinct regions of neurosecretory cells. These are specialised neurones which are at the same time both nerve cells and endocrine cells. They produce hormones (usually peptides) in the cell body, which are transported down the axon and released into the blood in response to nerve impulses.

Hormones of the Hypothalamus and Pituitary

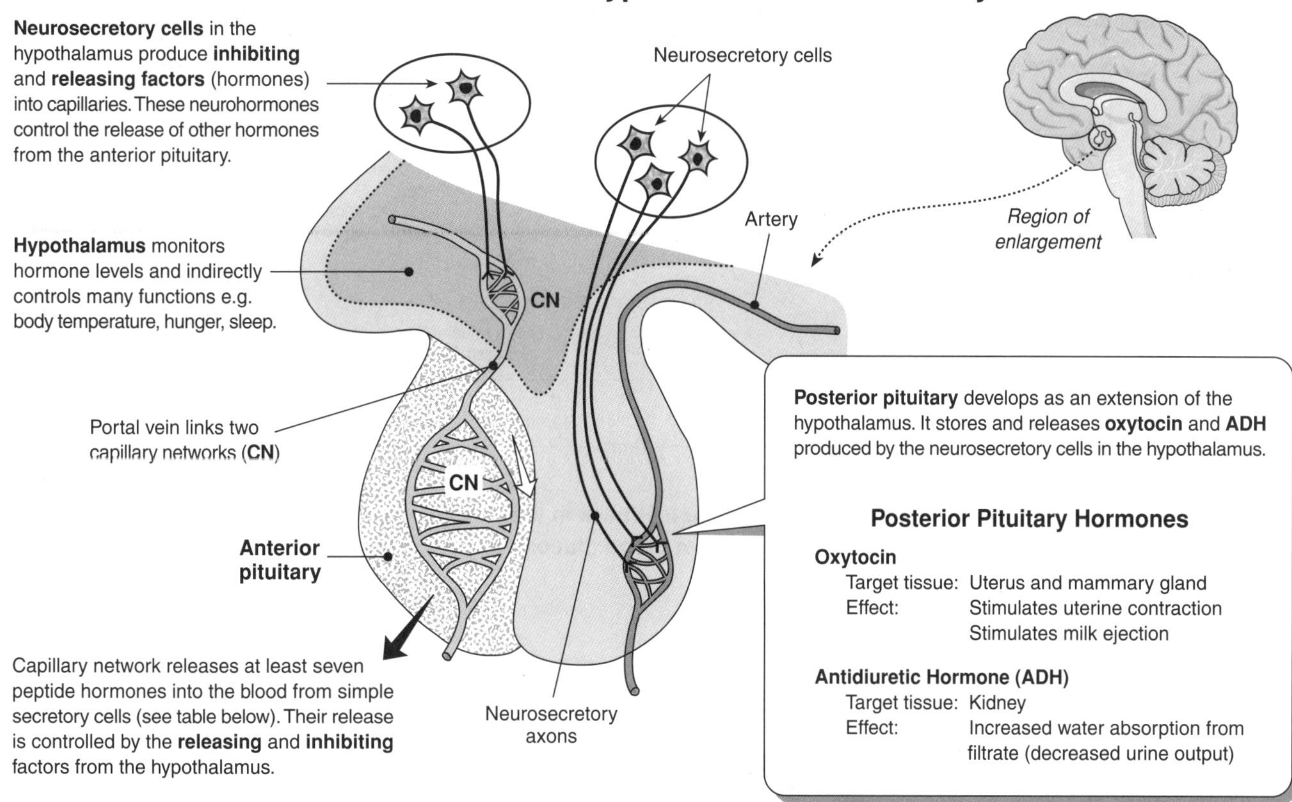

Posterior pituitary develops as an extension of the hypothalamus. It stores and releases **oxytocin** and **ADH** produced by the neurosecretory cells in the hypothalamus.

Posterior Pituitary Hormones

Oxytocin
- Target tissue: Uterus and mammary gland
- Effect: Stimulates uterine contraction; Stimulates milk ejection

Antidiuretic Hormone (ADH)
- Target tissue: Kidney
- Effect: Increased water absorption from filtrate (decreased urine output)

Anterior Pituitary Hormone	Target tissue	Primary action
Growth hormone (GH)	All tissues	Stimulates general tissue growth and protein synthesis
Prolactin	Mammary gland	Stimulates synthesis of milk protein, growth of mammary gland
Thyroid stimulating hormone (TSH)	Thyroid gland	Increases synthesis and secretion of thyroid hormones
Follicle stimulating hormone (FSH)	Seminiferous tubules (male), ovarian follicles (female)	Increases sperm production (male), stimulates follicle maturation (female)
Luteinising hormone (LH)	Interstitial cells in ovary (female), interstitial cells in testis (male)	Secretion of ovarian hormones, ovulation, formation of corpus luteum (female), androgen synthesis and secretion (male)
Melanophore-stimulating hormone	Melanophores and melanocytes	Increases melanin synthesis and dispersal (skin darkening)
Adrenocorticotrophin (ACTH)	Adrenal cortex	Increases synthesis and secretion of hormones from the adrenal cortex

1. Explain how the anterior and posterior pituitary differ with respect to their relationship to the hypothalamus:

2. Explain how the differences between the two regions of the pituitary relate to the nature of their hormonal secretions:

3. Explain how the release of TSH is regulated:

Control of Blood Glucose

The endocrine portion of the **pancreas**, the α and β cells of the **islets of Langerhans**, produces two hormones, **insulin** and **glucagon**. Together, these hormones mediate the regulation of blood glucose, maintaining a steady state through **negative feedback**. Insulin promotes a decrease in blood glucose through synthesis of glycogen and cellular uptake of glucose. Glucagon promotes an increase in blood glucose through the breakdown of glycogen and the synthesis of glucose from amino acids. Restoration of normal blood glucose level acts through negative feedback to stop hormone secretion. Regulating blood glucose to within narrow limits allows energy to be available to cells as needed. Extra energy is stored, as glycogen or fat, and is mobilised to meet energy needs as required. The liver is pivotal in these carbohydrate conversions.

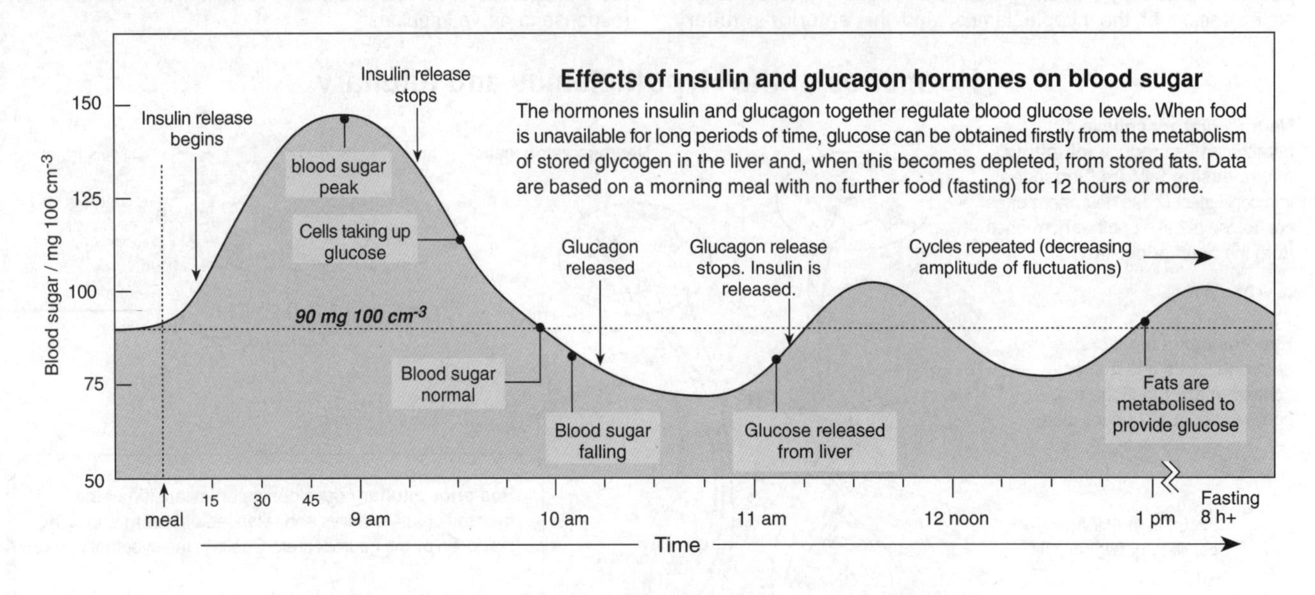

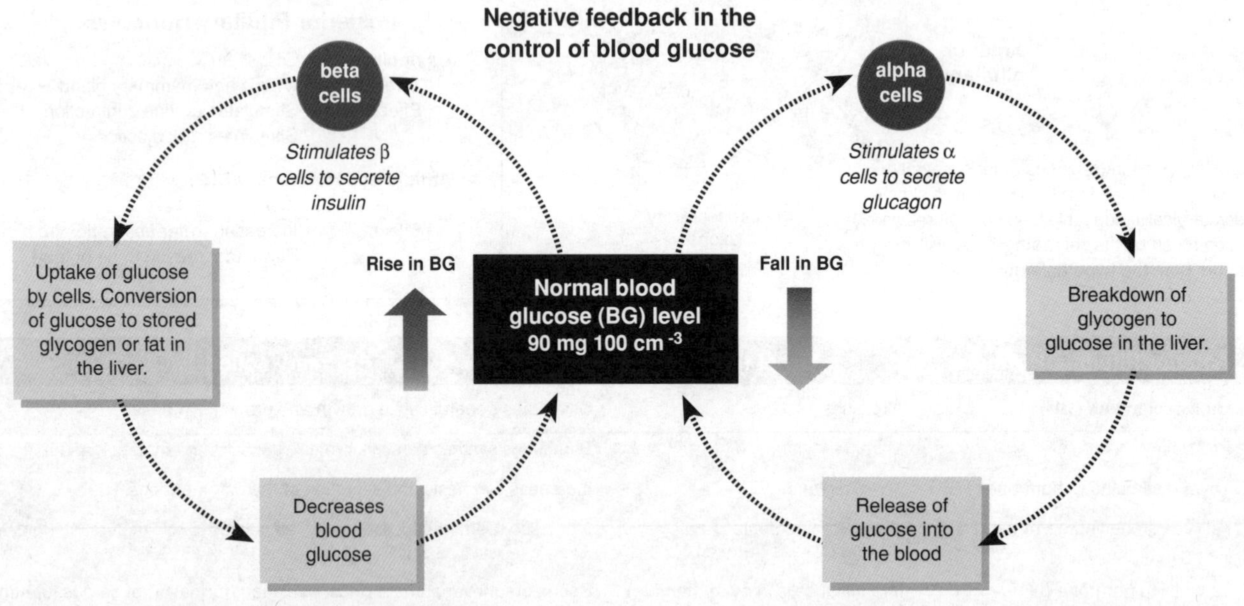

1. (a) Name the stimulus for the release of insulin: __________

   (b) Name the stimulus for the release of glucagon: __________

   (c) Explain how glucagon brings about an increase in blood glucose level: __________

   (d) Explain how insulin brings about a decrease in blood glucose level: __________

2. Outline the role of negative feedback in the control of blood glucose: __________

3. Explain why fats are metabolised after a long period without food: __________

Diabetes Mellitus

Diabetes is a general term for a range of disorders sharing two common symptoms: production of large amounts of urine and excessive thirst. Other symptoms may be present as well, depending on the type of diabetes. **Diabetes mellitus** is the most common form of diabetes and is characterised by high blood sugar. **Type I** has a juvenile onset while **Type II** affects adults.

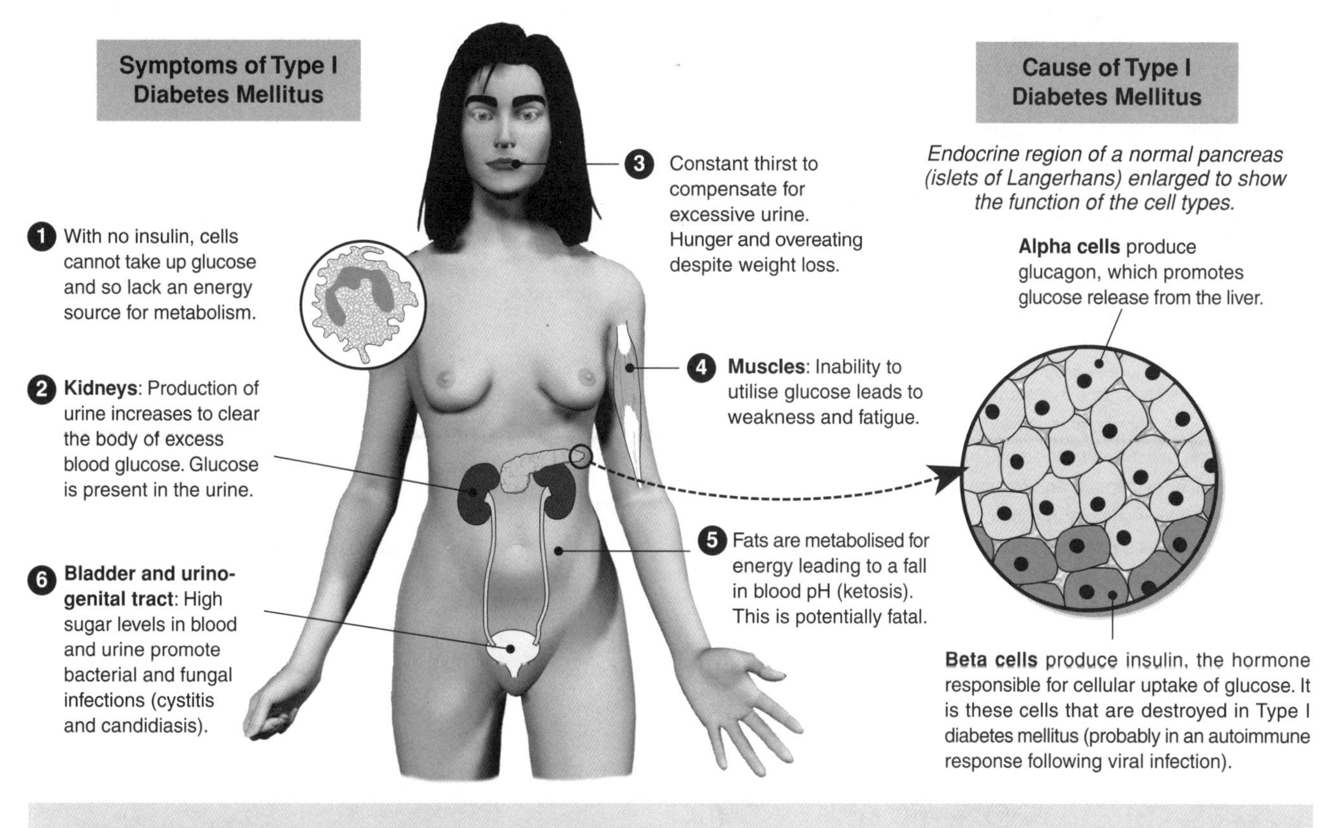

Type I Diabetes mellitus (insulin dependent)

Incidence: About 10-15% of all diabetics.

Age at onset: Early; often in childhood (often called juvenile onset diabetes).

Symptoms: Symptoms are severe. Insulin deficiency accelerates fat breakdown and leads to a number of metabolic complications: hyperglycaemia (high blood sugar), excretion of glucose in the urine, increased urine production, excessive thirst and hunger, weight loss, and ketosis.

Cause: Absolute deficiency of insulin due to lack of insulin production (pancreatic beta cells are destroyed in an autoimmune reaction). There is a genetic component but usually a childhood viral infection triggers the development of the disease. Mumps, coxsackie, and rubella are implicated.

Treatments

Present treatments: Regular insulin injections combined with dietary management to keep blood sugar levels stable. Blood glucose is monitored regularly with testing kits to guard against sudden falls in blood glucose (hypoglycaemia).

Until recently, insulin was extracted from dead animals. Now, genetically engineered yeast or bacterial cells containing the gene for human insulin are grown in culture, providing abundant, low cost insulin, without the side effects associated with animal insulin.

New treatments: Cell therapy involves the transplant of insulin producing islet cells. To September 2002, approximately 200 islet transplants had been performed by several groups around the world. Cell therapy promises to be a practical and effective way to provide sustained relief for Type I diabetics.

Future treatments: In the future, gene therapy, where the gene for insulin is inserted into the diabetic's cells, may be possible.

1. Describe the cause of Type I diabetes mellitus: ______

2. List the symptoms of Type I diabetes mellitus: ______

3. Summarise the treatments for Type I diabetes mellitus (list key words/phrases only):

   (a) Present treatment: ______

   (b) New treatments: ______

   (c) Future treatments: ______

The Liver's Homeostatic Role

The liver, located just below the diaphragm and making up 3-5% of body weight, is the largest homeostatic organ. It performs a vast number of functions including production of bile, storage and processing of nutrients, and detoxification of poisons and metabolic wastes. The liver has a **unique double blood supply** and up to 20% of the total blood volume flows through it at any one time. This rich vascularisation makes it the central organ for regulating activities associated with the blood and circulatory system. In spite of the complexity of its function, the liver tissue and the liver cells themselves are structurally relatively simple. Features of liver structure and function are outlined below. The histology of the liver in relation to its role is described opposite.

Homeostatic Functions of the Liver

The liver is one of the largest and most complex organs in the body. It has a central role as an organ of homeostasis and performs many functions, particularly in relation to the regulation of blood composition. General functions of the liver are outlined below. Briefly summarised, the liver:

1. Secretes bile, important in emulsifying fats in digestion.
2. Metabolises amino acids, fats, and carbohydrates (below).
3. Synthesises glucose from non-carbohydrate sources when glycogen stores are exhausted (gluconeogenesis).
4. Stores iron, copper, and some vitamins (A, D, E, K, B_{12}).
5. Converts unwanted amino acids to urea (urea cycle).
6. Manufactures heparin and plasma proteins (e.g. albumin).
7. Detoxifies poisons or turns them into less harmful forms.
8. Some liver cells phagocytose worn-out blood cells.
9. Synthesises cholesterol from acetyl coenzyme A.

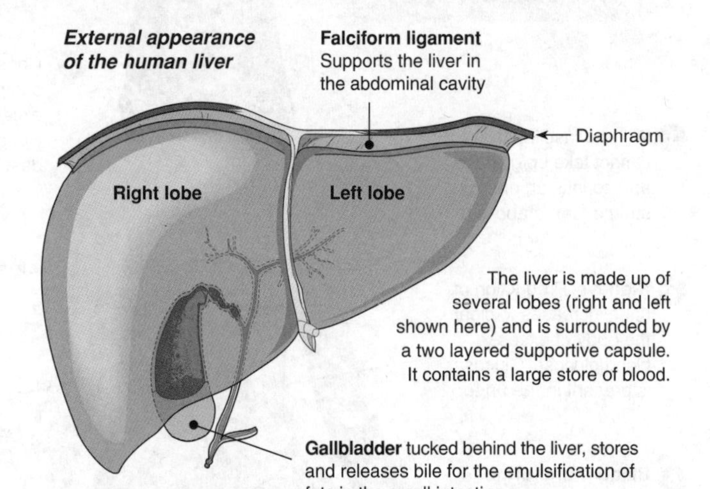

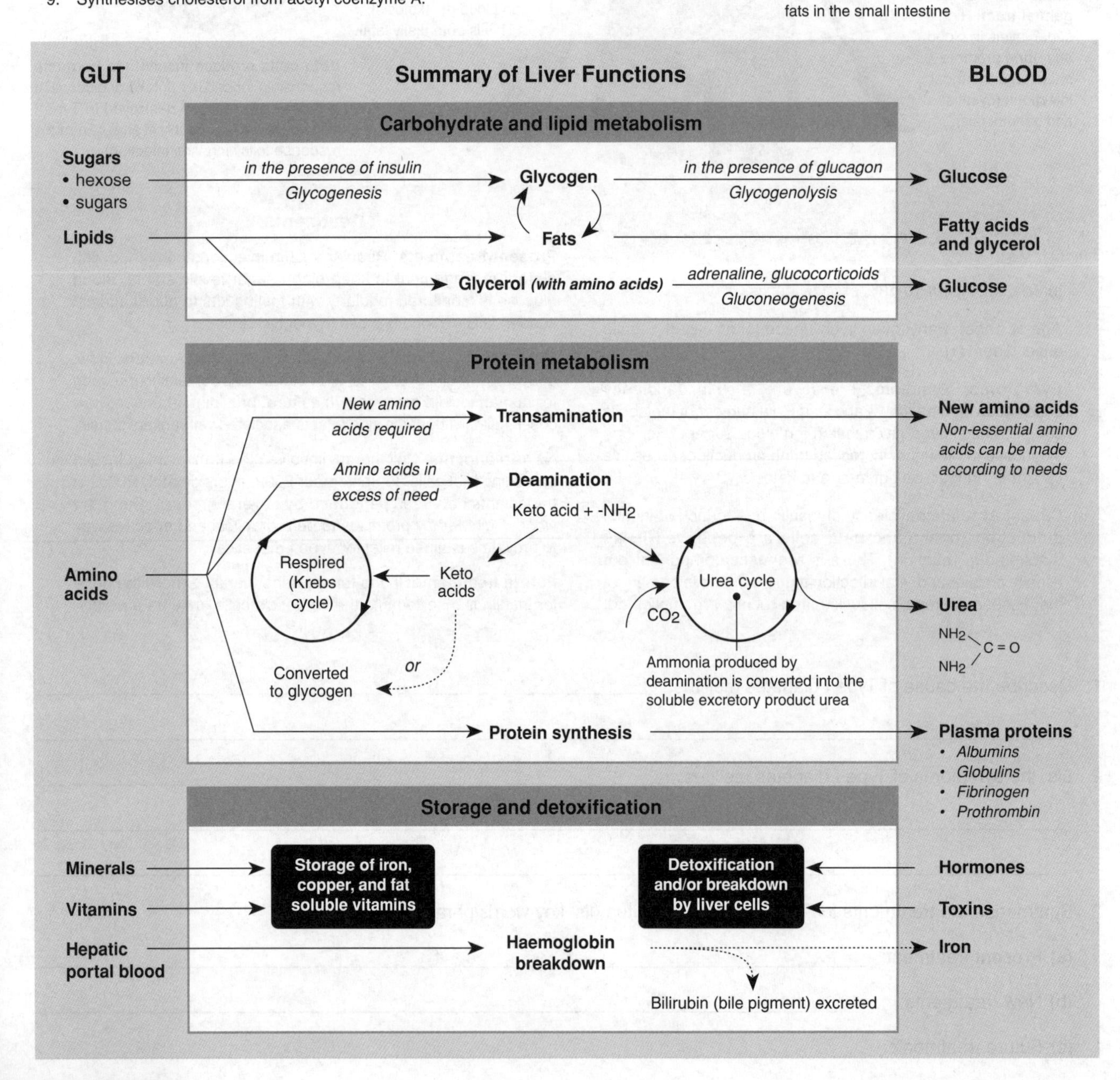

The Internal Structure of the Liver

Radiating cords of hepatocytes

CV

Schematic illustration of the arrangement of lobules and portal tracts in liver tissue

Bile ductule

Branch of hepatic portal vein

Branch of hepatic artery

Portal tract (triad)

Bile flow

Blood flows towards the central vein

CV

Central vein

Fibrous connective tissue capsule surrounds lobules

Lobule

The connective tissue capsule covering the liver branches through the tissue, dividing it into functional units called **lobules**. A lobule consists of rows (**cords**) of **hepatocytes** (liver cells) arranged in a radial pattern around a central vein. Between the cords are blood spaces called **sinusoids** and small channels through which the bile flows (the **bile canaliculi**). Between the lobules are branches of the hepatic artery, hepatic portal vein, and bile duct. These form a **portal tract** (triad). Lymphatic vessels and nerves are also found in this area (not shown). **The photograph above** shows most of a liver lobule in a human, illustrating the central vein, the cords of liver cells, and sinusoids (dark spaces).

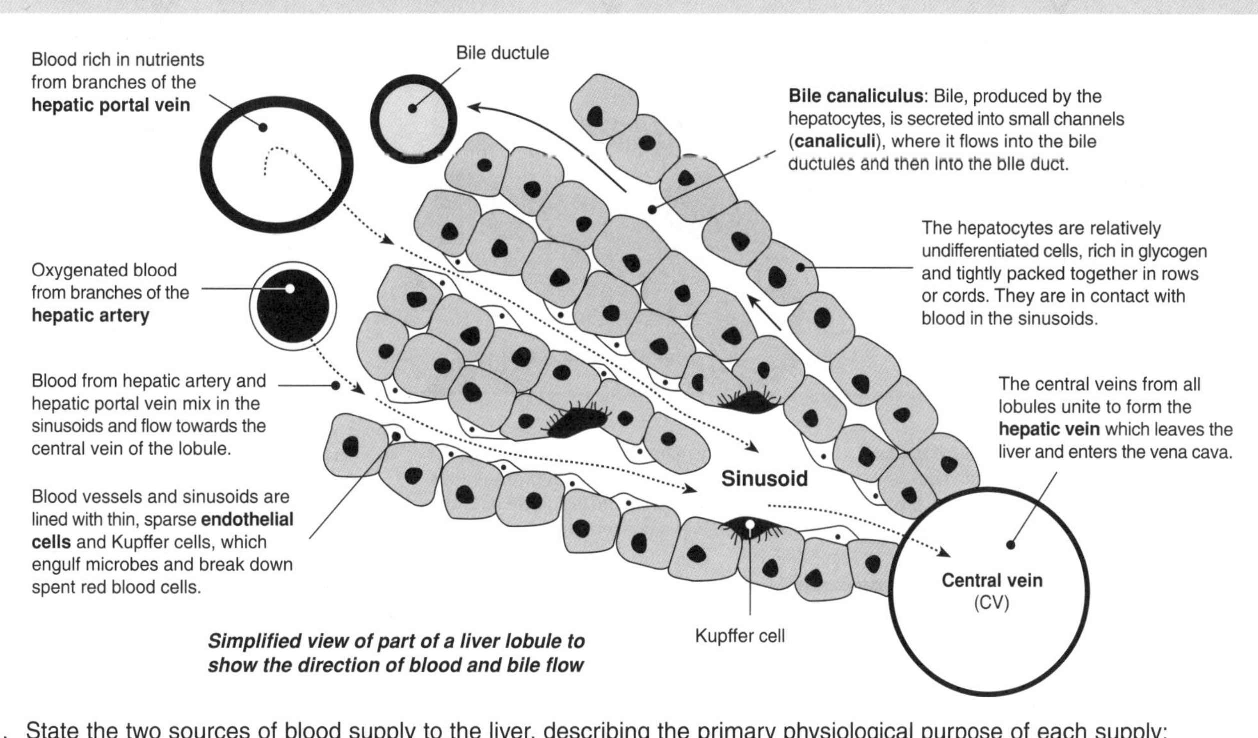

Simplified view of part of a liver lobule to show the direction of blood and bile flow

1. State the two sources of blood supply to the liver, describing the primary physiological purpose of each supply:

   (a) Supply 1: ______________________ Purpose: ______________________

   (b) Supply 2: ______________________ Purpose: ______________________

2. Briefly describe the role of the following structures in liver tissue:

   (a) Bile canaliculi: ______________________

   (b) Phagocytic Kupffer cells: ______________________

   (c) Central vein: ______________________

   (d) Sinusoids: ______________________

3. Briefly explain three important aspects of **either** protein metabolism **or** carbohydrate metabolism in the liver:

   (a) ______________________

   (b) ______________________

   (c) ______________________

Nitrogenous Wastes in Animals

Waste materials are generated by the metabolic activity of cells. If allowed to accumulate, they would reach toxic concentrations and so must be continually removed. Excretion is the process of removing waste products and other toxins from the body. Waste products include carbon dioxide and water, and the nitrogenous (nitrogen containing) wastes that result from the breakdown of amino acids and nucleic acids. The simplest breakdown product of nitrogen containing compounds is ammonia, a small molecule that cannot be retained for long in the body because of its high toxicity. Most aquatic animals excrete ammonia immediately into the water where it is washed away. Other animals convert the ammonia to a less toxic form that can remain in the body for a short time before being excreted via special excretory organs. The form of the excretory product in terrestrial animals (urea or uric acid) depends on the type of organism and its life history. Terrestrial animals that lay eggs produce uric acid rather than urea, because it is non-toxic and very insoluble. It remains as an inert solid mass in the egg until hatching.

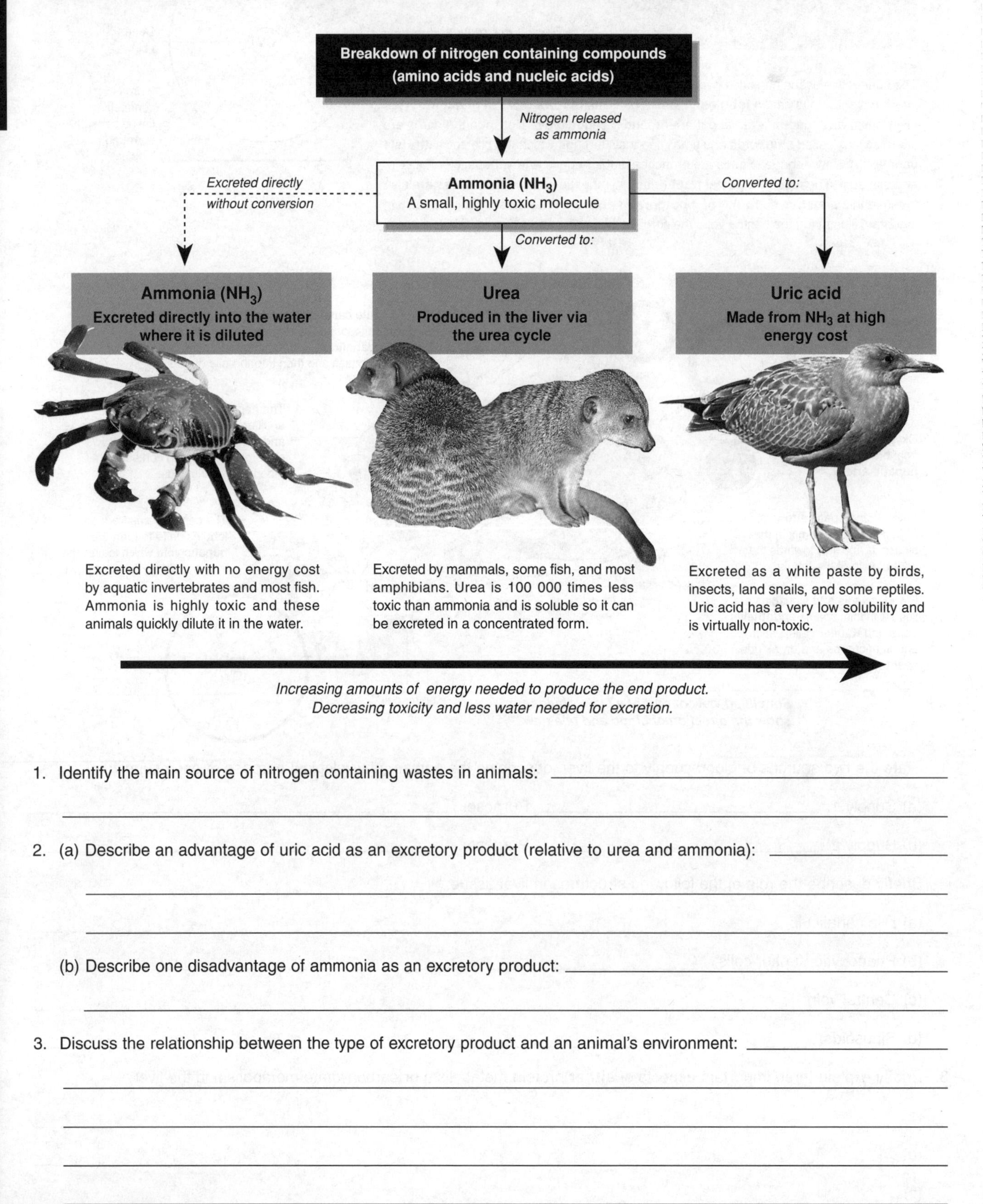

1. Identify the main source of nitrogen containing wastes in animals: ______________________

2. (a) Describe an advantage of uric acid as an excretory product (relative to urea and ammonia): ______________________

   (b) Describe one disadvantage of ammonia as an excretory product: ______________________

3. Discuss the relationship between the type of excretory product and an animal's environment: ______________________

Waste Products in Humans

In humans and other mammals, a number of organs are involved in the excretion of the waste products of metabolism: mainly the kidneys, lungs, skin, and gut. The liver is a particularly important organ in the initial treatment of waste products, particularly the breakdown of haemoglobin and the formation of urea from ammonia. Excretion should not be confused with the elimination or egestion of undigested and unabsorbed food material from the gut. Note that the breakdown products of haemoglobin (blood pigment) are excreted in bile and pass out with the faeces, but they are not the result of digestion.

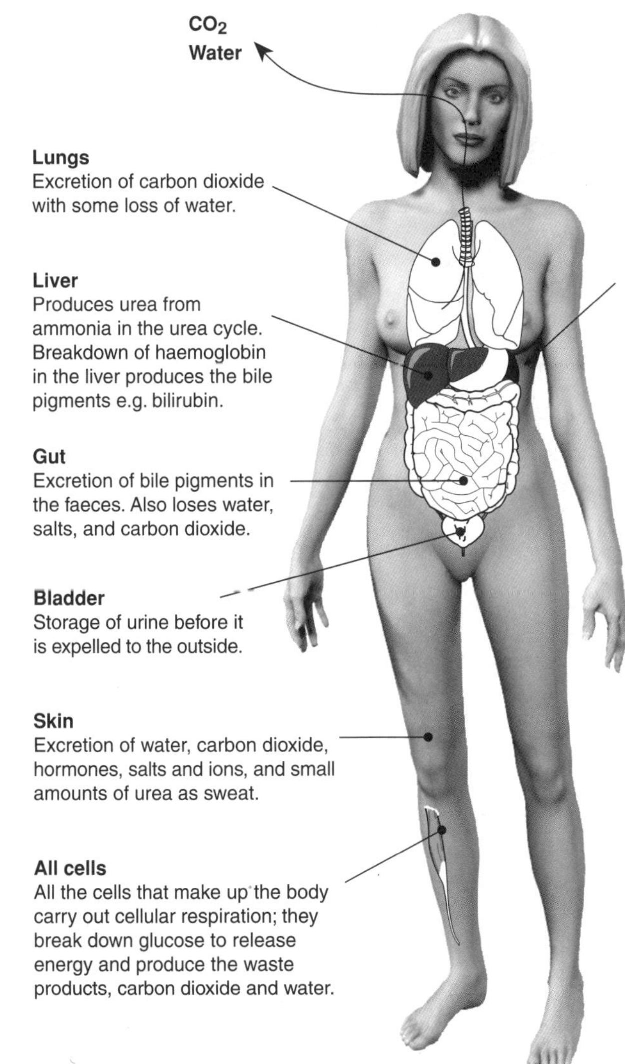

Excretion In Humans

In mammals, the kidney is the main organ of excretion, although the skin, gut, and lungs also play important roles. As well as ridding the body of nitrogenous wastes, the kidney is also able to excrete many unwanted poisons and drugs that are taken in from the environment. Usually these are ingested with food or drink, or inhaled. As long as these are not present in toxic amounts, they can usually be slowly eliminated from the body.

Substance	Origin*	Organ(s) of excretion
Carbon dioxide		
Water		
Bile pigments		
Urea		
Ions (K+, H+)		
Hormones		
Poisons		
Drugs		

* Origin refers to from where in the body each substance originates

1. In the diagram above, complete the table summarising the origin of excretory products and the main organ or organs of excretion involved for each substance.

2. Explain the role of the liver in excretion, even though it is not an excretory organ itself: ______

3. Tests for pregnancy are sensitive to an excreted substance in the urine. Suggest what this substance might be:

4. People sometimes suffer renal (kidney) failure, where the kidneys cease to operate and can no longer produce urine. Given that the kidney rids the body of excessive ions and water, as well as nitrogenous wastes, predict the probable effects of kidney failure. You may wish to discuss this as a group. HINT: Consider the effects of salt and water retention, and the effect of high salt levels on blood pressure and on the heart.

Water Balance in Animals

Many aspects of metabolism e.g. enzyme activity, membrane transport, and nerve conduction, are dependent on particular concentrations of ions and metabolites. To achieve this balance, the salt and water content of the internal environment must be regulated; a process called **osmoregulation**. The mechanisms for obtaining, retaining, and eliminating water and solutes (including excretion of nitrogenous wastes) in marine, freshwater and terrestrial organisms vary considerably. Differences reflect both the constraints of the environment and the evolutionary inheritance of the organism.

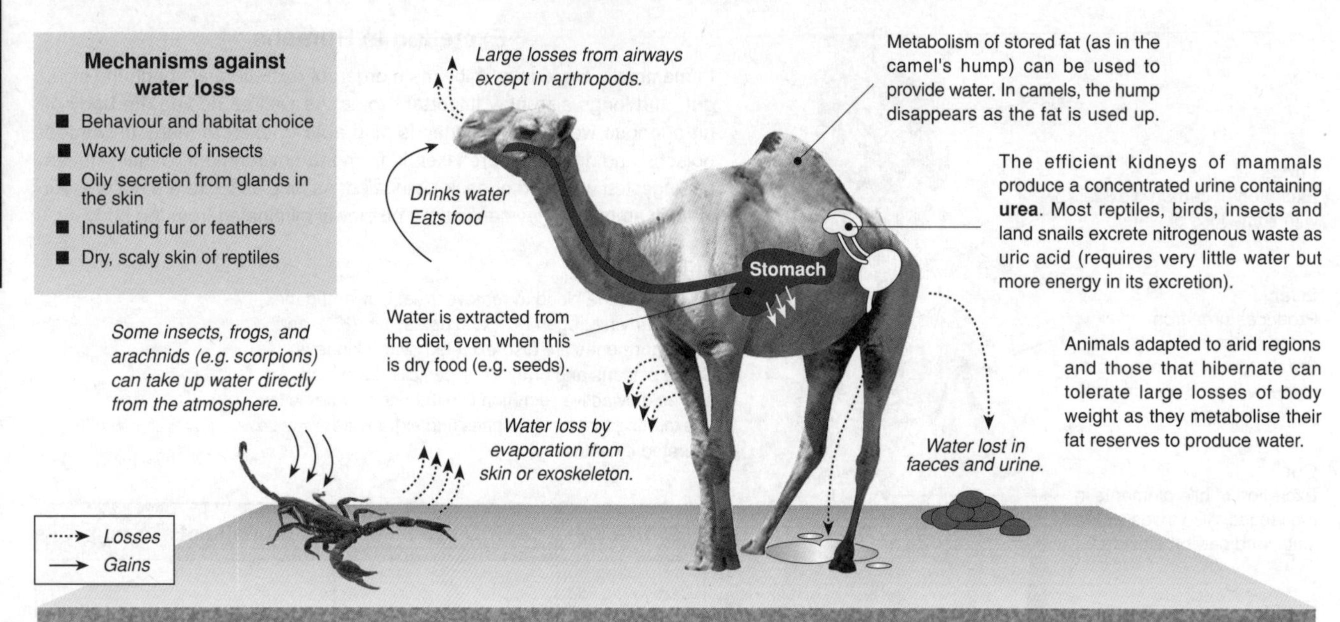

The primary water balance problem for terrestrial animals is water loss. Water is required for all metabolism, including the metabolism of food and the disposal of waste products. Mammals must **drink water** regularly, although some are able to survive for long periods without drinking by generating water from the metabolism of fats. Mammals have efficient **kidneys** and produce a concentrated urine high in urea. In most land arthropods, water is conserved by limiting losses to the environment. The **chitinous exoskeleton** itself does not reduce water loss much but the waxy cuticle of insects retards water loss very effectively. The respiratory structures of arthropods are chitinous, internal tubes and there is little loss from these. All animals show behavioural adaptations to limiting water loss by seeking out damper environments. This is particularly important for desert animals and for arthropods with little resistance to dehydration.

Tolerance of water loss and burrowing behaviour in desert frogs.

Behavioural adaptations, efficient kidneys, and thick fur in kangaroos

Humidity seeking behavioural adaptations in woodlice.

Chitinous exoskeleton with waxy, waterproof cuticle in insects.

1. (a) Briefly describe four ways in which animals obtain water: ____________________

   (b) Identify three ways in which water is lost: ____________________

2. Discuss structural and behavioural adaptations for reducing water loss in a named arthropod **or** a named mammal:

Marine Environments

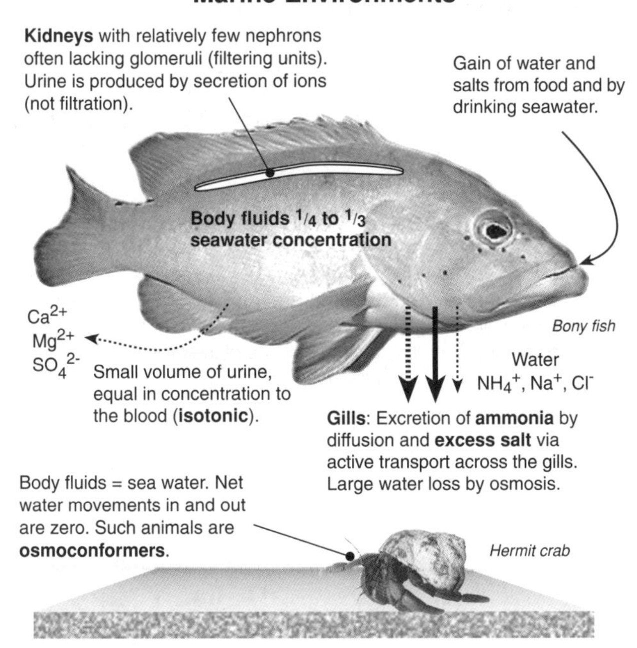

Freshwater Environments

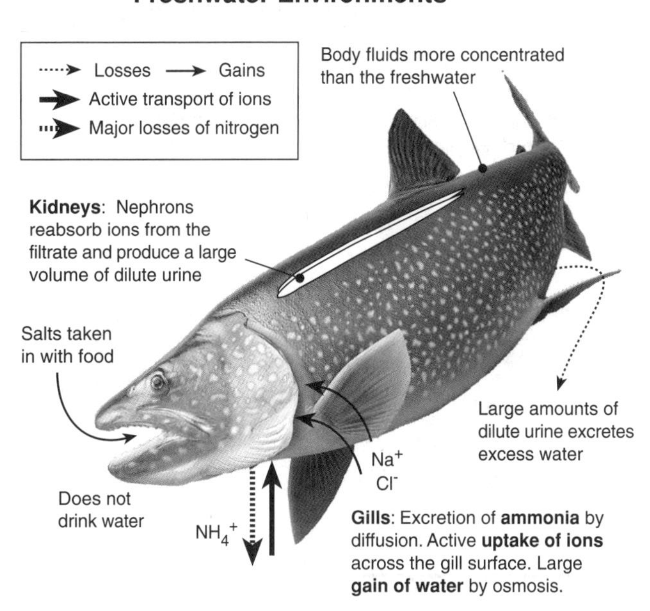

Most marine invertebrates do not regulate salt and water balance; they are **osmoconformers** and their body fluids fluctuate with changes in the environment. Animals, such as fish and marine mammals, that regulate their salt and water fluxes are termed **osmoregulatory**. Bony fish lose water osmotically and counter the loss by drinking salt water and excreting the excess salt. Marine elasmobranchs generate osmotic concentrations in their body fluids similar to seawater by tolerating high urea levels. Excess salt from the diet is excreted via a salt gland in the rectum. Marine mammals produce a urine that is high in both salt and urea. Some intertidal animals tolerate frequent dilutions of normal seawater and may actively take up salts across the gill surfaces to compensate for water gain and salt loss.

Freshwater animals have body fluids that are osmotically more concentrated than the water they live in. Water tends to enter their tissues by osmosis and must be expelled to avoid flooding the body. Simple protozoans use contractile vacuoles to collect the excess water and expel it. Other invertebrates expel water and nitrogenous wastes using simple nephridial organs. Bony fish and aquatic arthropods produce dilute urine (containing ammonia) and actively take up salts across their gills (in aquatic insects these are often non-respiratory, anal gills). The kidneys of **freshwater bony fish** (above) also reabsorb salts from the filtrate through active transport mechanisms. These ion gains are important because some loss of valuable ions occurs constantly as a result of the high urine volume.

Most marine invertebrates, like these sea anemones, are osmoconformers.

Elasmobranchs maintain an osmotic concentration similar to seawater.

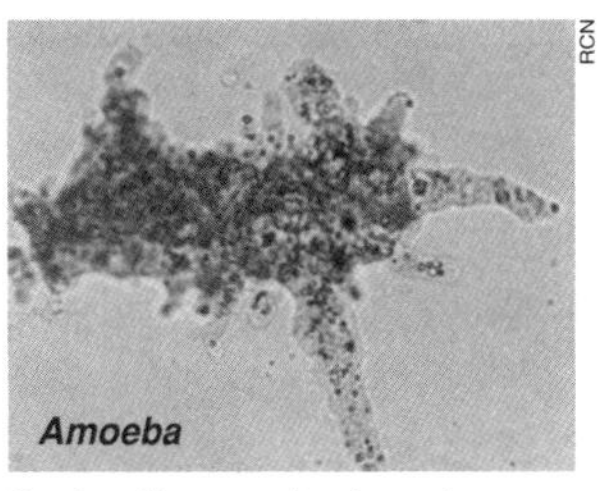

Contractile vacuoles in protozoans collect excess water and expel it.

In aquatic insect larvae, the gills actively take up salts from the water.

3. Describe how freshwater animals can compensate for salt losses that occur when they excrete large amounts of water:

4. (a) Explain what is meant by an **osmoregulator** and give an example:

   Example:

   (b) Explain what is meant by an **osmoconformer** and give an example:

   Example:

5. Freshwater and marine bony fish have contrasting excretion and osmoregulation problems. Explain why:

   (a) Marine bony fish drink vast quantities of salt water:

   (b) Freshwater bony fish do not drink water at all:

6. Describe the salt and water balance problems faced by migrating fish as they move from a marine environment to freshwater (as happens during spawning runs in salmon):

Water Budget in Mammals

Water loss is a major problem for most mammals. The degree to which urine can be concentrated (and water conserved) depends on the number of nephrons present in the kidney and the length of the loop of Henle. The highest urine concentrations are found in mammals from desert environments, such as kangaroo rats (below). Under normal conditions these animals will not drink water, obtaining most of their water from the metabolic break down of food instead.

Regulation of Water Balance in Humans

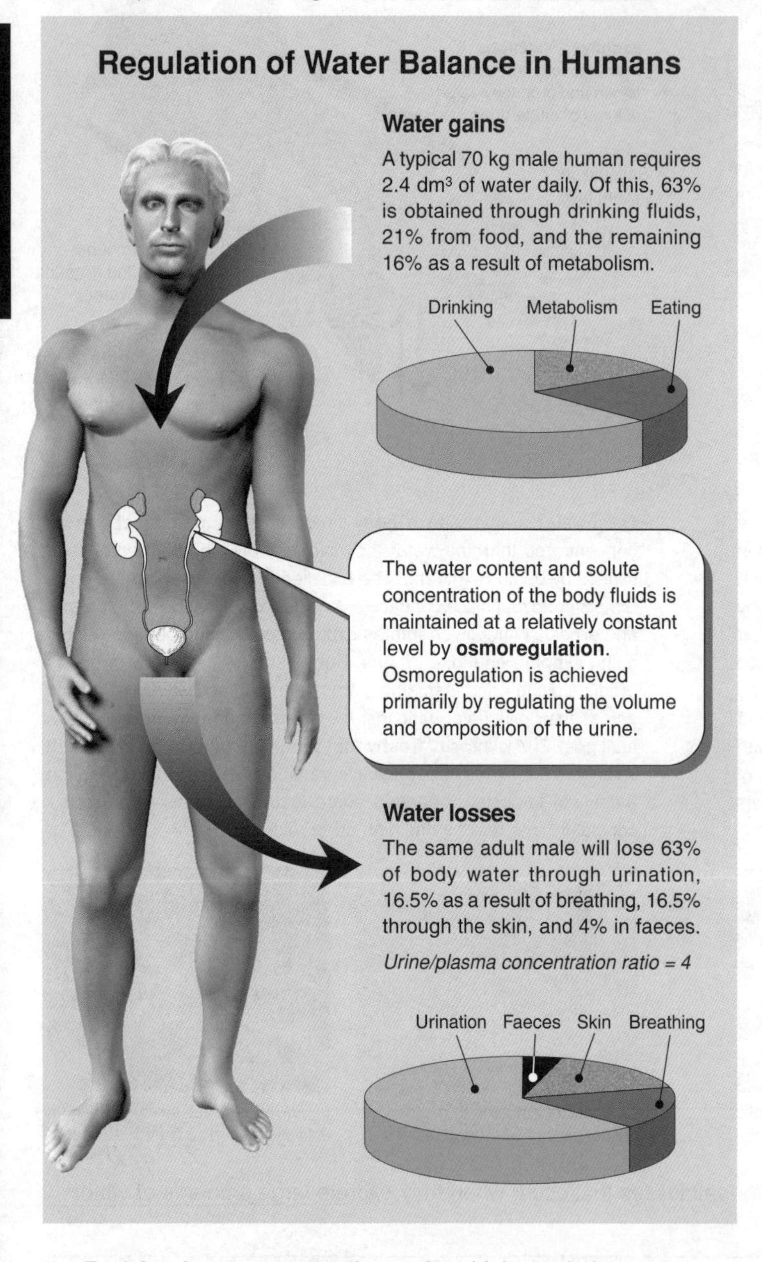

Water gains

A typical 70 kg male human requires 2.4 dm^3 of water daily. Of this, 63% is obtained through drinking fluids, 21% from food, and the remaining 16% as a result of metabolism.

The water content and solute concentration of the body fluids is maintained at a relatively constant level by **osmoregulation**. Osmoregulation is achieved primarily by regulating the volume and composition of the urine.

Water losses

The same adult male will lose 63% of body water through urination, 16.5% as a result of breathing, 16.5% through the skin, and 4% in faeces.

Urine/plasma concentration ratio = 4

Adaptations of Arid Adapted Rodents

Most desert-dwelling mammals are adapted to tolerate a low water intake. Arid adapted rodents, such as jerboas and kangaroo rats, conserve water by reducing losses to the environment and obtain the balance of their water needs from the oxidation of dry foods (respiratory metabolism). The table below shows the water balance in a kangaroo rat after eating 100 g of dry pearl barley. Note the high urine to plasma concentration ratio relative to that of humans.

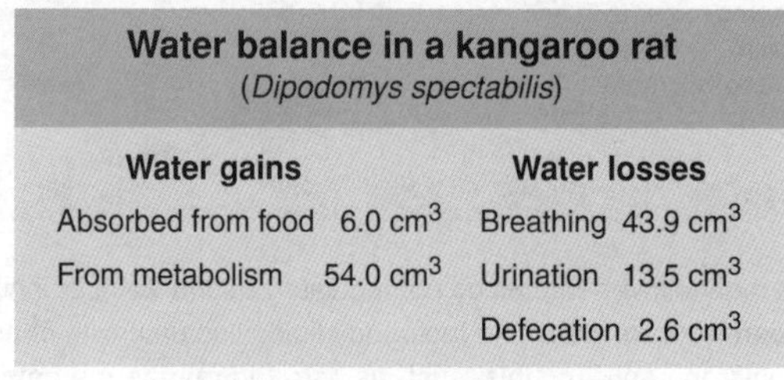

Water balance in a kangaroo rat (*Dipodomys spectabilis*)

Water gains		Water losses	
Absorbed from food	6.0 cm^3	Breathing	43.9 cm^3
From metabolism	54.0 cm^3	Urination	13.5 cm^3
		Defecation	2.6 cm^3

Urine/plasma concentration ratio = 17

Adaptations of kangaroo rats

Kangaroo rats, and other arid-adapted rodents, tolerate long periods without drinking, meeting their water requirements from the metabolism of dry foods. They dispose of nitrogenous wastes with very little output of water and they neither sweat nor pant to keep cool.

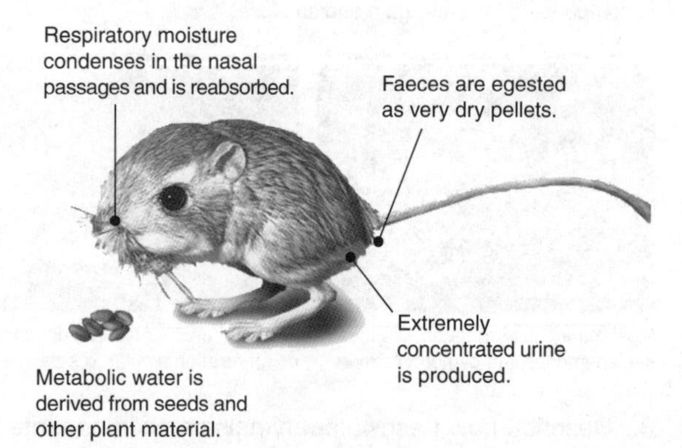

1. Explain why most mammals need to drink regularly: ______________________________

____

2. Using the tabulated data for the kangaroo rat (above), graph the water gains and losses in the space provided (below).

3. Describe three physiological adaptations of desert adapted rodents to low water availability:

(a) ______________________________

____

(b) ______________________________

____

(c) ______________________________

____

____

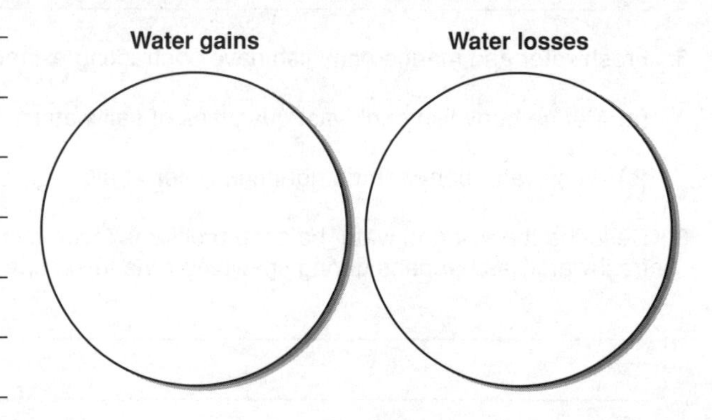

The Urinary System

The mammalian urinary system consists of the kidneys and bladder, and their associated blood vessels and ducts. The kidneys have a plentiful blood supply from the renal artery. The blood plasma is filtered by the kidneys to form urine. Urine is produced continuously, passing along the ureters to the bladder, a hollow muscular organ lined with smooth muscle and stretchable epithelium. Each day the kidneys filter about 180 dm^3 of plasma. Most of this is reabsorbed, leaving a daily urine output of about 1 dm^3. By adjusting the composition of the fluid excreted, the kidneys help to maintain the body's internal chemical balance. All vertebrates have kidneys, but their efficiency in producing a concentrated urine varies considerably. Mammalian kidneys are very efficient, producing a urine that is concentrated to varying degrees depending on requirements.

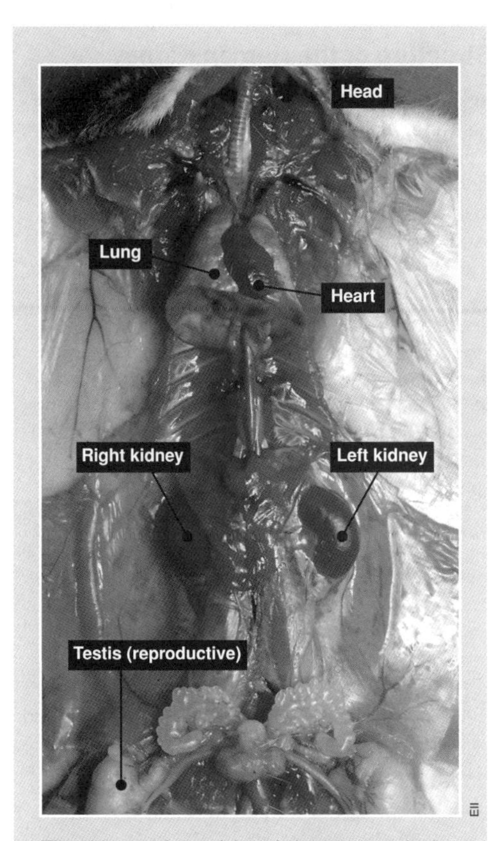

The kidneys of **rats** (above), humans, and other mammals are distinctive, bean shaped organs that lie at the back of the abdominal cavity to either side of the spine. The kidneys lie outside the peritoneum of the abdominal cavity and are partly protected by the lower ribs. Each kidney is surrounded by three layers of tissue. The inner-most renal capsule is a smooth fibrous membrane that acts as a barrier against trauma and infection. The two outer layers comprise fatty tissue and fibrous connective tissue. These act to protect the kidney and anchor it firmly in place.

The Human Urinary System

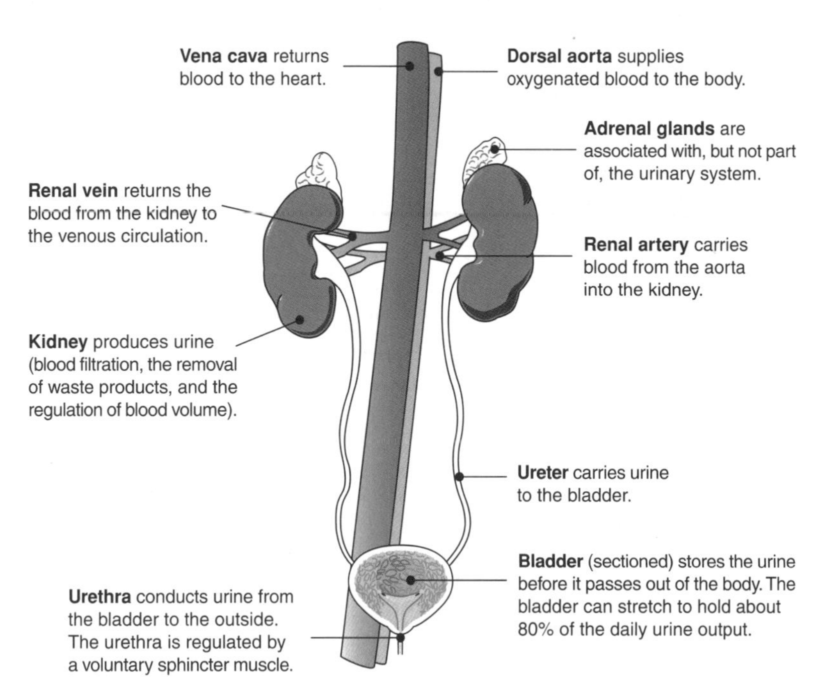

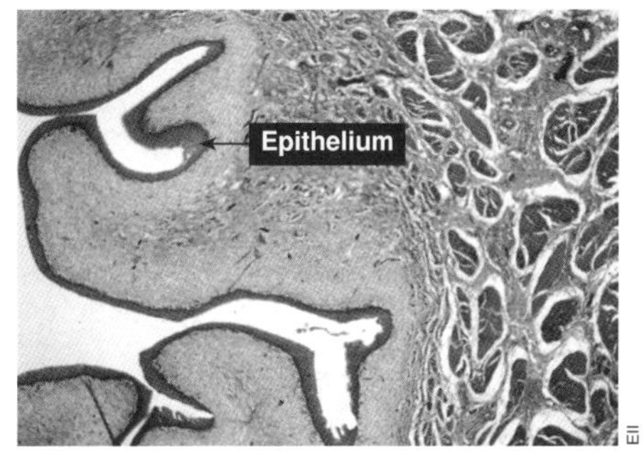

The bladder is lined with a **transitional epithelium**. This type of epithelium is layered, or stratified, and can be stretched without the outer cells breaking apart from each other. This a feature of hollow organs that are subjected to expansion from within. In the photograph, the lumen (cavity) of the bladder is to the left.

1. Identify the components of the urinary system and describe their functions: ______________________

____

____

____

2. Calculate the percentage of the plasma reabsorbed by the kidneys: ______________________

3. The kidney receives blood at a higher pressure than other organs. Suggest why this is the case: ______________________

____

____

4. Suggest why the kidneys are surrounded by fatty connective tissue: ______________________

____

The Physiology of the Kidney

The functional unit of the kidney, the **nephron**, is a selective filter element, comprising a renal tubule and its associated blood vessels. Filtration, i.e. forcing fluid and dissolved substances through a membrane by pressure, occurs in the first part of the nephron, across the membranes of the capillaries and the glomerular capsule. The passage of water and solutes into the nephron and the formation of the glomerular filtrate depends on the pressure of the blood entering the afferent arteriole (below). If it increases, filtration rate increases; when it falls, glomerular filtration rate also falls. This process is so precisely regulated that, in spite of fluctuations in arteriolar pressure, glomerular filtration rate per day stays constant. After formation of the initial filtrate, the **urine** is modified through secretion and tubular reabsorption according to physiological needs at the time.

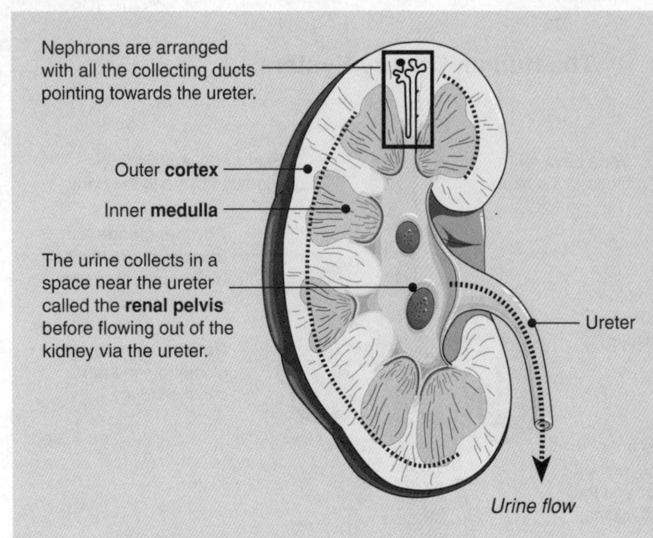

Internal Structure of the Human Kidney

Human kidneys are about 100-120 mm long and 25 mm thick. The functional unit of the kidney is the **nephron**. The other parts of the urinary system are primarily passageways and storage areas. The inner tissue of the kidney appears striated (striped), due to alignment of the nephrons and their surrounding blood vessels. It is the precise alignment of the nephrons in the kidney that makes it possible to fit in all the filtering units required. Each kidney contains more than 1 million nephrons. They are **selective filter elements**, which regulate blood composition and pH, and excrete wastes and toxins. The initial urine is formed by **filtration** in the glomerulus. Plasma is filtered through three layers: the capillary wall, and the basement membrane and epithelium of Bowman's capsule. The epithelium comprises very specialised epithelial cells called **podocytes**. The filtrate is modified as it passes through the tubules of the nephron and the final urine passes out the ureter.

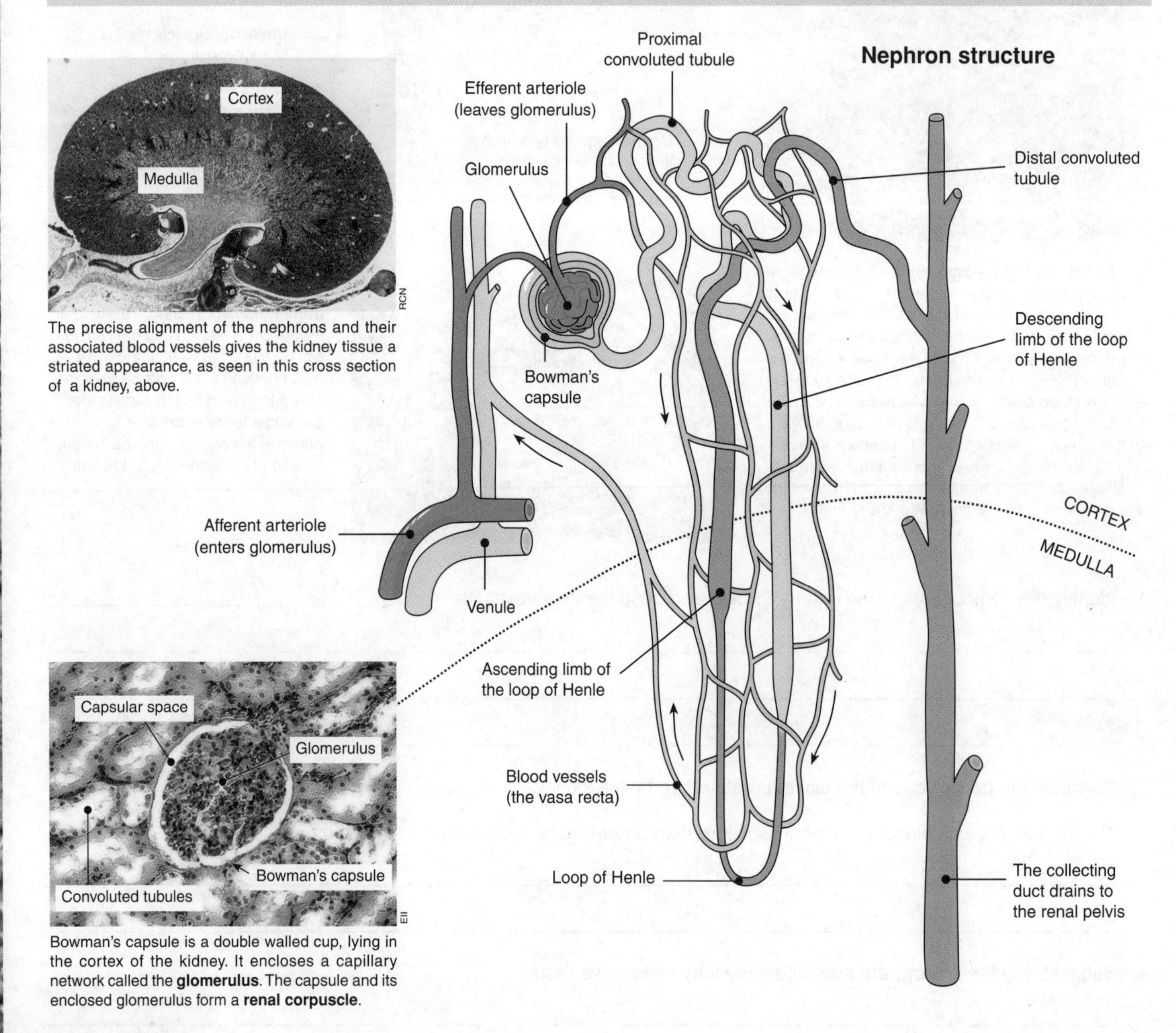

The precise alignment of the nephrons and their associated blood vessels gives the kidney tissue a striated appearance, as seen in this cross section of a kidney, above.

Bowman's capsule is a double walled cup, lying in the cortex of the kidney. It encloses a capillary network called the **glomerulus**. The capsule and its enclosed glomerulus form a **renal corpuscle**.

Summary of activities in the kidney nephron

Urine is formed in the kidney nephron by ultrafiltration of the blood and subsequent modification of the filtrate to add or remove substances (e.g. ions). The processes involved in urine formation are summarised below for each region of the nephron: glomerulus, proximal convoluted tubule, loop of Henle, distal convoluted tubule, and collecting duct.

Fluid is forced through the capillaries of the **glomerulus**, forming a filtrate similar to blood but lacking cells and proteins.

Blood flow in arteriole

Glomerulus

Filtration
H_2O
Salts
Glucose
Small proteins

Proximal convoluted tubule

Reabsorption of 90% of filtrate

Active transport	Passive transport	Osmosis
Glucose, Na^+, K^+, Mg^{2+}, Ca^{2+}	Cl^-	H_2O

Loop of Henle

Active transport
Salt transported from the ascending limb

Reabsorption
H_2O by osmosis from the descending limb

Passive transport
Na^+ and Cl^- from the thin part of the limb

*The **loop of Henle** has varying permeability to salt and water. The transport of salts and passive movement of water establish and maintain the salt gradient across the medulla necessary for the concentration of the urine in the collecting duct.*

Blood vessel

H_2O

H_2O

NaCl

Na^+
Cl^-

Increasing extracellular salt gradient in the medulla of the kidney

Distal convoluted tubule

Reabsorption
Na^+, Cl^-, Ca^{2+} by active transport
H_2O by osmosis

Secretion
H^+, K^+ by active transport
NH_3 by diffusion

H_2O

Water is carried away by blood capillaries and into the venous circulation so that the high interstitial salt gradient is maintained.

Collecting duct

Concentration of urine
H_2O leaves tubule by osmosis

The role of ADH
ADH promotes reabsorption of water from the collecting duct. When blood volume is low and more water is required, ADH promotes urine concentration.

Urine out to renal pelvis

1. Give a concise definition of a nephron and summarise its role in excretion: ______________________________

2. Explain the importance of the following in the production of urine in the kidney nephron:

   (a) Filtration of the blood at the glomerulus: ______________________________

   (b) Active secretion: ______________________________

   (c) Reabsorption: ______________________________

   (d) Osmosis: ______________________________

3. (a) Identify the purpose of the salt gradient in the kidney: ______________________________

   (b) Explain how this salt gradient is produced: ______________________________

Control of Kidney Function

Variations in salt and water intake, and in the environmental conditions to which we are exposed, contribute to fluctuations in blood volume and composition. The primary role of the kidneys is to regulate blood volume and composition (including the removal of nitrogenous wastes), so that homeostasis is maintained. This is achieved through varying the volume and composition of the urine. Two hormones, antidiuretic hormone and aldosterone, are involved in the process.

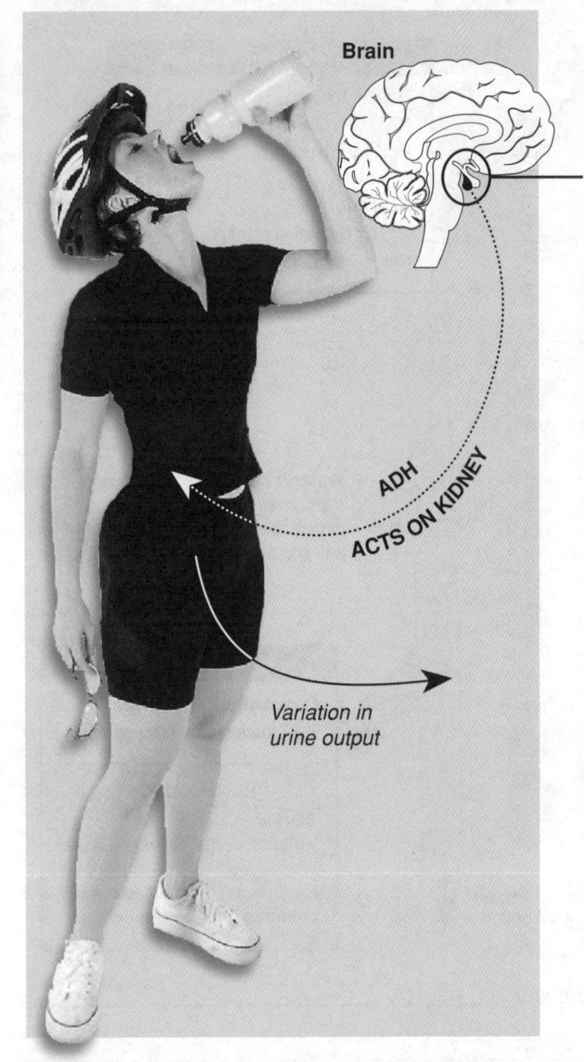

Control of Blood Volume

Osmoreceptors in the hypothalamus of the brain detect a fall in the concentration of water in the blood. They stimulate **neurosecretory cells in the hypothalamus** to synthesise and secrete the hormone ADH (antidiuretic hormone).

ADH passes from the hypothalamus to the posterior pituitary where it is released into the blood. ADH increases the permeability of the kidney collecting duct to water so that more water is reabsorbed and urine volume decreases.

Factors inhibiting ADH release

- High fluid intake
- High blood volume
- Low blood sodium levels
- Alcohol consumption

ADH levels decrease → Water reabsorption decreases. Urine output increases.

Factors causing ADH release

- Low fluid intake
- Low blood volume
- High blood sodium levels
- Nicotine and morphine

ADH levels increase → Water reabsorption increases. Urine output decreases.

Factors causing release of aldosterone

Low blood volumes also stimulate secretion of aldosterone from the adrenal cortex. This is mediated through a complex pathway involving the hormone renin from the kidney.

Aldosterone → Sodium reabsorption increases, water follows, blood volume restored.

1. (a) *Diabetes insipidis* is a type of diabetes, caused by a lack of ADH. Based on what you know of the role of ADH in kidney function describe the symptoms of this disease:

   (b) Suggest how this disorder might be treated:

2. Explain why alcohol consumption (especially to excess) causes dehydration and thirst:

3 (a) State the effect of aldosterone on the kidney nephron:

   (b) Explain the net result of this effect:

4. Explain how negative feedback mechanisms operate to regulate blood volume and urine output:

Kidney Dialysis

A dialysis machine is a machine designed to remove wastes from the blood. It is used when the kidneys fail, or when blood acidity, urea, or potassium levels increase much above normal. In kidney dialysis, blood flows through a system of tubes composed of semi-permeable membranes. Dialysis fluid (**dialysate**) has a composition similar to blood except that the concentration of wastes is low. It flows in the opposite direction to the blood on the outside of the dialysis tubes. Consequently, waste products like urea diffuse from the blood into the dialysis fluid, which is constantly replaced. The dialysis fluid flows at a rate of several 100 cm^3 per minute over a large surface area. For some people dialysis is an ongoing procedure, but for others dialysis just allows the kidneys to rest and recover.

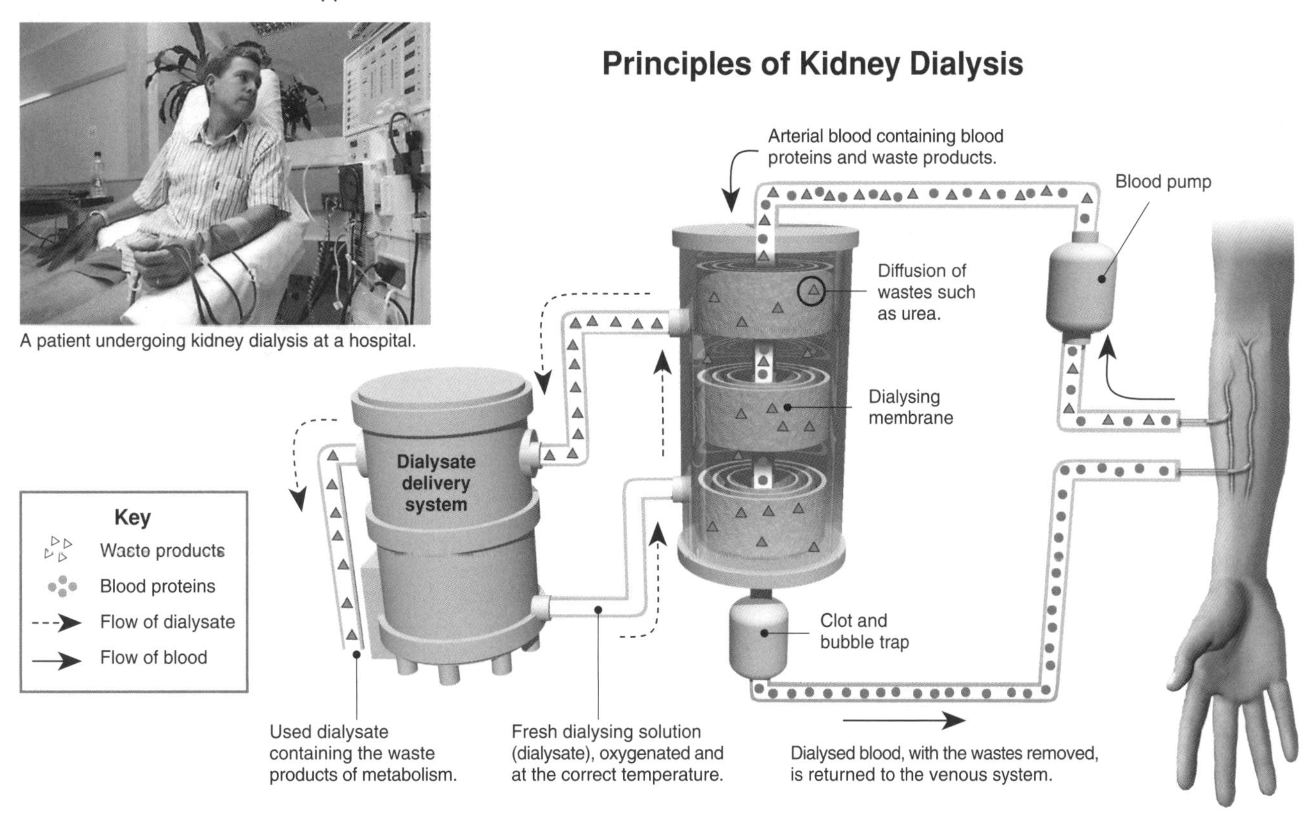

A patient undergoing kidney dialysis at a hospital.

1. In kidney dialysis, explain why the dialysing solution is constantly replaced rather than being recirculated:

2. Explain why ions such as potassium and sodium, and small molecules such as glucose, do not diffuse rapidly from the blood into the dialysing solution along with the urea:

3. Explain why the urea passes from the blood into the dialysing solution:

4. Describe the general transport process involved in dialysis:

5. Give a reason why the dialysing solution flows in the opposite direction to the blood:

6. Explain why a clot and bubble trap is needed after the blood has been dialysed but before it re-enters the body:

Responses and Coordination

AQA-A	AQA-B	CIE	Edexcel	OCR
Complete:	Complete:	Complete:	Complete:	Complete:
1-2, 8-9, 10(c), 11-24, 26-29, 34	1-3, 5, 8-10 (a)-(b), 11-22, 25-26, 28-29	1-2, 11-16, 18-21, 24-26, 37(a)	1-4, 11-16, 18-21, 22(a), 25-26, 29, 37	1-2, 11-16, 18-21, 24-26, 37(a)

Learning Objectives

☐ 1. Compile your own glossary from the **KEY WORDS** displayed in **bold type** in the learning objectives below.

The mammalian nervous system

☐ 2. With respect to structure and function, appreciate the differences between nervous and hormonal coordination (*see "Homeostasis", pp. 218-219*).

The central nervous system *(pages 238, 240-241)*

☐ 3. With respect to structure and function, distinguish between the two primary divisions of the mammalian nervous system: the **central nervous system** (CNS) and the **peripheral nervous system** (PNS).

☐ 4. Recognise main regions of the **brain** and their primary [illegible] **cerebral hemispheres** (cerebrum).

☐ 5. Describe the location and principal functions of the cerebral hemispheres with reference to the following:
- **Sensory** and **motor areas**.
- Association areas, e.g. the **visual association area**.
- The areas associated with speech.

Relate the size of a cerebral region to the complexity of its innervation and describe how each side of the body is controlled by the opposite hemisphere.

☐ 6. EXTENSION: Describe an example of brain malfunction, e.g. as occurs in **Alzheimer's disease**.

The peripheral nervous system *(pages 238-239)*

☐ 7. Explain the basic function of the two divisions of the PNS: **sensory division** and **motor division**.

☐ 8. Appreciate that the motor division of the PNS is further divided into the **autonomic** and **somatic nervous systems** and outline the role of each.

☐ 9. Recognise the components of the autonomic nervous system (**ANS**): the **parasympathetic** and **sympathetic nervous systems**. Describe their roles and recognise that they have generally opposite effects.

☐ 10. Describe examples of the effects of the autonomic nervous system with reference to the following:
(a) Control of pupil diameter and tear production.
(b) Control of bladder emptying.
(c) Control of heart rate and force of contraction.
(d) Control of digestive secretions and gut motility.

Understand that some reflex autonomic activity can be modified by conscious control, and provide an example.

Neurone structure and function *(pages 242-244)*

☐ 11. Understand the term **neurone**. Recognise the structural and functional features that distinguish different types of neurones: **sensory**, **effector** (**motor**), and **relay neurones** (**interneurones**).

☐ 12. Draw a labelled diagram to illustrate the structure of a **myelinated motor neurone**. Appreciate the role of the **Schwann cells** in the **myelination** of a motor neurone.

☐ 13. Explain what is meant by a **reflex**. Using a diagram, describe the functioning of a simple spinal reflex arc involving three neurones and identify the neurone types involved. Appreciate the adaptive value of reflexes.

☐ 14. Explain what is meant by the **resting potential** of a neurone and explain how it is established. Include reference to **ion pumps** and the movement of Na^+ and K^+, the differential permeability of the membrane, and the generation of an **electrochemical gradient**.

☐ 15. Describe the generation of the **action potential** (**nerve impulse**) with reference to the change in membrane permeability of the nerve leading to **depolarisation** [illegible] and the **all-or-nothing** nature of the [illegible]

[illegible] along a [illegible] **saltatory conduction**, identifying the roles of **myelin** and the **nodes of Ranvier**.

☐ 17. Contrast the nature of the nerve impulse in **myelinated** and **non-myelinated** fibres and explain reasons for the difference. Describe how myelination, axon diameter, and temperature affect speed of conductance.

☐ 18. Explain the nature and importance of the **refractory period** in producing discrete impulses.

Synapses *(pages 245-247)*

☐ 19. Identify the role of synapses in mammals. Describe the structure of a **cholinergic synapse** (as seen using electron microscopy). Recognise the **neuromuscular junction** as a specialised cholinergic synapse.

☐ 20. Outline the sequence of events involved in the action of a **cholinergic synapse**. Include reference to the arrival of the **action potential** at the **presynaptic terminal**, the role of Ca^{2+} and **acetylcholine**, and the generation of the action potential in the **postsynaptic neurone**.

☐ 21. Understand that transmission at chemical synapses involves **neurotransmitters** and understand how these differ functionally from hormones. Identify the roles of **acetylcholine** and **noradrenaline** at different synapses.

☐ 22. Describe the effects of drugs at synapses as shown by:
(a) Amplification at the synapse by mimicking the action of natural transmitters, e.g. **nicotine**, caffeine.
(b) Amplification at the synapse by preventing neurotransmitter breakdown, e.g. strychnine, nerve gas.
(c) Inhibition at the synapse by blocking transmission across the synapse, e.g. atropine, curare.

☐ 23. Understand the terms: **agonist** and **antagonist** with respect to the actions of drugs at synapses.

☐ 24. Appreciate the role of synapses in:
(a) **Unidirectionality**: impulses travel in one direction.
(b) Allowing the interconnection of nerve pathways.
(c) **Integration** through **summation** and **inhibition**.

Sensory reception in mammals *(pages 215, 248)*

☐ 25. Outline the need for communication systems to respond to environmental **stimuli**. Distinguish between internal and external stimuli and identify some of the types of **sense organs** that respond to these.

☐ 26. Appreciate the role of sensory receptors in converting different forms of energy into nerve impulses. With reference to specific examples, understand the following features of sensory receptors:
- (a) They respond only to specific stimuli.
- (b) They respond to **stimulation** by establishing generator potentials. Generator potentials that reach a **threshold** will produce action potentials.
- (c) The strength of receptor response is proportional to the stimulus strength.
- (d) They show **sensory adaptation**.

☐ 27. Describe the structure of a sensory receptor as illustrated by an appropriate example, e.g. a **Pacinian corpuscle** or a stretch receptor. Identify the stimulus for the response in the receptor, and explain how the response is brought about.

The mammalian eye *(pages 249-251)*

☐ 28. Describe the basic structure and properties of the mammalian **eye** with respect to the transmission and refraction of light, and the focusing of an image on the **retina**. Include reference to the following:
- (a) The structure and function of the iris.
- (b) The role of the cornea, lens, ciliary muscles, and suspensory ligaments (including **accommodation**).
- (c) The significance of the central fovea and blind spot.

☐ 29. Describe the structure and function of the photoreceptor cells (**rods** and **cones**) in the retina, including:
- (a) The role of the photosensitive pigments (e.g. **rhodopsin**) and their response to absorbed light.
- (b) The basis of monochromatic and trichromatic vision.
- (c) The creation of generator potentials in the rod or cone and action potentials in the optic nerve.
- (d) The reasons for differences in sensitivity and acuity.

☐ 30. In simple terms, explain how the nerve impulses from the eye are interpreted.

The mammalian ear *(page 252)*

☐ 31. Understand the nature of sound as a stimulus. Describe the basic structure of the mammalian **ear**, and outline the function of each of its main regions.

☐ 32. Describe how sound is received by the ear and transmitted to the sensory cells in the **organ of Corti**.

☐ 33. Explain how the sense cells produce a nerve impulse and identify the regions of the brain involved in processing auditory information.

Animal behaviour *(pages 253-258)*

☐ 34. Describe examples of simple responses as illustrated by **taxes** and **kineses**. Identify the adaptive role of these and other simple behaviours (e.g. reflexes).

☐ 35. Distinguish between **innate** and **learned behaviour**. Describe examples of learned behaviours, identifying their adaptive role in each case. Include reference to:
- **Habituation** and **imprinting**.
- **Classical** and **operant conditioning**.

☐ 36. Describe each of the following as they relate to successful breeding: **courtship behaviour**, including the role of hormones or pheromones, **species recognition**, and **pair bonding**. Appreciate the stereotyped nature of courtship displays, including the role of sign stimuli and innate releasing mechanisms.

Responses in plants *(pages 259-264)*

☐ 37. Appreciate the need for communication systems in flowering plants to respond to environmental stimuli. Describe responses in plants and their physiological basis, including, where appropriate, the commercial significance of these. Include reference to the following:
- (a) The role of **auxins**, **gibberellins**, and **abscisic acid** on plant growth and development.
- (b) The role of **cytokinins** and **ethene** on plant growth.
- (c) The role of **phytochrome pigments** in the detection and response to daylength.
- (d) The nature of **tropisms**, especially phototropism.
- (e) The nature of plant defence mechanisms.

Textbooks

See the 'Textbook Reference Grid' on pages 8-9 for textbook page references relating to material in this topic.

Supplementary Texts

See pages 4-6 for additional details of these texts:

- Adds, J. *et al.* 2001. **Respiration and Coordination** (NelsonThornes), pp. 14-33.
- Clegg, C.J., 1998. **Mammals: Structure and Function** (John Murray), pp. 58-69.
- Dockerty, M. & M. Reiss, 1999. **Behaviour** (CUP), entire text as required.
- Jones, M. & G. Jones, 2002. **Mammalian Physiology and Behaviour** (CUP), pp. 48-91.
- Murray, P. and N. Owens, 2001. **Behaviour and Populations** (Collins), pp. 6-27.
- Rowett, H.G.Q., 1999. **Basic Anatomy & Physiology** (John Murray), pp. 60-72.

See page 6 for details of publishers of periodicals:

STUDENT'S REFERENCE

- **All the Better to See You With** Biol. Sci. Rev., 8(2) Nov. 1995, pp. 30-33. *Eye structure & function.*
- **The Autonomic Nervous System** Bio. Sci. Rev. 11(4) March 1999, pp. 30-34. *The structure and function of the autonomic nervous system.*
- **What's Your Poison** Bio. Sci. Rev. 16(2) Nov. 2003, pp. 33-37. *The action of naturally derived poisons on synaptic transmission. This account includes an account of toxins and their actions.*
- **Synapses** Biol. Sci. Rev., 8(4) March 1996, pp. 26-29. *Types and functions of synapses including synaptic plasticity and the neural basis of learning.*
- **The Nervous System** (series) New Scientist, 10 June 1989, 11 Nov. 1989, 29 June 1991 (Inside Science). *Nervous system structure and function.*
- **Animals in Action** Biol. Sci. Rev., 7(2) Nov. 1994, pp. 31-35. *The study of animal behaviour, including simple behaviours (taxes and kineses).*
- **Serotonin - The Brain's Mood Modulator** Biol. Sci. Rev., 12(1), Sept. 1999, pp. 28-31. *The role of serotonin as a neurotransmitter.*
- **Making the Connection** Biol. Sci. Rev.,13(3) January 2001, pp. 10-13. *The central nervous system, neurotransmitters, and synapses.*
- **A Receptor** Biol. Sci. Rev., 12(2) Nov. 1999, pp. 6-7. *The nature of sensory receptors.*
- **A Pacinian Corpuscle** Biol. Sci. Rev., 12(3) Jan. 2000, pp. 33-34. *An account of the structure and operation of a common pressure receptor.*
- **Before your Very Eyes** New Scientist 15 March 1997 (Inside Science). *Excellent account of eye structure, & perception & processing of visual info.*
- **What the Nose Knows** Biol. Sci. Rev., 7(5) May 1995, pp. 2-5. *Physiochemical basis of smell, including scents and pheromones in life cycles.*
- **Sending Plants around the Bend** Biol. Sci. Rev., 12(4) March 2000, pp. 14-17. *How plants perceive and respond to stimuli around them.*

See pages 10-11 for details of how to access **Bio Links** from our web site: **www.biozone.co.uk**. From Bio Links, access sites under the topics:

GENERAL BIOLOGY ONLINE RESOURCES > Online Textbooks and Lecture Notes: • S-Cool! A level biology revision guide • Learn.co.uk • Mark Rothery's biology web site ... *and others*

ANIMAL BEHAVIOUR: • Animal behaviour • Innate behaviour • Ken's bioweb resources: Animal behaviour • Pavlovian conditioning • Sign stimuli and motivation... *and others*

ANIMAL BIOLOGY > Neuroscience: • Anatomy of the human ear • Basic neural processes • Cow's eye dissection • Nervous system • Seeing, hearing, smelling the world • The human eye • The effects of LSD on the human brain • Virtual tour of the ear: hearing mechanism • The secret of the brain... *and others*

PLANT BIOLOGY: • Kimball's plant biology lecture notes ... *and others* > **Hormones & Responses**: • Botany online: Plant hormones • Plant hormones • Plant hormones and growth regulators • Plant tropisms • Plants in motion

Software and video resources are provided on the Teacher Resource Handbook on CD-ROM

The Mammalian Nervous System

The **nervous system** is the body's control and communication centre. It has three broad functions: detecting stimuli, interpreting them, and initiating appropriate responses. Its basic structure is outlined below. Further detail is provided in the following pages.

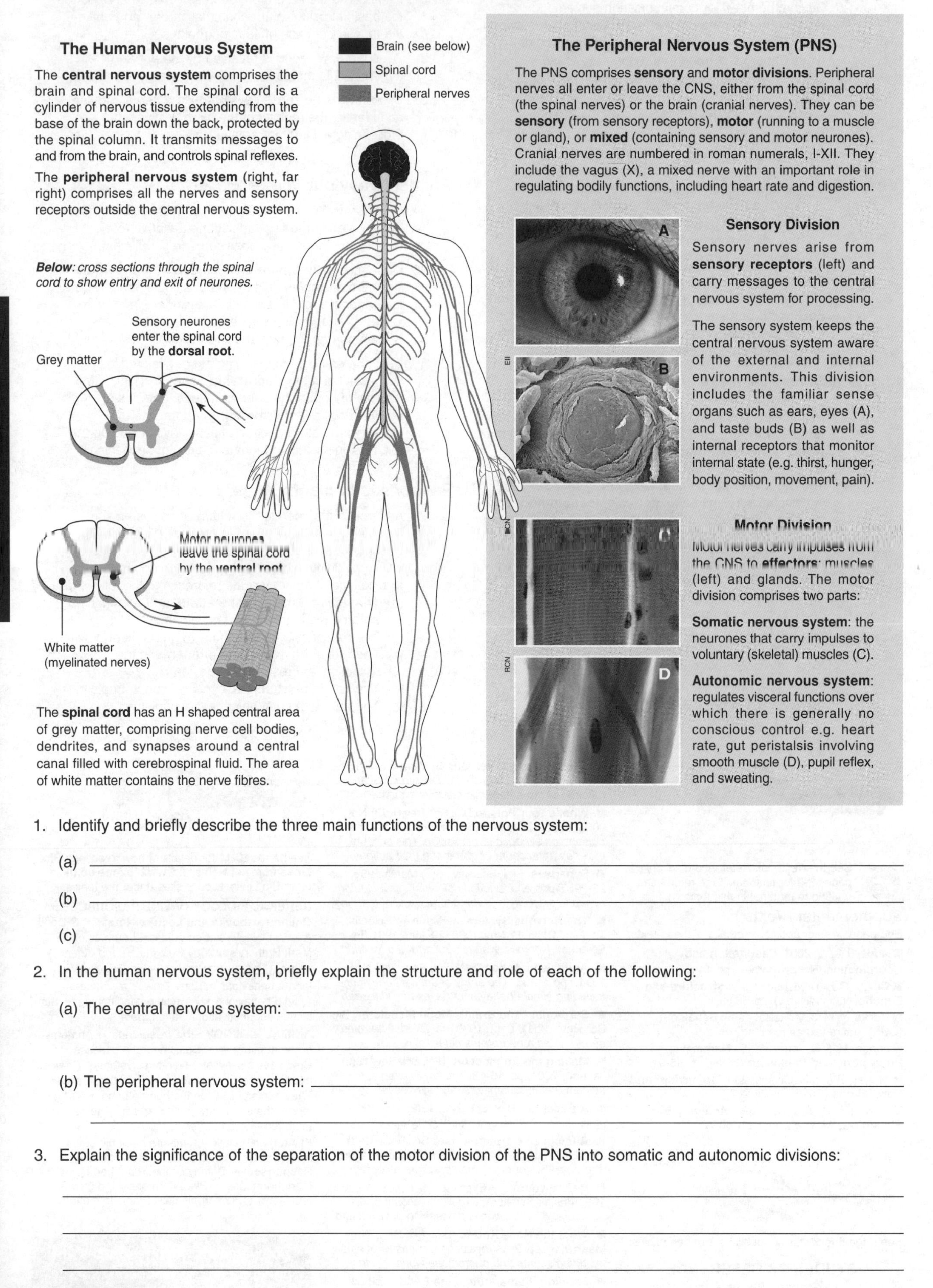

The Human Nervous System

The **central nervous system** comprises the brain and spinal cord. The spinal cord is a cylinder of nervous tissue extending from the base of the brain down the back, protected by the spinal column. It transmits messages to and from the brain, and controls spinal reflexes.

The **peripheral nervous system** (right, far right) comprises all the nerves and sensory receptors outside the central nervous system.

Below: *cross sections through the spinal cord to show entry and exit of neurones.*

The **spinal cord** has an H shaped central area of grey matter, comprising nerve cell bodies, dendrites, and synapses around a central canal filled with cerebrospinal fluid. The area of white matter contains the nerve fibres.

The Peripheral Nervous System (PNS)

The PNS comprises **sensory** and **motor divisions**. Peripheral nerves all enter or leave the CNS, either from the spinal cord (the spinal nerves) or the brain (cranial nerves). They can be **sensory** (from sensory receptors), **motor** (running to a muscle or gland), or **mixed** (containing sensory and motor neurones). Cranial nerves are numbered in roman numerals, I-XII. They include the vagus (X), a mixed nerve with an important role in regulating bodily functions, including heart rate and digestion.

Sensory Division

Sensory nerves arise from **sensory receptors** (left) and carry messages to the central nervous system for processing.

The sensory system keeps the central nervous system aware of the external and internal environments. This division includes the familiar sense organs such as ears, eyes (A), and taste buds (B) as well as internal receptors that monitor internal state (e.g. thirst, hunger, body position, movement, pain).

Motor Division

Motor nerves carry impulses from the CNS to **effectors**: muscles (left) and glands. The motor division comprises two parts:

Somatic nervous system: the neurones that carry impulses to voluntary (skeletal) muscles (C).

Autonomic nervous system: regulates visceral functions over which there is generally no conscious control e.g. heart rate, gut peristalsis involving smooth muscle (D), pupil reflex, and sweating.

1. Identify and briefly describe the three main functions of the nervous system:

   (a) ______________________________

   (b) ______________________________

   (c) ______________________________

2. In the human nervous system, briefly explain the structure and role of each of the following:

   (a) The central nervous system: ______________________________

   (b) The peripheral nervous system: ______________________________

3. Explain the significance of the separation of the motor division of the PNS into somatic and autonomic divisions:

____

The Autonomic Nervous System

The **autonomic nervous system** (ANS) regulates involuntary visceral functions by means of **reflexes**. Although most autonomic nervous system activity is beyond our conscious control, voluntary control over some basic reflexes (such as bladder emptying) can be learned. Most visceral effectors have dual innervation, receiving fibres from both branches of the ANS. These two branches, the **parasympathetic** and **sympathetic** divisions, have broadly opposing actions on the organs they control (excitatory or inhibitory). Nerves in the parasympathetic division release acetylcholine. This neurotransmitter is rapidly deactivated at the synapse and its effects are short lived and localised. Most sympathetic postganglionic nerves release noradrenaline, which enters the bloodstream and is deactivated slowly. Hence, sympathetic stimulation tends to have more widespread and long lasting effects than parasympathetic stimulation. Aspects of autonomic nervous system structure and function are illustrated below. The arrows indicate nerves to organs or ganglia (concentrations of nerve cell bodies).

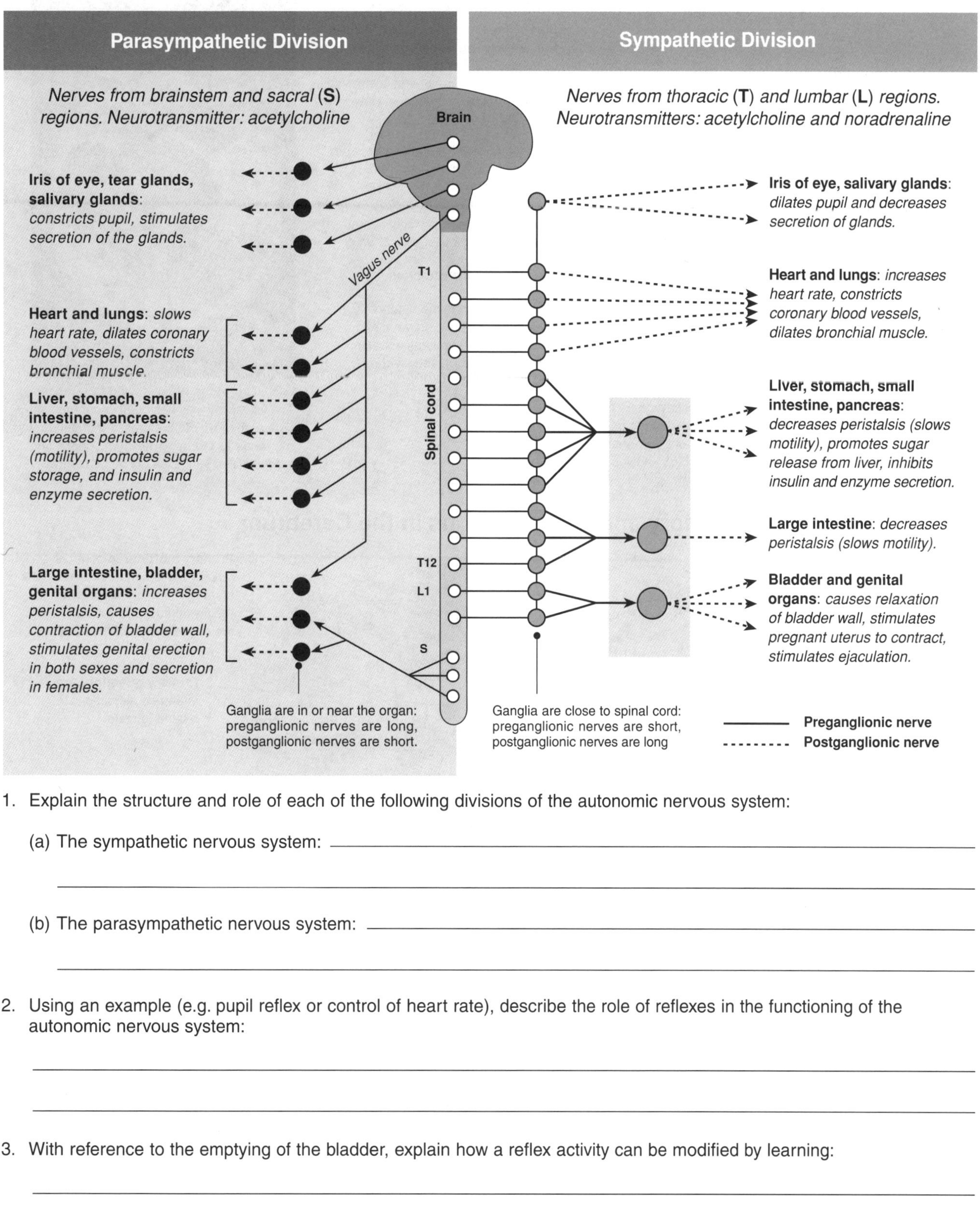

1. Explain the structure and role of each of the following divisions of the autonomic nervous system:

   (a) The sympathetic nervous system: ______

   (b) The parasympathetic nervous system: ______

2. Using an example (e.g. pupil reflex or control of heart rate), describe the role of reflexes in the functioning of the autonomic nervous system:

   ______

3. With reference to the emptying of the bladder, explain how a reflex activity can be modified by learning:

   ______

The Human Brain

The brain is one the largest organs in the body. It is protected by the skull, the meninges (membranous coverings), and the cerebrospinal fluid (CSF). The brain is the control centre for the body. It receives a constant flow of information from the senses, but responds only to what is important at the time. Some responses are very simple (e.g. cranial reflexes), whilst others require many levels of processing. The human brain is noted for its large, well developed cerebral region, and the region responsible for complex thought and reasoning. Each cerebral hemisphere is divided into four lobes by deep sulci or fissures. These lobes: temporal, frontal, occipital, and parietal, correspond to the bones of the skull under which they lie.

Primary Structural Regions of the Brain

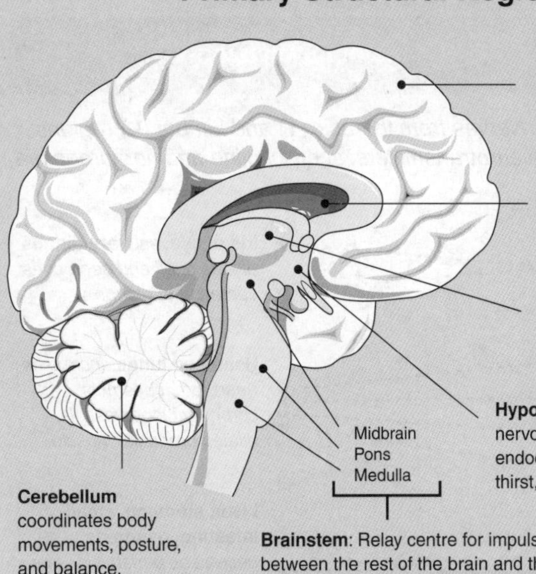

Cerebrum: the cerebrum is divided into two cerebral hemispheres and has many, complex roles. It contains sensory, motor, and association areas, and is involved in memory, emotion, language, reasoning, and sensory processing.

Ventricles: Cavities containing the CSF, which absorbs shocks and delivers nutritive substances.

Thalamus is the main relay centre for all sensory messages that enter the brain, before they are transmitted to the cerebrum.

Hypothalamus controls the autonomic nervous system and links nervous and endocrine systems. Regulates appetite, thirst, body temperature, and sleep.

Cerebellum coordinates body movements, posture, and balance.

Brainstem: Relay centre for impulses between the rest of the brain and the spinal cord. Controls breathing, heartbeat, and the coughing and vomiting reflexes.

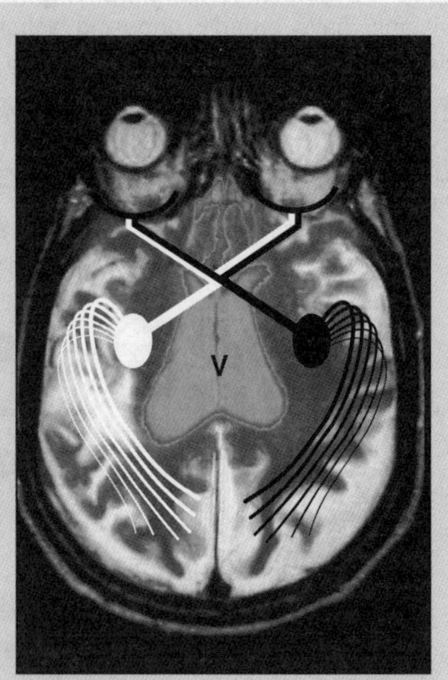

MRI scan of the brain viewed from above. The visual pathway has been superimposed on the image. Note the crossing of sensory neurones to the opposite hemisphere and the fluid filled ventricles (V) in the centre.

Sensory and Motor Regions in the Cerebrum

General sensory area receives sensations from receptors in the skin, muscles and viscera. Sensory information from receptors on one side of the body crosses over to the opposite side of the cerebral cortex where conscious sensations are produced. The size of the sensory region for different body parts depends on the number of receptors in that particular body part.

Primary motor area controls muscle movement. Stimulation of a point one side of the motor area results in muscular contraction on the opposite side of the body.

Parietal lobe

Primary gustatory area interprets sensations related to taste.

Frontal lobe

Occipital lobe

Temporal lobe

Visual areas within the occipital lobe receive, interpret, and evaluate visual stimuli. In vision, each eye views both sides of the visual field but the brain receives impulses from left and right visual fields separately (see photo caption above). The visual cortex combines the images into a single impression or **perception** of the image.

Auditory areas interpret the basic characteristics and meaning of sounds.

Language areas: The motor speech area (Broca's area) is concerned with speech production. The sensory speech area (Wernicke's area) is concerned with speech recognition and coherence.

1. For each of the following bodily functions, identify the region(s) of the brain involved in its control:

   (a) Breathing and heartbeat: ______________________

   (b) Memory and emotion: ______________________

   (c) Posture and balance: ______________________

   (d) Autonomic functions: ______________________

   (e) Visual processing: ______________________

   (f) Body temperature: ______________________

   (g) Language: ______________________

The Malfunctioning Brain: The Effects of Alzheimer's Disease

Alzheimer's disease is a disabling neurological disorder affecting about 5% of the population over 65. Its causes are largely unknown, its effects are irreversible, and it has no cure. Sufferers of Alzheimer's have trouble remembering recent events and they become confused and forgetful. In the later stages of the disease, people with Alzheimer's become very disorientated, lose past memories, and may become paranoid and moody. Dementia and loss of reason occur at the end stages of the disease.

Cerebral cortex: Conscious thought, reasoning, and language. Alzheimer's sufferers show considerable loss of function from this region.

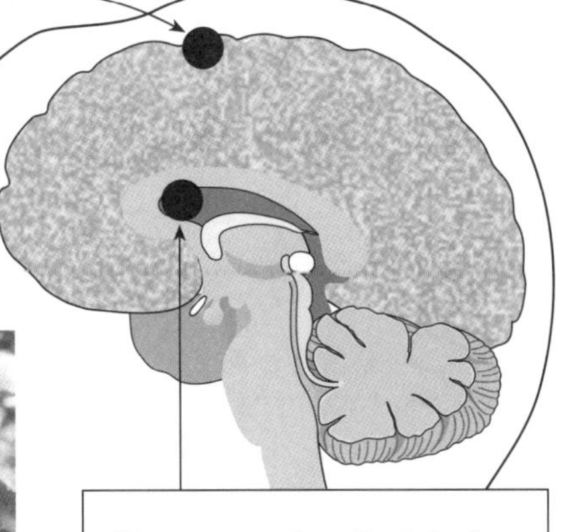

Hippocampus: A swelling in the floor of the lateral ventricle. It contains complex foldings of the cortical tissue and is involved in the establishment of memory patterns. In Alzheimer's sufferers, it is one of the first regions to show loss of neurones and accumulation of amyloid.

It is not uncommon for Alzheimer's sufferers to wander and become lost and disorientated.

Upper Brain	Lower Brain
Normal	
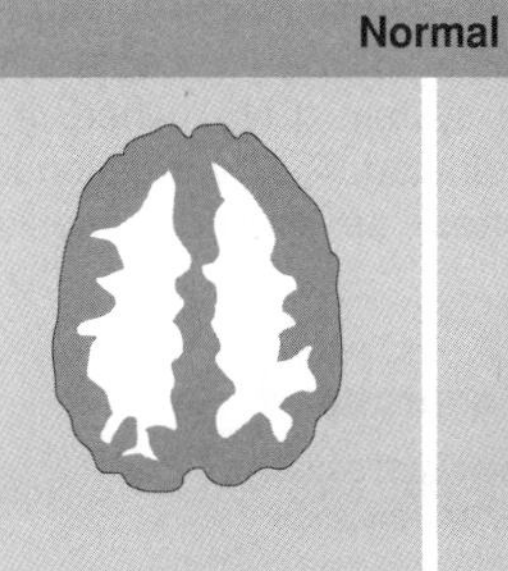	
Early Alzheimer's	
	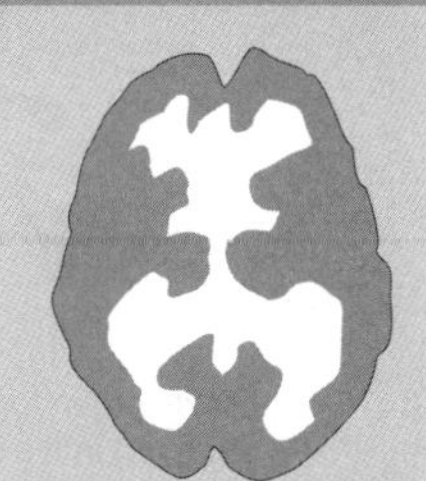
Late Alzheimer's	
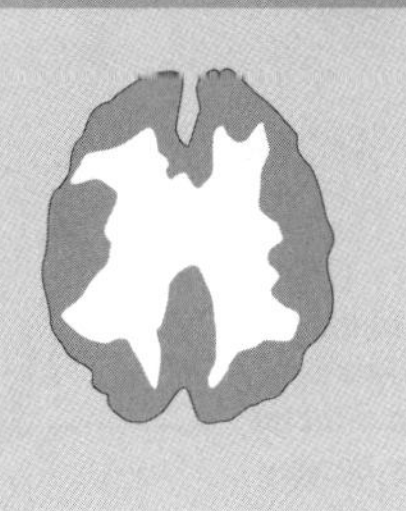	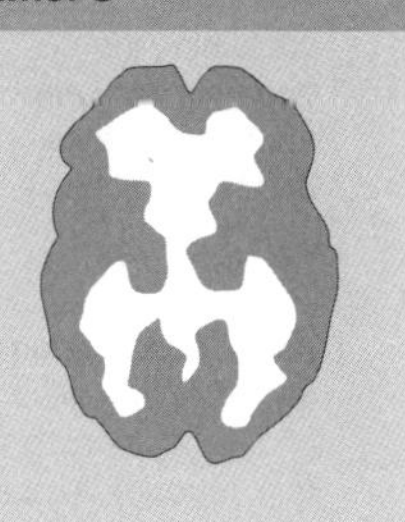

Alzheimer's is associated with accelerated loss of neurones, particularly in regions of the brain that are important for memory and intellectual processing, such as the cerebral cortex and hippocampus. The disease has been linked to abnormal accumulations of protein-rich **amyloid** plaques, which invade the brain tissue and interfere with synaptic transmission. The brain scans above show diminishing brain function in certain areas of the brain in Alzheimer's sufferers. Note, particularly in the scans to the left, how the brain has shrunk. Light areas indicate brain activity.

(h) Muscular contraction: ____________________

(i) Sensory processing related to taste: ____________________

(j) Sensory processing related to sound: ____________________

2. Describe the likely effect of a loss of function (through injury) to the primary motor area in the left hemisphere:

____

____

3. Describe the role of the ventricles of the brain: ____________________

____

____

4. Some loss of neurones occurs normally as a result of ageing. Identify the features distinguishing Alzheimer's disease from normal age related neurone loss:

____

____

____

Neurone Structure and Function

The nervous and endocrine systems are the body's regulatory and coordinating systems. Homeostasis depends on the ability of the nervous system to detect, interpret, and respond to, internal and external conditions. Sensory receptors relay information to the central nervous system (CNS) where it is interpreted and responses are coordinated. The information is transmitted along nerve cells (**neurones**) as electrical impulses. The speed of impulse conduction depends primarily on the axon diameter and whether or not the axon is myelinated (see below). Within the tolerable physiological range, an increase in temperature also increases the speed of impulse conduction: in cool environments, impulses travel faster in endothermic than in ectothermic vertebrates. Neurones typically consist of a cell body, dendrites, and an axon. Basic types are described below.

The Structure of Neurones

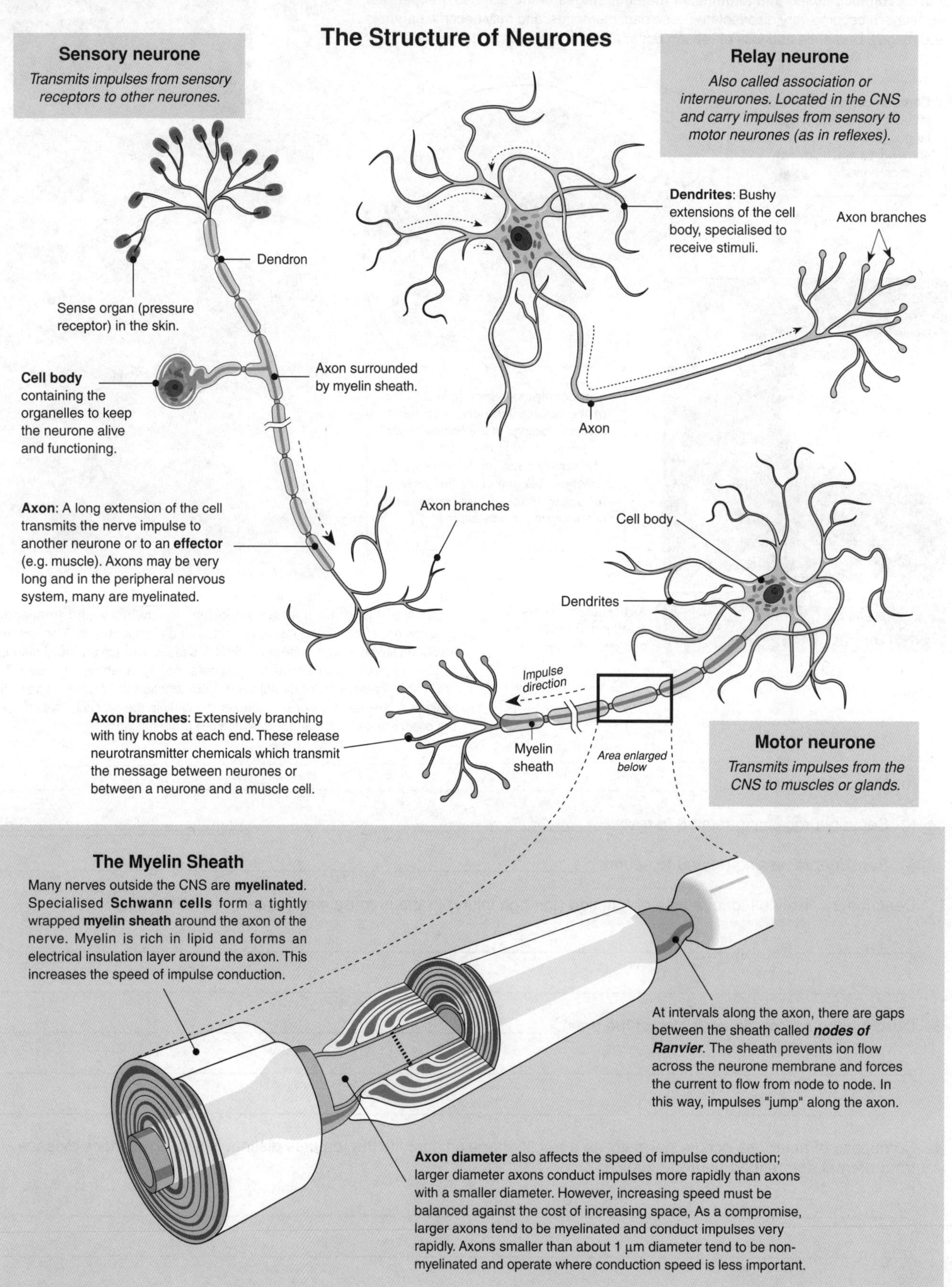

A Polysynaptic Reflex Arc: Pain Withdrawal

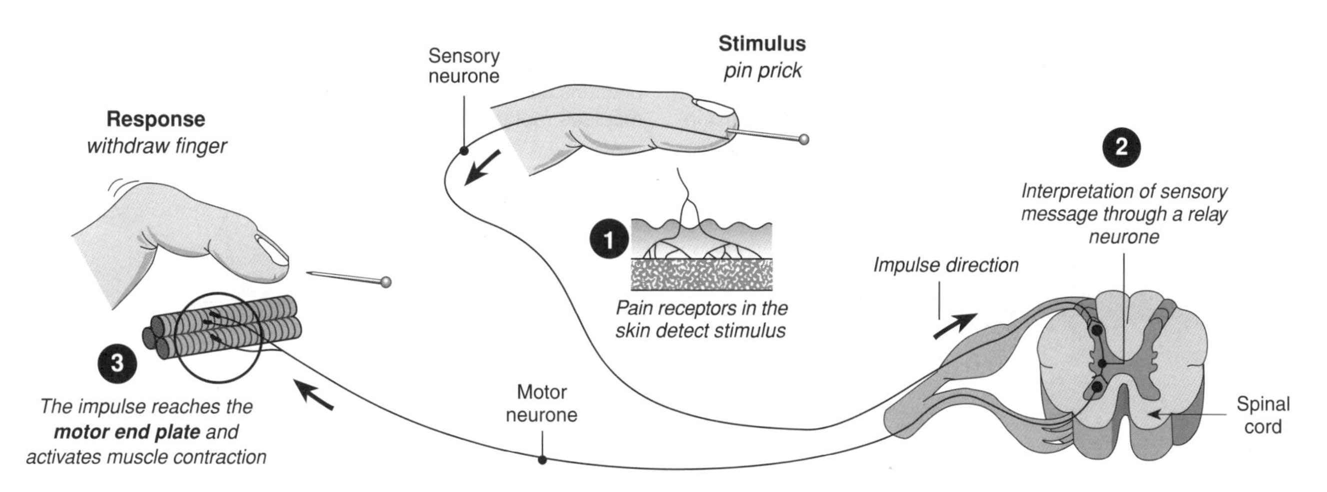

A reflex is an automatic response to a stimulus involving a small number of neurones and a central nervous system (CNS) processing point (usually the spinal cord, but sometimes the brain stem). This type of circuit is often called a **reflex arc**. Reflexes permit rapid responses to stimuli. They are classified according to the number of CNS synapses involved: *monosynaptic reflexes* involve only one CNS synapse (e.g. knee jerk), *polysynaptic reflexes* involve two or more (e.g. pain withdrawal reflex). Both are spinal reflexes. The pupil reflex (opening and closure of the pupil) is an example of a cranial reflex.

1. (a) Describe a structural difference between a motor and a sensory neurone: ______________________

   (b) Describe a functional difference between a motor and a relay neurone: ______________________

2. (a) Predict what would happen to the part of an axon if it is cut so that it is no longer connected to its nerve cell body:

   ____

   (b) Explain your prediction: ______________________

3. (a) Describe one way (other than insulation of the axon) in which impulse conduction speed could be increased:

   ____

   (b) Name one animal that uses this method: ______________________

   (c) Describe the adaptive advantage of faster conduction of nerve impulses: ______________________

4. (a) Briefly, describe the cause of the disorder, multiple sclerosis (MS): ______________________

   (b) Explain why MS impairs nervous system function even though axons are undamaged: ______________________

5. (a) Explain why higher reasoning or conscious thought are not necessary or desirable features of reflexes:

   ____

   (b) Explain when it might be adaptive for conscious thought to intervene and modify a reflex action: ______________________

6. Distinguish between a spinal and a cranial reflex and give an example of each: ______________________

Transmission of Nerve Impulses

Neurones, like all cells, contain ions or charged atoms. Those of special importance include sodium (Na^+), potassium (K^+), and negatively charged proteins. Neurones are **electrically excitable** cells: a property that results from the separation of ion charge either side of the neurone membrane. They may exist in either a resting or stimulated state. When stimulated, neurones produce electrical impulses that are transmitted along the axon. These impulses are transmitted between neurones across junctions called **synapses**. Synapses enable the transmission of impulses rapidly all around the body.

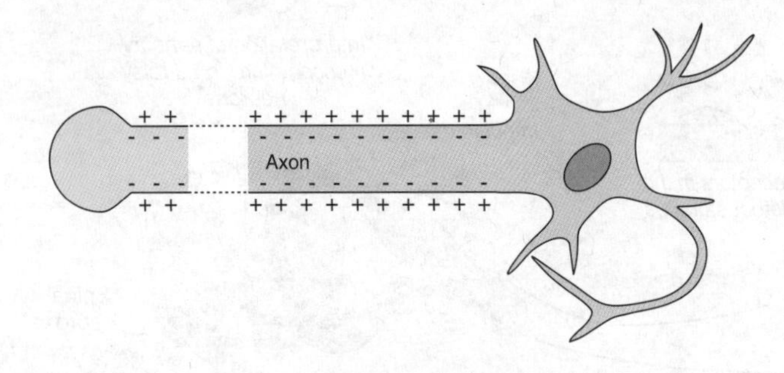

Impulse travels in this direction

Na^+ Na^+

Area of impulse

Next area to be stimulated

Area returning to resting state

The Resting Neurone

When a neurone is not transmitting an impulse, the inside of the cell is negatively charged compared with the outside of the cell. The cell is said to be electrically polarised, because the inside and the outside of the cell are oppositely charged. The potential difference (voltage) across the membrane is called the resting potential and for most nerve cells is about -70 mV. Nerve transmission is possible because this membrane potential exists.

The Nerve Impulse

When a neurone is stimulated, the distribution of charges on each side of the membrane changes. For a millisecond, the charges reverse. This process, called **depolarisation**, causes a burst of electrical activity to pass along the axon of the neurone. As the charge reversal reaches one region, local currents depolarise the next region. In this way the impulse spreads along the axon. An impulse that spreads this way is called an **action potential**.

The Action Potential

The depolarisation described above can be illustrated as a change in membrane potential (in millivolts). In order for an action potential to be generated, the stimulation must be strong enough to reach the **threshold** potential; the potential (voltage) at which the depolarisation of the membrane becomes "unstoppable" and the action potential is generated. The action potential is **all or none** in its generation. Either the **threshold** is reached and the action potential is generated or the nerve does not fire. The resting potential is restored by the movement of potassium ions (K^+) out of the cell. During this **refractory period**, the nerve cannot respond.

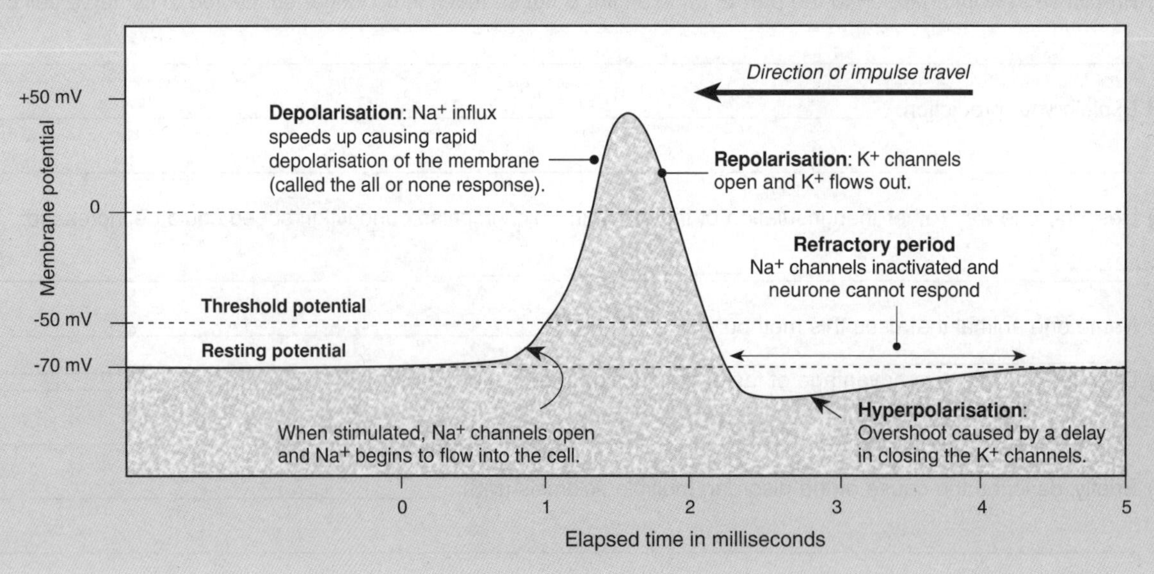

1. Explain how an action potential is able to pass along a nerve: ____________________

2. Explain how the refractory period influences the direction in which an impulse will travel: ____________________

3. Action potentials themselves are indistinguishable from each other. Explain how the nervous system is able to interpret the impulses correctly and bring about an appropriate response: ____________________

Chemical Synapses

Action potentials are transmitted between neurones across synapses: junctions between the end of one axon and the dendrite or cell body of a receiving neurone. **Chemical synapses** are the most widespread type of synapse in nervous systems. The axon terminal is a swollen knob, and a small gap separates it from the receiving neurone. The synaptic knobs are filled with tiny packets of chemicals called **neurotransmitters**. Transmission involves the diffusion of the neurotransmitter across the gap, where it interacts with the receiving membrane and causes an electrical response. The response of a receiving cell to the arrival of a neurotransmitter depends on the nature of the cell itself, on its location in the nervous system, and on the neurotransmitter involved. Synapses that release ACh are termed **cholinergic**. In the example below, acetylcholine (ACh) causes membrane depolarisation and the generation of an action potential (excitation).

The Structure of a Cholinergic Synapse

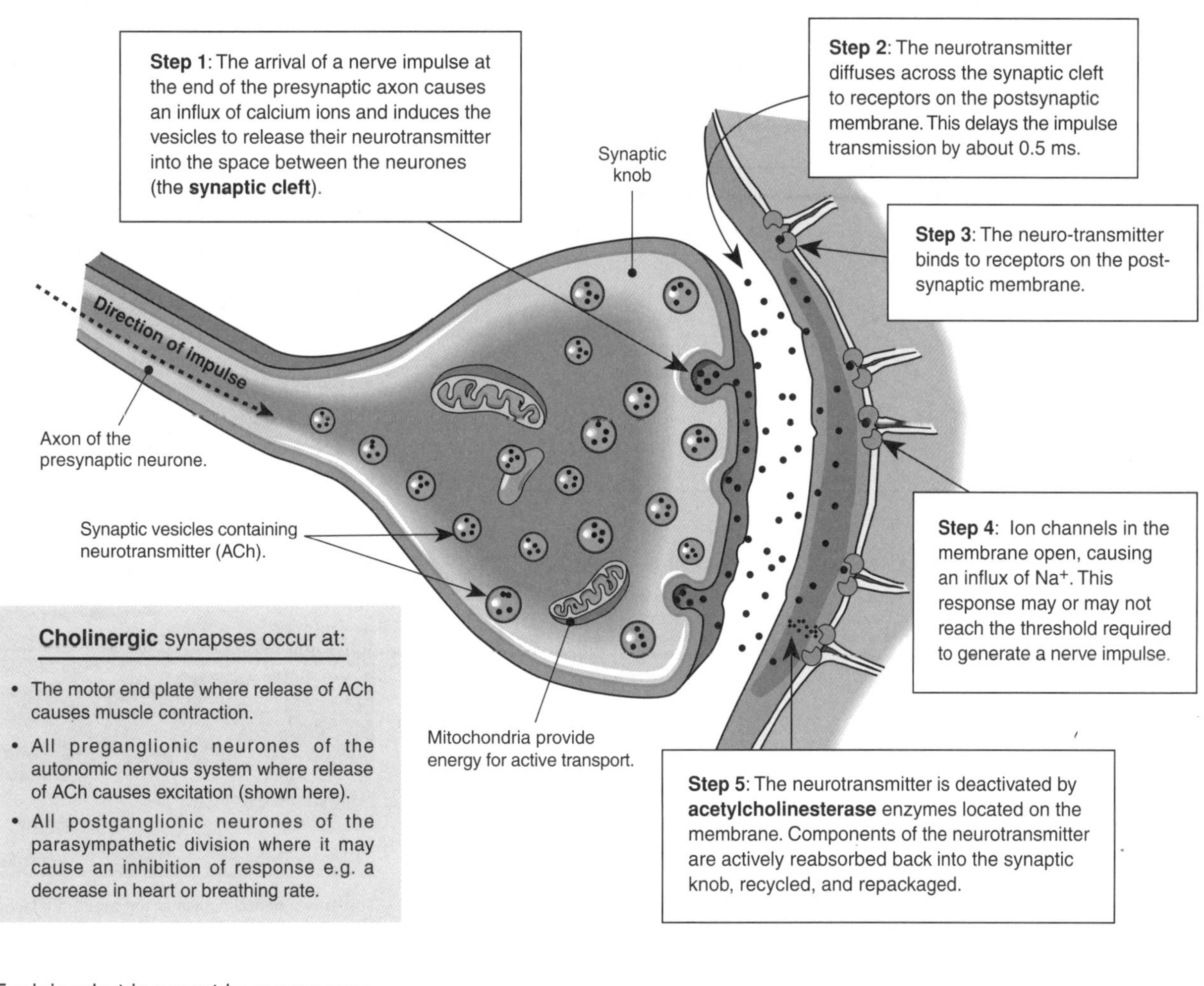

Cholinergic synapses occur at:

- The motor end plate where release of ACh causes muscle contraction.
- All preganglionic neurones of the autonomic nervous system where release of ACh causes excitation (shown here).
- All postganglionic neurones of the parasympathetic division where it may cause an inhibition of response e.g. a decrease in heart or breathing rate.

1. Explain what is meant by a synapse: ______

2. Explain what causes the release of neurotransmitter into the synaptic cleft: ______

3. State why there is a brief delay in transmission of an impulse across the synapse: ______

4. (a) State how the neurotransmitter is deactivated: ______

   (b) Explain why it is important for the neurotransmitter substance to be deactivated soon after its release: ______

5. Consult a reference source to identify one function of acetylcholine in the nervous system: ______

6. Suggest one factor that might influence the strength of the response in the receiving cell: ______

Integration at Synapses

Synapses play a pivotal role in the ability of the nervous system to respond appropriately to stimulation and to adapt to change. The nature of synaptic transmission allows the **integration** (interpretation and coordination) of inputs from many sources. These inputs need not be just excitatory (causing depolarisation). Inhibition results when the neurotransmitter released causes negative chloride ions (rather than sodium ions) to enter the postsynaptic neurone. The postsynaptic neurone then becomes more negative inside (hyperpolarised) and an action potential is less likely to be generated. At synapses, it is the sum of **all** inputs (excitatory and inhibitory) that leads to the final response in a postsynaptic cell. Integration at synapses makes possible the various responses we have to stimuli. It is also the most probable mechanism by which learning and memory are achieved.

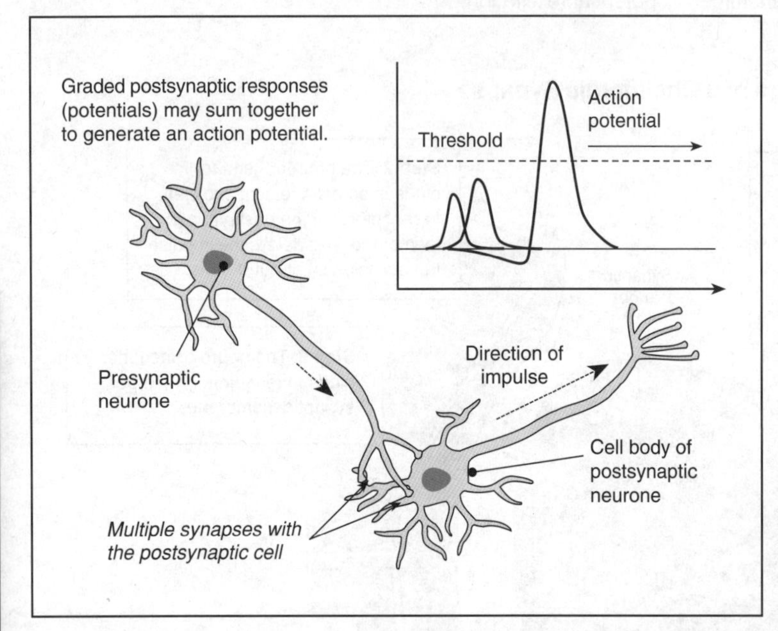

Synapses and Summation

Nerve transmission across chemical synapses has several advantages, despite the delay caused by neurotransmitter diffusion. Chemical synapses transmit impulses in one direction to a precise location and, because they rely on a limited supply of neurotransmitter, they are subject to fatigue (inability to respond to repeated stimulation). This protects the system against overstimulation.

Synapses also act as centres for the **integration** of inputs from many sources. The response of a postsynaptic cell is often graded; it is not strong enough on its own to generate an action potential. However, because the strength of the response is related to the amount of neurotransmitter released, subthreshold responses can sum to produce a response in the post-synaptic cell. This additive effect is termed **summation**. Summation can be **temporal** or **spatial** (below left and centre). A neuromuscular junction (photo below) is a specialised form of synapse between a motor neurone and a skeletal muscle fibre. Functionally, it is similar to any excitatory cholinergic synapse.

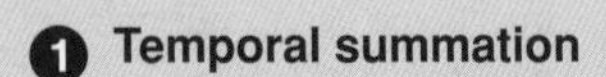

1 Temporal summation

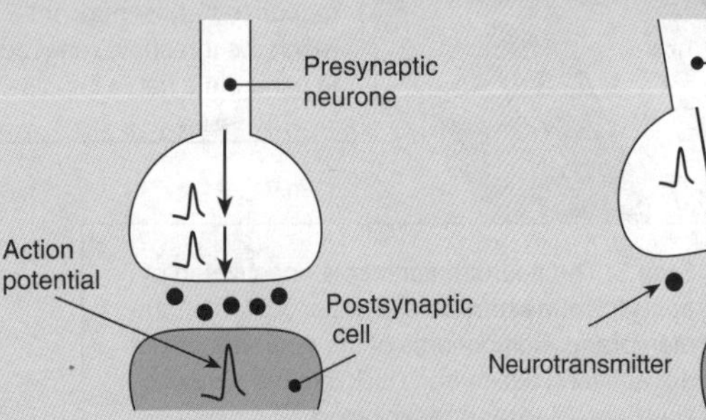

Several impulses may arrive at the synapse in quick succession from a single axon. The individual responses are so close together in time that they sum to reach threshold and produce an action potential in the postsynaptic neurone.

2 Spatial summation

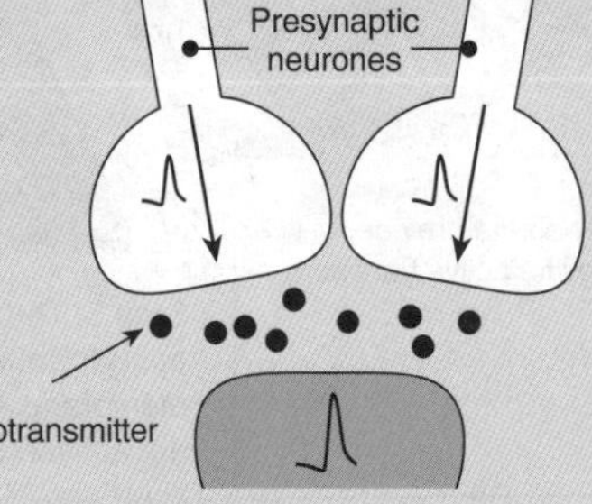

Individual impulses from spatially separated axon terminals may arrive **simultaneously** at different regions of the same postsynaptic neurone. The responses from the different places sum to reach threshold and produce an action potential.

3 Neuromuscular junction

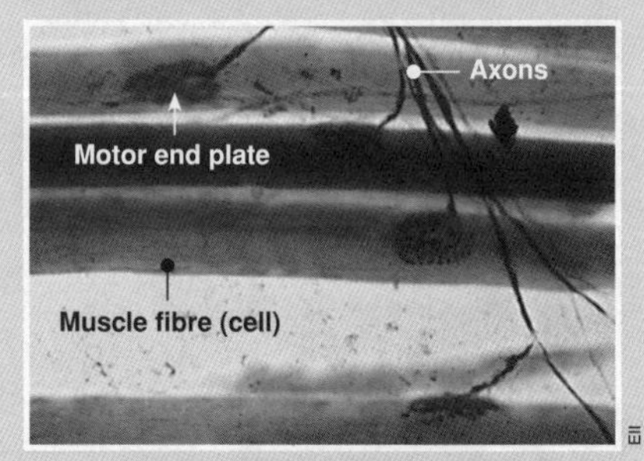

The arrival of an impulse at the neuromuscular junction causes the release of acetylcholine from the synaptic knobs. This causes the muscle cell membrane (sarcolemma) to depolarise, and an action potential is generated in the muscle cell.

1. Explain the purpose of nervous system integration: ______________________

____

2. (a) Explain what is meant by **summation**: ______________________

____

(b) In simple terms, distinguish between temporal and spatial summation: ______________________

____

____

3. Describe two ways in which a neuromuscular junction is similar to any excitatory cholinergic synapse:

(a) ______________________

(b) ______________________

Drugs at Synapses

Synapses in the peripheral nervous system are classified according to the neurotransmitter they release; **cholinergic** synapses release acetylcholine (**Ach**) while **adrenergic** synapses release adrenaline or noradrenaline. The effect produced by these neurotransmitters depends, in turn, on the type of receptors present on the postsynaptic membrane. Ach receptors are classified as nicotinic or muscarinic according of their response to nicotine or muscarine (a fungal toxin).

Adrenergic receptors are also of two types, alpha (α) or beta (β), classified according to their particular responses to specific chemicals. **Drugs** exert their effects on the nervous system by mimicking (**agonists**) or blocking (**antagonists**) the action of neurotransmitters at synapses. Because of the small amounts of chemicals involved in synaptic transmission, drugs that affect the activity of neurotransmitters, or their binding sites, can have powerful effects even in small doses.

Drugs at cholinergic synapses

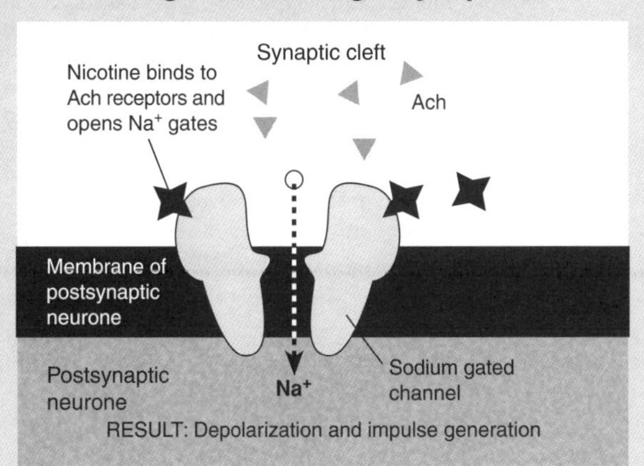

Nicotine acts as a **direct agonist** at nicotinic synapses. Nicotine binds to and activates acetylcholine (Ach) receptors on the postsynaptic membrane. This opens sodium gates, leading to a sodium influx and membrane depolarisation. Some agonists work indirectly at the synapse by preventing Ach breakdown. Such drugs are used to treat elderly patients with Alzheimer's disease.

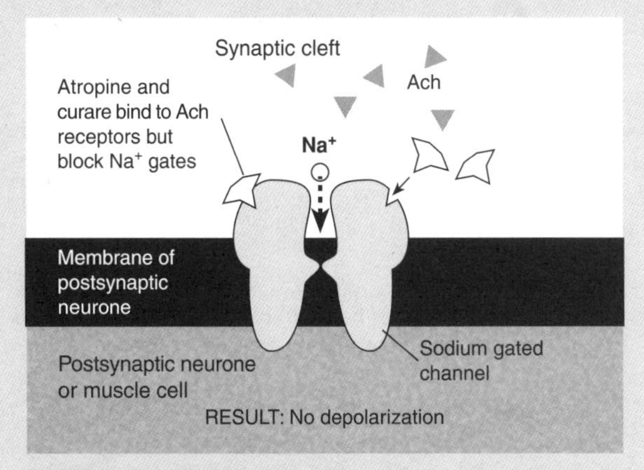

Atropine and **curare** act as antagonists at some cholinergic synapses. These molecules compete with Ach for binding sites on the postsynaptic membrane, and block sodium influx so that impulses are not generated. If the postsynaptic cell is a muscle cell, muscle contraction is prevented. In the case of curare, this causes death by flaccid paralysis.

Drugs at adrenergic synapses

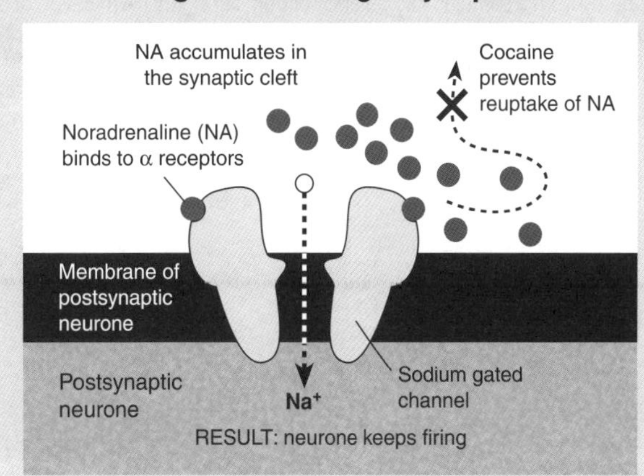

Under normal circumstances, the continued activity of the neurotransmitter noradrenaline (NA) at the synapse is prevented by reuptake of NA by the presynaptic neurone. **Cocaine** and **amphetamine** drugs act indirectly as agonists by preventing this reuptake. This action allows NA to linger at the synapse and continue to exert its effects.

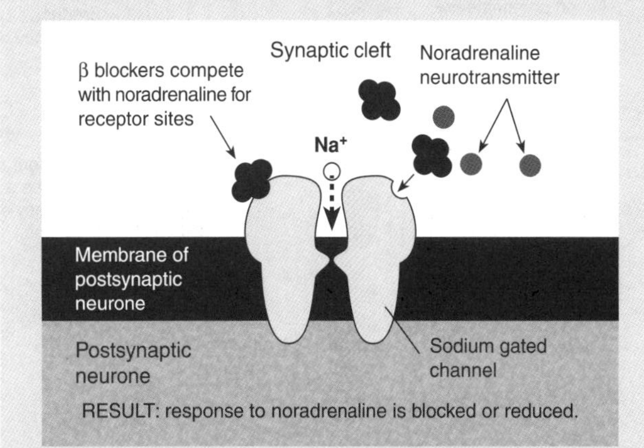

Therapeutic drugs called **beta** (β) **blockers** act as direct antagonists at adrenergic synapses (sympathetic nervous system). They compete for the adrenergic β receptors on the postsynaptic membrane and block impulse transmission. Beta blockers are prescribed primarily to treat hypertension and heart disorders because they slow heart rate and reduce the force of contraction.

1. Providing an example of each, outline two ways in which drugs can act at a cholinergic synapse:

   (a) ______________________________

   (b) ______________________________

2. Providing an example, outline one way in which drugs can operate at adrenergic synapses: ______________________________

3. Explain why atropine and curare are described as direct antagonists: ______________________________

4. Suggest why curare (carefully administered) is used during abdominal surgery: ______________________________

The Basis of Sensory Perception

All sensory receptors operate on the same basis: a **stimulus** is detected by the receptor cell, which triggers a change in the receptor membrane and the generation of an electrical response (below, left). The simplest sensory receptors consist of a single sensory neurone capable of detecting a stimulus and producing a nerve impulse (e.g. free nerve endings). More complex sense cells form synapses with their sensory neurones (e.g. taste buds). Sensory receptors are classified according to the stimuli to which they respond. The response of a **mechanoreceptor**, the Pacinian corpuscle, to a stimulus (pressure) is described below, right.

The Nature of Sensory Reception

Sensory receptors are specialised to detect stimuli and respond by producing an electrical discharge. In this way they act as **biological transducers**, converting the energy from a stimulus into an electrochemical signal. Stimulation of a sensory receptor cell results in an electrical response with certain, specific properties. These properties (outlined below) govern how receptors respond and adapt to stimuli, and how they encode information for interpretation by the central nervous system. The example below illustrates these properties in a stretch receptor.

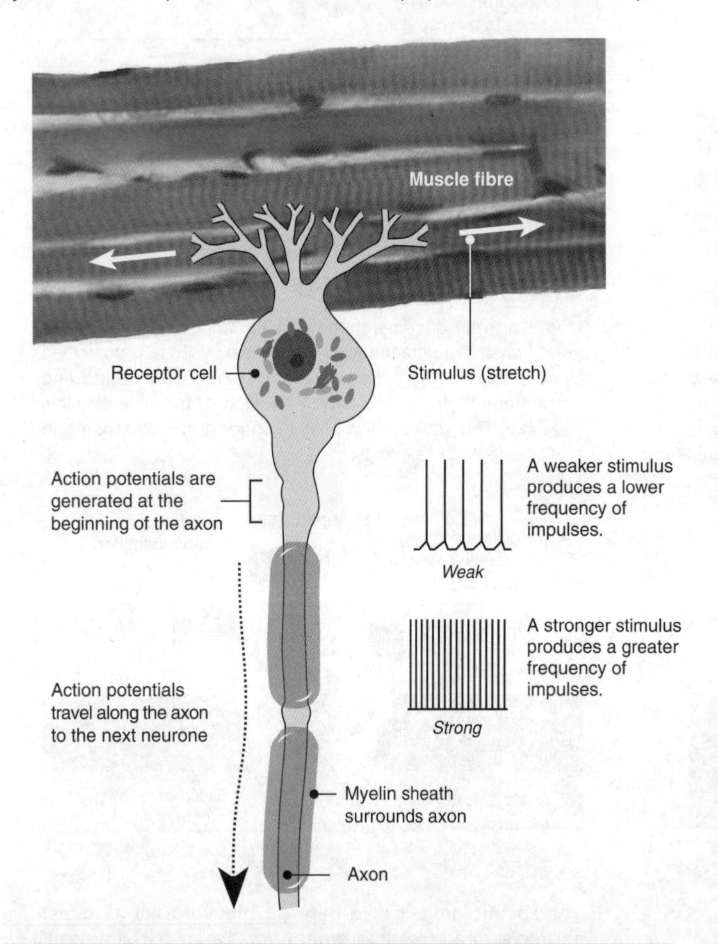

Properties of the Receptor Response

- The frequency of impulses in the receptor cell is directly proportional to the strength of the stimulus.
- Impulse frequency therefore provides information about the stimulus strength.
- Sensory receptors show **sensory adaptation** and will stop responding to a constant stimulus of the same intensity.

The Pacinian Corpuscle

Pacinian corpuscles are pressure receptors that occur deep within the skin all over the body. They are relatively large and simple in structure, consisting of a sensory nerve ending (dendrite) surrounded by a capsule of layered connective tissue. When pressure is applied to the skin, the capsule is deformed. The deformation stretches the nerve ending and leads to a localised depolarisation, called **generator potential**. Once the generator potential reaches or exceeds a **threshold** value, it causes an **action potential** to flow along the sensory pathway (axon).

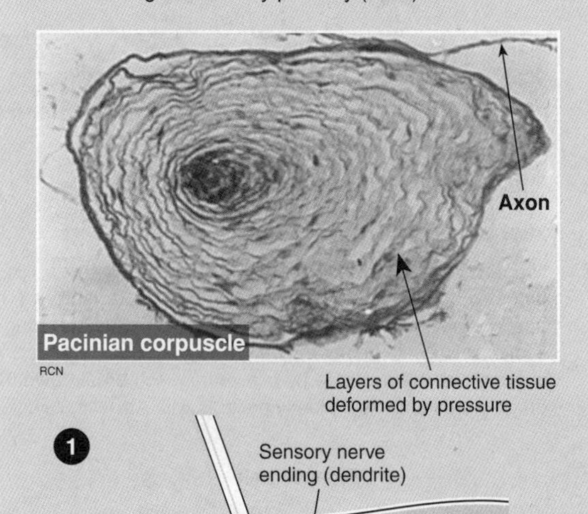

1. Sensory nerve ending (dendrite)

Na^+

Deforming the corpuscle leads to an increase in the permeability of the nerve to sodium. Na^+ diffuses into the nerve ending creating a localised depolarisation.

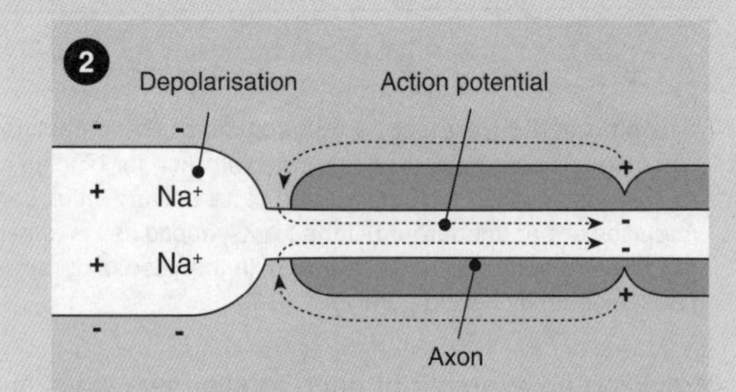

When the depolarisation (generator potential) reaches threshold, a volley of **action potentials** is triggered. These action potentials are conducted along the sensory axon.

1. Explain why sensory receptors are termed 'biological transducers': ______________________

2. Explain the significance of linking the magnitude of a sensory response to stimulus intensity: ______________________

3. Explain the physiological importance of sensory adaptation: ______________________

The Structure of the Eye

The eye is a complex and highly sophisticated sense organ specialised to detect light. The adult eyeball is about 25 mm in diameter. Only the anterior one-sixth of its total surface area is exposed; the rest lies recessed and protected by the **orbit** into which it fits. The eyeball is protected and given shape by a fibrous tunic. The posterior part of this structure is the **sclera** (the white of the eye), while the anterior transparent portion is the **cornea**, which covers the coloured iris.

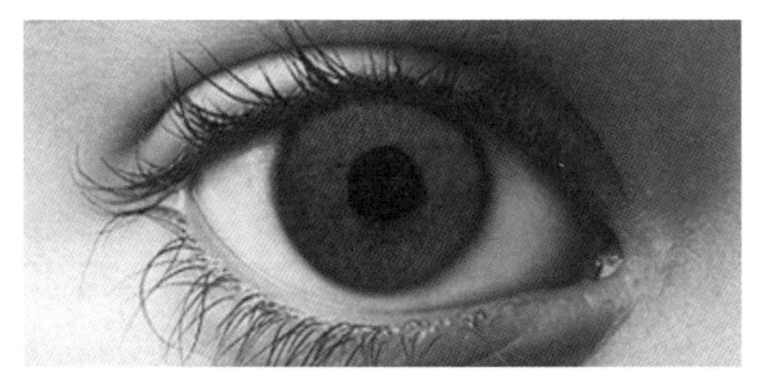

The Structure and Function of the Mammalian Eye

The human eye is essentially a three layered structure comprising an outer fibrous layer (the sclera and cornea), a middle vascular layer (the choroid, ciliary body, and iris), and inner **retina** (neurones and **photoreceptor cells**). The shape of the eye is maintained by the fluid filled cavities (aqueous and vitreous humours), which also assist in light refraction. Eye colour is provided by the pigmented iris. The iris also regulates the entry of light into the eye through the contraction of circular and radial muscles.

Forming a Visual Image

Before light can reach the photoreceptor cells of the retina, it must pass through the cornea, aqueous humour, pupil, lens, and vitreous humour. For vision to occur, light reaching the photoreceptor cells must form an image on the retina. This requires **refraction** of the incoming light, **accommodation** of the lens, and **constriction** of the pupil.

The anterior of the eye is concerned mainly with **refracting** (bending) the incoming light rays so that they focus on the retina (below left). Most refraction occurs at the cornea. The lens adjusts the degree of refraction to produce a sharp image. **Accommodation** (below right) adjusts the eye for near or far objects. Constriction of the pupil narrows the diameter of the hole through which light enters the eye, preventing light rays entering from the periphery.

The point at which the nerve fibres leave the eye as the optic nerve, is the **blind spot** (the point at which there are no photoreceptor cells). Nerve impulses travel along the optic nerves to the visual processing areas in the cerebral cortex. Images on the retina are inverted and reversed by the lens but the brain interprets the information it receives to correct for this image reversal.

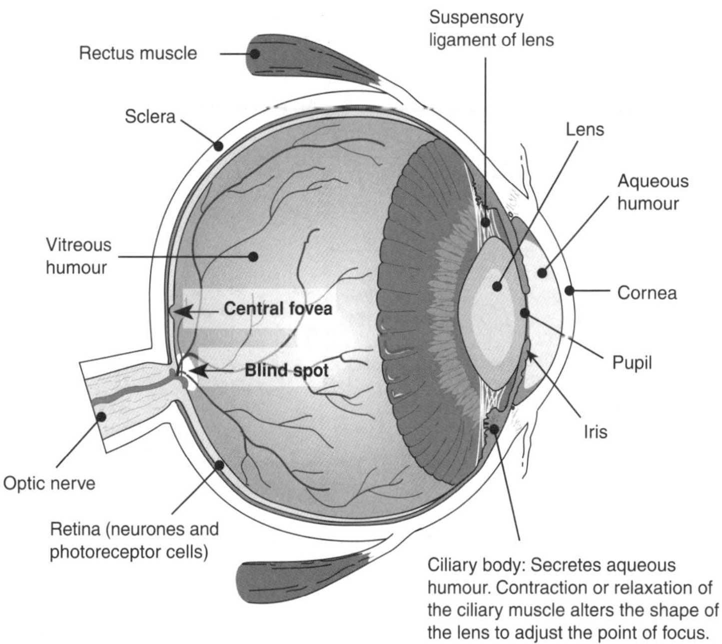

Normal vision

Focal point

Light

In normal vision, light rays from an object are bent sufficiently by the cornea and lens, and converge on the central fovea. A clear image is formed. Images are focused upside down and mirror reversed on the retina. The brain automatically interprets the image as right way up.

Accommodation for near and distant vision

Distant objects — Near parallel rays from a distant object — Flattened lens

Close objects — Divergent rays from a closer object — Rounded lens

The degree of refraction occurring at each surface of the eye is precise. The light rays reflected from an object 6 m or more away are nearly parallel to one another. Those reflected from near objects are divergent. The light rays must be refracted differently in each case so that they fall exactly on the central fovea. This is achieved through adjustment of the shape of the lens (**accommodation**). Accommodation from distant to close objects occurs by rounding the lens to shorten its focal length, since the image distance to the object is essentially fixed.

1. Identify the function of each of the structures of the eye listed below:

(a) Cornea: ______________________

(b) Ciliary body: ______________________

(c) Retina: ______________________

(d) Iris: ______________________

The Physiology of Vision

Vision involves essentially two stages: formation of the image on the retina (see previous activity), and generation and conduction of nerve impulses. When light reaches the retina, it is absorbed by the photosensitive pigments associated with the membranes of the photoreceptor cells (the rods and cones). The pigment molecules are altered by the absorption of light in such a way as to lead to the generation of nerve impulses. It is these impulses that are conducted via nerve fibres to the visual processing centre of the cerebral cortex (see the activity "*The Human Brain*" for the location of this region).

The Structure and Arrangement of Photoreceptor Cells in the Retina

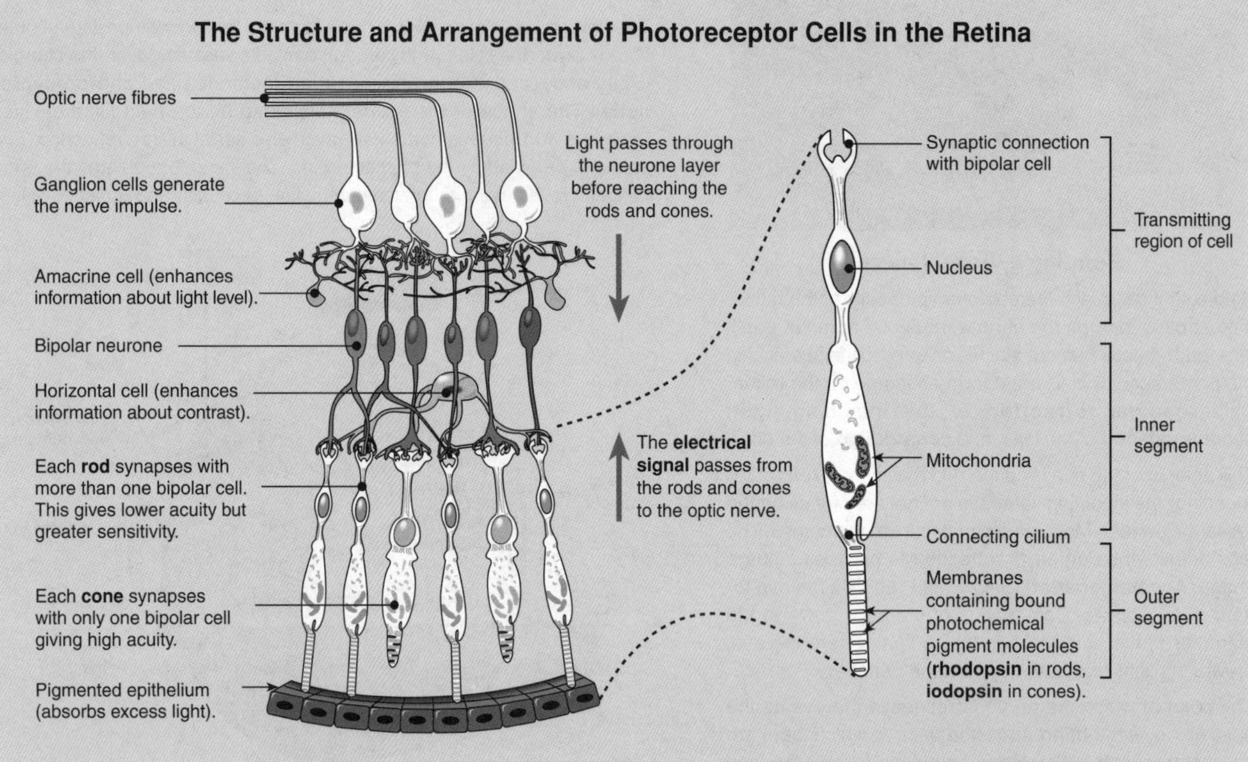

Arrangement of photoreceptors and neurones in the retina

Structure of a rod photoreceptor cell

The photoreceptor cells of the mammalian retina are the **rods** and **cones**. Rods are specialised for vision in dim light, whereas cones are specialised for colour vision and high visual acuity. Cone density and visual acuity are greatest in the **central fovea** (rods are absent here). After an image is formed on the retina, light impulses must be converted into nerve impulses. The first step is the development of **generator potentials** by the rods and cones. Light induces structural changes in the **photochemical pigments** (or photopigments) of the rod and cone membranes. The generator potential that develops from the pigment breakdown in the rods and cones is different from the generator potentials that occur in other types of sensory receptors because stimulation results in a **hyperpolarisation** rather than a depolarisation (in other words, there is a net loss of Na^+ from the photoreceptor cell). Once generator potentials have developed, the graded changes in membrane conductance spread through the photoreceptor cell. Each photoreceptor makes synaptic connection with a bipolar neurone, which transmits the potentials to the **ganglion cells**. The ganglion cells become **depolarised** and initiate nerve impulses which pass through the optic chiasma and eventually to the visual areas of the cerebral cortex. The frequency and pattern of impulses in the optic nerve conveys information about the changing visual field.

1. (a) The first stage of vision involves forming an image on the retina. Explain what this involves (see the previous activity):

   ___

   ___

   (b) Explain how accommodation is achieved (see the previous activity for help): ___

   ___

   ___

   ___

2. Contrast the structure of the blind spot and the central fovea: ___

   ___

   ___

   ___

The Basis of Trichromatic Vision

There are three classes of **cones**, each with a maximal response in either short (blue), intermediate (green) or long (yellow-green) wavelength light (below). The yellow-green cone is also sensitive to the red part of the spectrum and is often called the red cone. The differential responses of the cones to light of different wavelengths provides the basis of trichromatic colour vision.

Synaptic connection

Nucleus

Mitochondrion

Membranes containing bound **iodopsin** pigment molecules.

Each **cone** synapses with only one bipolar cell giving high acuity.

Cone response to light wavelengths

Violet | Blue | Green | Yellow | Orange

B G R

Electrical response

450 500 550 600 650 700

Wavelength /nm

3. Complete the table below, comparing the features of rod and cone cells:

Feature	Rod cells	Cone cells
Visual pigment(s):		
Visual acuity:		
Overall function:		

4. Account for the differences in acuity and sensitivity between rod and cone cells:

5. (a) Explain clearly what is meant by the term photochemical pigment (photopigment):

   (b) Identify two photopigments and their location:

6. In your own words, explain how light is able to produce a nerve impulse in the ganglion cells:

7. Explain the physiological basis for colour vision in humans:

Hearing

Most animals respond to sound and so have receptors for the detection of sound waves. In mammals, these receptors are organised into hearing organs called ears. Sound is produced by the vibration of particles in a medium and it travels in waves that can pass through solids, liquids, or gases. The distance between wave 'crests' determines the frequency (pitch) of the sound. The absolute size (amplitude) of the waves determines the intensity or loudness of the sound. Sound reception in mammals is the role of **mechanoreceptors**: tiny hair cells in the cochlea of the inner ear. The hair cells are very sensitive and are easily damaged by prolonged exposure to high intensity sounds. Gradual hearing loss with age is often caused by the cumulative loss of sensory hair cell function, especially at the higher frequencies. Such hearing loss is termed perceptive deafness.

The Human Ear

Ear canal
Ear drum
Vestibular apparatus (senses balance and body position)
Auditory nerve
Cochlea
Round window
Oval window
Eustachian tube
Malleus (hammer)
Incus (anvil)
Stapes (stirrup)
Ear ossicles (bones)
Pinna focuses sound waves

In mammals, sound waves are converted to pressure waves in the inner ear. The ears of mammals use mechanoreceptors (sensory hair cells) to change the pressure waves into nerve impulses. The mammalian ear contains not only the organ of hearing, the **cochlea**, but all the specialised structures associated with gathering, directing, and amplifying the sound. The cochlea is a tapered, coiled tube, divided lengthwise into **three fluid filled canals**. The cochlea is shown below, unrolled to indicate the way in which sound waves are transmitted through the canals to the sensory cells. The mechanisms involved in hearing are outlined in a simplified series of steps. In mammals, the inner ear is also associated with the organ for detecting balance and position (the vestibular apparatus), although this region is not involved in hearing.

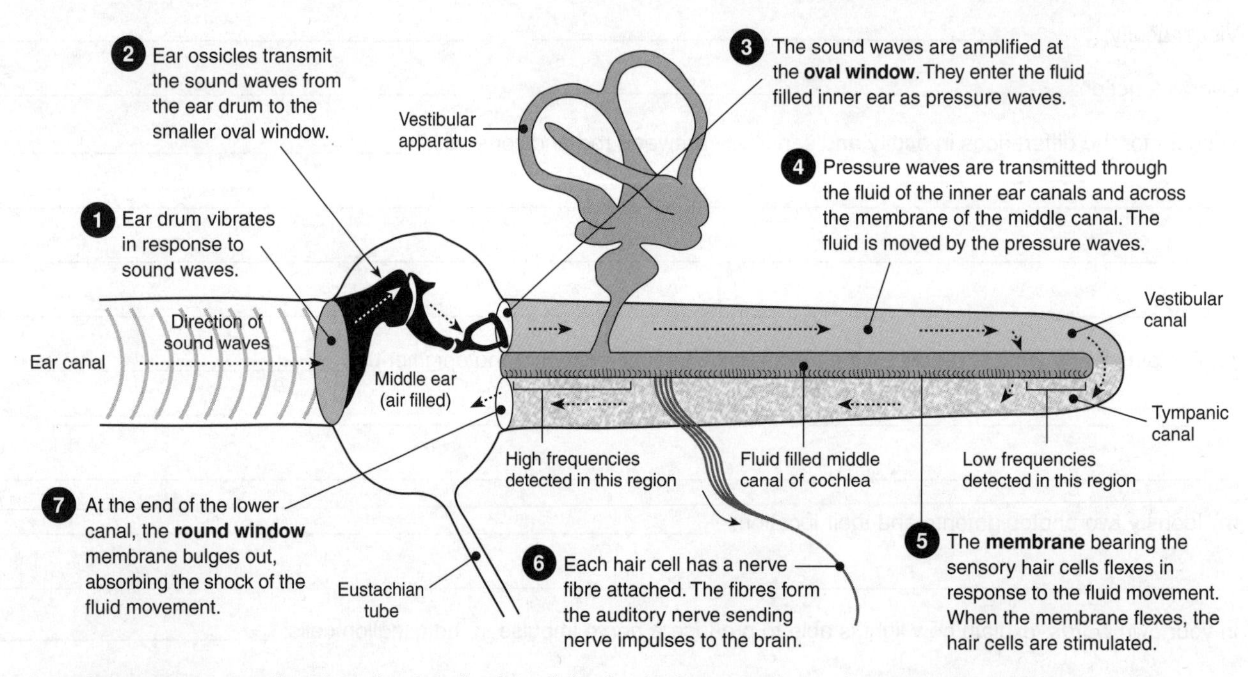

1. In a short sentence, outline the role of each of the following in the reception and response to sound:

   (a) The ear drum: ______

   (b) The ear ossicles: ______

   (c) The oval window: ______

   (d) The sensory hair cells: ______

   (e) The auditory nerve: ______

2. Explain the significance of the inner ear being fluid filled: ______

The Components of Behaviour

Behaviour in animals can be attributed to two components; **innate behaviour** that has a genetic basis, and **learned behaviour**, which results from the experiences of the animal. Together they combine to produce the total behaviour exhibited by the animal. It should also be noted that experience can modify innate behaviour. Animals behave in 'automatic ways' in many situations. The innate behaviour follows a classical pathway called a **fixed-action pattern** (FAP) where an innate behavioural programme is activated by a stimulus or **releaser** to direct some kind of behavioural response. Innate behaviours are generally adaptive and are performed for a variety of reasons. Learning, which involves the modification of behaviour by experience, occurs in various ways.

Innate Behaviours

Reflex behaviour
Simplest type of animal behaviour. A sudden stimulus induces an automatic involuntary and stereotyped response. Many reflexes are protective.

Kinesis
Random movement of an animal in which the rate of movement is related to the intensity of the stimulus, but not to its direction.

Taxis
A movement in response to the direction of a stimulus. Movement towards a stimulus are positive while those away from a stimulus are negative.

Stereotyped behaviour
Occurs when the same response is given to the same stimulus on different occasions. This behaviour shows fixed patterns of coordinated movements called fixed action patterns.

The complex behaviour patterns exhibited by an animal

Learned Behaviours

Classical conditioning
Animals may associate one stimulus with another.

Habituation
Response to a stimulus wanes when it is repeated with no apparent effect.

Insight behaviour
Correct behaviour on the first attempt where the animal has no prior experience.

Imprinting behaviour
During a critical period, an animal can adopt a behaviour by latching on to its first stimulus.

Operant conditioning
Also called trial and error learning; an animal is rewarded or punished after chance behaviour.

Fixed Action Pattern

A **releaser** (sign stimulus) triggers the operation of an **innate behavioural programme** in the brain that results in a **fixed-action pattern** (FAP), i.e. a predictable behavioural response.

Releaser → Innate behavioural programme →

1. Distinguish between innate and learned behaviours: ______________________________

____

2. (a) Explain the role of releasers in innate behaviours: ______________________________

____

(b) Name a releaser for a fixed action pattern and the animal involved, and describe the behaviour elicited:

____

____

Learned Behaviour

Imprinting occurs when an animal learns to make a particular response only to one type of animal or object. Imprinting differs from most other kinds of learned behaviour in that it normally can occur only at a specific time during an animal's life. This **critical period** is usually shortly after hatching (about 12 hours) and can last for several days. While a critical period and the resulting imprinted behaviour are normally irreversible, they are not considered rigidly fixed. There are examples of animals that have had abnormal imprinted behaviours revert to the 'wild type'. There are two main types of imprinting using visual and auditory stimuli: filial and sexual imprinting. Breeding ground imprinting uses olfactory (smell) stimuli.

Filial (Parent) Imprinting

Filial imprinting is the process by which animals develop a social attachment. It differs from most other kinds of learning (including other types of imprinting), in that it normally can occur only at a specific time during an animal's life. This **critical period** is usually shortly after hatching (about 12 hours) and may last for several days. Ducks and geese have no innate ability to recognise *mother* or even their own species. They simply respond to, and identify as mother, the first object they encounter that has certain characteristics.

Breeding Ground Imprinting

Salmon undertake long migrations in the open ocean where they feed, grow and mature after hatching in freshwater streams. Some species remain at sea for several years, after which each fish returns to its exact home stream to spawn. Research has shown that this ability is based on **olfactory imprinting** where the fish recognise the chemical odours of their specific stream and swim towards its source. Other animals (bears, humans, etc.) have learned to exploit this behaviour.

Sexual Identity Imprinting

Individuals learn to direct their sexual behaviour at some stimulus objects, but not at others. Termed **sexual imprinting**, it may serve as a species identifying and species isolating mechanism. The mate preferences of birds have been shown to be imprinted according to the stimulus they were exposed to (other birds) during early rearing. Sexual imprinting generally involves longer periods of exposure to the stimulus than filial imprinting (*see left*).

Habituation

Habituation is a very simple type of learning involving a loss of a response to a repeated stimulus when it fails to provide any form of reinforcement (reward or punishment). Habituation is different to fatigue, which involves loss of efficiency in a repeated activity, and arises as a result of the nature of sensory reception itself. An example of habituation is the waning response of a snail attempting to cross a platform that is being tapped at regular time intervals. At first, the snail retreats into its shell for a considerable period after each tap. As the tapping continues, the snail stays in its shell for a shorter duration, before resuming its travel.

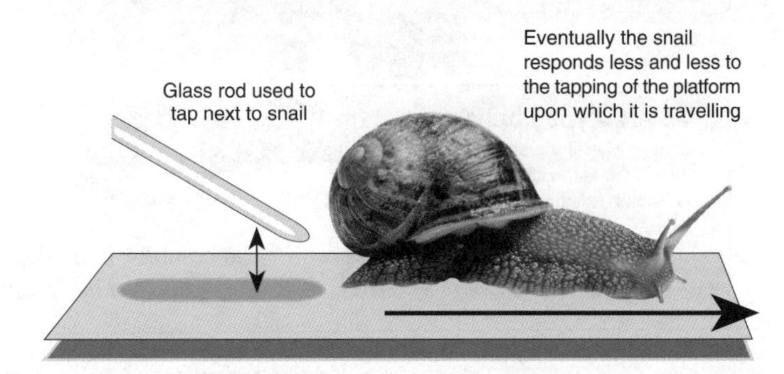

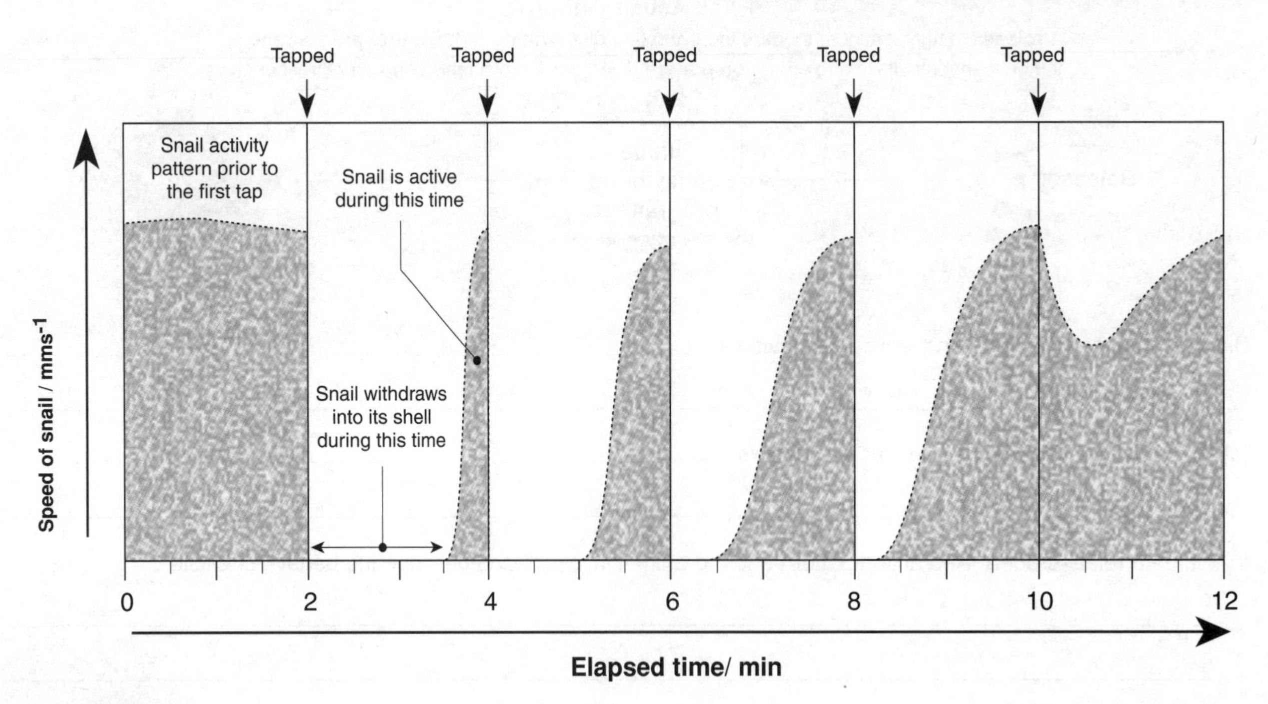

Classical (Pavlovian) Conditioning

Classical conditioning, founded by **Ivan Pavlov**, describes a type of **associative learning** in which behaviour that is normally triggered by a certain stimulus comes to be triggered by a substitute stimulus that previously had no effect on the behaviour. Between 1890 and 1900, Pavlov noticed that the dogs he was studying would salivate when they knew they were to be fed. It was determined that the dogs were alerted by a bell that rung every time the door into the lab was opened. Through experimentation, Pavlov discovered that the ringing of the bell initially brought about no salivation, but the dogs could be **conditioned** to relate the ringing of the bell to the presentation of food. Eventually the ringing of the bell elicited the same salivation response as the presentation of food, indicating that the dog was conditioned to associate the two stimuli.

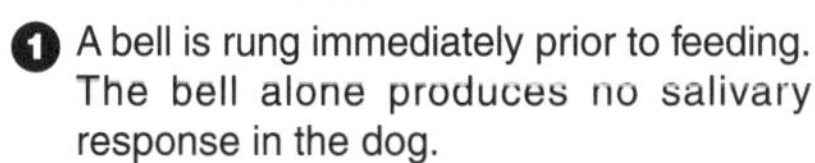

1. A bell is rung immediately prior to feeding. The bell alone produces no salivary response in the dog.

2. Food is introduced after the bell has rung. Steps one are two are repeated a number of times (association of bell and food).

3. Eventually the dog becomes conditioned to salivate whenever the bell is rung, even when no food is presented.

Operant Conditioning

Operant conditioning is used to describe a situation where an animal learns to associate a particular behavioural act with a **reward** (as opposed to a stimulus in classical conditioning). This behaviour determines whether or not the reward appears. **Burrhus Skinner** studied operant conditioning using an apparatus he invented called a **Skinner box** *(see right)*. Skinner designed the box so that when an animal (usually a pigeon or rat) pushed a particular button it was rewarded with food. The animals learned to associate the pushing of the button with obtaining food (the reward). The behavioural act that leads the animal to push the button in the first place is thought to be generated spontaneously (by accident or curiosity). This type of learning is also called **instrumental learning** because the spontaneous behaviour is instrumental in obtaining the reward. Operant conditioning is the predominant learning process found in animals.

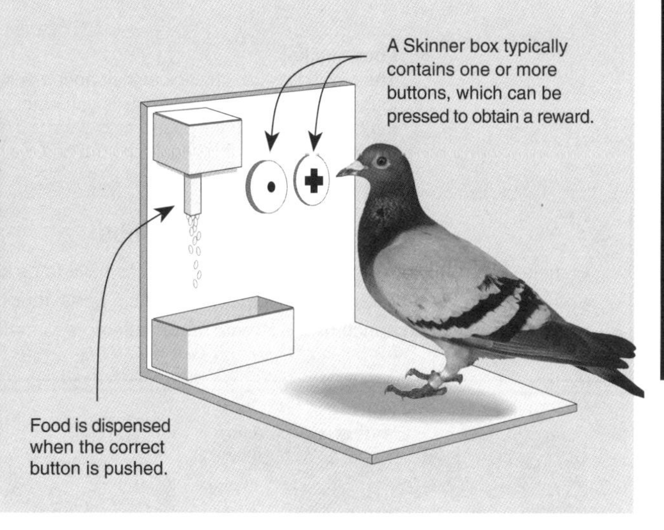

1. Explain what **filial imprinting** is: ___

2. For the example of **filial imprinting** (opposite), identify which parts of the behaviour can be attributed to:

   (a) Innate behaviour: ___

   (b) Learned behaviour: ___

3. In relation to human behaviour, describe an example of the following:

   (a) Habituation: ___

   (b) Imprinting: ___

   (c) Classical conditioning: ___

   (d) Operant conditioning: ___

Breeding Behaviour

Many of the behaviours observed in animals are associated with reproduction, reflecting the importance of this event in an individual's life cycle. Many types of behaviour are aimed at facilitating successful reproduction. These include **courtship** behaviours, which may involve attracting a mate to a particular breeding site (often associated with high availability of resources such as food or nesting sites). Courtship behaviours are aimed at reducing conflict between the sexes and are often **stereotyped** or ritualistic. They rely on **sign stimuli** to elicit specific responses in potential mates. Other reproductive behaviours are associated with assessing the receptivity of a mate, defending mates against others, and rearing the young.

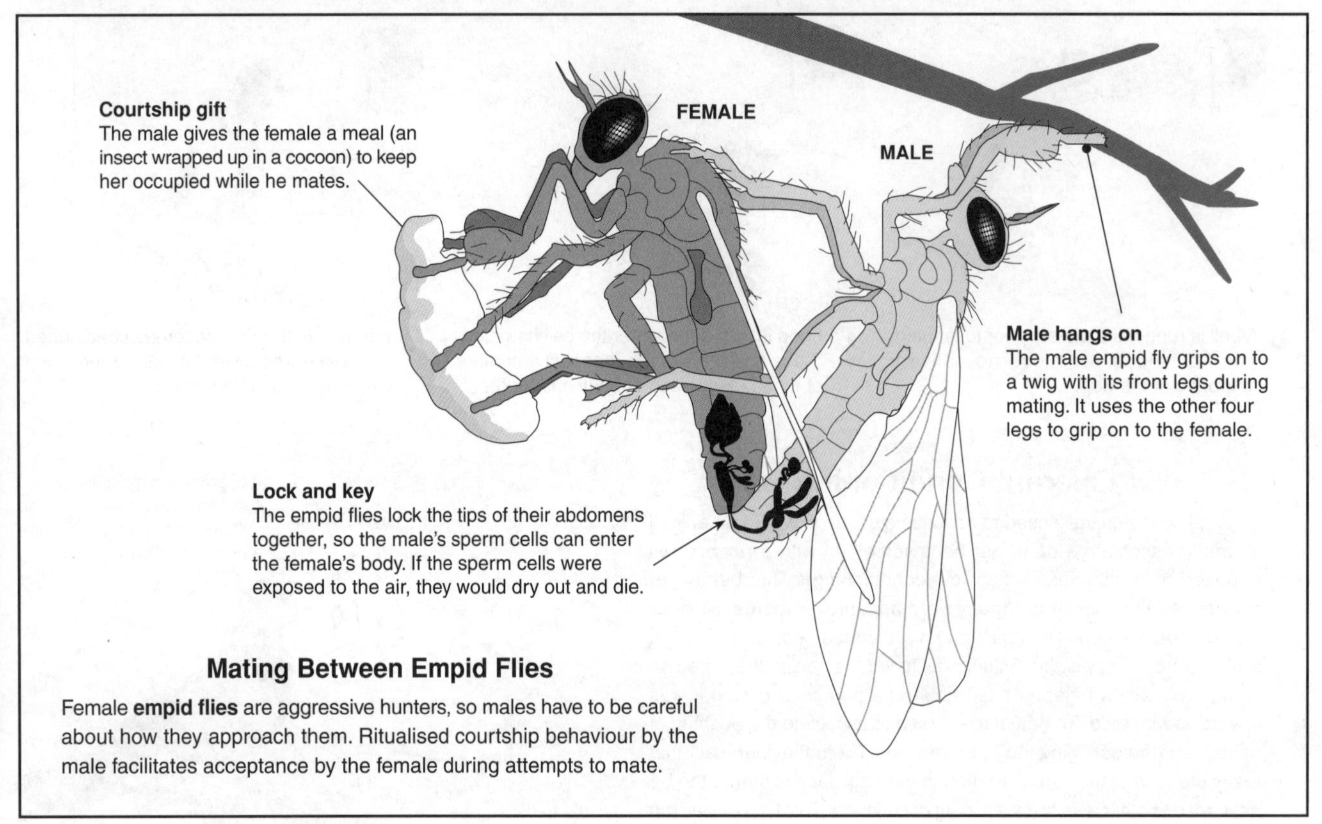

Mating Between Empid Flies

Female **empid flies** are aggressive hunters, so males have to be careful about how they approach them. Ritualised courtship behaviour by the male facilitates acceptance by the female during attempts to mate.

Courtship behaviour occurs as a prelude to mating and is most common in vertebrates and insects. One function of courtship is to synchronise the behaviours of the male and female so that mating can occur, and to override attack or escape behaviour. Here, a male greater frigatebird calls, spreads its wings, and inflates its throat pouch to court a female.

It is common in some birds (and many arthropods) for the male to provide an offering, such as food or nesting material, to the female. These **rituals** reduce aggression in the male and promote appeasement behaviour by the female. For some **monogamous** species, e.g. the blue-footed boobies mating above, the pairing begins a long term breeding partnership.

Many marine birds, such as the emperor penguins above, form breeding colonies where large numbers of birds come together to lay their eggs and raise their young. The adults feed at sea, leaving the young unattended for varying lengths of time. Penguin chicks congregate in large, densely packed groups to conserve body heat while their parents are away.

1. (a) Explain what is meant by **courtship behaviour**: ______________________________

   ____

   (b) Suggest a reason why courtship behaviour may be necessary prior to mating: ______________________________

   ____

   (c) Explain why courtship behaviour is often ritualised and involves stereotyped displays: ______________________________

   ____

   ____

Territorial Behaviour and Courtship Displays

A territory is any area that is defended against members of the same species. Territories are usually defended by clear acts of aggression or ritualised signals (e.g. vocal, visual or chemical signals).

Birds often engage in complex courtship behaviour. Males display in some way and females select their mates. These courtship displays are usually species specific and include ritualised acts such dancing, feeding, and nest-building.

Resource availability often determines territory size and the population becomes spread out accordingly. Gannet territories are relatively small, with hens defending only the area they can reach while sitting on their nest.

(d) Describe one example of a courtship behaviour: ______________________

2. Using an example, describe an advantage of cooperation during breeding: ______________________

3. (a) Explain the nature and purpose of the courtship display in empid flies: ______________________

(b) Identify the **sign stimulus** in this behaviour: ______________________

4. (a) Expain what is meant by the formation of a **pair-bond**: ______________________

(b) Using an example, describe an advantage of pair-bonding behaviour: ______________________

5. (a) Explain what is meant by a **territory**: ______________________

(b) Explain why a male will defend a particular breeding territory against other males: ______________________

(c) Suggest how a female in this situation would select a mate: ______________________

6. Discuss the advantages and disadvantages of maintaining a territory: ______________________

Simple Behaviours

Taxes and kineses are examples of **orientation behaviours**. Such behaviours describe the way in which motile organisms (or gametes) **position** themselves and move in response to external cues (stimuli). Common stimuli are gravity, light, chemicals, and temperature. Some animals and many protozoa respond to certain stimuli simply by changing their rate of movement or by randomly turning without actually orientating to the stimulus. These movements are called **kineses**. In contrast, **taxes** involve orientation and movement directly to or away from one or more stimuli, such as temperature. Taxes often involve moving the head (which carries the sensory receptors) from side to side until the sensory input from both sides is equal (a klinotaxic response). Note that many taxic responses are complicated by a simultaneous response to more than one stimulus. For example, fish orientate dorsal side up by responding to both light and gravity. Male moths orientate positively to pheromones but use the wind to judge the direction of the odour source (the female moth).

When confronted with a vertical surface, snails will reorientate themselves so that they climb vertically upwards.

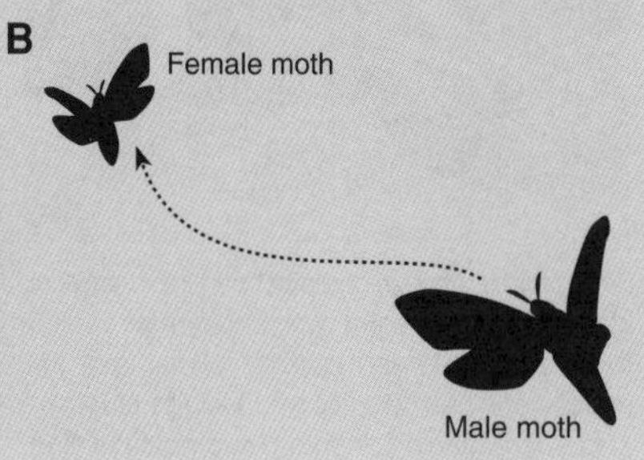

A flying male moth, encountering an odour (pheromone) trail left by a female, will turn and fly upwind until it reaches the female.

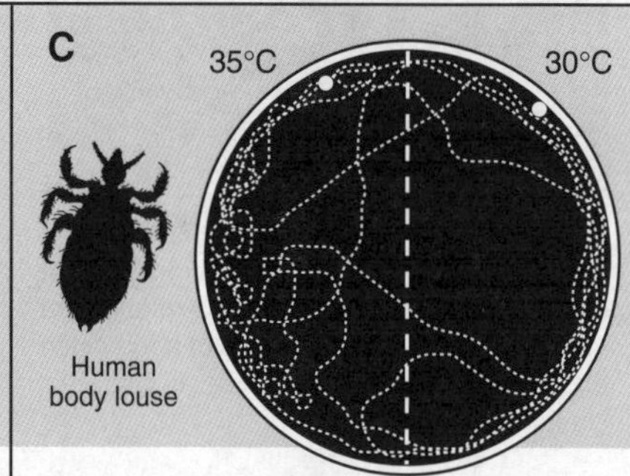

In a circular chamber, lice make relatively few turns at 30°C, but many random turns at 35°C.

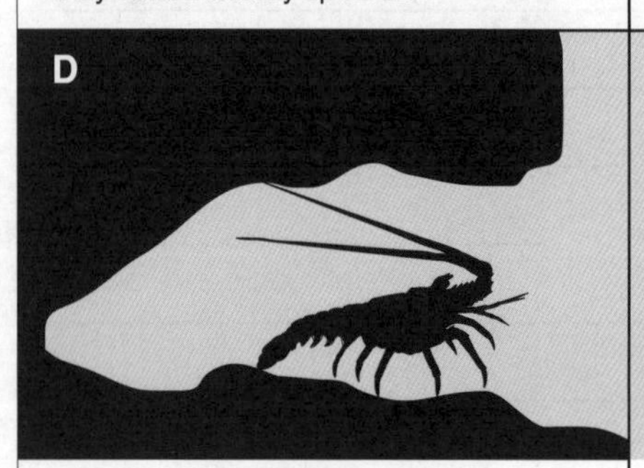

Spiny lobsters will back into tight crevices so that their body is touching the crevice sides. The antennae may be extended out.

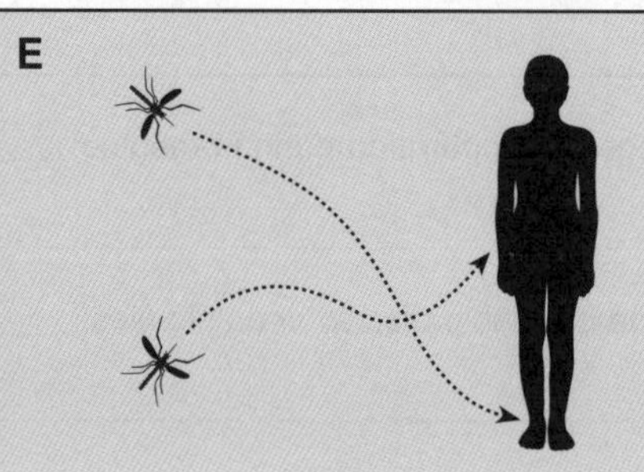

At close range, mosquitoes use the temperature gradient generated by the body heat of a host to home in on exposed flesh.

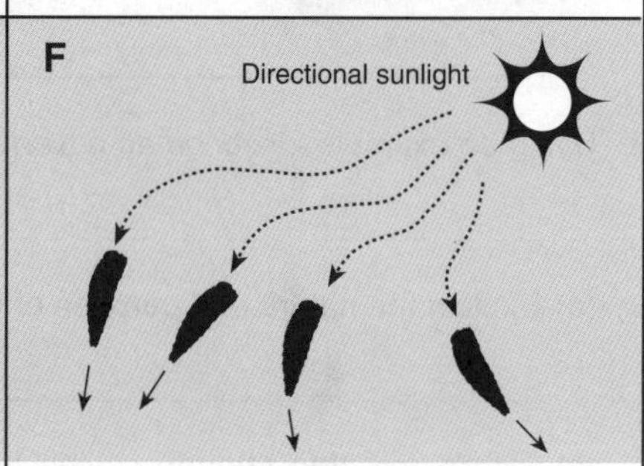

Blowfly maggots will turn and move rapidly away from a directional light source.

1. Define the following terms used to describe orientation responses in animals.

   (a) Kinesis: __________

   (b) Taxis: __________

2. Comment on the adaptive (survival) value of simple behaviours such as kineses: __________

3. Name the physical stimulus for each of the following **prefixes** used in naming orientation responses:

   (a) Geo- _____ (b) Hydro- _____ (c) Thigmo- _____

   (d) Photo- _____ (e) Chemo- _____ (f) Thermo- _____

4. For each of the above examples (A-F), describe the orientation response. Indicate whether the response is positive or negative (e.g. positive phototaxis):

   (a) **A:** _____ (b) **B:** _____

   (c) **C:** _____ (d) **D:** _____

   (e) **E:** _____ (f) **F:** _____

5. Suggest what temperature body lice "prefer", given their response in the chamber (in C): _____

Plant Responses

Even though most plants are firmly rooted in the ground, they are still capable of responding and making adjustments to changes in their external environment. This ability is manifested chiefly in changing patterns of growth. These responses may involve relatively sudden physiological changes, as occurs in flowering, or a steady growth response, such as a tropism. Other responses made by plants include nastic movements, circadian rhythms, photoperiodism, dormancy, and vernalisation.

TROPISMS
Tropisms are growth responses made by plants to directional external stimuli, where the direction of the stimulus determines the direction of the growth response. A tropism may be positive (towards the stimulus), or negative (away from the stimulus). Common stimuli for plants include light, gravity, touch, and chemicals.

LIFE CYCLE RESPONSES
Plants use seasonal changes in the environment as cues for the commencement or ending of particular life cycle stages. Such changes are mediated by **plant growth factors**, such as phytochrome and gibberellin. Examples include flowering and other **photoperiodic responses**, dormancy and germination, and leaf fall.

RAPID RESPONSES TO ENVIRONMENTAL STIMULI
Plants are capable of quite rapid responses. Examples include the closing of **stomata** in response to water loss, opening and closing of flowers in response to temperature (photo, below right), and **nastic responses** (photos, below left). These responses often follow a circadian rhythm.

PLANT COMPETITION AND ALLELOPATHY
Although plants are rooted in the ground, they can still compete with other plants to gain access to resources. Some plants produce chemicals that inhibit the growth of neighbouring plants. Such chemical inhibition is called **allelopathy**. Plants also compete for light and may grow aggressively to shade out slower growing competitors.

PLANT RESPONSES TO HERBIVORY
Many plant species have responded to grazing or browsing pressure with evolutionary adaptations enabling them to survive constant cropping. Examples include rapid growth to counteract the constant loss of biomass (grasses), sharp spines or thorns to deter browsers (acacias, cacti), or toxins in the leaf tissues (eucalyptus).

RCN

Some plants, such as *Mimosa* (above), are capable of **nastic responses**. These are relatively rapid, reversible movements, such as leaf closure in response to touch. Unlike tropisms, nastic responses are independent of stimulus direction.

The growth of tendrils around a support is a response to a mechanical stimulus, and is called thigmomorphogenesis.

FA

The opening and closing of this tulip flower is temperature dependent. The flowers close when it is cool at night.

1. Identify the stimuli to which plants typically respond: ______________________

2. Explain how plants benefit by responding appropriately to the environment: ______________________

____

____

Auxins, Gibberellins, and ABA

Auxin was the first substance to be identified as a plant hormone. Charles Darwin and his son Francis were first to recognise its role in stimulating cell elongation, while Frits W. Went isolated this growth-regulating substance, which he called auxin. **Indole-acetic acid** (IAA) is the only known naturally occuring auxin. Shortly after its discovery it was found to have a role in suppressing the growth of lateral buds. This inhibitory influence of a shoot tip or apical bud on the lateral buds is called **apical dominance**. Two Japanese scientists isolated **gibberellin** in 1934, eight years after the isolation of auxin, and more than 78 gibberellins have now been identified. Gibberellins are involved in stem and leaf elongation, as well as breaking dormancy in seeds. Specifically, they stimulate cell division and cell elongation, allowing stems to 'bolt' and the root to penetrate the testa in germination. During the 1960s, Frederick T. Addicott discovered a substance apparently capable of accelerating **abscission** in leaves and fruit (which he called abscisin), and which is now called **abscisic acid** (ABA). Although it now seems that ABA has very little to do with leaf abscission, it is a growth inhibitor and also stimulates the closing of stomata in most plant species. It is also involved in preventing premature germination and development of seeds.

Auxins, Gibberellins, and Abscisic Acid (ABA) and Plant Responses

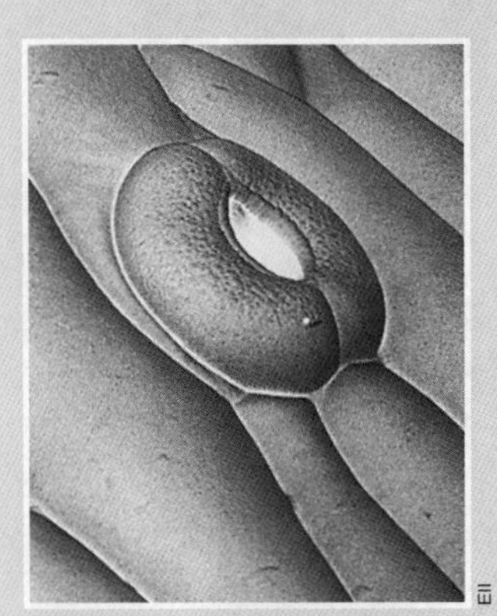

ABA stimulates the closing of stomata in most plant species. Its synthesis is stimulated by water deficiency (water stress).

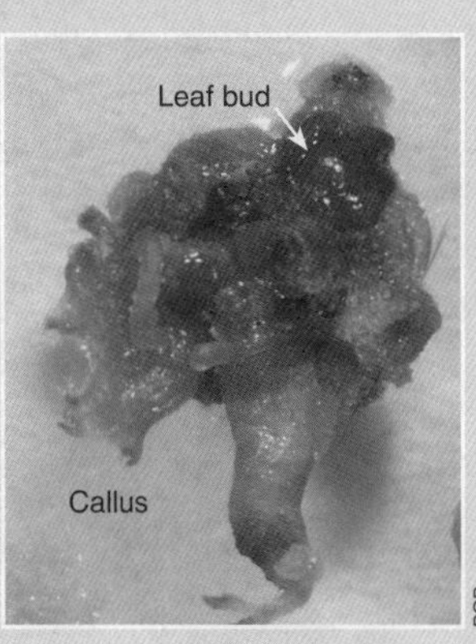

Cytokinins promote the development of leaf buds in calluses (above). In contrast, increased auxin levels will promote the development of roots.

Gibberellins are responsible for breaking dormancy in seeds and promote the growth of the embryo and emergence of the seedling.

ABA promotes seed dormancy. It is concentrated in senescent leaves, but it is probably not involved in leaf abscission except in a few species.

Gibberellins cause stem and leaf elongation by stimulating cell division and cell elongation. They are responsible for **bolting** in brassicas.

Gibberellins are used to hasten seed germination and ensure germination uniformity in the production of barley malt in brewing.

ABA is produced in ripe fruit and induces fruit fall. The effects of ABA are generally opposite to those of cytokinins.

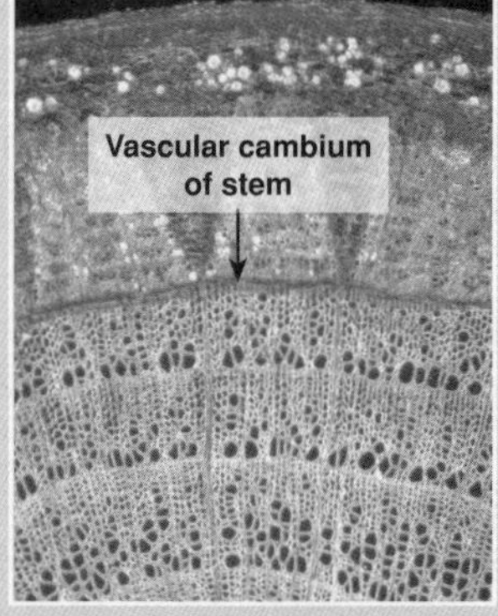

Auxins promote the activity of the vascular cambium (above), stem length, differentiation of tissues, and **apical dominance** .

1. Describe the role of **auxins** in apical dominance: ______________________

____

____

2. Describe the role of **gibberellins** in stem elongation and in the germination of grasses such as barley:

____

____

____

3. Describe the role of abscisic acid in closure of stomata: ______________________

____

Flowering and Dormancy

Photoperiodism is the response of a plant to the relative lengths of daylight and darkness. Flowering is one of the activities of plants that is photoperiodic, and plants can be classified according to how light affects their flowering. Photoperiodism is controlled through the action of a pigment called **phytochrome**, which acts like an alarm button for some biological clocks in plants. Phytochrome is also involved in other light initiated responses in plants, such as seed germination, leaf growth, and chlorophyll synthesis. Plants do not grow at the same rate all of the time. In temperate regions, many perennial and biennial plants start to shut down growth as autumn approaches and the days grow shorter. During the unfavourable seasons, they limit their growth or cease to grow altogether. This condition of arrested growth, or **dormancy**, enables plants to survive periods of water shortage or low temperature. The plant's buds will not resume growth until there is a convergence of environmental cues in early spring. Short days and long, cold nights (as well as dry, nitrogen deficient soils) are strong cues for dormancy. Seasonal changes in temperature also influence many plant responses, including seed germination and flowering. In many plants the dormancy breaking process and flowering are triggered by a specific period of exposure to low winter temperatures. This low-temperature stimulation of flowering or seed germination is called **vernalisation**.

Photoperiodism

Photoperiodism is based on a system that monitors the day/night cycle. The photoreceptor involved in this, and a number of other light-initiated plant responses, is a blue-green pigment called **phytochrome**. Phytochrome is universal in vascular plants and has two forms: active and inactive. On absorbing light, it readily converts from the inactive form (**P_r**) to the active form (**P_{fr}**). **P_{fr}** predominates in daylight, but reverts spontaneously back to the inactive form in the dark. The plant measures daylength (or rather night length) by the amount of phytochrome in each form.

Summary of phytochrome related activities in plants

Process	Effect of daylight	Effect of darkness
Conversion of phytochrome	Promotes $P_r \rightarrow P_{fr}$	Promotes $P_{fr} \rightarrow P_r$
Seed germination	Promotes	Inhibits
Leaf growth	Promotes	Inhibits
Flowering - long day plants	Promotes	Inhibits
Flowering - short day plants	Inhibits	Promotes
Chlorophyll synthesis	Promotes	Inhibits

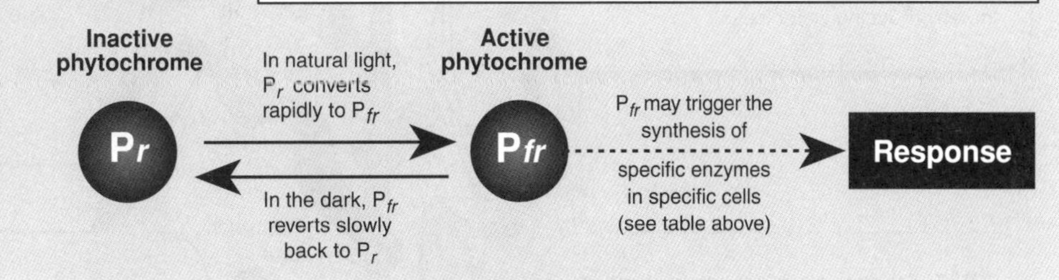

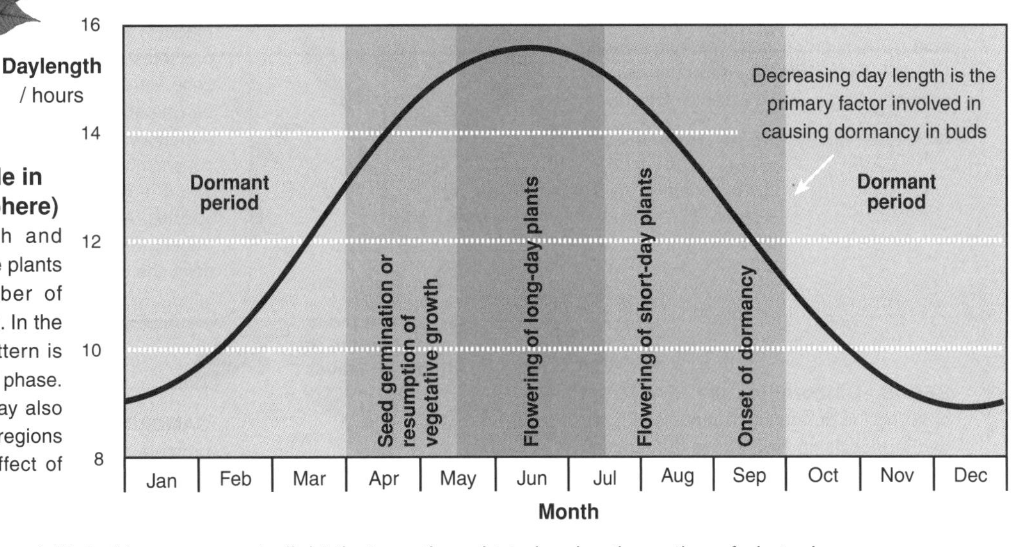

Day length and life cycle in plants (Northern Hemisphere)
The cycle of active growth and dormancy shown by temperate plants is correlated with the number of daylight hours each day (right). In the southern hemisphere, the pattern is similar, but is six months out of phase. The duration of the periods may also vary on islands and in coastal regions because of the moderating effect of nearby oceans.

1. List three plant responses initiated by exposure to light that are thought to involve the action of phytochrome:

2. State what causes phytochrome to change from its inactive to active form:

3. Explain the adaptive role of **vernalisation** in a plant life cycle:

4. Suggest why dormancy is advantageous to plants living in temperate climates:

5. Describe three strong environmental cues that may cause the onset of dormancy in plants:

Applications of Plant Hormones

Like animals, plants use hormones to regulate their growth and development. Plant hormones (**phytohormones**) are organic compounds produced in one part of the plant and transported to another part, where they produce a growth response. Hormones are effective in extremely small amounts. There are five groups of phytohormones: **auxins** (indolacetic acid or IAA), **cytokinins**, **gibberellins**, **ethene**, and **abscisic acid** (ABA). Together they control growth and development of the plant at various stages. Synthetic analogues of IAA have been produced for commercial use and are applied as growth promoters in rooting powders, and as inducers for fruit production. Some analogues (e.g. 2-4-5-T) even act as growth inhibitors and are used as selective herbicides.

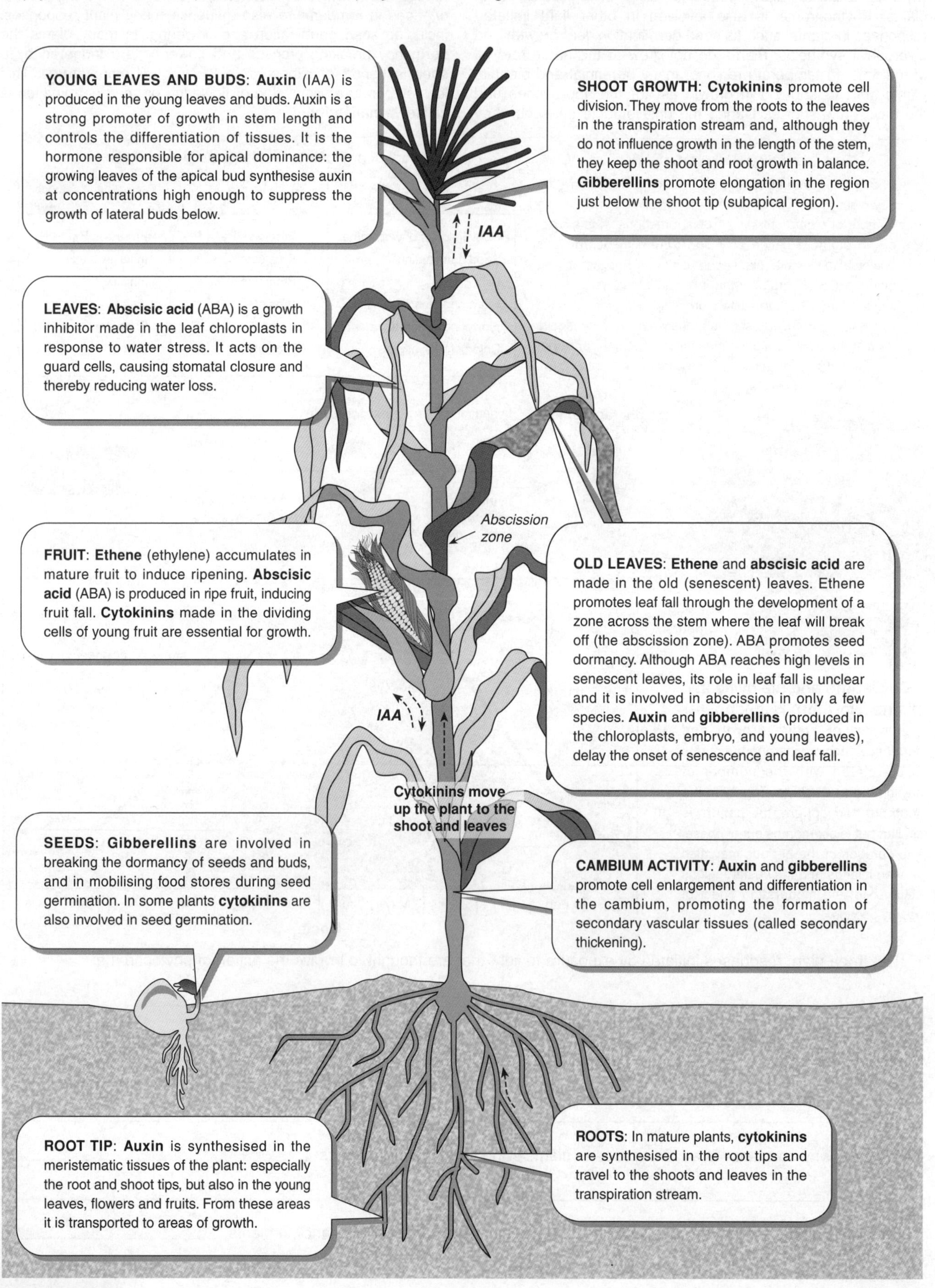

1. Explain the difference between plant hormones and animal hormones:

2. Outline one major effect of each of the hormones listed below:

   (a) Gibberellins:

   (b) Cytokinins:

   (c) Ethene:

   (d) Abscisic acid:

3. Explain the role of auxin (IAA) in the following plant growth processes:

   (a) Apical dominance:

   (b) Stem growth:

   (c) Secondary growth:

4. Outline why pruning (removing the central leader) induces bushy growth in plants:

5. Describe how a horticulturist could utilise hormones in each of the following situations:

   (a) To promote root development in plant cuttings:

   (b) To induce ripening in fruit:

   (c) To act as weed killer:

   (d) To promote seed germination:

   (e) To encourage seed dormancy prior to storage:

Plant Defence Mechanisms

Plants possess various biochemical and structural defence mechanisms, which protect them from infection and the activities of herbivorous animals that graze on plants or suck their sap. Some defence mechanisms are always present as part of the plant's basic make-up, while others are activated in response to an attack. **Passive defences** take the form of physical and chemical barriers. **Active defences** are produced in direct response to an infection or physical attack and act more specifically against the pathogen. Some plants produce chemicals that inhibit the growth others nearby (**allelopathy**).

Passive Defences

Passive defences are always present and are not the result of contact with pathogens or grazers. Passive defences may be **physical** or **chemical**. Physical barriers (e.g. hairs, spines) help to prevent pathogen entry or grazer attack. In some plants, the particular growth form (e.g. divarication in coprosmas) reduces browser damage. Despite physical barriers, plants are still damaged by the pathogens and grazers that penetrate their defences. If this occurs, chemical defences or active cellular defence mechanisms (right) are used to protect the plant against further damage.

Examples of physical barriers

Hairs on the leaf surface may deter pathogens.

Thorns or **spines** on the plant surface may deter grazers.

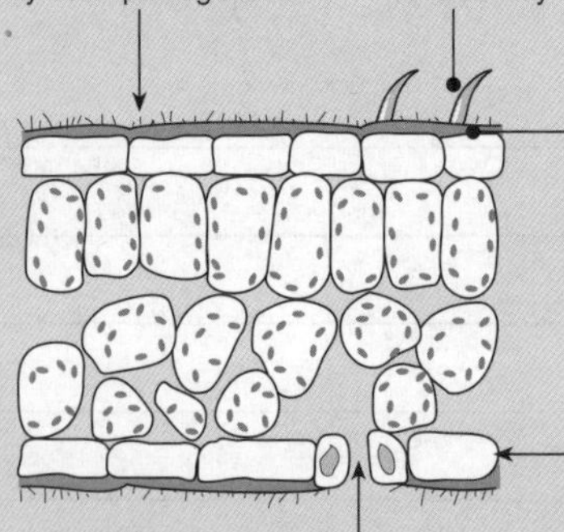

A thick, **waxy cuticle** forms a barrier against degrading enzymes used by pathogens to gain entry to the host.

Thickened **cell walls** reduce the ability of a pathogen to invade a plant.

Stomata can be a point of entry for some pathogens. Plants may use hairs to guard these openings, or the stomata may be small enough to exclude some larger pathogens.

Examples of chemical barriers

Some plants contain pungent, volatile compounds that deter grazers and **inhibit** the development of **pathogens**.

Many plants have developed distasteful or toxic chemicals as a defence to **deter** insects and other **grazers**. Examples include tannins, alkaloids, pyrethrins, and phenols.

Bracken (*Pteridium aquilinum*) contains hormones that disrupt the development of insect predators.

Bracken also contains a powerful carcinogen that is toxic to livestock when eaten.

Active Defences

Once infected, a plant needs to respond actively to prevent any further damage. **Active defences** are invoked only after the **pathogen** has been recognised, or after wounding or attack by a herbivore. Many plant defences contribute to slowing pathogen growth without necessarily stopping it.

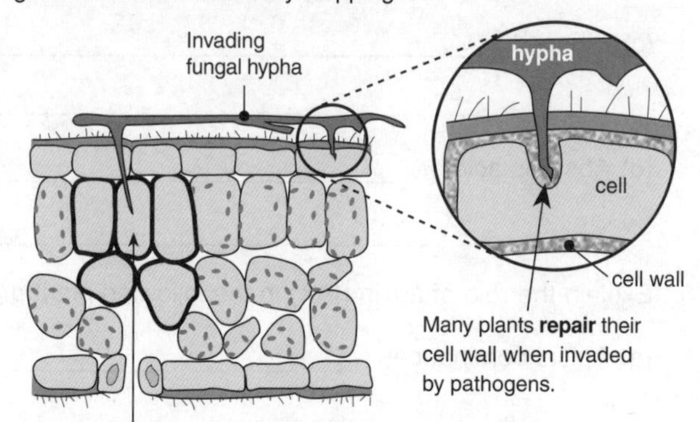

Many plants **repair** their cell wall when invaded by pathogens.

Many plants produce a **hypersensitive response** when invaded by pathogens. This enzyme activated response leads to the production of reactive nitric oxide and **cell death**. Cell death in the infected region limits the spread of the pathogen.

Other cellular active defences

Phytoalexins	Antibiotic-like substances are secreted to destroy a range of pathogens.
Reactive oxygen levels	An increase in reactive oxygen levels in cell membranes kills microorganisms.
Wound repair	Infected areas are sealed off by layers of thickened cells called **cork cells**.

Sealing off infected areas gives rise to abnormal swellings called **galls** (oak gall, above left; bullseye galls on a maple leaf, above right). These galls limit the spread of the parasite or the infection in the plant.

1. Distinguish between **passive** and **active** defence mechanisms:

   (a) Passive: ______________________

   (b) Active: ______________________

2. Describe two purposes of chemical defences in plants:

   (a) ______________________

   (b) ______________________

3. Name the interaction between plants when a chemical from one inhibits growth of the other: ______________________

Index